Geophysical Monograph 26

Heterogeneous Atmospheric Chemistry

David R. Schryer
Editor

American Geophysical Union
Washington, D.C.
1982

Heterogeneous Atmospheric Chemistry

Library of Congress Cataloging in Publication Data
Main entry under title:

Heterogeneous atmospheric chemistry.

(Geophysical monograph; 26)
Includes bibliographies.
1. Atmospheric chemistry--Addresses, essays, lectures.
I. Schryer, David R. II. Series.
QC879.6.H47 1982 551.5'112 82-11451
ISBN 0-87590-051-8

Printed in the United States of America.

CONTENTS

PREFACE

In the past few years it has become increasingly clear that heterogeneous, or multiphase, processes play an important role in the atmosphere. Unfortunately the literature on the subject, although now fairly extensive, is still rather dispersed. Furthermore, much of the expertise regarding heterogeneous processes lies in fields not directly related to atmospheric science. Therefore, it seemed desirable to bring together for an exchange of ideas, information, and methodologies the various atmospheric scientists who are actively studying heterogeneous processes as well as other researchers studying similar processes in the context of other fields.

This was accomplished in the summer of 1981 at the Conference on Multiphase Processes - Including Heterogeneous Catalysis - Relevant to Atmospheric Chemistry (originally entitled Workshop/Conference on Heterogeneous Catalysis - Its Importance to Atmospheric Chemistry). The conference was held June 29 to July 3, 1981, at Albany, N.Y. It was sponsored by the National Science Foundation and the National Aeronautics and Space Administration and was hosted by the Atmospheric Sciences Research Center (ASRC) of the State University of New York at Albany. The conference chairman was Volker A. Mohnen of ASRC. Over 100 researchers from a wide variety of fields attended the conference, which brought to attention the broad scope of ongoing research on various phenomena related to heterogeneous processes.

During the organization of the conference it was decided that, in addition to the conference itself, a book was needed which would present, in a single source, an effective introduction to the present state of knowledge of heterogeneous processes of relevance to the atmosphere. It was soon decided, however, that the book should not be just a collection of conference papers. Rather, additional papers not received in time for the conference would be included and all papers would be individually refereed and edited.

This has now been accomplished and the result is the present monograph. The majority of the papers presented herein were presented at the Albany conference, although many of these have been revised to reflect the latest research results. Several papers presented at the conference are, for a number of reasons, not included in the monograph, and a few papers not presented at the conference are presented here. The attempt has been made, within individual papers as well as among the papers, to present a mix between review of the various fields covered and presentation of new research results. A similar mix has been sought between papers dealing directly with atmospheric science and papers from other fields dealing with phenomena and/or methodologies which are potentially applicable to atmospheric science but are not directly related to it.

It is hoped that this monograph will prove useful not only to researchers actively engaged in the study of heterogeneous processes in the atmosphere but also to researchers who might profitably consider such processes but who have not done so, either for lack of knowledge of their potential importance or because of the dispersed nature of the relevant literature. If this monograph contributes to a broader consideration of heterogeneous processes in the study of atmospheric chemistry then the editor's goal will have been realized.

Use of trade names or names of manufacturers in this monograph does not constitute an official endorsement of such products or manufacturers, either expressed or implied, by the sponsoring organizations.

David R. Schryer
National Aeronautics and Space Administration
Langley Research Center
Hampton, Virginia
March 1, 1982

Clusters, Microparticles, and Particles

COMMON PROBLEMS IN NUCLEATION AND GROWTH, CHEMICAL KINETICS, AND CATALYSIS

Howard Reiss

Department of Chemistry, University of California at Los Angeles, Los Angeles, California 90024

The editor of this monograph has asked me to make a few remarks about nucleation and growth in the atmosphere in the context of the goals of this document. The task, it turns out, is a formidable one. The difficulty lies in the interdisciplinary character of our mission and, in the end, encounters the problem which has characterized the field of nucleation and particulate formation for as long as I can remember. The field seems, always, to fall "between the cracks". From time to time it has attracted the attention of such diverse specialists as metallurgists, engineers, chemists, physicists, astronomers, aerodynamicists, meteorologists, and many more. These individuals have brought with them such diversities of languages and pyschologies of innovation that they have really experienced difficulty in communicating with one another.

Although this is frustrating it does illustrate one cardinal point. The problem is of immense scientific and technological importance. A list of those subjects to which nucleation and/or growth have contributed would have to include metallurgy, high energy physics, atmospheric science, planetary physics and cosmochemistry, vulcanism, aerodynamics, turbine technology, cavitation, boiling, air pollution, crystal growth, magnetism, and chemical processing. Now, we are asked to consider the relationship between atmospheric particulates and catalysis. We will be adding one more discipline, and another group of specialists, to an already top heavy collection. It therefore seems appropriate to consider some of the examples from the past in which confusion was generated among capable people solely because of their diverse orientations. With such knowledge we might avoid similar difficulties because of our own diverse orientations.

The process of nucleation involves the escape of a system from a metastable state, and, often to the development of a new (daughter) phase within an outwardly homogeneous mother phase. The student of this phenomenon is therefore faced with spanning a long sequence of events which stretches all the way from the molecular to the macroscopic domain. In different stages of this sequence he may be forced to use different approximate methods in order to arrive at tractable solutions to the problem. The "interfaces" between these methods of approximation have only rarely preserved continuity of concept, and this lack of continuity has invariably led to confusion, some of which is not even resolved today.

For example, the earliest practitioners in the field of nucleation were experimentalists (physical chemists, metallurgists, or engineers) who tended to view the process more in terms of its finished product, e.g., a macroscopic fragment or drop of the daughter phase. Reasoning then traveled along the path directed from the macroscopic to the molecular level, and, through all of this, the drop remained a drop until it became evident that one could not proceed further without accounting for molecular detail. The strategy then became one in which the drop was retained while attempts were made to graft molecular-like behavior onto its continuum personality.

For example, the drop was now regarded as a large molecule and as such had to be described by a molecular partition function. Since the calculation of this partition function, beginning truly at the molecular level, represented a very difficult task, the issue was avoided by appealing to the relation between the partition function of the drop and its free energy. The free energy could be evaluated from the bulk thermodynamic properties of the liquid if satisfactory corrections were made for surface free energy, a quantity that had to be considered in view of the drop's large ratio of surface to volume. This procedure is the well-known capillarity approximation.

But then it was noted that not all the free energy was accounted for by this device. For example, a drop suspended from a capillary tube does not seem to be translating (as a molecule would) throughout the laboratory. Furthermore, it does not seem to be rotating (as a molecule would). Corrections were therefore made for these omissions. These corrections which endeavor to take account of the fact that the necessary degrees of freedom were nonetheless present in the drop (but not as translations and rotations) are known as "replacement free energy", and lead to the prediction that calculated rates of nucleation might be in error by a factor as large as 10^{18}. Later work took account of the fact that translational degrees of freedom refer to the motion of the center of mass of the drop and that a drop suspended from a capillary tube in the laboratory still had a center of mass which was able to translate (fluctuate). So the original drop did have some translational free energy. Furthermore, since a liquid cannot rotate rigidly, a drop cannot rotate in the conventional manner as a rigid body. However, the molecules in the drop still had angular momentum, even if not concerted. But the same nonconcerted angular momentum was present in the bulk liquid. Thus, very little replacement free energy was needed for the case of rotation. This brought the factor of predicted error down from approximately 10^{18} to about 10^4. Although I think that the confusion has been resolved, it still exists in the minds of many.

This example is a good one because the confusion was generated by the need to maintain a continuity of concept at an interface between the stages of reasoning associated with the molecular level on the one hand and the macroscopic point of view on the other hand. As soon as individuals, more accustomed to thinking about things on the molecular level, became involved they began to make progress but also generated confusion since they crossed the interface in the other direction, traveling along the path from the molecular to the macroscopic level. They also started from several different points.

A popular but erroneous starting point was the theory of so-called "mathematical" clusters which, among other things, are useful for the statistical mechanical treatment of imperfect gases. Unfortunately, the mathematical clusters imply just what their name says; they are merely bookkeeping devices for keeping track of the details of a complicated mathematical development. What was really needed was a treatment using "physical" clusters such as the elementary aggregates of molecules which are incipient fragments of the new phase. However, it immediately became apparent that the "convention" by means of which a physical cluster was defined was central to the whole problem. For example, if one was interested in developing a theory (an equation of state) for an imperfect gas, making use of physical clusters, any number of conventions were suitable as long as they were dealt with consistently, but all of them made a theory far more difficult than with the simple device of the mathematical cluster.

In addition, the problem of nucleation and subsequent growth is a kinetic problem, not an equilibrium one. In this case the physical cluster had to be chosen so that it was, at once treatable, and yet corresponded to the fragments of the new phase which are the products of the nucleation process. Even today, no thoroughly satisfactory solution to this dilemma has been devised.

Several partial solutions have involved the use of large computers, appealing to molecular dynamics or Monte Carlo techniques. In one example the cluster is treated as a molecule having internal vibrational degrees of freedom. Since the center of mass is chosen consistently in this approach, the problem of replacement free energy never even makes its debut. We are then left with another discontinuity of concept - this time originating at the molecular rather than the macroscopic side of the conceptual interface. Other conventions involve confining molecules inside large spherical regions with perfectly hard walls, the center of the sphere being located at the center of mass of the assembly of molecules contained within it. The free energy of this "physical" cluster is then evaluated by Monte Carlo means. This convention is more "droplike", except that in a drop the center of mass is not confined to the geometric center so that another discontinuity of concept arises, and an adjustment has to be made.

There is another kind of dichotomy which arises in this field. This has to do with the distinction between equilibrium (thermodynamics) and kinetic processes. In recent years photochemists and more conventional chemical kineticists have become interested in nucleation for a number of reasons. Among these is a strong interest in the photochemistry of atmospheric processes, especially those which lead to photochemically generated particles and pollutants. In addition, nucleation and growth are being used for detection and amplification in the study of the mechanisms of some chemical reactions. This constitutes another reason. Now, although most chemical kineticists are well schooled in the modern disciplines of chemical rate theory, and are even well into the subject of "selected state" chemistry made possible by the advent of the laser, the molecular beam, and other modern instrumentation, they have shown a tendency to misunderstand the kinetic problems in nucleation theory.

In many, but not all, nucleation processes a quasi-equilibrium theory does fairly well. There are several good examples of this kind of theory within the discipline of standard chemical kinetics. One example concerns unimolecular decomposition. The old Lindemann mechanism first published in 1922 is a case in point. In this process two molecules which we denote by A enter into a reversible reaction as follows:

$$A + A \rightleftarrows A^* + A \qquad (1)$$

in which A^* represents some sort of excited molecule, capable of undergoing the decomposition represented by the following reaction:

$$A^* \rightarrow B + C \qquad (2)$$

where B and C represent decomposition products. To the forward reaction in equation (1) we assign the specific rate constant k_1, while to the reverse reaction we assign the constant k_{-1}. At equilibrium in the reaction specified by equation (1), we then have the following relation:

$$k_1[A]^2 = k_{-1}[A^*][A] \qquad (3)$$

where bracketed quantities represent concentrations.

If the reaction of equation (2) occurs relatively slowly compared to the rate at which the equilibrium of equation (3) is established, then it perturbs that equilibrium only slightly. In fact, the rate associated with equation (2) can be accounted for by the term

$$k_2[A^*] \tag{4}$$

where k_2 is again a specific rate constant. The rate of change of the concentration of A^* is then given by the net rate of production represented by equation (3) with a minus sign replacing the equal sign, from which we subtract the loss represented by expression (4). Thus, we have

$$\frac{d[A^*]}{dt} = k_1[A]^2 - k_{-1}[A^*][A] - k_2[A^*] \tag{5}$$

However, the concentration of A^* is assumed to be so small that it reaches a quasi-steady state, and the derivative on the left side of equation (5) can be set to zero. Then one can solve for the concentration of A^* in the resulting equation and substitute it into expression (4) to obtain an expression for the rate of appearance of the decomposition products, or, what is the same thing, the rate of disappearance of the reactants A. The result is

$$\frac{-d[A]}{dt} = \frac{k_1k_2[A]^2}{k_{-1}[A] + k_2} \tag{6}$$

We also note from equation (3) that an equilibrium constant K can be defined as follows:

$$K = \frac{[A^*][A]}{[A]^2} = \frac{[A^*]}{[A]} = \frac{k_1}{k_{-1}} \tag{7}$$

so that equation (6) may be written in the form

$$\frac{-d[A]}{dt} = \frac{k_{-1}k_2K[A]^2}{k_{-1}[A] + k_2} \tag{8}$$

In the case that the concentration or pressure of A is high, the first term in the denominator of equation (8) dominates the second term to the extent that the second may be ignored. In that case equation (8) reduces to

$$-\frac{d[A]}{dt} = k_2K[A] \tag{9}$$

This is a first-order reaction even though the overall process is not really unimolecular, and this famous result shows how a reaction can appear to be unimolecular. Of course at low pressures one must use the full equation (8) and the appearance of the kinetics changes. The important feature of equation (9) is the appearance of the equilibrium constant K. This reminds us that the result is made possible by having one of the steps of the overall process being virtually in equilibrium.

What is not always perceived by kineticists, who often tend to regard the process of nucleation as a kind of chain polymerization affair, is that most nucleating systems resemble the case of unimolecular decomposition (a case with which they are very familiar). A cluster of a new phase is built up by the stepwise addition of molecules to smaller clusters. Each of these steps is a reaction much like equation (1), and, for the most part, most of the early steps are in virtual equilibrium, again like equation (1). Eventually a step is reached (at the so-called nucleus size) where the cluster can "disappear", this time, not by decomposition, but by rapidly growing into a macroscopic drop. The rate of this process, which is the rate of nucleation, is basically the rate of this last step. If the nucleus contains n molecules, and if we denote it by the formula A_n, then the concentration of nuclei will be given by

$$[A_n] = K'[A]^n \tag{10}$$

in which K' is simply the equilibrium constant characterizing the equilibrium between single molecules (monomers) A and nuclei A_n. This expression is the analog of equation (7), and to complete a crude theory of nucleation we require the analog of expression (4). This is simply the rate at which monomers encounter and join the nucleus, so that it quickly grows to a macroscopic drop. We can express this rate as

$$J = k_n[A_n][A] \tag{11}$$

where again k_n is a specific rate constant and J is the rate of nucleation. Substituting equation (10) into equation (11) gives

$$J = k_nK'[A]^{n+1} \tag{12}$$

From this we see that the rate may depend very critically on the concentration of monomers, since n may be a fairly large number and a slight change in the concentration of A may lead to an enormous change in J. Since the concentration of A will be roughly proportional to its pressure, the nucleation rate will be very sensitive to pressure or so-called supersaturation. If, for example, there are of the order of 100 molecules in the nucleus, n will be of the order of 100, and a slight change in pressure will have an enormous effect on J.

Now the quasi-equilibrium nature of the theory, embodied in the use of K', has prompted workers to attempt to evaluate K' by estimating the standard free energy of reaction (as with the case of any good equilibrium constant), and what better means for doing this than by evaluating the free energy of the drop destined to represent the nucleus. Thus, the capillarity approximation enters the picture, and the drop becomes the model for a cluster. Unfortunately, only the drop representing the nucleus can be considered to be in equilibrium with the metastable phase, and the use of a drop to represent smaller clusters becomes immediately suspect. However, if one truly uses an equilibrium theory as is the case with equation (12), this inconsistency is avoided, since smaller clusters do not have to be considered.

Unfortunately, no theory of rate can be a truly equilibrium theory. When adjustments are made to take into account contributions of nonequilibrium phenomena, more problems are created because one must then deal with clusters having sizes smaller than the nucleus. Incidentally, the primary effect of these nonequilibrium corrections is to "perturb" the equilibrium distribution of clusters, which for the nucleus would be represented by equation (10), in view of the fact that clusters are being "drained" away by the process represented by equation (9). This is not an uncommon effect in other kinetic examples; it occurs, for example, when one considers the process of effusion of a gas through a pin hole where the distribution of velocities can no longer be considered Maxwellian near the pin hole because molecules with selected velocities are being drained out of the distribution. In fact, chemical kineticists are also familiar with this process in connection with chemical rate theories. For example, the theory of absolute reaction rates must be adjusted for the fact that activated complexes are not truly in equilibrium with reactants simply because they are being drained away by the reaction whose rate one is trying to estimate.

All of this has been meant to emphasize the similarities between the problems which appear in the kinetics of phase transitions and more conventional kinetics associated with chemical processes. In spite of this, there is ample evidence that many sophisticated chemical kineticists have not fully penetrated the subtleties of nucleation theory. Of course the reverse is also true. Sometimes, because of a concentration on thermodynamics in the evaluation of K', workers seem to think that nucleation is a branch of thermodynamics rather than kinetics. Of course this is not true.

Incidentally, an interesting goal in the theory of nucleation is the use of modern chemical kinetic theory. We are approaching the point where this may be feasible. For example, it may be possible to produce such high states of supersaturation (e.g., by rapid expansion of a gas through a nozzle) so that the nucleus is a dimer. This may be possible in the case of supersaturated argon vapor. In that case the process represented by equation (11) would involve no more than three molecules, the two molecules in the nucleus and the third molecule needed to further the process of condensation. With only three molecules, and the knowledge of the intermolecular potential, it becomes possible to perform "trajectory analyses", in the modern sense, for the noncollinear case, and to treat the nucleation process as an ordinary chemical reaction. So here again, we have the union of two diverse disciplines, and it is important to carry out this union in as smooth a manner as possible.

In this monograph we are considering the possible importance of heterogeneous catalysis on atmospheric particulate surfaces. Here again we are bringing together individuals from different fields who already have worked on problems, each in their respective fields, which happen to be isomorphic, but which may be couched in different languages. For example, in the case of heterogeneous nucleation (or in the growth of particulates) atoms or molecules adsorbed on a particle are limited by steric factors, surface energies, and processes of migration which take place on the surface itself. In the field of nucleation and growth of particulates, these factors have frequently been considered. Perhaps one of the most famous examples, pioneered in the Albany/Schenectady region, is that of silver iodide which is capable of nucleating ice crystals, presumably because, in some way, the silver iodide surface and the ice crystal surface can be made to register. Heterogeneous nucleation is itself an example of catalysis. In this case the nucleus catalyzes the decay of the metastable state.

Cluster calculations have been relevant to both fields. I have already mentioned the treatment of nuclei as though they were large molecules having internal degrees of vibrational freedom. Fairly ambitious computer experiments have been carried out in this direction. But clusters have also been used by theorists interested in exploring the quantum mechanical aspects of catalytic surfaces. For example, studies have been performed on clusters of as many as 36 beryllium atoms. In these studies a fairly complete treatment of the electrons in the cluster is performed. The surface properties of the cluster seem to become fairly stable even when only 26 atoms are involved. Alternatively, surface theorists have used the strategy of embedding a cluster in a surface, treating the rest of the surface in some average way (by means of boundary conditions), and learning something about the electronic properties of the surface through the calculated behavior within the embedded cluster. Sometimes instead of embedding the clusters the "dangling" bonds are "capped" by satisfying them with hydrogen atoms.

I could, in fact, go much further. For example, the catalyzing of heterogeneous processes on a particular surface could itself affect the nature of the particulate, its size and configuration, and in so doing affect the growth. There is a real physical symbiosis between the fields we are considering and a potential intellectual symbiosis between the workers therein. One of the primary objectives of this monograph is to explore these symbioses and to determine what effect they have on the atmospheric processes that affect all of us.

EFFECT OF THE MECHANISM OF GAS-TO-PARTICLE CONVERSION ON THE EVOLUTION OF AEROSOL SIZE DISTRIBUTIONS

John H. Seinfeld and Mark Bassett

Department of Chemical Engineering, California Institute of Technology, Pasadena, California 91125

Abstract. The evolution of the size distribution of an aerosol undergoing growth by gas-to-particle conversion is investigated theoretically when growth occurs by any of three mechanisms, vapor phase diffusion, reaction of adsorbed vapor species on the particle surface, and reaction of dissolved vapor species in the particle volume.

Introduction

Atmospheric aerosols evolve in size by coagulation and gas-to-particle conversion. To interpret the evolution of a size spectrum it is necessary to understand the influences of these two phenomena. It has been found that aerosols in the size range 0.01 μm to 1.0 μm diameter grow principally by gas-to-particle conversion, the process by which vapor molecules diffuse to the surface of a particle and subsequently are incorporated into the particle. Considerable work has been carried out to identify the chemical pathways of incorporation of vapor species into atmospheric particles. Much of that effort has focused on elucidating gas-phase reaction mechanisms that lead to condensable vapor species such as sulfuric acid, ammonium sulfate, ammonium nitrate, and organic acids and nitrates (Hidy et al., 1980; Seinfeld, 1980), whereas additional studies have concentrated on reactions that occur on the surface of or within particles between particulate phase components such as metals and carbon, and adsorbed or absorbed vapor molecules (Judeikis and Siegel, 1973; Novakov et al., 1974; Peterson and Seinfeld, 1979, 1980; Seinfeld, 1980).

The rate-controlling step in gas-to-particle conversion may be a result of one or a combination of three mechanisms: the rate of diffusion of the vapor molecule to the surface of the particle; the rate of a surface reaction involving the adsorbed vapor molecule and the particle surface; and the rate of a reaction involving the dissolved species occurring uniformly throughout the volume of the particle. The particle growth rates that result in the three cases can be referred to as diffusion-controlled, surface reaction-controlled, and volume reaction-controlled growth, respectively. In fact, it has been suggested that information about possible chemical conversion mechanisms can be inferred from data on the evolution of an aerosol size distribution (Seinfeld and Ramabhadran, 1975; Heisler and Friedlander, 1977; Gelbard and Seinfeld, 1979; McMurry and Wilson, 1982). By calculating growth rates for particles of different sizes, the functional dependence of growth rate on particle size can be determined and compared with theoretical expressions relating particle growth rate to particle size, so-called growth laws. In this way it is possible to suggest chemical mechanisms that are consistent with the data. Carefully executed laboratory studies of this type have not yet been reported, although at least two are currently in progress.

The main object of this paper is to theoretically compare aerosol size spectra evolving by the mechanisms of diffusion-, surface reaction-, and volume reaction-controlled growth. The results will provide a basis for the interpretation of atmospheric and laboratory aerosol size spectra with respect to the governing growth mechanisms.

Growth of a Single Component Aerosol

Let $n(m,t)dm$ be the number of particles per unit volume of air having mass in the range m, $m + dm$ and let $I_m(m,t)$ be the rate of change of the mass of a particle of mass m due to gas-to-particle conversion. When the only process occurring is gas-to-particle conversion, the size distribution function $n(m,t)$ is governed by

$$\frac{\partial n}{\partial t} + \frac{\partial}{\partial m}(I_m(m,t)n) = 0 \qquad (1)$$

To study the evolution of an aerosol from an initial distribution,

$$n(m,0) = n_0(m) \qquad (2)$$

under different modes of gas-to-particle conversion requires the solution of equation (1) for the forms of $I_m(m,t)$ corresponding to the modes of conversion.

Equation (1) can be placed in dimensionless

form by defining the dimensionless time and particle mass,

$$\left.\begin{aligned} \tau &= \frac{tI_m^r}{\rho\lambda^3} \\ \mu &= \frac{m}{\rho\lambda^3} \end{aligned}\right\} \quad (3)$$

where t is time, ρ is the density of the particle, λ is the mean free path of the air, and I_m^r is a reference value of I_m. By then defining the reference size distribution function n^r and the dimensionless growth rate, $I(\mu,\tau) = I_m(m,t)/I_m^r$, we obtain the dimensionless form of equation (1),

$$\frac{\partial N}{\partial \tau} + \frac{\partial}{\partial \mu}(I(\mu,\tau)N) = 0 \quad (4)$$

where $N(\mu,\tau) = n(m,t)/n^r$.

The object of this work is to examine solutions of equation (4) for several forms of the growth law $I(\mu,\tau)$. In particular, three forms of I will be studied corresponding to three different rate-controlling mechanisms for gas-to-particle conversion.

Diffusion-Controlled Growth

The rate of change of the mass of a particle resulting from diffusion of vapor molecules of species A to the particle can be expressed as (Friedlander, 1977; Seinfeld, 1980)

$$I_m = \left(\frac{48\pi^2 m}{\rho}\right)^{1/3} \frac{\mathcal{D}_A M_A}{RT} (p_A - p_{A_s}) f(Kn) \quad (5)$$

where $\mathcal{D}_A$ is the molecular diffusivity of A in air, M_A is the molecular weight of A, R is the ideal gas constant, T is the absolute temperature, p_A is the partial pressure of A in the air, p_{A_s} is the vapor pressure of A just above the particle surface, and

$$f(Kn) = \frac{1 + Kn}{1 + 1.71Kn + 1.333Kn^2} \quad (6)$$

where Kn is the Knudsen number, the ratio of the mean free path of the air λ to the particle radius r. The term f(Kn) accounts for the transition between the two limiting cases:

$$f(Kn) = \begin{cases} 1 & Kn \to 0 \quad \text{continuum limit} \\ \dfrac{3}{4Kn} & Kn \to \infty \quad \text{free molecule limit} \end{cases} \quad (7)$$

The vapor pressure of A just above the surface can be related to the particle mass through the Kelvin equation,

$$\frac{p_{A_s}}{p_{A_0}} = \exp\left[\left(\frac{32\pi}{3}\right)^{1/3} \frac{\sigma\rho^{1/3}\bar{v}}{RTm^{1/3}}\right] \quad (8)$$

where σ is the surface tension of the particle, $\bar{v}$ is the molar volume of condensed A, and p_{A_0} is the vapor pressure of A over a flat interface at temperature T.

We define the saturation ratio $S = p_A/p_{A_0}$ and the reference growth rate,

$$I_m^r = (48\pi^2)^{1/3} \frac{\lambda \mathcal{D}_A M_A p_{A_0}}{RT} \quad (9)$$

Then the dimensionless growth rate corresponding to equations (5) and (8) is

$$I(\mu,\tau) = \mu^{1/3}[S(\tau) - \exp(K\mu^{-1/3})]f(Kn) \quad (10)$$

where the dependence on time τ enters through S and where

$$K = \left(\frac{32\pi}{3}\right)^{1/3} \frac{\sigma\bar{v}}{\lambda RT} \quad (11)$$

and

$$Kn = \left(\frac{3\mu}{4\pi}\right)^{-1/3} \quad (12)$$

In the case of so-called perfect absorption, $p_{A_s} = 0$, and equation (10) reduces to

$$I(\mu,\tau) = \mu^{1/3} S\, f(Kn) \quad (13)$$

The continuum and free molecule limits can be examined with reference to equation (13). We find

$$I(\mu,\tau) = \begin{cases} S\mu^{1/3} & Kn \to 0 \\ \left(\dfrac{81}{256\pi}\right)^{1/3} S\mu^{2/3} & Kn \to \infty \end{cases} \quad (14)$$

Surface Reaction-Controlled Growth

The second case we consider is that of surface reaction-controlled growth, namely when the rate of particle growth is controlled by the rate at which adsorbed A on the particle surface is converted to another species B. Thus, we take, as the simplest representation of such a situation, the sequence,

$$A(g) \rightleftarrows A(s) \rightleftarrows B$$

where A(s) denotes an adsorbed vapor molecule A on the surface that subsequently is converted to species B.

If the concentration of adsorbed A on the surface is c_s, and the rate of conversion to B is first order, with rate constant k_s, the rate of gain of particle mass due to the surface reaction is $4\pi r^2 M_B k_s c_s$. At steady state this rate must equal the rate of diffusion of molecules of A to the surface, which is given by equation (5). Thus,

$$4\pi r^2 M_B k_s c_s = \left(\frac{48\pi^2 m}{\rho}\right)^{1/3} \frac{\mathcal{D}_A M_A}{RT}(p_A - p_{A_s}) f(Kn) \tag{15}$$

Let us assume that adsorption equilibrium can be expressed by a relation of the form,

$$p_{A_0} = H_s c_s \tag{16}$$

Then c_s can be determined from equations (15), (16), and (8) as

$$c_s = \left[\left(\frac{48\pi^2 m}{\rho}\right)^{1/3} \frac{\mathcal{D}_A M_A}{RT} p_A f(Kn)\right] \left\{4\pi r^2 M_B k_s + \left(\frac{48\pi^2 m}{\rho}\right)^{1/3} \frac{\mathcal{D}_A M_A H_s}{RT} \exp\left[\left(\frac{32\pi}{3}\right)^{1/3} \frac{\sigma \rho^{1/3} \bar{v}}{RTm^{1/3}}\right] f(Kn)\right\}^{-1} \tag{17}$$

When the rate-determining step is surface reaction, the second term in the denominator of equation (17) dominates the first term and equation (17) reduces to

$$c_s \simeq \frac{p_A}{H_s} \exp\left[-\left(\frac{32\pi}{3}\right)^{1/3} \frac{\sigma \rho^{1/3} \bar{v}}{RTm^{1/3}}\right] \tag{18}$$

The corresponding rate of particle growth is

$$I_m = 4\pi r^2 M_B k_s \frac{p_A}{H_s} \exp\left[-\left(\frac{32\pi}{3}\right)^{1/3} \frac{\sigma \rho^{1/3} \bar{v}}{RTm^{1/3}}\right] \tag{19}$$

Defining the reference growth rate,

$$I_m^r = (36\pi)^{1/3} M_B k_s \frac{p_{A_0}}{H_s} \tag{20}$$

we obtain the surface reaction-controlled dimensionless growth rate as

$$I(\mu,\tau) = S\mu^{2/3} \exp(-K\mu^{-1/3}) \tag{21}$$

<u>Volume Reaction-Controlled Growth</u>

Finally we consider the case in which the rate of growth is controlled by the conversion of dissolved A to a second species B. The sequence can be depicted as

$$A(g) \rightleftarrows A(\ell) \rightleftarrows B(\ell)$$

If the concentration of dissolved A is c_v and the rate of conversion to B is first order, with rate constant k_v, the rate of gain of particle mass due to volume reaction is $\frac{4}{3}\pi r^3 M_B k_v c_v$. At steady state this rate must equal the rate of diffusion of molecules of A to the particle. Thus,

$$\frac{4}{3}\pi r^3 M_B c_v = \left(\frac{48\pi^2 m}{\rho}\right)^{1/3} \frac{\mathcal{D}_A M_A}{RT}(p_A - p_{A_s}) f(Kn) \tag{22}$$

As before, we assume that the equilibrium can be expressed by

$$p_{A_0} = H_v c_v \tag{23}$$

Then c_v can be determined from equations (22), (23), and (8) as

$$c_v = \left[\left(\frac{48\pi^2 m}{\rho}\right)^{1/3} \frac{\mathcal{D}_A M_A}{RT} p_A f(Kn)\right] \left\{\frac{4}{3}\pi r^3 M_B k_v + \left(\frac{48\pi^2 m}{\rho}\right)^{1/3} \frac{\mathcal{D}_A M_A H_v}{RT} \exp\left[\left(\frac{32\pi}{3}\right)^{1/3} \frac{\sigma \rho^{1/3} \bar{v}}{RTm^{1/3}}\right] f(Kn)\right\}^{-1} \tag{24}$$

When the rate-determining step is volume reaction, the second term in the denominator of equation (24) dominates the first term and equation (24) reduces to

$$c_v \simeq \frac{p_A}{H_v} \exp\left[-\left(\frac{32\pi}{3}\right)^{1/3} \frac{\sigma \rho^{1/3} \bar{v}}{RTm^{1/3}}\right] \tag{25}$$

The corresponding rate of particle growth is

$$I_m = \frac{4}{3}\pi r^3 M_B k_v \frac{p_A}{H_v} \exp\left[-\left(\frac{32\pi}{3}\right)^{1/3} \frac{\sigma \rho^{1/3} \bar{v}}{RTm^{1/3}}\right] \tag{26}$$

Defining the reference growth rate,

$$I_m^r = \lambda^3 M_B k_v \frac{p_{A_0}}{H_v} \tag{27}$$

we obtain the volume reaction-controlled dimensionless growth rate as

$$I(\mu,\tau) = S\mu \exp(-K\mu^{-1/3}) \tag{28}$$

Dimensionless Size Spectra Evolution

We now consider size spectra $N(\mu,\tau)$ corresponding to the three growth cases just devel-

TABLE 1. Dimensionless Growth Laws

Rate-Controlling Mechanism	$I(\mu)$
Diffusion	$\mu^{1/3}[S - \exp(K\mu^{-1/3})]f(Kn)^a$
Diffusion (perfect absorption)	$\mu^{1/3}S\ f(Kn)$
Surface reaction	$S\mu^{2/3}\exp(-K\mu^{-1/3})$
Volume reaction	$S\mu\exp(-K\mu^{-1/3})$

$^a Kn = \left(\frac{3\mu}{4\pi}\right)^{-1/3};\quad f(Kn) = \frac{1 + Kn}{1 + 1.71Kn + 1.333Kn^2}$

oped. The initial distribution $N_0(\mu)$ was adapted from one measured in a power plant plume. (See case d in Table 3 of Eltgroth and Hobbs, 1979.) The dimensionless size distributions are presented in terms of the dimensionless mass distribution $M(\log_{10} D_p,\tau)$, where $(\rho\lambda^3)^2 n_r M(\log_{10} D_p,\tau)\, d\log_{10} D_p$ is the mass of particles having logarithm of diameter in the range $(\log_{10} D_p, \log_{10} D_p + d\log_{10} D_p)$, where $\log_{10} D_p$ is understood as $\log_{10}(D_p/1\ \mu m)$. Thus, M is related to N by

$$M(\log_{10} D_p,\tau) = 6.9\mu^2 N(\mu,\tau) \qquad (29)$$

For the initial distribution used, the value of n_r was chosen so that the maximum value of $M(\log_{10} D_p,0)$ is 1.0. This value of n_r is $1.645 \times 10^{14}\ \mu g^{-1}\text{-cm}^{-3}$.

The dimensionless growth laws are summarized in Table 1. It is necessary to specify the parameters S and K. The saturation ratio S was set at 2.878 corresponding to a critical diameter $D_p^* = 4\sigma\bar{v}/RT\ \ell n\ S$ of 0.01 μm for a sulfuric acid/water aerosol at 25°C. The Kelvin parameter K is then equal to 0.1282. Since S is taken as independent of time τ, the growth laws I are functions of μ only.

Figure 1 shows the evolution of the dimensionless mass distribution at τ = 0, 0.2, 0.4, and 0.6 for case (1), diffusion-controlled growth with perfect absorption. We see that the smaller particles grow proportionally faster than the larger ones. The dependence of I on particle diameter gradually shifts from D_p^2 for the smallest particles (free molecule regime) to D_p for the largest particles (continuum regime), as indicated in equation (14).

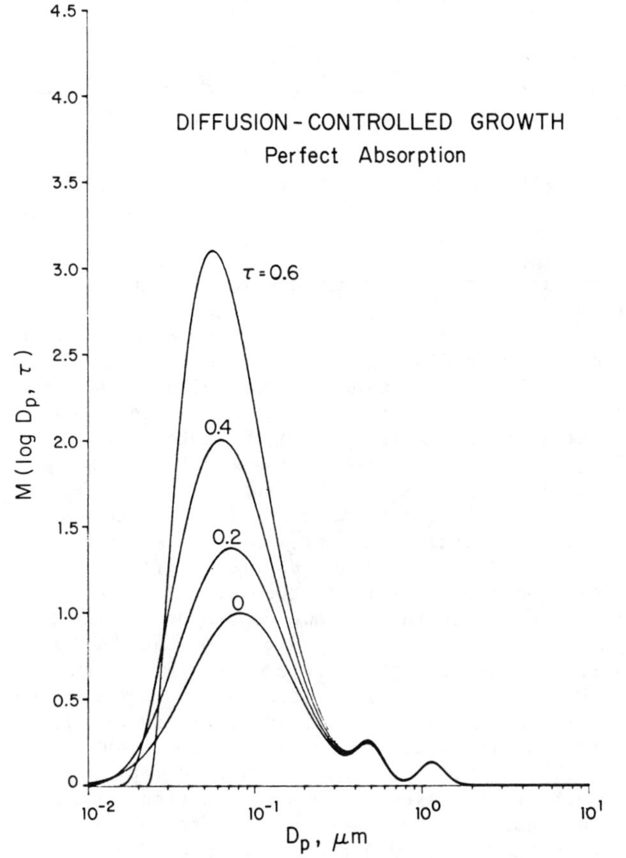

Fig. 1. Dimensionless mass distribution for diffusion-controlled growth with perfect adsorption.

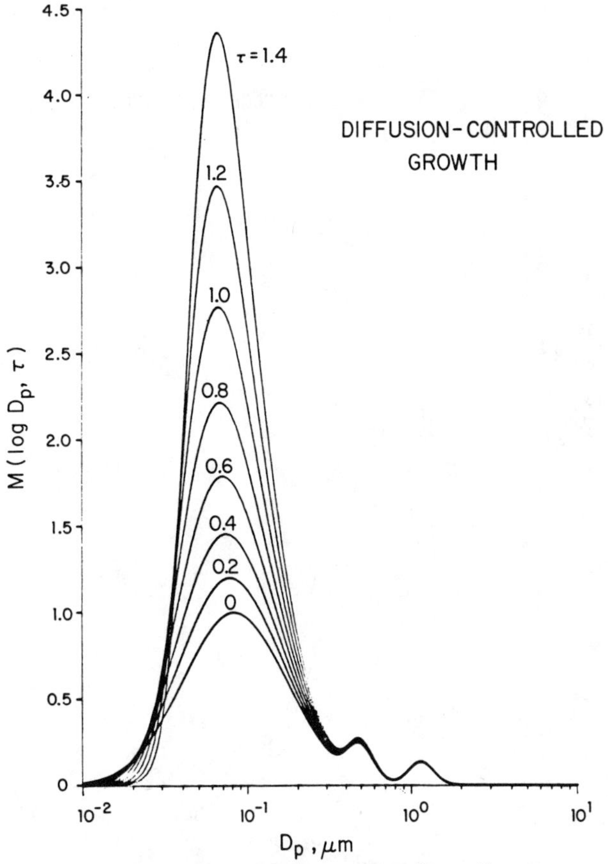

Fig. 2. Dimensionless mass distribution for diffusion-controlled growth.

In case (2), diffusion-controlled growth with a non-zero vapor pressure over the particle surface, as shown in Figure 2, addition of a vapor pressure leads to a much slower growth of the smaller particles than in case (1) in Figure 1. This, in turn, reduces the tendency of the major mode in the mass distribution to steepen. To understand why this is so consider equation (10) which may be expressed in terms of D_p^* as

$$I = \alpha D_p \left[S - \exp(K/\alpha D_p^*) \exp \frac{K(D_p^* - D_p)}{\alpha D_p^* D_p} \right] f(Kn) \quad (30)$$

where $\alpha = (\pi/6)^{1/3}\lambda^{-1}$. Now consider a particle of diameter close to D_p^*, i.e., for which $(D_p/D_p^* - 1) << \alpha D_p/K$. Then the second exponential in equation (30) can be expanded to give

$$I \simeq \frac{KS(D_p - D_p^*)}{D_p^*} f(Kn) \quad (31)$$

Thus, particles close to the critical diameter grow at a rate proportional to $(D_p - D_p^*)$. This situation can be contrasted with that in case (1) in which $I \sim D_p^2$.

Figures 3 and 4 show the evolution of the dimensionless mass distribution in cases (3) and (4), surface and volume reaction-controlled growth, respectively. In the case of surface reaction-controlled growth (Figure 3), we note that the large particles, for which the Kelvin effect is negligible, grow at a rate proportional to D_p^2. Recall that in diffusion-controlled growth the smallest particles also grow at a rate proportional to D_p^2. The larger particles growing by diffusion grow at a rate proportional to D_p. Thus, we expect that of two continuum regime particles, one growing by diffusion and one by surface reaction, the particle growing by surface reaction does so at a greater rate. When gas-to-particle conversion occurs by volume reaction, large particles grow at a rate proportional to D_p^3. Thus, these particles grow more rapidly than in either diffusion or surface reaction cases.

In Figure 5, the distributions from all three mechanisms at $\tau = 0.4$ are shown, providing a summary of the effects previously mentioned. With the current non-dimensionalization, the surface reaction is faster than diffusional

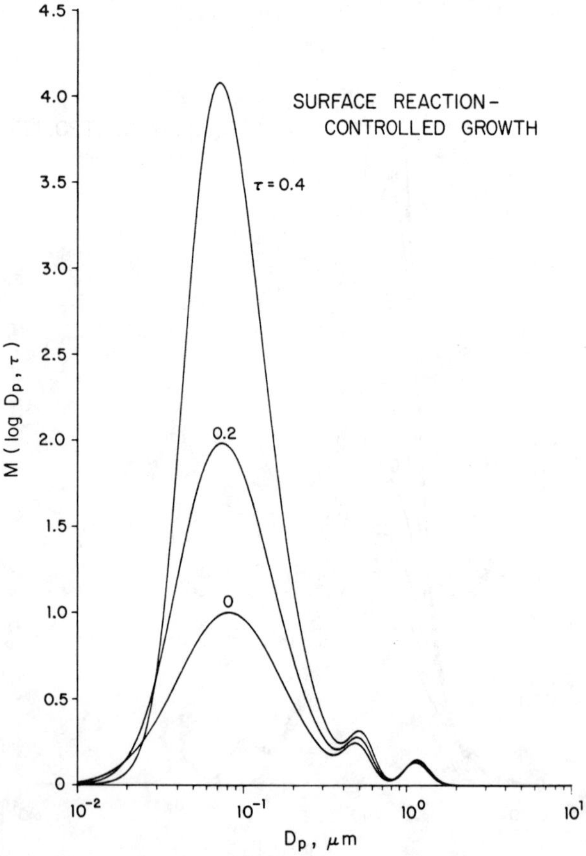

Fig. 3. Dimensionless mass distribution for surface reaction-controlled growth.

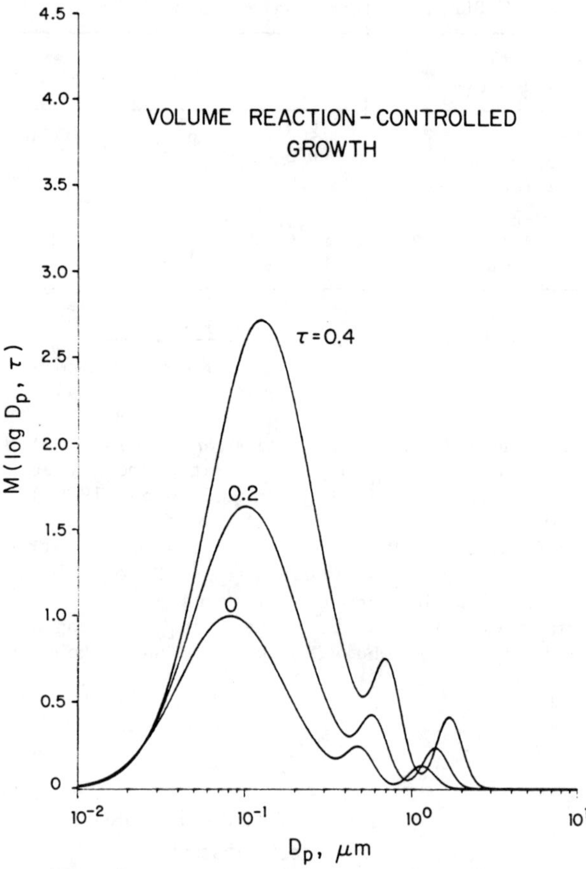

Fig. 4. Dimensionless mass distribution for volume reaction-controlled growth.

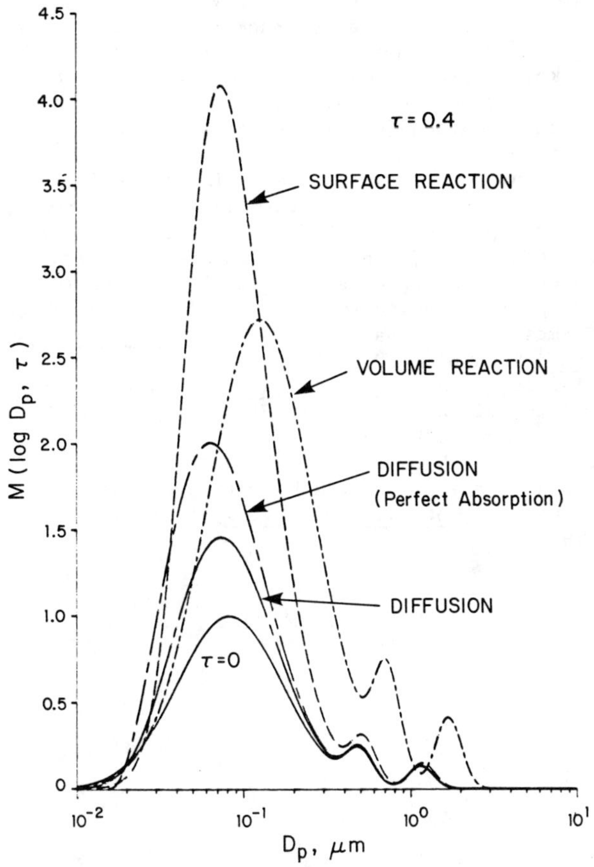

Fig. 5. Dimensionless mass distributions for the three growth mechanisms at τ = 0.4.

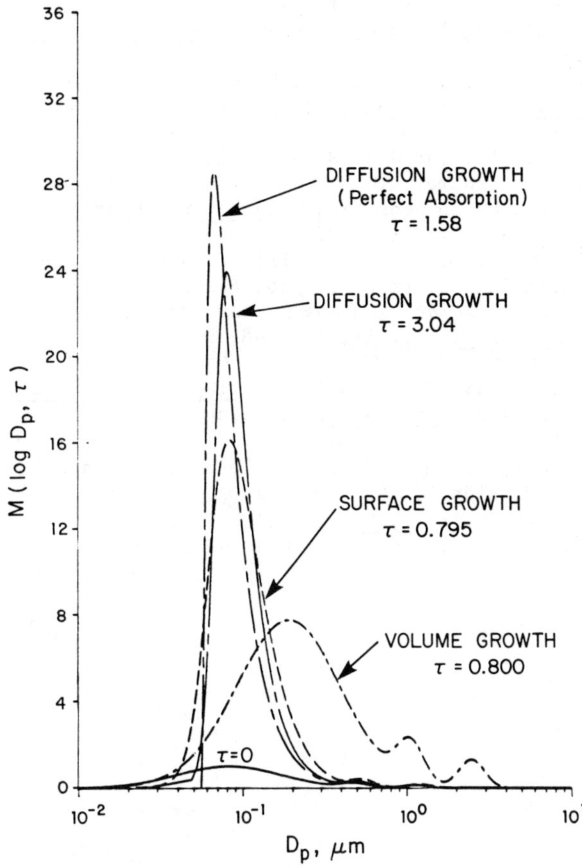

Fig. 6. Dimensionless mass distributions for the three growth mechanisms when the ratio of mass added by gas-to-particle conversion to initial mass is 7.

growth. However, with a different non-dimensionalization, diffusional growth could be faster.

The location of the main peak provides an indication of the relative importance of the growth of small and large particles. For the volume reaction case, where the ratio of large particle growth to small particle growth is the greatest, the particle diameter at which the peak is located is the greatest. On the other hand, for diffusional growth with no vapor pressure, where the growth of small particles is the most important, the particle diameter at which the peak is located is smaller than for any of the other cases.

Application to Identification of Growth Mechanism

The results of the previous section indicate that the mechanism of growth of an aerosol can be inferred from the evolution of its size spectrum. Figure 6 shows three size distributions at times when the total mass added to the particulate phase is the same (seven times the initial aerosol mass). Thus, the different size distributions are solely the result of the manner in which the different mechanisms distribute mass among the different particle sizes. This approach is useful, because in general, there may be uncertainties in the parameters needed to relate τ to t.

The distribution for the volume reaction case has significantly more large particles than any other. Thus, it should be fairly easy to distinguish experimentally between a volume reaction and the other mechanisms. On the other hand, it would probably be difficult to distinguish between the surface reaction case and diffusion growth without a vapor pressure. It would probably be impossible to distinguish experimentally between diffusion growth with a vapor pressure and a surface reaction.

Acknowledgement. This work was supported by U. S. Environmental Protection Agency grant R806844.

References

Eltgroth, M. W., and P. V. Hobbs, Evolution of particles in the plumes of coal-fired power plants - II. A numerical model and comparisons with field measurements, Atmos. Environ., 13, 953-976, 1979.

Friedlander, S. K., Smoke, Dust and Haze: Fundamentals of Aerosol Behavior, Wiley, New York, 1977.

Gelbard, F., and John H. Seinfeld, Exact solution of the general dynamic equation for aerosol growth by condensation, J. Colloid Interface Sci., 68, 173-183, 1979.

Heisler, S. L., and S. K. Friedlander, Gas-to-particle conversion in photochemical smog; Growth laws and mechanisms for organics, Atmos. Environ., 11, 158-168, 1977.

Hidy, G. M., P. K. Mueller, D. Grosjean, B. R. Appel, and J. J. Wesolowski, The Character and Origins of Smog Aerosols, Wiley, New York, 1980.

Judeikis, H. S., and S. Siegel, Particle catalyzed oxidation of atmospheric pollutants, Atmos. Environ., 7, 619-631, 1973.

McMurry, P. H., and J. C. Wilson, Growth laws for the formation of secondary ambient aerosols: Implications for chemical conversion mechanisms, Atmos. Environ., 16, 121-134, 1982.

Novakov, T., C. S. Chang, and A. B. Harker, Sulfates as pollution particles: Catalytic formation on carbon (soot) particles, Science, 186, 256-261, 1974.

Peterson, T. W., and John H. Seinfeld, Calculation of sulfate and nitrate levels in a growing, reacting aerosol, AIChE J., 25, 831-838, 1979.

Peterson, T. W., and John H. Seinfeld, Heterogeneous condensation and chemical reaction in droplets - Application to the heterogeneous atmospheric oxidation of SO_2, Adv. Environ. Sci. Technol., 10, 125-180, 1980.

Seinfeld, J. H., Lectures in Atmospheric Chemistry, AIChE Monogr. Ser. 12, Am. Inst. Chem. Eng., New York, 1980.

Seinfeld, J. H., and T. E. Ramabhadran, Atmospheric aerosol growth by heterogeneous condensation, Atmos. Environ., 9, 1091-1097, 1975.

NEUTRAL AND CHARGED CLUSTERS IN THE ATMOSPHERE: THEIR IMPORTANCE AND POTENTIAL ROLE IN HETEROGENEOUS CATALYSIS

A. W. Castleman, Jr.

Department of Chemistry and Chemical Physics Laboratory, CIRES*, University of Colorado, Boulder, Colorado 80309

Introduction

A major aspect of atmospheric chemistry for which there is currently a paucity of basic information concerns the interconversion of atmospheric molecules between various physical and chemical states, and the influence which this has on their incorporation into the aerosol phase. The transfer of molecules from the gaseous to condensed state removes them from further participation in direct gas-phase reactions, but at the same time introduces them to a site where they may be more catalytically active (i.e., adsorption on a catalytic surface or absorption in a reactive aqueous phase surrounding an aerosol particle). The transport and removal processes of various atmospheric constituents are greatly influenced by incorporation into an aerosol in contrast to being in the gas phase. Additionally, particles that are formed by a gas-to-particle conversion process often serve as condensation nuclei, which can have an important bearing on their climatic consequences.

While there is presently a considerable data base for homogeneous chemical reactions among atmospherically important molecules, it is well recognized that one of the least understood aspects of atmospheric transformations concerns heteromolecular and related heterogeneous processes. In particular, there is currently very little basic information available on the detailed mechanisms of gas-to-particle conversion. Furthermore, even the general nature and potential importance of heterogeneous reactions on aerosol surfaces and within the droplet phase surrounding them are largely unknown.

Gas-to-particle conversion processes are known to be initiated via a sequence of clustering reactions. These ultimately lead to the production of aerosol particles either by a direct gas-to-particle conversion process or through the formation of critical size embryos which eventually undergo phase transformation. Despite the lack of knowledge in the area of heterogeneous chemistry, it is recognized that the reactivity of constituents associated with an aerosol or adsorbed on a particle surface can be understood in terms of their interaction with neighboring molecules analogous to those involved in a cluster. Interactions among salts and acids present in the aqueous phase surrounding an aerosol can likewise be most easily studied as isolated ion clusters. The results of studies of these interactions also find direct application in assessing the energy barrier to nucleation. An attractive model of a catalytic surface, which enables a more detailed understanding of the factors responsible for its catalytic activity, is one in which the molecules are viewed as forming localized clusters with specific electronic properties. Likewise, it is becoming increasingly evident that even homogeneous reactions which have been well characterized in the laboratory may be influenced if the reactants are present in a clustered state in the free atmosphere.

Based on the foregoing examples, it is evident that small clusters play an important role in atmospheric processes and an understanding of their states and reactivity is essential in developing a fundamental understanding and basis for assessing the importance of heterogeneous processes. The purpose of this paper is to survey briefly the current knowledge in these various fields and to recommend studies needed for further progress in quantifying aerosol formation and catalytic reactivity.

Clusters Between Ions and Molecules and Among Neutral Molecules

Although it is well recognized that ionic processess play an important role in the upper atmospheric (Ferguson, 1975) the presence of ions in the troposphere and the middle atmosphere is often neglected. Nevertheless, they play an important role in fair-weather electricity and in

*The Cooperative Institute for Research in Environmental Sciences is jointly sponsored by the University of Colorado and the National Oceanic and Atmospheric Administration.

the charging of atmospheric aerosols, and as discussed in the next section are believed to be responsible for a gas-to-particle conversion process under certain conditions. Ions are ubiquitous throughout the Earth's environment, and mechanisms leading to their origin are generally well understood (Whitten and Poppoff, 1971). In high regions of the atmosphere, galactic cosmic rays as well as ultraviolet light from the Sun ensure partial ionization from the atmosphere at all levels. Near the ground the decay of radon and its progeny contributes substantially to processes initiating the ionization of air molecules (Bricard and Pradel, 1966). Excellent reviews of the various steps in the mechanisms of ion interconversion are found in recent references by Ferguson (1975), Smith and Adams (1980), and Mohnen (1971a, b). As a result of the extensive work of Ferguson and Smith and their coworkers, the rates and mechanisms of ion formation in the upper atmosphere are now relatively well in hand; they have been generally substantiated by rocket measurements made by the groups of Narcisi (1969) and Arnold (Arnold et al., 1981a, b). As a result of the difficulty in making ion measurements at the higher pressures existing in lower regions of the atmosphere as well as in unraveling the many complex reactions which can take place between numerous complex ionic species, the details of ion initiation and loss processes are less well known in the stratosphere and especially in the troposphere. Their elucidation is currently the subject of extensive research in numerous laboratories, including our own.

In order to investigate the bonding of molecules to ions, various techniques are employed. These include the flowing and stationary afterflow, ion cyclotron resonance, and high-pressure mass spectrometric sampling of ion-drift cells. The first three are widely used to measure reaction rates, and the fourth is the major technique for measuring bond energies and entropies. Its advantage lies in the fact that measurements can be made over a wide range of pressures, temperatures, and electric field parameters whereby the energy between the colliding pairs can be varied. Thereby, the attainment of equilibrium, requisite in determining thermodynamic properties of clusters, can be ensured.

With careful attention to the various parameters influencing equilibrium (Castleman, 1979c), measured ion intensities and knowledge of the partial pressures of clustering neutrals in the high-pressure reaction cells can be employed to determine equilibrium constants, $K_{n,n+1}$, for reactions of the type

$$A^{\pm}\cdot B_n + B\,(+M) \rightleftarrows A^{\pm}\cdot B_{n+1}\,(+M) \qquad (1)$$

Here, $A^{\pm}$ designates a positive or negative ion, B the clustering neutral, and M the third body necessary for collisional stabilization of the complex. Taking the standard state to be 1 atm, and making the usual assumptions (Castleman, 1979c) concerning ideal gas behavior and the proportionality of the chemical activity of an ion cluster to its measured intensity, it follows that

$$\ell n K_{n,n+1} = \ell n\left[\frac{I_{n+1}}{I_n p_B}\right] = -\frac{\Delta G^o_{n,n+1}}{RT}$$

$$= -\frac{\Delta H^o_{n,n+1}}{RT} + \frac{\Delta S^o_{n,n+1}}{R} \qquad (2)$$

Here, I_{n+1} and I_n represent the respective measured ion intensities, p_B the pressure of clustering molecules B, $\Delta G^o_{n,n+1}$, $\Delta H^o_{n,n+1}$, and $\Delta S^o_{n,n+1}$ the standard Gibbs free energy, enthalpy, and entropy, respectively, R the gas-law constant, and T absolute temperature.

A typical van't Hoff plot of data derived from experiments of the clustering of H_2O to NO_3^- is shown in Figure 1. In order to obtain the enthalpies and entropies of clustering from van't Hoff plots, the following well-known relationship is employed: $\Delta G^o = \Delta H^o - T\Delta S^o$. Clearly, the slopes provide values of enthalpies and the entropies are determined from the intercepts via a

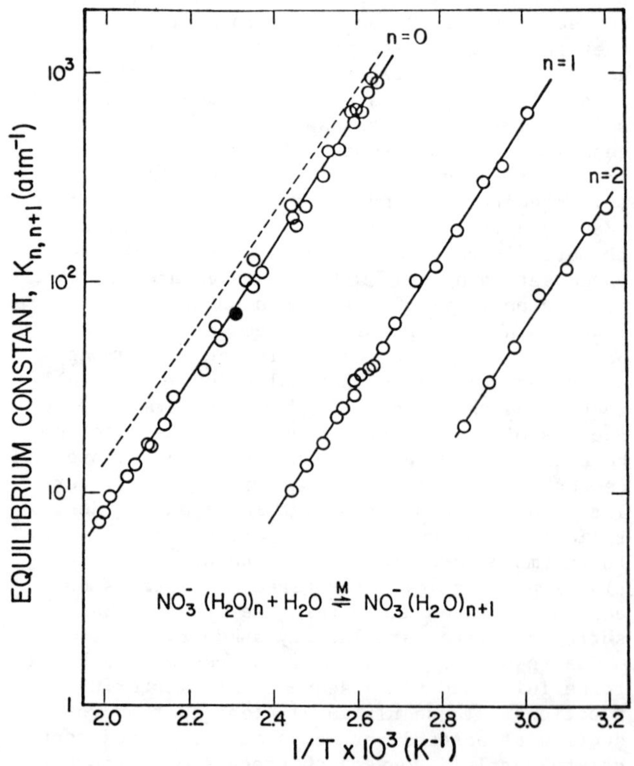

Fig. 1. van't Hoff plots of equilibrium constants for gas-phase reactions $NO_3^-(H_2O)_n + H_2O \rightleftarrows NO_3^-(H_2O)_{n+1}$; dashed line is from Kebarle's work (Payzant et al., 1971).

TABLE 1. Bond Energies Recently Measured in the Author's Laboratory for Selected Anion/Ligand Systems: $A^- B_n + B \overset{M}{\rightleftharpoons} A^- B_{n+1}$

		$-\Delta H^o_{n,n+1}$ (kcal/mol)				$-\Delta S^o_{n,n+1}$ (gibb/mol)			
Ligand	Ion	(0,1)	(1,2)	(2,3)	(3,4)	(0,1)	(1,2)	(2,3)	(3,4)
H_2O	Cl^-	14.9	12.6	11.5	10.9	19.7	20.5	22.4	24.8
	I^-	11.1	9.9	9.3		19.3	20.3	21.0	
	CO_3^-	14.1	13.6	13.1		25.2	29.6	32.5	
	HCO_3^-	15.7	14.9	13.6	13.4	24.1	29.1	30.2	33.3
	NO_2^-	15.2	13.6	11.7	11.6	23.8	26.4	25.8	29.0
	NO_3^-	14.6	14.3	13.8		25.0	30.3	33.2	
CO_2	Cl^-	8.0				19.6			
	I^-	5.6				18.2			
	CO_3^-	7.1				21.8			
	NO_2^-	9.3				24.2			
	SO_3^-	6.5				20.7			
	K^+	8.5				15.2			
SO_2	Cl^-	21.8	12.3	10.0	8.6	23.2	22.7	23.1	23.2
	NO_2^-	25.9	9.0	6.6		36.8	16.8	13.4	
	SO_2^-	24.0				33.8			
	SO_3^-	13.3				18.9			
	I^-	12.9	10.1	9.2		20.2	21.6	24.7	
HNO_2	NO_2^-	33	21.3						
HNO_3	NO_3^-		17.7	16.0					
HCl	I^-	14.2				22.7			

least-squares analysis of the data. Analogous data were obtained recently for a number of negative and positive ion systems and have been used to demonstrate that the breakdown in the classical liquid-drop formulation of nucleation theory is based on the inability of the model to properly account for the configurational entropy of small clusters (structure features) rather than its inability to describe bonding adequately (Castleman et al., 1978a; Castleman, 1979a). The application of recent laboratory data in explaining observations of negative ion cluster distributions in the lower ionosphere is discussed in Keesee et al. (1979b).

Recent data for selected negative ion ligand systems are shown in Table 1. Interestingly, the relative pattern of bond energies for the association of one neutral molecule to a negative ion was found to be the same for water (a protic polar molecule), sulfur dioxide (an aprotic polar molecule), and carbon dioxide (a nonpolar molecule). Theoretical calculations based on the approximate quantum mechanical procedure CNDO/2 (complete neglect of differential overlap) are successful in explaining the relative bonding of higher order clusters of H_2O about HCO_3^-, CO_3^-, NO_2^-, and NO_3^- (Keesee et al., 1979a; Lee et al., 1980b).

The results of the present studies have led to an understanding of factors controlling the strength of the bonding of various neutrals to ions with regard to the role of both the ion and the neutral. The results show that the strength of the bonding of the ions with the three neutrals CO_2, H_2O, and SO_2 is consistent with the inequality

$$OH^- > F^-,\ O^- > O_2^- > NO_2^- > Cl^- > NO_3^-$$

$$> CO_3^- > SO_4^- \approx SO_3^- \approx Br^- > I^- \qquad (3)$$

This relationship appears to parallel the order of the gas-phase basicity of the negative ions where the strongest bases exhibit the largest bond dissociation energies (Keesee et al., 1980).

Another interesting comparison is that of the relative bond dissocation energies, $D(A^- - B)$ for a given ion, with the different ligands SO_2, H_2O, and CO_2. The hydrogen, sulfur, and carbon are the centers which are attracted to a negative charge. Water has a slightly larger dipole moment than sulfur dioxide (1.85 versus 1.63 D) (McDaniel and Mason, 1973), but the quadrupole moment of water is repulsive when the dipole is attractive to a negative ion. The sign of the quadrupole interaction for SO_2 is not known.

Carbon dioxide has no dipole but does have a significant quadrupole moment. Considering only charge-dipole (or charge-multipole) interactions, the bond strength for a ligand to a given ion would be expected to be on the order of $H_2O \approx SO_2 > CO_2$.

For a weakly basic or large ion like I^-, the more or less expected order $SO_2 > H_2O > CO_2$ is observed. But as the ions become smaller or more basic, SO_2 binds relatively more strongly than water. With O_2^- the enthalpy change for addition of CO_2 becomes comparable to that of H_2O. Finally, for a small ion like O^-, the order $SO_2 > CO_2 > H_2O$ actually occurs. Hence, the simple correlation of the bond dissociation energies $D(A^- - B)$ being related to the gas-phase acidity of the neutral which was suggested heretofore (Yamdagni and Kebarle, 1971) fails in this more general case.

Interestingly, the mean polarizabilities for SO_2, CO_2, and H_2O follow the same order as their relative bond energies to small ions: 3.78, 2.59, and 1.45 $Å^3$, respectively (McDaniel and Mason, 1973). Polarization energies are known to be relatively more important for smaller ions, due to the ability of the neutral to approach close to the ion and thereby become more influenced by the ionic electric field, with the attendant result of a larger induced dipole. Qualitatively, consideration of the polarizabilities partially explains the increased bonding strength of SO_2 and CO_2 over that of water in clustering to smaller ions.

In some cases the bonds are strong enough to be considered as chemical instead of merely being weak electrostatic ones. In other words, the bond may be of a covalent nature, which is equivalent to stating that significant charge transfer occurs between the original ion and the clustering neutral. Such an occurence would invalidate the application of classical electrostatics to estimate bond energies (Spears and Ferguson, 1973; Spears, 1972). An example is $OH^- \cdot CO_2$, which is more properly considered as HCO_3^-.

In terms of charge transfer, the electron affinity of the clustering neutral is a relevant factor to consider in assessing the relative bonding trends. Indeed, SO_2 has an electron affinity (Franklin and Harland, 1974) of 1.0 eV and forms a stable gas-phase negative ion, whereas the negative ion of water has not been observed (Caledonia, 1975). The negative ion of CO_2 has been detected in high-energy processes in which the linear neutral can be bent to form the ion (Paulson, 1970), although CO_2^- is short-lived and autodetaches. Nevertheless, $(CO_2)_2^-$ is apparently stable (Klots and Compton, 1977; Rossi and Jordan, 1979). Collective effects are evidently important in the formation and stabilization of certain negative ion complexes.

It is recognized that dispersion of charge, created by partial charge transfer to a neutral ligand upon the first association step, can lead to a significant decrease in the bond energy for succeeding steps. A surprising finding from some of the recent data is the fact that the bond energy value for a small cluster can actually be lower than the bulk-phase value for the heat of vaporization of a particular solvent molecule from the ion cluster.

Shown in Table 2 is a listing of the bond energies for a number of molecules clustered to positive ions, including the alkali metal series, certain transition metal ions, alkaline earth ions, protons, and NH_4^+ (Castleman, 1978; Tang and Castleman, 1972, 1974, 1975; Dzidic and Kebarle, 1970; Castleman et al., 1978b, 1981a; Tang et al., 1976; Holland and Castleman, 1982; Kebarle et al., 1967; Payzant et al., 1973; Munson, 1965; Staley and Beauchamp, 1975; Payzant and Kebarle, 1972). Because of its high proton affinity, NH_4^+ is expected to be an important terminal positive ion in the atmosphere whenever trace concentrations of ammonia are present. The thermodynamic parameters derived for positive-ion/neutral complexes have provided unique information on the strength of bonding and structure of these complexes as well. In addition, structural computations using quantum mechanical and electrostatic theories have provided further insight into the nature of bonding and arrangement of ligands about the central ion. Details of the solvation process of various ions (Lee et al., 1980a; Castleman et al., 1982), evidence to support proposed identifications of cluster species in the atmosphere (Keesee et al., 1979b), and extended understanding of transition metal complexes have resulted from the investigations (Holland and Castleman, 1982). In addition to the aforementioned, the results have been useful in developing an interpretation of the formation mechanisms of similar cluster species that have been observed in surface analysis studies utilizing a number of currently accessible surface bombardment techniques.

In comparing strengths of first ligand-ion bonds for the case of alkali metal ions, a strong radial dependence is observed. This would be expected for systems having essentially only electrostatic contributions to their bonding. The first ligand is found to bind to cations of the transition metal series much more strongly. As borne out by extended Hückel calculations (Castleman, 1978), this is evidence for the presence of partial covalent hybrid bonding. The larger enthalpies of ammoniation relative to hydration of alkali metals have been traced to the greater polarizability of the ammonia molecule, which more than overcomes the contribution due to the large dipole moment of water.

A number of important findings have been derived from studies of successive ligand clustering. The observation of breaks in plots of successive enthalpy changes as a function of the number of ligands illustrates the formation of coordination shells and the presence of analogous solvation breaks (Castleman, 1979c; Holland and Castleman, 1980b, 1982; Tang and Castleman,

TABLE 2. Selected Bond Energies for Cation/Ligand Systems: $A^+ B_n + B = A^+ B_{n+1}$

Cation	n=0	n=1	n=2	n=3	n=4	n=5	Reference
			$-\Delta H^o_{n,n+1}$ (Hydration, kcal/mol)				
H^+	165[a]	31.6	19.5	17.5	15.3	13	Kebarle et al. (1967), Payzant and Kebarle (1972)
NH_4^+	17.3	14.7	13.4	12.2	9.7	-	Payzant et al. (1973)
Li^+	34.0	25.8	20.7	16.4	13.9	12.1	Dzidic and Kebarle (1970)
Na^+	24.0	19.8	15.8	13.8	12.3	10.7	Dzidic and Kebarle (1970)
K^+	17.9	16.1	13.2	11.8	10.7	10.0	Dzidic and Kebarle (1970)
Rb^+	15.9	13.6	12.2	11.2	10.5	-	Dzidic and Kebarle (1970)
Cs^+	13.7	12.5	11.2	10.6	-	-	Dzidic and Kebarle (1970)
Sr^+	34.5	30.5	25.7	22.3	20.6	18.3	Tang et al. (1976)
Bi^+	22.8	17.7	14.0	12.0	10.5	-	Tang and Castleman (1974)
Pb^+	22.4	16.9	12.2	10.8	10.0	9.6	Tang and Castleman (1972)
Cu^+	-	-	16.4	16.7	14.0	-	Holland and Castleman (1982)
Ag^+	33.3	25.4	15.0	14.9	13.7	13.3	Castleman et al. (1981a)
			$-\Delta H^o_{n,n+1}$ (Ammoniation, kcal/mol)				
NH_4^+	25.4	17.3	14.2	11.8	-	-	Tang and Castleman (1975)
Li^+	38.8[b]	33.1	21.0	16.5	11.1	9.3	Castleman et al. (1978b)
Na^+	29.1	22.9	17.1	14.7	10.7	9.7	Castleman et al. (1978b)
K^+	20.1	16.3	13.5	11.6	-	-	Castleman (1978)
Rb^+	18.7	15.2	13.1	11.4	10.2	-	Castleman (1978)
Bi^+	35.5	23.2	13.4	9[c]	-	-	Castleman (1978)
Cu^+	-	-	14.0	12.8	12.8	-	Holland and Castleman (1982)
Ag^+	~47[d]	36.9	14.6	13.0	12.8	-	Castleman et al. (1981a)

[a]Munson (1965).
[b]Staley and Beauchamp (1975).
[c]Extrapolated.
[d]Estimated.

1975). The preferential binding of ammonia over water to copper and silver ions reverses after the addition of the first two ligands, a cross-over phenomenon explaining the observed solution complexes of these ions (Holland and Castleman, 1982; Fisher and Hall, 1967).

The stability of small gas-phase clusters is influenced by entropy as well as bonding. There are three major contributions to the entropy change for a clustering reaction. These are translational, rotational, and vibrational. A configurational contribution may also need to be considered when distinguishable cluster structures having similar energies exist (Bauer and Frurip, 1977). However, this contribution is essentially related to the number of isomers of comparable energy and should be negligible for small ion clusters.

In the case of an ion-neutral association reaction, the translational contribution is dominant since the combining of two particles into one results in an overall negative sign in the entropy change. The magnitude of the translational contribution to ΔS^o is generally -35 to -48 cal/K·mol. The entropy changes for the various reactions differ primarily because of the rotational and vibrational contributions. Except in the case of the first ligand attachment to an atomic ion, the rotational contribution is of the same sign as the translational one (Castleman et al., 1978a). These two contributions are partially offset by the vibrational frequencies. Consequently, the entropy changes should be similar for ions of similar geometry and clustering energetics. This was found to be the case for data obtained for the hydration of CO_3^- as well as NO_3^- and is generally observed for other ion-ligand systems as well. An accurate calculation of entropy changes, employing standard statistical mechanical procedures, requires information on the structure and vibration frequencies of the ion clusters; unfortunately, in most cases the properties are not reliably known.

As clustering proceeds, the motions of the ligands are constricted due to crowding by neighboring ones. As a result, internal rotational and bending frequencies become higher, but this is countered by a decrease in the stretching frequencies due to weaker bonding. Experimentally, the $\Delta S^o_{n-1,n}$ values are often observed at first to become more negative upon successive clustering. This is an indication that crowding is overcompensating for the effect of weaker bonding. Such trends are instructive regarding an understanding of structural changes upon clustering. Referring to Table 1, the particularly small entropy values for the second and third additions of SO_2 onto NO_2^- may indicate that sulfur dioxide is not directly adding to the ion but instead is avoiding crowding by forming a chain-like cluster.

The Bonding and Stability of Small Neutral-Neutral Complexes and Related Charge Transfer Complexes

In recent years, ever increasing attention has been directed to investigations of van der Waals molecules (see, for instance, Amirav et al. (1981)). However, most of these have involved studies of weakly bound clusters often comprised of noble gas molecules. While data from such systems provide basic information on the formation and stability of van der Waals complexes, they are of limited value in interpreting those of importance in the atmosphere or as prototypes for elucidating the catalytic behavior of surfaces (Messmer, 1979; Borel and Buttet, 1981). The interested reader is referred to a recent review (Hobza and Zahradnik, 1980).

In a recent publication Calo and Narcisi (1980) have made an estimate of the equilibrium number densities for certain van der Waals molecules in an attempt to assess their atmospheric importance. Those considered are: $N_2{\cdot}N_2$, $N_2{\cdot}O_2$, $O_2{\cdot}O_2$, $N_2{\cdot}CO_2$, $O_2{\cdot}CO_2$, $N_2{\cdot}H_2O$, $O_2{\cdot}H_2O$, $CO_2{\cdot}CO_2$, $H_2O{\cdot}H_2O$, $Ar{\cdot}Ar$, $Ar{\cdot}CO_2$, and $Ar{\cdot}H_2O$ for the altitude range 5 to 90 km. It was concluded by the authors that number densities of atmospheric dimer species can be appropriately given by an equilibrium estimate, and that in some cases their concentrations are of the same order of magnitude as those of trace gas constituents. The authors have assessed possible roles of van der Waals molecules in neutral and ionic termolecular reactions as well as in infrared adsorption and emission in the natural atmosphere.

In order to quantify the formation and properties of van der Waals molecules and similar weakly bound clusters, we have conducted a study of the continuous course of change of a number of systems from the gaseous toward the condensed state. The goals of the work have been to investigate the dynamics of cluster formation and the resultant changing properties through studies of bonding of analogous compounds via molecular-beam electric deflection experiments. A number of systems have been chosen for investigation in order to compare the behavior of ones which are stabilized by induced-dipole/induced-dipole interactions with other complexes which involve simple hydrogen bonding, and finally with ones which can form charge transfer complexes upon progressive clustering.

A systematic investigation has been made of the distributions of clusters formed in a freely expanding jet comprised of CO_2, SO_2, NH_3, H_2O, C_6H_6, CH_3OH, C_2H_5OH, HNO_3, HCl, H_2SO_4, and also of some of these involving clusters with water. Particular insight into the dynamics of cluster formation and growth were provided by studies of the isotope effects in hydrogen-bonded complexes. The results have yielded evidence that clusters are vibrationally hot upon formation, and grow progressively following cooling collisions with carrier-gas molecules in the beam expansion. Isotopic enhancement during cluster growth has been observed. The observed cluster distributions are consistent with a simple RRK (Rice, Ramsperger, and Kassel) picture accounting for cluster lifetime and growth due to the energetics involved during the expansion, cooling, and clustering processes. This is a statistical model of unimolecular decomposition in which the rate is taken to be proportional to a function of the excess energy above threshold and the number of vacillations in the molecule (or cluster) (Forst, 1973).

Studies of cluster distributions resulting from association reactions have been particularly valuable in elucidating the changing properties of molecular clusters (Hermann et al., 1981). Results of investigations of the formation of water clusters are of particular interest since they provide insight into the nature of hydrogen-bonded systems existing not only in the gaseous state but also as liquid water and ice. A mass spectrometric analysis of the resulting cluster distribution has revealed a number of interesting features corresponding to weakly bound clusters containing 6, 8, 10, and 12 molecules. The results show that the formation of small clusters occurs through well-defined ring closures, with structures being exhibited which are similar to those occurring in bulk ice crystals. Electron impact ionization of larger clusters showed the existence of special features at clusters $H^+(H_2O)_n$ where n = 21, 26, 28, and 30. Each of these corresponds to known unit cell structures for clathrates in the condensed phase, and it is speculated that what is actually observed in the gas phase is a cluster having an analogous structure and which is stabilized by a mobile proton (Holland and Castleman, 1980a).

Based on this model, the cluster involving 21 water molecules about an H^+ ion is a pentagonal dodecahedron. The findings and related explanation are especially intriguing since they account for the observations of Searcy and Fenn (1974), R. Burke (private communication, 1978), and Rabalais and coworkers (Lancaster et al., 1979). The work in the first two references is based on studies of the expansion of a weakly ionized gas. The results of Rabalais and coworkers give evidence of the formation of similar species during the ion bombardment of ice surfaces.

More recent experiments have been conducted (A. W. Castleman, Jr. and B. D. Kay, unpublished data, 1982; J. Muenter, University of Rochester, private communication, April 1981) which suggest the stability of SO_2-water complexes. These have been found to have an appreciable dipole moment, as evidenced by electric deflection experiments performed in our laboratory. Calculations (Holland and Castleman, 1981a) of the optimized geometry of the $SO_2 \cdot H_2O$ adduct (Figure 2) show the plane of the water molecule to be tilted back from the S-O intermolecular axis, thus projecting a lone electron pair of water at the sulfur of SO_2. This configuration suggests that the adduct can be described as Lewis acid-base pair, analogous to that of the trimethylamine/sulfur-dioxide complex reported in the literature (Good, 1975). Although there is no stable sulfurous acid structure for the $SO_2 \cdot H_2O$ adduct to which it can become rearranged, comparison with the results for other calculations of an $SO_3 \cdot H_2O$ adduct is useful (Holland and Castleman, 1978). The CNDO/2 results give a dipole moment of 6.94 D for the optimized $SO_2 \cdot H_2O$ adduct, compared to 7.28 for the $SO_3 \cdot H_2O$ adduct.

A system of considerable importance is the interaction of SO_3 and water to form the $SO_3 \cdot H_2O$ adduct and the subsequent rearrangement to sulfuric acid. Studies made in our laboratory suggest that SO_3 and water readily cocluster, but the rearrangement requires substantial time. The energy barrier to rearrangement from $SO_3 \cdot H_2O$ to H_2SO_4 is about 10 kcal/mol (Holland and Castleman, 1978). Electric deflection experiments (A. W. Castleman, Jr., and B. D. Kay, unpublished data, 1982) prove that sulfuric acid dimer is a head-to-tails configuration with no permanent dipole moment. The fact that cluster interactions with other molecules are governed by a dipole/induced-dipole term suggests that the clustering reaction may be an especially slow one. This might give rise to a rather low rate of nucleation, or rate of aerosol formation in the case of this particular species.

There is an urgent need for additional studies of cluster structure and stability in order to further elucidate problems involving gas-to-particle conversion among atmospherically important molecules.

Considerations of Gas-to-Particle Conversion

There are two generally recognized mechanisms by which aerosols might be formed in the atmosphere via gas-to-particle conversion. One is a direct nucleation mechanism, by which a gas-phase molecule undergoes a phase transformation involving a heteromolecular (bonding to a dissimilar molecule or to an ion), homogeneous, or heterogeneous process. The second is visualized to occur by interaction between the various preexisting quasi-stable clusters of the aerosol-forming species. It is generally recognized that homogeneous nucleation does not operate in the atmosphere. Likewise, direct heterogeneous processes are predominant in regions where there is a large

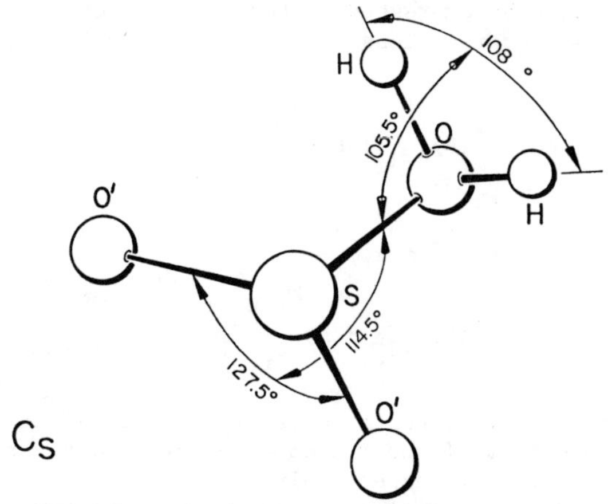

Fig. 2. CNDO/2 optimized geometry of the adduct of sulfur dioxide and water. The calculated interatomic distances are O-H = 1.04 Å, S-O = 1.75 Å, and S-O^1 = 1.52 Å. The symmetry of the complex is C_s.

concentration of preexisting aerosol.

In both the upper atmosphere and the relatively unpolluted troposphere aerosol formation processes involving cluster-cluster interactions provide an important and attractive hypothesis of formation (Castleman and Keesee, 1981; Keesee and Castleman, 1982). The direct heteromolecular mechanisms have many features related to the cluster-cluster model, and much of the data gained in studying one is of direct value in assessing the other. In particular, studies of the nature and stability of quasi-stable cluster species provide important new information needed in formulating theoretical models to account for both processes. The suggestions made by Mohnen (Coffey and Mohnen, 1971) and E. E. Ferguson (private communication, 1977) are extended by F. Arnold (private communication, 1980; Arnold and Fabian, 1980) to include the growth of multiple ion clusters into a particle have implications in both the upper atmosphere and in the troposphere, as do neutral cluster interaction models (Friedlander, 1978; McMurry, 1977; Friend and Vasta, 1980).

As a result of the extensive data coming from laboratory studies, the role of cluster stability in gas-to-particle conversion processes involving a single ionic species has been clarified. When heteromolecular nucleation proceeds under conditions where there is a finite energy barrier (Castleman et al., 1978a; Castleman, 1979b), the formulation may be developed in terms of a steady-state sequence of reactions. For example, ion-induced nucleation may be viewed as proceeding via a sequence of clustering reactions commencing with the third-order reaction shown in equation (1). Whether or not the kinetics eventually become second-order involves a consideration of cluster lifetimes preceding subsequent collision with a third body. In any event, the kinetics can always be expressed in terms of effective second-order reactions:

$$A^+(B)_{n-1} + B \underset{k_{r,n}}{\overset{k_{f,n-1}}{\rightleftharpoons}} A^+(B)_n \tag{4}$$

$$\vdots \qquad \vdots \qquad \vdots$$

$$A^+(B)_{n-1^*} + B \underset{k_{r,n^*}}{\overset{k_{f,n^*-1}}{\rightleftharpoons}} A^+(B)_{n^*} \tag{5}$$

where n* designates the cluster of critical size and subscripts f and r denote forward and reverse rate constants, respectively.

In order to demonstrate the essential nature of the energy barrier to nucleation, the rate of formation of a stable condensed phase, R, can be expressed by (Castleman et al., 1978a; Castleman, 1979a)

$$R = k_{n^*}[B]^{n^*+1} \prod_{n=1}^{n^*} K_{n-1,n} \tag{6}$$

with

$$\prod_{n=1}^{n^*} K_{n-1,n}[B]^n = e^{-\Delta\varphi/kT} \tag{7}$$

In arriving at the above, the fictitious equivalence between equilibrium and steady state has been assumed, and clustering beyond the critical size has been assumed to be very fast and effectively unidirectional. Of course in practice equilibrium does not exist at each cluster, and in the case of finite nucleation rates the steady-state concentrations of the clusters will depart from equilibrium values. In such cases, the well-known Zel'dovich preexponential factor (Zettlemoyer, 1969) must be included in the rate expressions. The energy barrier defined above, $\Delta\varphi$, is referenced to a standard state of 1 atm; account must be taken of the vapor pressure of the condensing phase in order to relate the numerical values to conventional supersaturation ratios.

At the present time there are no general theories of association reactions which can be used to treat the many-body interactions responsible for the formation of a critical size nucleus. Therefore, in cases where the phase transformation is kinetically controlled, recourse has been taken to employing molecular dynamics calculations (McGinty, 1972) or Monte Carlo techniques (Abraham et al., 1976; Mrusik et al., 1976; Bauer et al., 1978). Where a finite energy barrier exists, the nucleation rate can be expressed in terms of (1) a preexponential factor accounting for both the effective collision rate of a vapor with a prenucleation cluster of n molecules (designated an n-mer) and the departure of the distribution from one of equilibrium, and (2) an exponential involving the energy barrier to nucleation. The problem is to formulate an appropriate expression for the free energy of the barrier.

In principle, the energy barrier can be computed with knowledge of the free energy of formation of n-mer, which in turn can be derived from the relationship

$$-\frac{\Delta G^\circ}{RT} = \frac{n\mu_1^\circ - \mu_n^\circ}{RT} = \ln\left[\frac{1}{n!}\,\frac{Qn/v}{(Q_1/v)^n}\right] \tag{8}$$

The symbols μ_1° and μ_n° correspond to the chemical potential of the monomer vapor and n-mer having respective partition functions Q_1 and Q_n, and v is volume. In the case of condensation to the solid rather than the liquid phase, n! is to be replaced by the symmetry number of the crystal. Major difficulty, and hence the origin of the uncertainty, is associated with properly formulating Q_n. Further consideration of these aspects is beyond the scope of this paper; the interested reader is referred to several sources which detail the theoretical difficulties (Lee et al., 1980a; Bauer and Frurip, 1977; Reiss, 1977; Binder and Stauffer, 1976; Abraham, 1974).

In order to avoid the exceedingly difficult task of calculating Q_n from basic principles of equilibrium statistical mechanics, it has become common to evaluate the total energy barrier to homogeneous nucleation in terms of an expression for the influence of surface tension on vapor pressure, i.e., the Kelvin equation (Abraham, 1974; Mason, 1971; Holland and Castleman, 1981b; Heicklen, 1976). The appropriate equation for evaluating the free energy of formation of a charged droplet is referred to as the Thomson equation and is given by

$$\Delta G^\circ_{n-1,n} = NkT\ln p^\circ + \sigma\left[\frac{32\pi NM^2}{3\rho^2 n}\right]^{1/3} - \frac{q^2}{8\pi}\left[1-\frac{1}{\varepsilon}\right]\left[\frac{4\pi N}{3n}\right]^{4/3}\left[\frac{\rho}{M}\right]^{1/3} \tag{9}$$

where p° is the vapor pressure of the condensing molecules at the temperature of the system, N is Avogadro's number, M is the molecular weight, ρ is the bulk density of the clustering molecules, σ is the surface tension, and q is the elementary charge.

In the case of all ions of atmospheric interest, the Thomson equation has been shown to represent well the bonding for the successive clustering of molecules about ions of both simple atomic and complex molecular configuration. A model developed by Castleman and coworkers (Lee et al., 1980a) now enables the bonding and related enthalpy change for clustering to be evaluated for all cluster sizes, ranging from a few hydrations up to and including the bulk phase. The validity of the correlation for both positive and negative ions is shown in Figures 3 and 4. Referring to the figures, it can be seen that the ratio of the ΔH^o_{solv} (standard heat of solvation) to the summed gas-phase enthalpies $\Delta H^o_{0,n}$ rapidly approaches a value which is independent of the nature of the ion for both positive and negative systems. The relationship which leads to this correlation is intimately related to the Born equation, which expresses the energy necessary to place a charge in the cavity of a dielectric medium.

In those situations where the classical formulations do not hold, it has been found that the failure of the equation is due to its inability to appropriately account for the structure of the evolving prenucleation embryo (Castleman, 1979b). In contrast to the assumptions in the theoretical model, the data suggest that small ion clusters have a relatively ordered structure, and this provides a qualitative picture of the reasons why the Thomson equation is not valid for certain systems. It is interesting to note that no variation in the dielectric constant can give accord between the predicted and experimental values. However, an empirical adjustment of the size of the ion has been shown to give much better agreement between the experimentally measured entropy and enthalpy values and those predicted for the Thomson equation (Holland and Castleman, 1981b). A recent modification of the Fisher droplet model for nucleation, developed by Chan and Mohnen

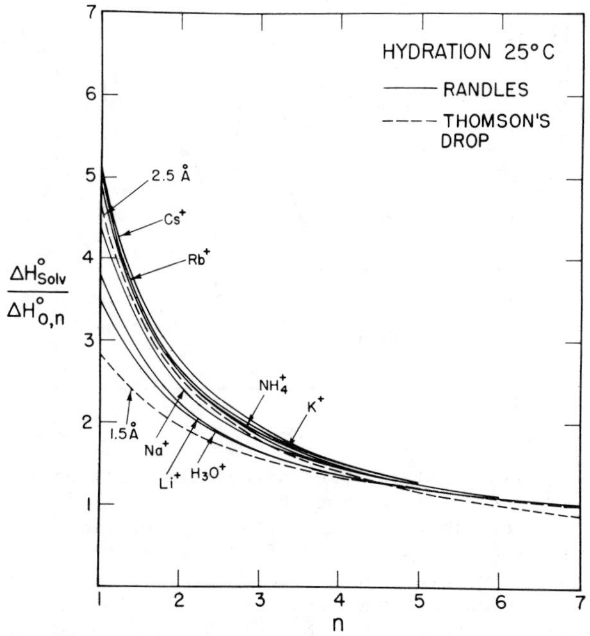

Fig. 3. Ratio of Randles' total enthalpy of solvation to the partial gas-phase enthalpy of hydration for positive ionic cluster size n.

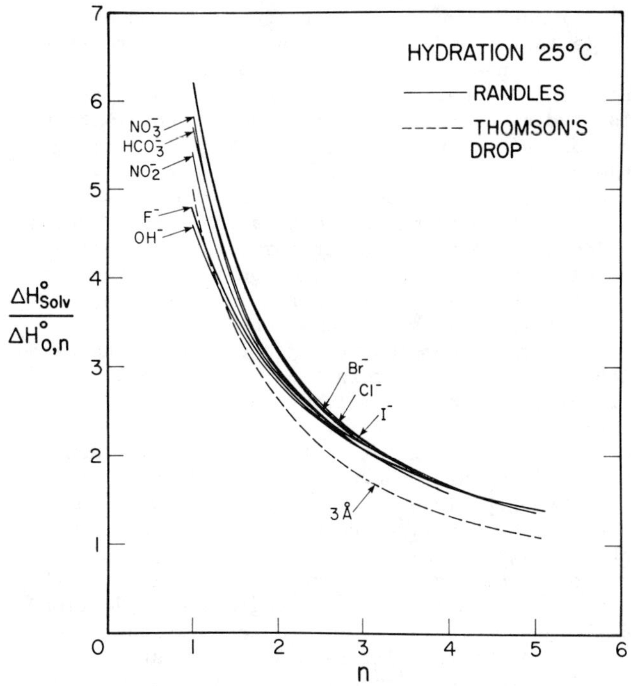

Fig. 4. Ratio of Randles' total enthalpy of solvation to the partial gas-phase enthalpy of hydration for negative ionic cluster size n.

(1980), displays a similar result. Their model is presently the most satisfactory one for accounting for the general process of ion-induced nucleation in terms of a semiempirical theory. A major need in this field is for more work on the structure of small clusters.

The alternate mechanism of gas-to-particle conversion proposed by Mohnen (Coffey and Mohnen, 1971) and independently by E. E. Ferguson (priate communication, 1977), which accounts for particle formation via the interaction of ionic species of unlike sign in the atmosphere, is especially attractive. Ferguson's hypothesis is based on a consideration of the ionization potentials and electron affinities of small clusters for situations that might arise where, after sufficient clustering to an ion, charge neutralization would not be possible. Yet, because of the opposite charges of the interacting ionic species, they would tend to become associated in a small complex, perhaps initiating nucleation through the formation of a small electrolyte-like droplet. This idea has been extended by F. Arnold (private communication, 1980; Arnold and Fabian, 1980) to include the growth of multiple ion clusters in a particle.

At the present time there are no definitive laboratory results to establish the fact that positive and negative ion complex interactions can lead to the formation of stable "charged" condensation embryos. The work of Smith and Adams

(1980) has shown that the recombination coefficients for negative and positive ions are remarkably independent of the complexity of the reacting ion pairs. What is most urgently needed is studies of the products of such recombination and an assessment of their stability in forming the proposed charge-transfer complexes.

Research is in progress in our laboratory on the formation of mixed clusters involving species which form charge-transfer complexes, and ultimately ion pairs, upon solvation in the condensed phase. This especially intriguing question is directly related to the one concerning the products of the recombination of positive and negative ion clusters upon interaction. It would be electrically neutral overall but have separated charges contained within.

Recent ab initio calculations by Kollman and Kuntz (1976) have shown that the hydration of NH_4F by only six water molecules should be sufficient to stabilize the solvated ion pair $NH_4^+(H_2O)_4F^-$. This suggests that other ionic compounds such as strong acids might also undergo ion pair formation at small degrees of hydration. Although no similar ab initio calculations have been done for the nitric-acid/water system, it is possible to make reasonable estimates of whether ion pair formation can occur when one nitric acid is hydrated with five water molecules. A calculation of the separation distance for H^+ and NO_3^- in such a cluster was made by determining the point at which the reaction becomes thermoneutral (Kay et al., 1981; Castleman et al., 1981b):

$$HNO_3 + 5H_2O \rightleftharpoons H^+(H_2O)_5NO_3^- \qquad (10)$$

Employing data obtained in our laboratory and utilizing other data available in the literature, the coulomb energy for two charges separated by a given distance can be made in a straightforward fashion if, as they are brought into close proximity, the hydrogen bonding between the water molecules on the two charges is neglected. In actual fact hydrogen bonding of the ligands on the individual charges has been accounted for since actual gas-phase thermochemical data are used in the calculation. Consequently, the magnitude of the error made in this approximation is expected to be less than the energy equivalent to the addition of one more water molecule. Based on the above, it is found that the thermal neutral distance is approximately 4.1 Å, thereby suggesting the existence of a large dipole moment in such clusters.

In order to provide information on the nature of interactions between ion pairs in solution and related problems in terms of these cluster types, we have undertaken studies of clustering between nitric acid and water molecules. Clusters corresponding to $(HNO_3)_x(H_2O)_y$ (where $1 \leq x \leq 6$ and y ranged to 20) were observed (Kay et al., 1981; Castleman et al., 1981b). For large values of y and in situations where x equals unity, species of the type H_3O^+ and NO_3^- coclustered with water (analogous to the recombination products of these two hydrated ion species) will eventually form.

Experimental studies of the neutral cluster distribution of nitric acid in water revealed an intriguing finding concerning the lack of stability of certain mixed clusters of particular composition. While all sizes of clusters produced by expanding water vapor in a carrier gas are found to be stable, experiments in which two or more nitric acid molecules were present in the aqueous cluster revealed instability in cases where the nitric acid to water ratios exceeded certain critical values (see Figure 5). The general shape of the envelope of the observed

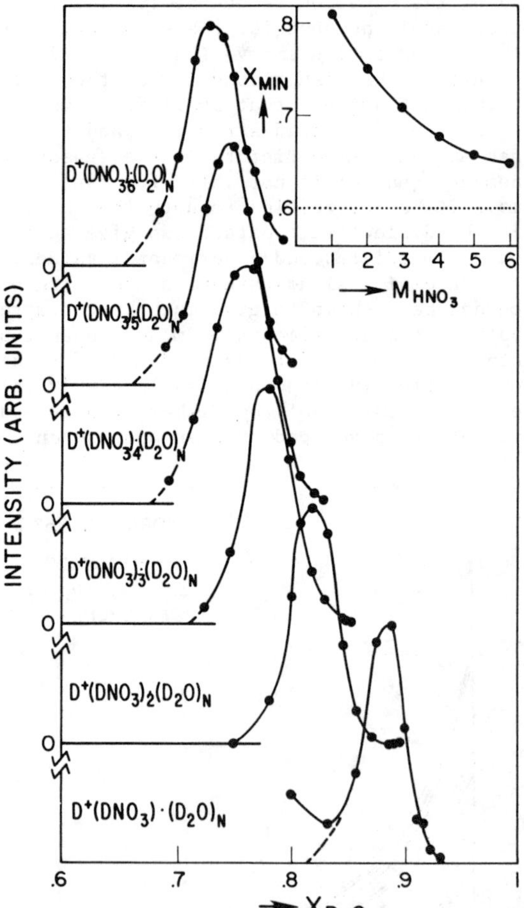

Fig. 5. Linear plot of normalized intensity versus mole-fraction of D_2O (X_{H_2O}) for cluster distributions containing between one and six DNO_3 molecules. The insert in the upper right corner displays a plot of the extrapolated onset mole-fraction (X_{MIN}) versus the number of DNO_3 molecules contained in the cluster. The horizontal dashed line corresponds to the bulk-phase azeotropic composition of nitric acid/water solutions (from Kay et al., 1981).

cluster distributions was found to depend on the expansion characteristics of the molecular beam source. However, the size of cluster for which onset of a particular clustering sequence (involving a cluster with a specific number of nitric acid molecules) begins was found to be independent of all experimentally accessible parameters.

In order to interpret the results, the equivalent concentration corresponding to the first stable cluster in a given sequence was converted to a mole fraction. The results are plotted versus the number of nitric acid molecules in the cluster in the insert in the upper right corner of Figure 5. When plotted in this fashion, the composition corresponding to the first stable clusters is seen to approach the bulk phase composition (indicated by the dashed line) at which binary water/nitric-acid solutions, more enriched in HNO_3, are found to be both photochemically and thermally unstable. Here the composition x refers to that of the parent cluster. Such solutions decompose to give NO_2 and oxygen. These findings strongly suggest that small clusters begin to display properties normally associated with the bulk liquid phase. Furthermore, they indicate that ion pair formation is occurring in these clusters in a manner analogous to that anticipated from the suggestions of Mohnen (Coffey and Mohnen, 1971), E. E. Ferguson (private communication, 1977), and F. Arnold (private communication, 1980; Arnold and Fabian, 1980).

The presence of a very large dipole resulting from charge separation should lead to enhanced clustering rates. Such effects are indicated in the distributions measured for nitric-acid/water clusters as shown by the maxima in Figure 5, again providing further evidence that such clusters having large dipoles may be important in the formation of hydrated complexes.

Recently, unique mixed-cluster species were observed in our laboratory during hydration studies of alkali metal ions (Märk et al., 1980). The direct observation of water clustered to sodium dimer ions and of alkali metal hydroxides clustered to alkali metal ions confirms the stability of such species, which could be considered as being equivalent to multi-ion complexes suggested as being potential aerosol-contributing species in the upper atmosphere.

Research in this field is only just beginning, and there is an urgent need for attention to virtually all systems for which hydration complexes might potentially lead to the formation of charge transfer entities.

In order to provide additional information on the various interrelated processes associated with the large dipole moments expected for certain mixed clusters, we have commenced experiments employing both first- (electrostatic hexapole) and second-order (electrostatic quadrupole) Stark focusing in order to investigate the differing dipole moments, and therefore structures, associated with clusters of varying size and composition (A. W. Castleman, Jr. and B. D. Kay, unpublished data, 1982). In agreement with theory and earlier experiments, the water dimer was found to have a large dipole moment, while higher-order clusters display no focusing up to clusters comprised of nine molecules. Other recent focusing experiments made in our laboratory have shown that the molecule $HNO_3 \cdot H_2O$ is polar. Higher order clusters involving HNO_3 and the HNO_3 dimer clustered to water, up to and including the seventh water molecule, display no first- or second-order focusing, possibly due to the presence of a mobile proton on the clusters. Cluster focusing experiments have also been made for $(NH_3)_n$, $(CH_3OH)_n$, $(CD_3OD)_n$, $(C_6H_6)_n$, $(SO_2)_n$, $(H_2SO_4)_n$, HCl, and some mixed clusters with H_2O (A. W. Castleman, Jr. and B. D. Kay, unpublished data, 1982).

Implications of Cluster Research to Other Problems in the Field of Atmospheric Science

Work on atmospherically important clusters is important in understanding many other related problems in the atmospheric sciences. Included among these are questions concerning the fate of the products of ion-cluster/ion-cluster recombination and related problems of aerosol charge acquisition, the influence of small clusters and aerosols on the propagation of electromagnetic radiation, and the role of clustered species in homogeneous gas-phase reactions.

A significant number of atmospheric reactions are believed to be termolecular, and it is clear that if one of the reactants is coclustered, this could appreciably influence reaction rates since only one particle in the complex could be released and thereby remove energy during the reaction step. Such effects have been suggested in the transformation of NO^+ to H_3O^+ in the upper atmosphere, and may be important in other clustering reactions as well. In particular, Hamilton and Naleway (1976) have suggested that the free radical HO_2 may cocluster easily with water vapor and ammonia, thereby influencing reaction rates involving this particular species.

It has been suggested that electron donor-acceptor complexes of SO_2 are involved in its photooxidation (Richards et al., 1976). In this regard, other systems for which data are needed in order to interpret their atmospheric behavior and assess their possible influence on reactivity due to the extent of clustering include SO_2 with alkenes, alkynes, aromatic hydrocarbons, ammonia, amines, and NO_2 (Pitts, 1978). Regarding (1) potential clustering with H_2O, and (2) understanding ultimate solvation in the aqueous phase, data are needed on H_2CO (Gordon et al., 1975), which is an important intermediate in the oxidation of methane and the NO_x's, including NO_2 and HNO_2.

Electronic States of Small Clusters Bonding, Surface Studies, and Charge Acquisition

Based on numerous papers presented at a recent international conference devoted to the subject of small inorganic clusters (Borel and Buttet, 1981), it is clear that investigations of the states of aggregates of small inorganic compounds are useful in order to further an understanding of the role of small particulates in catalyzing surface reactions. Numerous calculations (Messmer, 1979; Robinson, 1976; Slater and Johnson, 1974; Hamilton and Baetzold, 1979) have demonstrated the importance of information on the electronic states of small clusters as a model for understanding surface reactivity. Important work in this area has been initiated by Schumacher and coworkers (Hermann et al., 1978) on alkali metal systems. Investigations of coclustered ligands on metal aggregates including transition metals are urgently needed. Work in this area is commencing in our laboratory.

In conjunction with the investigations referred to here that will lead to information on ionization potentials as well as changing electronic properties with cluster size, data on the attachment of ligands to small clusters can be obtained from investigations of the appearance potentials of metal-ligand complexes. Consequently, the same studies which are relevant in determining the ionization potential and therefore the charging of small clusters and aerosols can in some cases be combined with the results of the ion-ligand bonding to deduce the requisite neutral-cluster bond energies.

Aerosol particles can be altered by subsequent condensation of additional vapor or by coagulation with small charged and uncharged molecular clusters and with other aerosols. It is generally believed that aerosols can acquire a charge by interaction with atmospheric ions, but the mechanisms of charge acquisition and transfer are not well studied. In this context, there is need for basic studies of ionization potential as a function of cluster size and composition.

Conclusions

Nucleation from the gaseous to the aerosol state commences via the formation of clusters among molecules participating in the phase-transformation process. Nucleation may proceed in some cases via the formation of prenucleation embryos which subsequently evolve through the energy barrier and undergo phase transformation. In other cases, cluster-cluster interaction among neutral particles or stagewise building of alternate-sign ion clusters may be important in the gas-to-particle conversion process.

An extensive investigation of the bonding of atmospherically important molecules to various ions has been made in our laboratory, and the results have clarified the overall gas-to-particle conversion process involving a single ionic species. In particular, the results have provided direct information on processes of atmospheric importance as well as giving basic insight into the theory of nucleation. By carefully selecting the ions and polar molecules clustering about them and conducting the experiments with systems which can be most readily treated theoretically, a great deal of information on the basic phenomenon was obtained. The results have led to an understanding of the limitations of the conventional theoretical methods often used to treat nucleation and aerosol formation. The data have also found application in interpreting the nature of the interactions between ions and molecules, e.g., those which occur in microdroplets or the solution phase constituting the liquid part of an aerosol particle.

At this point no definitive conclusions can be reached regarding the importance of single ion species in effecting nucleation in the atmosphere. Based on considerations of the energy barrier to nucleation, it is likely, at least in the case of the perturbed atmosphere, that ion-induced nucleation does represent some contribution to aerosol formation processes. Calculations by Mohnen and coworkers (Chan et al., 1978) demonstrate that species such as $H_3O^+(H_2O)_n \cdot (H_2SO_4)_x$ can form optimal cluster combinations under conditions existing in the stratosphere. These ions seem to be likely candidates for the formation of aerosols by a single ionic-type species. The findings of Mohnen (Mohnen and Kadlecek, 1980) regarding the unusual stability of the species $NO_2^-(HNO_2)$ suggest that this complex may also need to be considered in any nucleation scheme. Its stability has also been verified by the work of Castleman and coworkers (Keesee et al., 1980), which determined the NO_2^--HNO_2 bond energy to be 33 kcal/mol.

The reactive nature of constituents both in the aerosol phase and adsorbed on aerosol surfaces is influenced largely by the nature of the electronic interactions with neighboring molecules; these can also be viewed as a type of cluster-cluster interaction. For this reason there is considerable interest in studies of clusters among molecules of atmospheric importance, both those which may exist in the gas phase as well as those formed between gas- and condensed-phase partners. There is also growing evidence that even atmospheric homogeneous chemical conversions are influenced by the clustered nature of certain reactant molecules, especially those involving free-radical species. Consequently, the results of such studies will have an impact on the area of homogeneous as well as heterogeneous chemistry. Investigations of reactions among cluster ions as well as the interconversion of ionic species in such clusters provide further information on the nature of reactions which may proceed in the droplet phase surrounding aerosol particles.

The state of matter midway between the gaseous and the condensed is one for which there is growing interest, yet a paucity of basic information.

Future research in this area should be fruitful in unraveling important problems in the basic field of chemical physics, with many important applications in the field of atmospheric sciences.

Acknowledgements. Support of the Atmospheric Sciences Section of the National Science Foundation under Grant No. ATM 79-13801, the National Aeronautics and Space Administration under Grant No. NSG-2248, the Department of Energy under Grant No. DE-AC02-78EV04776, and the U.S. Army Research Office under Grant No. DAAG-29-79-C-0133 is gratefully acknowledged.

References

Abraham, F. F., Homogeneous Nucleation Theory: The Pretransition Theory of Vapor Condensation, Supplement No. 1 to Advances in Theoretical Chemistry, Academic Press, Inc., N.Y., 1974.

Abraham, F. F., M. R. Mrusik, and G. M. Pound, The thermodynamics and structure of hydrated halide and alkali ions, Faraday Disc. Chem. Soc., 61, 34-47, 1976.

Amirav, A., U. Even, and Joshua Jortner, Microscopic solvation effects on the excited state energetics and dynamics of aromatic molecules in large van der Waals complexes, J. Chem. Phys., 6, 2489, 1981.

Arnold, F., and R. Fabian, Changes in the optical light curve of the Crab pulsar between 1970 and 1977, Nature, 283, 50-51, 1980.

Arnold, F., R. Fabian, E. E. Ferguson, and W. Joos, Mass spectrometric measurements of fractional ion abundances in the stratosphere II: Negative ions, Planet. Space Sci., 29, 195, 1981a.

Arnold, F., G. Henschen, and E. E. Ferguson, Mass spectrometric measurements of fractional ion abundances in the stratosphere I: Positive ions, Planet. Space Sci., 29, 185-193, 1981b.

Bauer, S. H., and D. J. Frurip, Homogeneous nucleation in metal vapors. 5. A self-consistent kinetic model, J. Phys. Chem., 81, 1015-1024, 1977.

Bauer, S. H., C. F. Wilcox, Jr., and S. Russo, Stochastic simulation of homogeneous condensation, J. Phys. Chem., 82, 59-62, 1978.

Binder, K., and D. Stauffer, Statistical theory on nucleation, condensation and coagulation, Adv. Phys., 25, 343-396, 1976.

Borel, J.-P., and J. Buttet (Eds.), Small particles and inorganic clusters, Surface Sci., 106 (1-3), 1981.

Bricard, J., and J. Pradel, Electric charge and radioactivity of naturally occurring aerosols, in Aerosol Science, edited by C. N. Davies, p. 87, Academic Press, London, 1966.

Caledonia, G. E., Survey of the gas-phase negative ion kinetics of inorganic molecules. Electron attachment reactions, Chem. Rev., 75, 333-351, 1975.

Calo, J. M., and R. S. Narcisi, Van der Waals molecules - possible roles in the atmosphere, Geophys. Res. Lett., 7, 289-292, 1980.

Castleman, A. W., Jr., The properties of clusters in the gas phase: Ammonia about bismuth (1+), rubidium (1+), and potassium (1+), Chem. Phys. Lett., 53, 560-564, 1978.

Castleman, A. W., Jr., Advances in colloid and interface science, in Nucleation, edited by A. Zettlemoyer, pp. 73-128, Elsevier Press, Oxford, 1979a.

Castleman, A. W., Jr., A reconsideration of nucleation phenomena in light of recent findings concerning the properties of small clusters and a brief review of some other particle growth processes, Astrophys. Space Sci., 65, 337-349, 1979b.

Castleman, A. W., Jr., Studies of ion clusters: Relationship to understanding nucleation and solvation phenomena, in Kinetics of Ion-Molecule Reactions, edited by P. Ausloos, Plenum Press, N.Y., pp. 295-321, 1979c.

Castleman, A. W., Jr., P. M. Holland, B. D. Kay, R. G. Keesee, T. D. Märk, K. I. Peterson, F. J. Schelling, and B. L. Upschulte, Studies of small clusters: Relationship to understanding nucleation and surfaces, J. Vac. Sci. Technol., 18, 586, 1981a.

Castleman, A. W., Jr., P. M. Holland, and R. G. Keesee, The properties of ion clusters and their relationship to heteromolecular nucleation, J. Chem. Phys., 68, 1760-1767, 1978a.

Castleman, A. W., Jr., P. M. Holland, and R. G. Keesee, Ion association processes and ion clustering: Elucidating transitions from the gaseous to the condensed phase, Radiat. Phys. Chem., in press, 1982.

Castleman, A. W., Jr., P. M. Holland, D. M. Lindsay, and K. I. Peterson, The properties of clusters in the gas phase. 2. Ammonia about metal ions, J. Am. Chem. Soc., 100, 6039-6045, 1978b.

Castleman, A. W., Jr., B. D. Kay, V. Hermann, P. M. Holland, and T. D. Märk, Studies of the formation and structure of homomolecular and heteromolecular clusters, Surface Sci., 106, 179, 1981b.

Castleman, A. W., Jr., and R. G. Keesee, Nucleation and growth of stratospheric aerosols, Ann. Rev. Earth Planet. Sci., 9, 227-250, 1981.

Chan, L. Y., J. A. Kadlacek, V. A. Mohnen, and J. DelSanto, Ion Molecule Interactions of Atmospheric Importance, Interim Report to National Science Foundation, ASRC Publication No. 681, State Univ. of N.Y., Albany, 1978.

Chan, L. Y., and V. A. Mohnen, Ion nucleation theory, J. Atmos. Sci., 37, 2323-2331, 1980.

Coffey, P., and V. A. Mohnen, Ion-induced large cluster formation, paper presented at 24th Annual Gaseous Electronics Conference, Univ. of Florida, Gainesville, October 1971.

Dzidic, I., and P. Kebarle, Hydration of the alkali ions in the gas phase. Enthalpies and entropies of reactions $M^+(H_2O)_{n-1} + H_2O \rightarrow M^+(H_2O)_n$, J. Phys. Chem., 74, 1466-1485, 1970.

Ferguson, E. E., Ionospheric ion-molecule reactions, in Interactions Between Ions and Molecules, edited by P. Ausloos, p. 313, Plenum Press, New York, 1975.

Fisher, J. F., and J. L. Hall, Polarographic study of copper (III) complexes of mono-, di-, and triethanolamine, Anal. Chem., 39, 1550, 1967.

Forst, W., Theory of Unimolecular Reactions, Academic Press, New York, 1973.

Franklin, J. L., and P. W. Harland, Gaseous negative ions, Ann. Rev. Phys. Chem., 25, 485-526, 1974.

Friedlander, S. K., Theory of new particle formation in the presence of an aerosol, in Proc. Gesellschaft fuer Aerosolforschung (GAF) Conf., Vienna, 1978.

Friend, J. P., and R. Vasta, Nucleation by free radicals from the photooxidation of sulfur dioxide in air, submitted to J. Phys. Chem., 84, 2423, 1980.

Good, A., Third-order ion-molecule clustering reactions, Chem. Rev., 75, 561-583, 1975.

Gordon, M. S., D. E. Tallman, C. Monroe, M. Steinbach, and J. Armbrust, Localized orbital studies of hydrogen bonding. II. Dimers containing H_2O, HN_3, HF, H_2CO, and HCN, J. Am. Chem. Soc., 97, 1326-1333, 1975.

Hamilton, E. J., and C. A. Naleway, Theoretical calculation of strong complex formation by the HO_2 radical: $HO_2 \cdot H_2O$ and $HO_2 \cdot NH_3$, J. Phys. Chem., 80, 2037-2040, 1976.

Hamilton, J. F., and R. C. Baetzold, Catalysis by small metal clusters. Size-dependent catalytic activity can be correlated with changes in physical properties, Science, 205, 1213-1220, 1979.

Heicklen, J., Colloid Formation and Growth, A Chemical Kinetics Approach, Academic Press, New York, 1976.

Hermann, A., S. Leutwyler, E. Schumacher, and L. Woeste, On metal-atom cluster. IV. Photoionization thresholds and multiphoton ionization spectra of alkali metal molecules, Helv. Chim. Acta, 61, 453-487, 1978.

Hermann, V., B. D. Kay, and A. W. Castleman, Jr., Evidence for the existence of structures in gas-phase homomolecular clusters of water, submitted to Chem. Phys., 1981.

Hobza, P., and R. Zahradnik, Van der Waals systems: Molecular orbitals, physical properties, thermodynamics of formation and reactivity, in Topics in Current Chemistry, vol. 93, pp. 53-90, Springer-Verlag, Berlin/Heidelberg, 1980.

Holland, P. M., and A. W. Castleman, Jr., Gas-phase complexes: Considerations of the stability of clusters in the sulfur trioxide-water system, Chem. Phys. Lett., 56, 511-514, 1978.

Holland, P. M., and A. W. Castleman, Jr., A model for the formation and stabilization of charged water clathrates, J. Chem. Phys., 72, 5984-5990, 1980a.

Holland, P. M., and A. W. Castleman, Jr., Gas-phase complexes of Cu^+ and Ag^+ via thermoionic emission sources, J. Am. Chem. Soc., 102, 6174-6175, 1980b.

Holland, P. M., and A. W. Castleman, Jr., Structure and stability of the gas-phase complex of SO_2 with H_2O, J. Photochem., 16, 347, 1981a.

Holland, P. M., and A. W. Castleman, Jr., The Thomson equation revisited: Formulations of the energy barrier to nucleation in light of ion clustering experiments, submitted to J. Phys. Chem., 1981b.

Holland, P. M., and A. W. Castleman, Jr., The thermochemical properties of transition metal ion complexes, J. Chem. Phys., in press, 1982.

Kay, B. D., V. Hermann, and A. W. Castleman, Jr., Studies of gas-phase clusters: The solvation of HNO_3 in microscopic aqueous clusters, Chem. Phys. Lett., 80, 469, 1981.

Kebarle, P., S. K. Searles, A. Zolla, J. Scarborough, and M. Arshadi, The solvation of the hydrogen ion by water molecules in the gas phase. Heats and entropies of solvation of individual reactions: $H^+(H_2O)_{n-1} + H_2O \rightarrow H^+(H_2O)_n$, J. Am. Chem. Soc., 89, 6393-6399, 1967.

Keesee, R. G., and A. W. Castleman, Jr., The chemical kinetics of aerosol formation, in Stratospheric Aerosols, edited by R. C. Whitten, Springer-Verlag, Berlin, Heidelberg, New York, in press, 1982.

Keesee, R. G., N. Lee, and A. W. Castleman, Jr., Properties of clusters in the gas phase. 3. Hydration complexes of CO_3^- and HCO_3^-, J. Am. Chem. Soc., 101, 2599-2604, 1979a.

Keesee, R. G., N. Lee, and A. W. Castleman, Jr., Atmospheric negative ion hydration derived from laboratory results and comparison to rocket-borne measurements in the lower ionosphere, J. Geophys. Res., 84, 3719-3722, 1979b.

Keesee, R. G., N. Lee, and A. W. Castleman, Jr., Properties of clusters in the gas phase: V. Complexes of neutral molecules onto negative ions, J. Chem. Phys., 73, 2195-2202, 1980.

Klots, C. E., and R. N. Compton, Electron attachment to carbon dioxide clusters in a supersonic beam, J. Chem. Phys., 67, 1779-1780, 1977.

Kollman, P., and I. Kuntz, Hydration of NH_4F, J. Am. Chem. Soc., 98, 6820-6825, 1976.

Lancaster, G. M., F. Honda, Y. Fukuda, and J. W. Rabalais, Secondary ion mass spectrometry of molecular solids. Cluster formation during ion bombardment of frozen water, benzene, and cyclohexane, J. Am. Chem. Soc., 101, 1951-1958, 1979.

Lee, N., R. G. Keesee, and A. W. Castleman, Jr., On the correlation of total and partial enthalpies of ion solvation and the relationship to the energy barrier to nucleation, J. Colloid Inter. Sci., 75, 555-565, 1980a.

Lee, N., R. G. Keesee, and A. W. Castleman, Jr., The properties of clusters in the gas phase. IV. Complexes of H_2O and HNO_x clustering on NO_x^-, J. Chem. Phys., 72, 1089-1094, 1980b.

Märk, T. D., K. I. Peterson, and A. W. Castleman, Jr., New gas phase inorganic ion cluster species and their atmospheric implications, Nature, 285, 392-393, 1980.

Mason, B. J., The Physics of Clouds, 2nd ed., Clarendon Press, Oxford, 1971.

McDaniel, E. W., and E. A. Mason, Mobility and Diffusion of Ions in Gases, John Wiley and Sons, New York, 1973.

McGinty, D. J., Single-configuration approximation in the calculation of the thermodynamic properties of microcrystalline clusters, Chem. Phys. Lett., 13, 525-528, 1972.

McMurry, P. H., On the relationship between aerosol dynamics and the rate of gas-to-particle conversion, Ph. D. Thesis, Calif. Inst. Tech., Pasadena, 1977.

Messmer, R. P., Cluster model theory and its application to metal surface-adsorbate systems, in The Nature of Surface Chemical Bond, edited by T. N. Rhodin and G. Ertl, p. 51, North Holland Publishing Co., Amsterdam, 1979.

Mohnen, V. A., Discussion of the formation of major positive and negative ions up to the 50 km level, Pure Appl. Geophys., 84, 141, 1971a.

Mohnen, V. A., Discussion of the formation of major positive and negative ions up to the 50 km level, in Mesospheric Models and Related Experiments, pp. 210-219, D. Reidel Publ. Co., Dordrecht, Holland, 1971b.

Mohnen, V. A., and J. A. Kadlacek, Nature, mobility and pysico-chemical reactivity of ions in the lower atmosphere (less than 40 km), paper presented at VIth International Conf. on Atmospheric Electricity, Manchester, England, July 1980.

Mrusik, M. R., F. F. Abraham, and D. E. Schreiber, A Monte Carlo study of ion-water clusters, J. Chem. Phys., 64, 481-491, 1976.

Munson, M. S. B., Proton affinities and the methyl inductive effect, J. Am. Chem. Soc., 87, 2332-2336, 1965.

Narcisi, R. S., On water cluster ions in the ionospheric D region, in Planetary Electrodynamics, edited by S. C. Coroniti and J. Hughes, p. 447, Gordon Breach, New York, 1969.

Paulson, J. F., Some negative ion reactions with CO_2, J. Chem. Phys., 52, 963-964, 1970.

Payzant, J. D., A. J. Cunningham, and P. Kebarle, Kinetics and rate constants of reactions leading to hydration of NO_2^- and NO_3^- in gaseous oxygen, argon, and helium containing traces of water, Can. J. Chem., 51, 2230-2235, 1973.

Payzant, J. D. and P. Kebarle, A kinetic study of the proton hydrate $H^+(H_2O)_n$ equilibria in the gas phase, J. Am. Chem. Soc., 94, 7627-7632, 1972.

Payzant, J. D., R. Yamdagni, and P. Kebarle, Hydration of CN^-, NO_2^-, and OH^- in the gas phase, Can. J. Chem., 49, 3308-3314, 1971.

Pitts, J. N., Jr., Mechanisms, models, and myths: Fiction and fact in tropospheric chemistry, in Man's Impact on the Troposphere, edited by Joel S. Levine and David Schryer, NASA RP-1022, pp. 1-25, 1978.

Reiss, H., The replacement free energy in nucleation theory, Adv. Colloid Interface Sci., 7, 1-66, 1977.

Richards, J., D. L. Fox, and P. C. Reist, The influence of molecular complexes on the photooxidation of sulfur dioxide, J. Atmos. Environ., 10, 211-217, 1976.

Robinson, A. L., Heterogeneous catalysis: Can surface science contribute, Science, 185, 772-774, 1976.

Rossi, A. R., and K. D. Jordan, Comment on the structure and stability of $(CO_2)_2^-$, J. Chem. Phys., 70, 4422-4424, 1979.

Searcy, J. Q., and J. B. Fenn, Clustering of water on hydrated protons in a supersonic free jet expansion, J. Chem. Phys., 61, 5282-5288, 1974.

Slater, J. C., and K. H. Johnson, Quantum chemistry and catalysis, Phys. Today, 27, 34-41, 1974.

Smith, D., and N. G. Adams, Elementary plasma reactions of environmental interest, in Topics in Current Chemistry, vol. 89, Springer-Verlag, Berlin/Heidelberg, 1980.

Spears, K. G., Ion-neutral bonding, J. Chem. Phys., 57, 1850-1858, 1972.

Spears, K. G., and E. E. Ferguson, Termolecular and saturated termolecular kinetics for Li^+ and F^-, J. Chem. Phys., 59, 4174-4183, 1973.

Staley, R. H., and J. L. Beauchamp, Intrinsic acid-base properties of molecules. Binding energies of Li^+ to π^- and n-donor bases, J. Am. Chem. Soc., 97, 5920, 1975.

Tang, I. N., and A. W. Castleman, Jr., Mass spectrometric study of the gas-phase hydration of the monovalent lead ion, J. Chem. Phys., 57, 3638-3644, 1972.

Tang, I. N., and A. W. Castleman, Jr., Mass spectrometric study of gas-phase clustering reactions: Hydration of the monovalent bismuth ion, J. Chem. Phys., 60, 3981-3986, 1974.

Tang, I. N., and A. W. Castleman, Jr., Gas-phase solvation of the ammonium ion in ammonia, J. Chem. Phys., 62, 4576-4578, 1975.

Tang, I. N., M. S. Lian, and A. W. Castleman, Jr., Mass spectrometric study of gas-phase clustering reactions: Hydration of the monovalent strontium ion, J. Chem. Phys., 65, 4022-4027, 1976.

Whitten, R. C., and I. G. Poppoff, Fundamentals of Aeronomy, John Wiley and Sons, New York, 1971.

Yamdagni, R., and P. Kebarle, Hydrogen-bonding energies to negative ions from gas-phase measurements of ionic equilibriums, J. Am. Chem. Soc., 93, 7139-7143, 1971.

Zettlemoyer, A. C. (Ed.), Nucleation, Marcel Dekker, New York, 1969.

ROLE OF IONS IN HETEROMOLECULAR NUCLEATION: FREE ENERGY CHANGE OF HYDRATED ION CLUSTERS

S. H. Suck, T. S. Chen, R. W. Emmons, D. E. Hagen, and J. L. Kassner, Jr.

Department of Physics and Graduate Center for Cloud Physics Research, University of Missouri-Rolla, Rolla, Missouri 65401

Abstract. The presence of foreign particles (i.e., ions) in vapor is known to enhance nucleation. Recently, less-biased molecular dynamics calculations of hydrated ion clusters were made by other researchers to compare with the classical electrochemical theory of Thomson. Using a theory developed earlier, we present here the computed excess free energies of the hydrated ion (Cs^+ and F^-) clusters of opposite sign. This theory yields good agreement with the numerical molecular dynamics method in the estimation of the excess free energy. However, Thomson's theory predicts much smaller (in absolute magnitude) excess free energies at all cluster sizes than both ours and theirs. In view of a good agreement with the curvature variation of the excess free energy obtained by the accurate molecular dynamics method, their theory is promising particularly due to its merit of simplicity and analyticity. Further, good agreement with the less-biased molecular dynamics method (which requires the knowledge of only interaction potentials) leads us to a seemingly obvious but important conclusion: the difference in the excess free energies between the two oppositely charged ion clusters is largely due to the difference in the electrostatic interaction energy between the ion and water molecules since both the volume and surface energy terms remained unchanged in our calculations.

Introduction

Heteromolecular nucleation is more relevant to atmospheric problems than homomolecular nucleation. It is known that the presence of foreign particles in vapor enhances the nucleation process. Ions in the atmosphere are unceasingly produced due to constant earth radioactivity and cosmic rays. The created ions are instantaneously subject to ion-molecule collision processes and start to form heteromolecular clusters. Since the number concentration of the created ions is relatively small compared to vapor phase molecules, they are likely to form the cluster made of one ion and its surrounding molecules. These ions and ion clusters are important in two respects; atmospheric electricity and ion nucleation. The size and concentration of these ion clusters are important for determining ion mobilities which are immediately relevant to atmospheric electricity. For such studies of both ion nucleation and atmospheric electricity, the free energy change (or excess free energy) is of utmost importance. In our present study, we examine the free energy changes of the ion clusters of opposite sign, with the illustrations of Cs^+ and F^- hydrated ion clusters, based on an earlier theory of Suck (1981).

Castleman et al. (1972, 1978) have made extensive studies on the roles and properties of small ion clusters by considering the changes of enthalpy and entropy. Such studies are important for examining the source of failures in various theories. For example, their studies of the entropy change in 1978 revealed that the actual clusters have more ordered structure than predicted by the classical liquid drop formulations. Without an accurate estimation of the contribution of the electrostatic interaction energy between the ion and its surrounding substance, it will in general be difficult to find where the main source of errors occurs. Thomson's ion-liquid drop model may not be adequate with the electrostatic energy term due to its neglect of relatively strong intermolecular interactions between the ion and surrounding molecules. This may be one of the greatest sources which predicts less ordered structures.

Recently, Chan and Mohnen (1980) developed a semimolecular model theory to estimate the free energy change. Their analytic theory predicted qualitatively excellent agreement with the numerical molecular dynamics results of Briant and Burton (1976a, b) in the excess free energy (free energy change) and with the experimental values of Scharrer (1939) in the saturation ratio. Briant and Burton found that the classical electrothermodynamic theory (Wilson, 1899a, b; Frenkel, 1946) of Thomson did not agree with their molecular dynamics. Disagreement between the two in the excess free energy estimations is generally serious at all sizes of the ion clusters. It is known that the classical ion-liquid drop theory often does not predict energies agreeable with experiments in both small prenucleation clusters and larger clusters. However, the source of such disagreement with both the theories and experiments mentioned above is still to be discovered. Besides the volume and surface energy terms, the classical ion-liquid drop theory may not adequately describe the electrostatic energy term as mentioned earlier. For this reason, here we give primary consideration to the contribution of the electrostatic interaction energy between the ion and surrounding substance in the cluster.

Electrostatic Energy Contribution to the Total Free Energy Change

The airborne ions in the atmosphere are subject to ion-molecule reactions. Here we consider only the hydration of the ions via bimolecular addition reactions involving a third-body M as a catalysis to form stabilized product ion clusters:

$$I + H_2O + M \rightleftarrows I(H_2O) + M$$

$$I(H_2O)_1 + H_2O + M \rightleftarrows I(H_2O)_2 + M$$

$$\vdots$$

$$I(H_2O)_{i-1} + H_2O + M \rightleftarrows I(H_2O)_i + M \qquad (1)$$

$$\vdots$$

The hydrated ion cluster $I(H_2O)_i$ of size i is assumed to have been formed with the ion I at its center at sufficiently large sizes. In classical theories, the isomerization of the clusters is not taken into account. For an analytic description of the free energy change with respect to the size i, such simplification seems to be inevitable.

The classical electrothermodynamic free energy change $\Delta\Phi$ per hydrated ion cluster is written

$$\Delta\Phi = \Delta\Phi_v + \Delta\Phi_s + \Delta\Phi_E \qquad (2)$$

where the volume energy term is given by

$$\Delta\Phi_v = -\frac{4}{3}\pi R^3 nkT \ln S \qquad (3)$$

the surface energy term,

$$\Delta\Phi_s = 4\pi R^2\sigma \qquad (4)$$

and the electrostatic energy term,

$$\Delta\Phi_E = -\frac{1}{2}\left(1 - \frac{1}{\varepsilon}\right)\left(\frac{1}{R_I} - \frac{1}{R}\right)q_I^2 \qquad (5)$$

The symbol definitions are as follows: R is the radius of the ion cluster, n is the number of molecules/cm^3 in the cluster, k is the Boltzmann constant, T is the absolute temperature, S is the saturation ratio, σ is the surface energy, ε is the dielectric constant, R_I is the ionic radius, and q_I is the ionic charge. The electrostatic energy term here refers to the interaction energy between the central ion and the surrounding substance.

On a molecular level, the polarization energy is to be added to the usual electrostatic interaction energy when the intermolecular interaction due to the induced dipole moment is not negligible. Considering this and the importance of the surrounding dipolar molecules and foreign center (with zero dipole moment for atomic ions), the electrostatic energy expression (Suck, 1981) becomes

$$\Delta\Phi_E = -\frac{9}{2}\frac{\varepsilon(\varepsilon-1)}{[(2\varepsilon+1)(\varepsilon+2)]}\alpha\left[\left(\frac{1}{R_I} - \frac{1}{R}\right)q_I^2 + \frac{2}{3}\left(\frac{1}{R_I^3} - \frac{1}{R^3}\right)\mu_I^{o2}\right] + \frac{4\pi n\alpha_I\varepsilon}{(2\varepsilon+1)}\left(\frac{1}{R_I^3} - \frac{1}{R^3}\right)\mu_s^{o2} \qquad (6)$$

where μ_I^o in the second term of the first bracketed term is the permanent dipole moment of the foreign center in the cluster, α_I is the electronic polarizability of the foreign center, μ_s^o is the permanent dipole moment of the surrounding molecules, and α is the scaling factor.

If we ignore all the higher electric moment (dipole moment) terms, we simply have

$$\Delta\Phi_E = -\frac{9}{2}\{\varepsilon(\varepsilon-1)/[(2\varepsilon+1)(\varepsilon+2)]\}\,\alpha\left(\frac{1}{R_I} - \frac{1}{R}\right)q_I^2 \qquad (7)$$

The radial coordinate r dependent dielectric constant (Glueckauf, 1964; Chan and Mohnen, 1980) is given by

$$\varepsilon = n_r^2 + (\varepsilon_o - n_r^2)\frac{3}{\beta s}\left(\frac{1}{\tan h\ \beta s} - \frac{1}{\beta s}\right) \qquad (8)$$

where

$$s = -\frac{q_I}{\varepsilon r^2} \qquad (9)$$

and

$$\beta = -0.126/T \qquad (10)$$

Here n_r is the refractive index ($n_r = 1.33$) of water at long wavelengths and ε_o is the dielectric constant in the absence of the electric field.

The total number i of water molecules in the hydrated ion cluster is approximated by

$$i \simeq \frac{4}{3}\pi R^3 n \qquad (11)$$

and the number density is given by

$$n = \frac{A}{M}\rho \qquad (12)$$

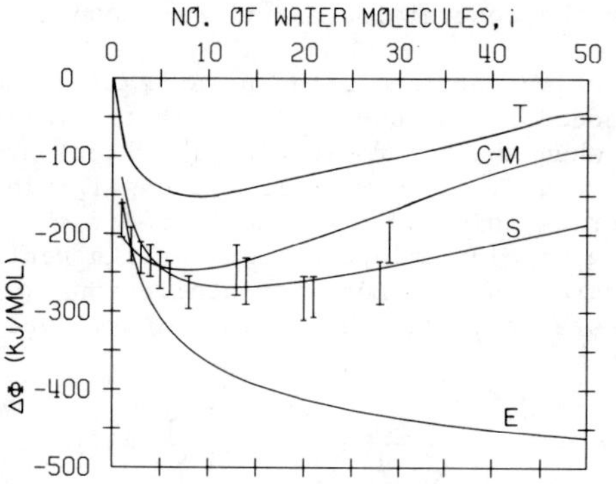

Fig. 1. Excess free energy of $Cs^+(H_2O)_i$ vs. size i at T = 300 K and at S = 1. The curve E is to denote the electrostatic energy that appears in equation (13), the curve S the excess free energy (eq. (13)), the bars 'I' the molecular dynamics result of Briant and Burton (1976), the curve C-M the theory of Chan and Mohnen (1980), and the curve T Thomson's theory.

where A is the Avogadro's number, M is the molecular weight, and ρ is the mass density of the cluster. The approximation (11) is more valid for larger clusters. We did not exclude the volume occupied by the ion in the equation. To see the difference between Thomson's theory and ours only in the contribution of the electrostatic energy, we retain the same volume and surface energy terms in the expression of the free energy change ΔΦ. The introduction of equations (11) and (12) into equations (3), (4), and (7) leads to the excess free energy ΔΦ,

$$\Delta\Phi = -ikT \ln S + (36\pi)^{1/3}\left(\frac{M}{A\rho}\right)^{2/3} i^{2/3}\sigma$$

$$- \frac{9}{2}\,\frac{\varepsilon}{2\varepsilon+1}\,\frac{(\varepsilon-1)}{(\varepsilon+2)}\,\alpha\left[\frac{1}{R_I} - \left(\frac{4\pi A\rho}{3Mi}\right)^{1/3}\right] q_I^{\,2} \quad (13)$$

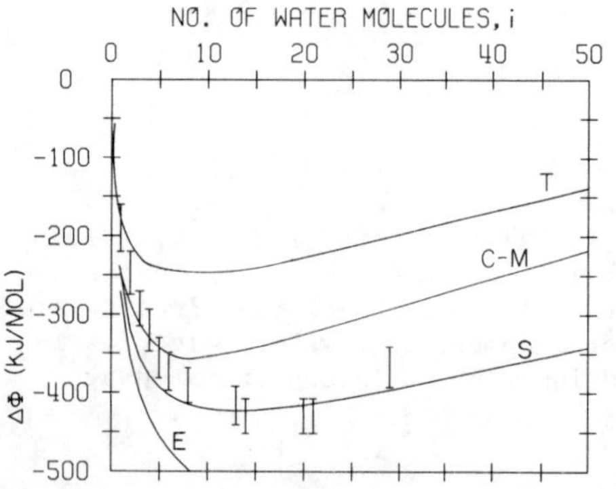

Fig. 2. Excess free energy of $F^-(H_2O)_i$ vs. size i at T = 300 K and at S = 1. The symbols associated with the curves are the same as in Figure 1.

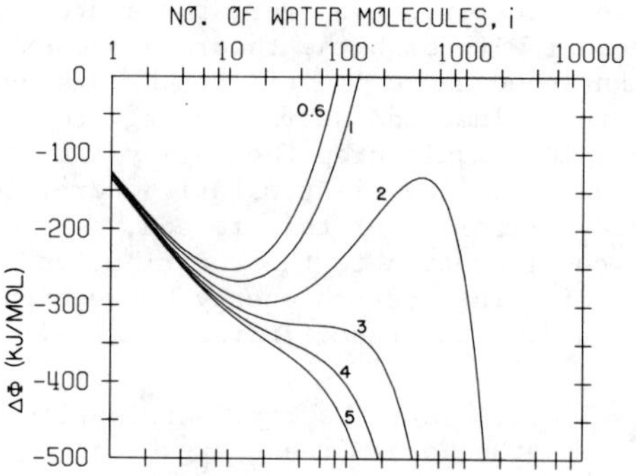

Fig. 3. Excess free energies of $Cs^+(H_2O)_i$ vs. size i at T = 300 K and at S = 0.6, 1.0, 2.0, 3.0, 4.0, and 5.0.

Here ε is approximately treated as a size R dependent parameter. As in the classical Thomson's electrothermodynamic equation (Frenkel, 1946), equation (13) above is relatively insensitive to the variation of ε.

Computed Results and Discussions

For purposes of direct comparisons with available theoretical results (Chan and Mohnen, 1980; Briant and Burton, 1976a, b), we select two hydrated ion clusters of $Cs^+(H_2O)_i$ and $F^-(H_2O)_i$ in our present study. Our selection of physical parameters that appear in equations (6) and (13) is as follows: (1) for $C_s{}^+$, R_I = 1.67 Å and α_I = 3.02 $Å^3$ (Kittel, 1971), (2) for F^-, R_I = 1.33 Å and α_I = 0.652 $Å^3$ (Kittel, 1971), (3) for H_2O, μ_o = 1.84 Debye and α = 1.44 $Å^3$ (Eisenberg and Kauzmann, 1969), (4) σ = 71.74 erg/cm^2 (interpolated value from Weast, 1975) at T = 300 K, (5) ε_o = 77.648 (Eisenberg and Kauzmann, 1969), (6) ρ = 1 g/cm^3, and (7) α = 1/3. However, higher electric moment terms add relatively small contributions. The computer-plotted excess free energies shown in Figures 1 through 4 are the results which contain only the monopole contribution in the electrostatic energy term.

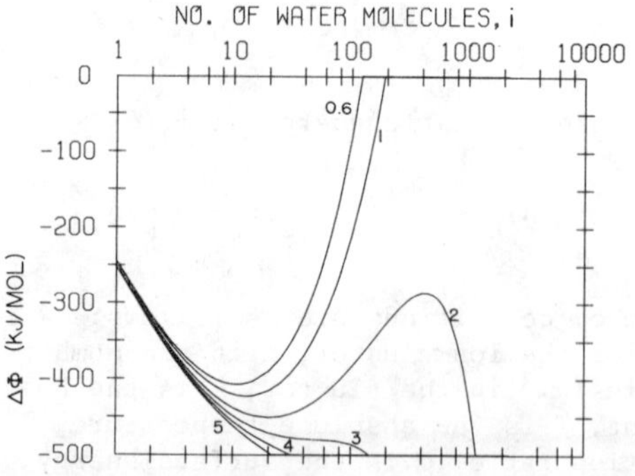

Fig. 4. Excess free energies of $F^-(H_2O)_i$ vs. size i at T = 300 K and at S = 0.6, 1.0, 2.0, 3.0, 4.0, and 5.0.

Figures 1 and 2 show the computed excess free energies $\Delta\Phi$ of $Cs^+(H_2O)_i$ and $F^-(H_2O)_i$ vs. size i at T = 300 K and S = 1. Our results (curve labeled as S) are in good agreement with the molecular dynamics calculations of Briant and Burton (1976a, b). Our predicted minimum excess free energy occurred at the cluster size of i = 13, showing a reasonable agreement with the $F^-(H_2O)_i$ results (vertical bars in Figures 1 and 2) of Briant and Burton. On the other hand, the theory (labeled as C-M in the same figures) of Chan and Mohnen (1980) predicted the minimum around i = 7. Thus, their quasi-stable cluster sizes of the hydrated ion clusters are not in good agreement with the molecular dynamics results. This is due to a distinctively different curvature of the excess free energy predicted by their theory. In general, our theory (Suck, 1981) predicts good agreement with the molecular dynamics results in the size region between i = 1 and 29. The difference in the excess free energies $\Delta\Phi$ between the oppositely charged clusters $Cs^+(H_2O)_i$ and $F^-(H_2O)_i$ largely comes from the difference in the ionic radii, as is clearly predicted by the theory. This is a consequence of the squared value of the charge in equations (6) and (13). It is of note that the energy minimum positions predicted by Chan and Mohnen (1980) are close to Thomson's values (labeled as T in Figures 1 and 2). It is to be noted that the ionic radii of ours (R_I = 1.67 Å for Cs^+ and R_I = 1.33 Å for F^- (Kittel, 1971)) are slightly different from their molecular dynamics effective radii (R_I = 1.3 ± 0.1 Å for F^- and R_I = 1.6 ± 0.1 Å for Cs^+ (Briant and Burton, 1976a, b)).

In the same figures, the electrostatic energy (labeled as E) is also plotted. The difference between the S-curve (excess free energy) and E-curve (electrostatic energy) defines the sum of volume and surface energy terms. This is the excess free energy of the uncharged homomolecular cluster. The "near" convergence between the S-curve and E-curve at small cluster sizes indicates that the excess free energy at small sizes is mostly determined by the electrostatic interaction energy between the ion and the surrounding molecules. To state it otherwise, the initial ion cluster growth (at small sizes) is largely determined by the strong attractive electrostatic interaction between the ion and surrounding vapor phase molecules. Without the contribution of the attractive electrostatic energy, the local minimum in the excess free energy will not occur. The clusters in this "well" are relatively stable. However, at saturation ratios greater than 1, thermodynamic fluctuations will cause the growth and decay of the quasi-stable clusters to reach the birth of critical state (transition state) clusters. The lower the energy barrier height, the easier the phase transition. Thus, the study of the curvature variation of the excess free energy depending on ions is of great importance. As is seen in Figures 1 and 2, the curvatures between the two atomic ions are strikingly similar, although the magnitude of the excess free energies is markedly different. This is not surprising as can be easily understood from equation (13). However, for molecular ions such similarity is likely to be lessened due to the increased importance of the higher-order terms in equation (6).

Figures 3 and 4 show the variation of the excess free energy with the saturation ratio S. As is expected, the energy barrier height is lowered with the increase of S. Our results show relatively lower barrier heights than the results of Chan and Mohnen (1980), which are not shown in the figures in order to avoid crowdedness. Their "critical" (referring to the zero slope at an inflection point) saturation ratios were in better agreement with the experimental values of positive ions and negative ions (Scharrer, 1939). The experimental "critical" saturation ratios were S_c = 4.87 for the positive ions, and S_c = 4.14 for the negative ions. On the other hand, ours were near S_c = 3.0 for both ions with the negative ion clusters having a slightly smaller value. Further, our critical cluster size i near the "critical" saturation was predicted to be 65 for both ions. The theory of Chan and Mohnen also predicted nearly the same but larger critical cluster sizes for both ions. In both their and our theories, the predicted barrier heights at S = 2 are nearly the same between the two oppositely charged ions (Cs^+ and F^-). Their energy barrier heights were in the vicinity of 270 kJ/mol compared to our 165 kJ/mol. It is expected that, in general, homomolecular nucleation will require larger saturation ratios than the ion nucleations considered here. Thus, compared to the homogeneous nucleation of water, the hydrated ion nucleation will have lower "critical" (referring to the zero slope at an inflection point) saturation ratios.

Conclusion

Our theory presented here has the merit of simplicity and analyticity over molecular dynamics or Monte-Carlo methods, while these methods are more realistic. Excellent agreement between our analytic theory and the less-biased numerical molecular dynamics method of Briant and Burton (1976a, b) is encouraging. Their results are less-biased in that all the isomerizations of a given size cluster are effectively included once intermolecular interaction potentials are properly taken into account.

Due to the charge-squared dependent term, the major difference in the excess free energy will come from the ionic radius difference between two oppositely charged ions. However, for oppositely charged but nearly same size atomic ions, the energy difference will result from the difference in polarizabilities, as can be seen from equation (6). In such cases, a preferential charge sign does not exist for enhancing nucleation. Further, the difference in the excess free energy is largely determined by the ionic radius, but not by the charge sign. Thus, there is no strict rule to determine the charge sign preference.

For both the hydrated Cs^+ and F^- clusters, we have used the same volume and surface energy terms. Thus, the difference in the excess free energy will occur through the discrepancy in the electrostatic energies of the two ions. Thus, in view of good agreement with the less-biased molecular dynamics method in the size regime of prenucleation clusters, we conclude that the electrostatic interaction energy between the ion and its surrounding water molecules in the hydrated ion cluster is mostly responsible for the large difference in the excess free energies between the two oppositely charged ion clusters of Cs^+ and F^-, as can be seen from Figures 1 and 2.

Acknowledgement. This material is based on work supported by the Division of Atmospheric Sciences, National Science Foundation, under Grant ATM 79-19480.

References

Briant, C. L., and J. J. Burton, A molecular model for the nucleation of water on ions, J. Atmos. Sci., 33, 1357, 1976a.

Briant, C. L., and J. J. Burton, Molecular dynamics study of the effects of ions on water microclusters, J. Chem. Phys., 64, 2888, 1976b.

Castleman, A. W., Jr., P. M. Holland, and R. G. Keesee, The properties of ion clusters and their relationship to heteromolecular nucleation, J. Chem. Phys., 68, 1760, 1978.

Castleman, A. W., Jr., and I. N. Tang, Role of small clusters in nucleation about ions, J. Chem. Phys., 57, 3629, 1972.

Chan, L. I., and V. A. Mohnen, Ion nucleation theory, J. Atmos. Sci., 37, 2323, 1980.

Eisenberg, D., and W. Kauzmann, The Structure and Properties of Water, Oxford, New York, 1969.

Frenkel, J., Kinetic Theory of Liquids, Oxford, London, 1946.

Glueckauf, E., Heats and entropies of ions in aqueous solution, Trans. Faraday Soc., 60, 572, 1964.

Kittel, C., Solid State Physics, John Wiley, New York, 1971.

Scharrer, L., Kondensation von übersättigten dampfen an ionen. Ann. Phys., 35, 619, 1939.

Suck, S. H., Change of free energy on heteromolecular nucleation: Electrostatic energy contribution, J. Chem. Phys., 75, 5090-5096, 1981.

Weast, R. C. (Ed.), Handbook of Chemistry and Physics, p. F-43, C.R.C. Press, Cleveland, Ohio, 1975.

Wilson, C. T. R., Condensation of water vapor in the presence of dust-free air and other gases, Phil. Trans. Roy. Soc. London, Ser. A, 189, 265, 1899a.

Wilson, C. T. R., On the comparative efficiency as condensation nuclei of positively and negatively charged ions, Phil. Trans. Roy. Soc. London, Ser. A, 193, 289, 1899b.

STRUCTURAL STUDIES OF ISOLATED SMALL PARTICLES USING MOLECULAR BEAM TECHNIQUES

Sang Soo Kim and Gilbert D. Stein

Gasdynamics Laboratory, Department of Mechanical and Nuclear Engineering, Northwestern University, Evanston, Illinois 60201

Abstract. Small particles or clusters have been shown to be of importance in catalysis. Specially designed Laval nozzles are used as a continuous source of clusters of Ar, Kr, or Xe in a crossed molecular beam/40-keV electron beam experiment. Each of these noble gases is expanded adiabatically through the nozzle as a small mole fraction in He, thus permitting one to vary cluster concentration, mean size, and temperature. Because the clusters have random orientation, the electron diffraction patterns are of the Debye-Scherrer type. They reveal that the clusters are crystalline, with temperatures in the range 15 to 60 K and mean size from g = 50 to 1000 atoms per cluster. As g decreases below 500 a progressive change in structure from bulk face-centered cubic is observed. Theoretical diffraction patterns are calculated using a multishell icosahedral structure in addition to that of the bulk. Similar structure changes occur in all three species. Since particles in this size range can be present in the atmosphere, especially due to photochemical production, the possibility exists that they may possess nonbulk properties in general and enhanced catalytic activity in particular.

Introduction

The chemistry of the atmosphere alone is a multifaceted problem encompassing many physical phenomena and a myriad of chemical species and processes. Aerosols and particles can exist in the atmosphere over time scales ranging from relatively short all the way to permanent residence due to wind currents and Brownian motion. Their size can range from automobile size and larger during meteorite showers, volcano eruptions, hurricanes, tornados, and the like, down to molecular-size aggregates or clusters as small as a few tenths of a nanometer.

The sources for aerosols include the Earth, oceans, and space as well as the atmosphere itself and can be both natural and anthropogenic. The flux of subatomic particles and photons from beyond the Earth (mainly, of course, from the Sun) has a substantial effect on the atmosphere in general and on aerosols in particular. Due to their much higher atmospheric mole fraction and greatly increased residence times, it is the lower end of this size spectrum which most influences the processes in our gaseous envelope.

Aerosol particles are important for the chemistry of the atmosphere because they can potentially function as an efficient third body when there are molecular species chemisorbed or physisorbed onto their surfaces. Particle structure, at the surface in particular, is deemed important in the freezing process for supercooled cloud water droplets.

The primary thrust of the research in our laboratory has been the study of cluster structure and other physical properties in the size range between monomer (g = 1) and the upper limit wherein all properties are those characteristic of matter in bulk (Stein, 1979). Depending on the chemical species in question and the particular physical property of interest, estimates of this upper bulk limit vary from as little as g = 5 up to 10 000 atoms (or molecules) per cluster. It is interesting to note that the critical size cluster ($g = g^*$) for nucleating a phase change, whether a homogeneous, heterogeneous, or binary process, virtually always occurs with $g^* < 500$. This is a regime in which bulk properties do not obtain or are at least suspect.

Specifically, with regard to the topic of heterogeneous catalysis, the activity of some catalysts is greatly enhanced as particle size is reduced into this interesting transition size regime (Burton, 1974). As the cluster size gets smaller the structure begins to change, as do numerous other properties. Thus a crystallite having bulk-band electronic structure will degenerate into discrete energy levels as size decreases, which in all probability will differ from that of isolated monomer. As a consequence the bonding between cluster and adsorbate will change, and under some circumstances can promote a specific reaction on the surface.

Molecular beam techniques, in conjunction with sources designed to produce very small particles or clusters, have been employed in our research (DeBoer et al., 1979). The resultant cluster beam is of a low enough density and traverses a vacuum

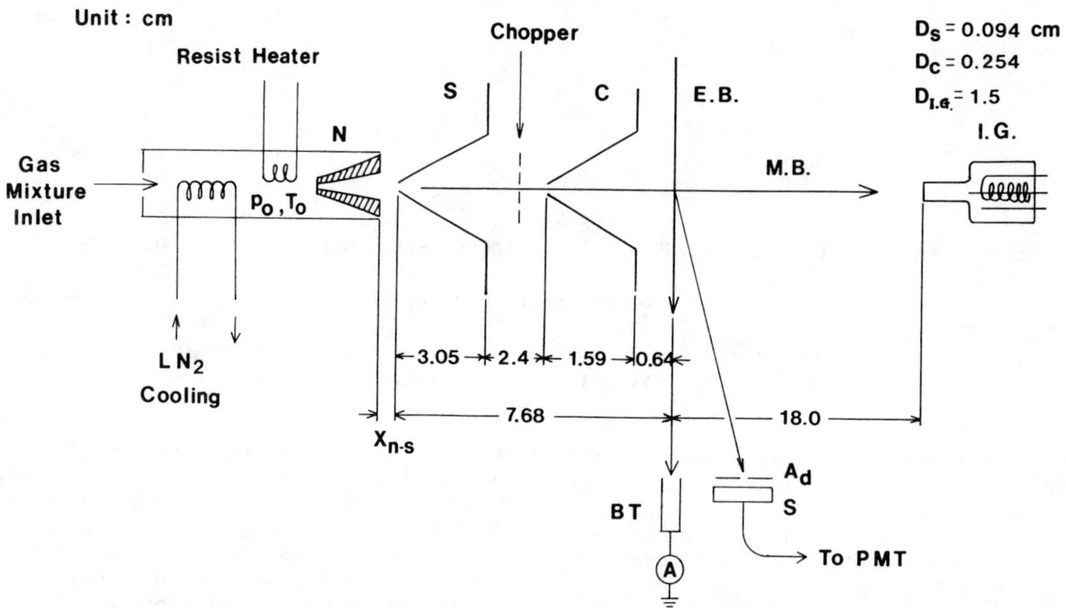

Fig. 1. A schematic of the molecular beam system is shown where N = nozzle, S = skimmer, C = collimator, E.B. = electron beam, M.B. = molecular beam, I.G. = ionization gauge, BT = beam trap, A_d = detector aperature, Sc = scintillator, and PMT = photomultiplier. The distance x_{n-s} between nozzle exit and skimmer orifice is variable.

chamber with low enough background density that the clusters do not interact with each other or with any support media such as surfaces, solvents, matrix, etc. There is an experimental price to be paid for this low cluster density, and that is a much lower signal and often lower signal-to-noise ratio than with other materials.

The results reported here are for noble gas clusters of Ar, Kr, or Xe formed in He carrier-gas expansions through supersonic, diverging Laval nozzle sources (Kim and Stein, 1982).

Description of Experiment

The molecular beam apparatus is of a conventional, concentric design, differentially pumped in three stages using a 10-cm ring-jet booster pump, a 15-cm diffusion pump, and two 15-cm diffusion pumps. The cluster sources for this work are diverging Laval nozzles with inlet throat (i.e., minimum) diameters $0.09 > D_o > 0.05$ mm (see Figure 1 and Table 1 for details). The inlet flow is subsonic and not of importance for cluster nucleation. The nozzles used here, numbers 11 to 13, are designed specifically for the high kinematic viscosity of He and have exit Mach numbers (i.e., flow velocity/sound speed) in the range from 2 to 10 depending on the starting pressure P_o and the gas mixture used (Kim et al., 1981). Even with viscous dissipation due to boundary layers inside these nozzles, the resultant beam intensities are orders of magnitude greater than free jet sources of the same diameter.

The flow passes through the skimmer and collimator into the final or scattering chamber where it is crossed by a 40-keV electron beam (de Broglie wavelength λ = 0.0061 nm). The flow is essentially collisionless by the time it reaches the skimmer with, therefore, no further changes in the cluster beam until it collides with the walls of the ionization gauge. The noble gas clusters are so weakly bound that they fall apart upon wall impact and are counted as gas atoms in the gauge.

Due to the small scattering cross section of high-energy electrons with atoms, most of the electron beam traverses the cluster beam with no deflection and is trapped in the beam stop. The small fraction that is scattered is detected by a single-channel scintillation detection system. The cluster crystallites, having random orientation in the beam, produce axisymmetric Debye-Scherrer diffraction patterns. The molecular beam is chopped in order to implement synchronous detection methods. These patterns are analyzed to obtain the cluster structure, size, and temperature.

TABLE 1. Nozzle Geometry

Nozzle Dimensions		Nozzle Number		
		11	12	13
Inlet diam.	D_o(cm)	0.0089	0.0089	0.0056
Exit diam.	D_e(cm)	0.47	0.47	0.47
Length	L_e(cm)	1.18	1.36	1.28

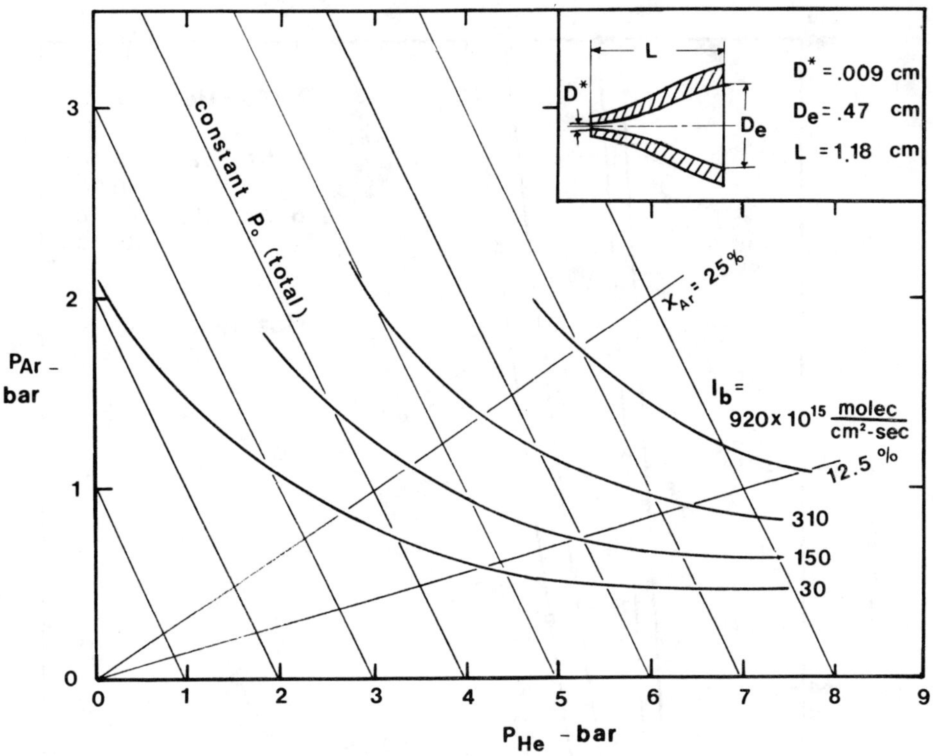

Fig. 2. Molecular beam intensity contours (constant I_b at the electron beam intersection) are presented as a function of the partial pressure of Ar and He for nozzle 11 at $T_o \simeq 295$ K.

The gas properties that are varied are p_o, temperature T_o, and condensable mass fraction ω_o (or mole fraction χ_o). Also, there is a choice of several nozzles. The distance between nozzle exit and skimmer entrance x_{ns} was optimized for maximum cluster beam intensity and not adjusted thereafter.

The diffraction detection system is controlled electronically to accumulate signal counts at a given angle θ by a synchronous up-down counting scheme for a preset number of chopper cycles, to eliminate the effects of background gas scattering. The detector is then stepped to the next increment in angle and the process repeated until a complete pattern is recorded.

Cluster Formation

The Ar, Kr, or Xe, as a small mole fraction χ_o in He, cluster during the adiabatic expansion after crossing their respective vapor-solid equilibrium lines and supersaturating to the nucleation onset point (Wu et al., 1978). The reason the Laval sources are so much more effective than free jets in nucleating and growing clusters is due to their controlled rate of expansion, particularly from the point of onset to the skimmer location where the flow becomes collisionless. The uncontrolled free-jet expansion moves quickly through this region, thus minimizing the production and growth of the condensed phase.

The mole fraction χ_o has been varied in an effort to (1) maximize cluster production with minimum mass flow rate from the nozzle, (2) control the final cluster temperature T_c (lower χ_o should give lower T_c), (3) determine what premixed mole fractions are most suitable, and (4) minimize the use of expensive gases such as Xe. The results for Ar in He are shown in Figure 2. Since the specific heat ratio $\gamma = 5/3$ for both condensable and He, there is no γ change with χ, which affects the nozzle gasdynamics for molecular condensable species (Abraham et al., 1981b). As a result of tests varying χ_o, most of the experiments could be undertaken with values as low as $\chi_o = 0.06$ (see Figure 2).

The most convenient procedure consists of operating with one nozzle for a series of experiments and fixing χ_o and T_o while varying p_o. Temperature is controlled with liquid N_2 and fine-tuned using resistance heating. An example of cluster beam intensity I_b for nozzle 11 and $\chi_o = 0.06$ is shown in Figure 3. As expected, onset of condensation occurs in the order Xe, Kr, and Ar as p_o increases, since the mixtures saturate and reach onset at the end of the adiabatic expansion in this order. Lowering T_o causes saturation and thus onset at lower p_o. When $I_b \geq 100 \times 10^{-6}$ torr on the cluster beam ionization gauge, the beam density is high enough to obtain diffraction data. (To convert from torr to atoms/cm^2·sec at the electron beam location use $I_b = KI_b$(torr) with

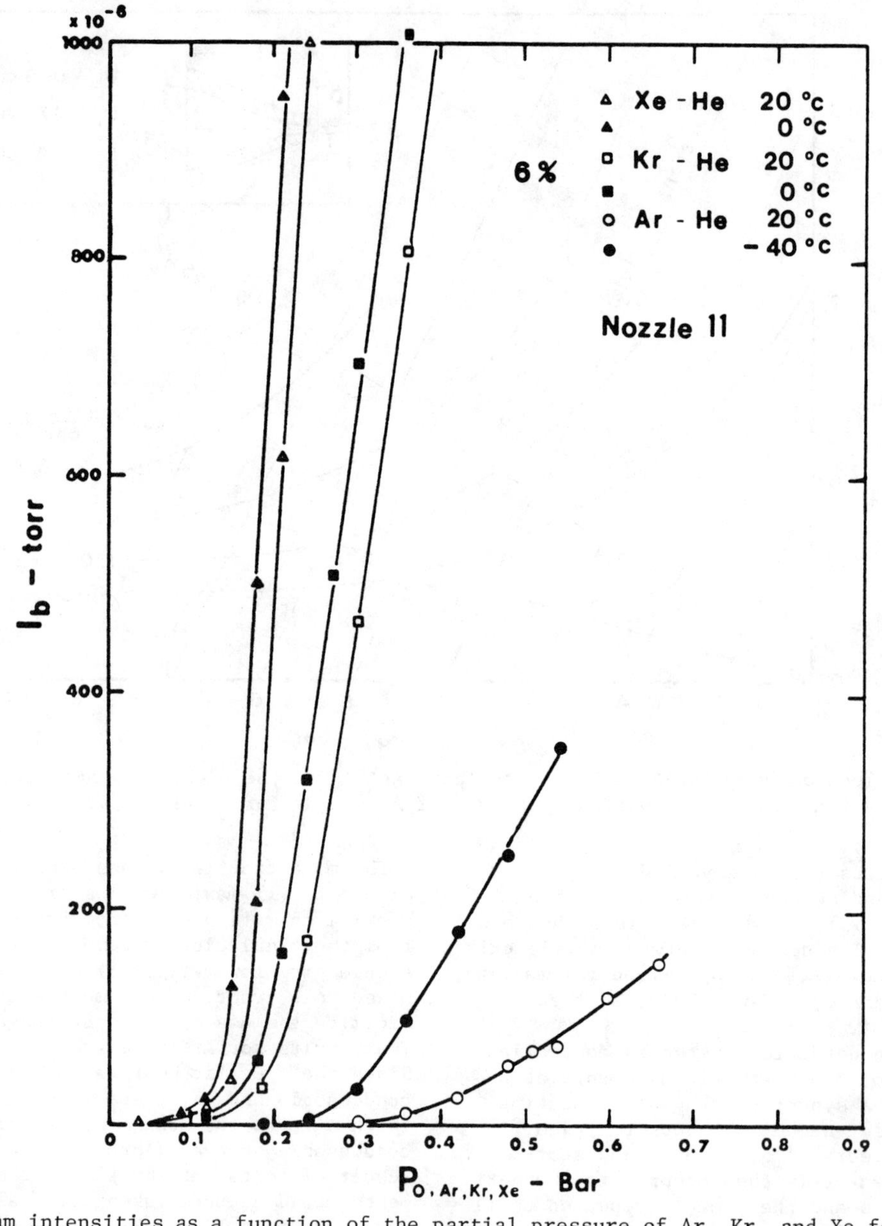

Fig. 3. Beam intensities as a function of the partial pressure of Ar, Kr, and Xe for nozzle 11 are shown at $T_o \simeq 20°$, $0°$, and $-40°C$.

$K = 3 \times 10^{21}$, 1.4×10^{21}, and 8×10^{20} for Ar, Kr, and Xe, respectively.)

Cluster Electron Diffraction

The solution of the Schrödinger equation for high-energy electrons scattering from N clusters whose positions and orientations are uncorrelated but with the same size and structure is given by (Pirenne, 1946)

$$I_\theta = \frac{NI_o}{L^2}\left[\sum_i^g \sum_j^g f_i f_j \frac{\sin(sr_{ij})}{sr_{ij}} \exp\left(\frac{-s^2\langle u^2\rangle}{3}\right) + \frac{2}{a_H}\sum_i^g \frac{S_i}{s^4}\right] \quad (1)$$

where I_θ and I_o are the scattered and incident electron beam intensities and L is the distance from the two-beam intersection to the detector. The electron elastic scattering factor is $f_i = (2/a_H)(Z_i - F_i)/s^2$ where Z_i is the atomic number and F_i is the X-ray scattering factor of atom i, $a_H = h^2/(4\pi^2 m_e e^2)$ is the classical Bohr radius with h as Planck's constant, m_e and e are the electron mass and charge, $s = (4\pi/\lambda)\sin\theta/2$ is the scattering parameter, S_i is the inelastic scattering

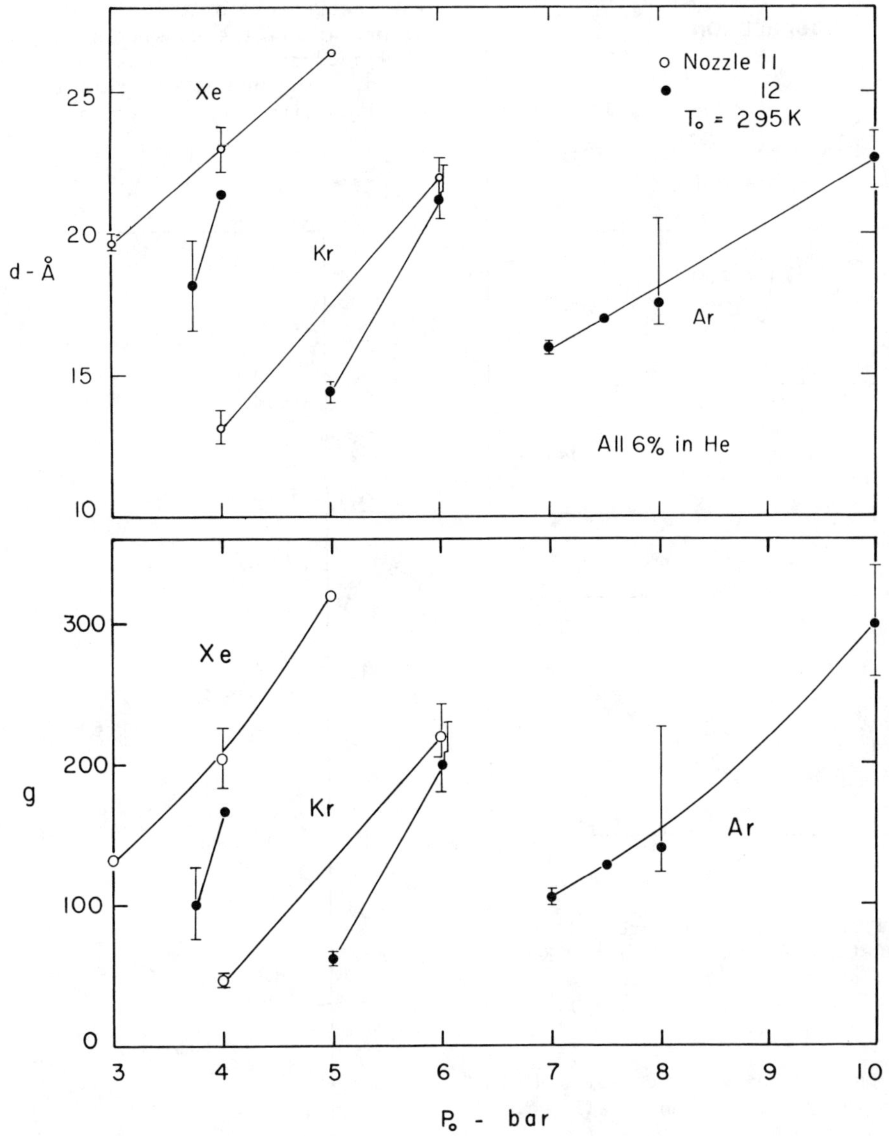

Fig. 6. Cluster diameter d and atoms per cluster g as a function of p_o for Ar, Kr, and Xe from nozzles 11 and 12 are obtained from diffraction experiments and equations (4) and (5).

factor of atom i, and r_{ij} is the distance from atom i to atom j.

The exponential $\exp(-s^2\langle u^2\rangle/3)$ is the Debye-Waller factor which is due to the thermal vibration, $\langle u^2\rangle$ being the average square displacement of the atoms about their equilibrium position. This displacement is related to the cluster temperature and results in decreased peak height with increasing s (i.e., increasing scattering angle Θ) or with increasing temperature T_c,

$$\langle u^2\rangle = \frac{424}{m}\frac{T_c}{\Theta^2}\left[\frac{\Theta}{4T_c} + \phi(\Theta/T_c)\right] \qquad (2)$$

where m is the atomic mass, Θ is the Debye temperature, and the function ϕ is tabulated in the International Tables for X-Ray Crystallography (MacGillary and Rieck, 1962; Ibers and Hamilton, 1973).

For clusters large enough to form bulk cubic structure, the Bragg law of reflection gives the interplane spacing $d_i = \lambda/2\sin(\theta_i/2)$. Using the relation between the spacing and the Miller indices, $1/d_i^2 = (h_i^2 + k_i^2 + l_i^2)/a^2$ where a is the lattice parameter, and for small scattering angle θ_i: $\sin(\theta_i/2) \simeq \tan(\theta_i/2) = R_i/2L$ one obtains

$$\frac{R_i}{\lambda L} = \frac{\left(h_i^2 + k_i^2 + l_i^2\right)^{1/2}}{a} \qquad (3)$$

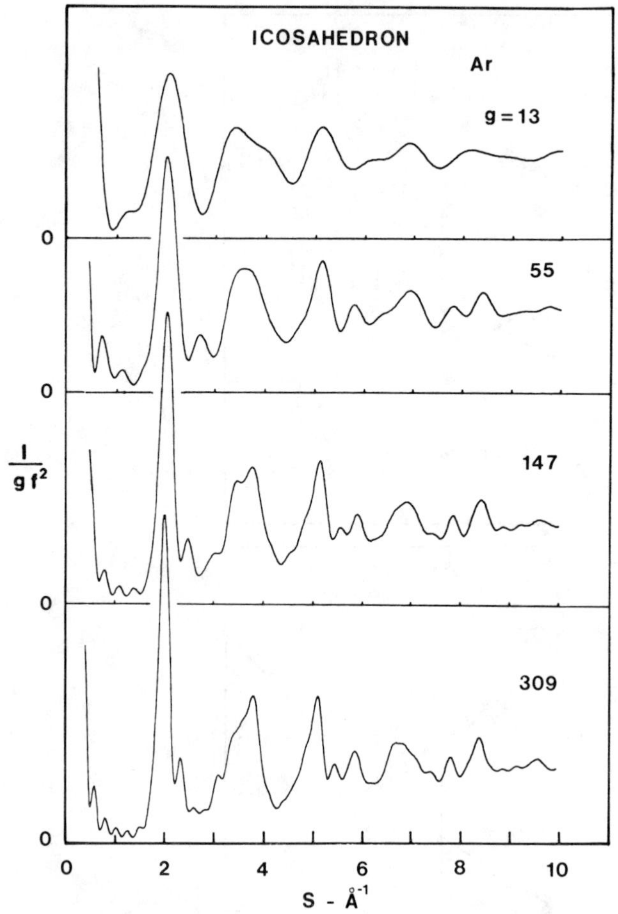

Fig. 4. Diffraction functions are calculated using equation (1) for Ar at T = 0 K for 1-, 2-, 3-, and 4-shell icosahedra (g = 13, 55, 147, and 309, respectively).

which relates the Debye-Scherrer ring radius R_i to the Miller indices and yields the unit cell parameter for the bulk structure.

Due to the finite extent of the microcrystal the diffraction ring peaks are broadened and are related to size ℓ (diameter d if approximately spherical) by the Scherrer equation (Scherrer, 1918)

$$\ell = \lambda L/(B^2 - B_o^2)^{\frac{1}{2}} \tag{4}$$

where B and B_o are the half-width of the cluster peak and that due to instrument broadening, respectively. (The use of this relationship is qualitative at best even for a single size crystal structure. Since there is always a size distribution in the molecular beam this relation cannot yield an exact cluster size. Moreover, it is interesting to note that it has the correct trend, i.e., narrower peak width for larger cluster size, even for the nonperiodic structure of the theoretical multishell icosahedra shown in Figure 4. If the cluster structure were amorphous or liquid the use of equation (4) may not be valid.) The number of atoms in a cluster is obtained with the expression

$$g = n(d^3/a^3) \tag{5}$$

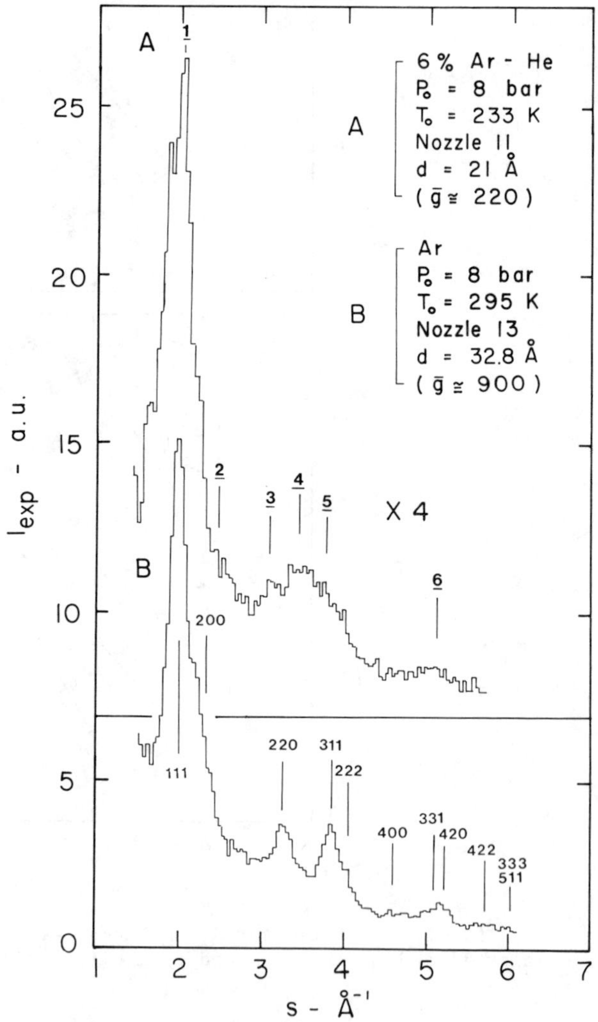

Fig. 5. Electron diffraction patterns for Ar clusters are obtained with the single-channel detection system. The beam was chopped at 100 Hz and the net signal accumulated for 5 sec at each discrete detection angle ($s = (4\pi/\lambda)\sin(\theta/2)$). The intensity is plotted in arbitrary units with I_{exp}(counts/sec amp electron beam) = 6×10^7 I_{exp}(a.u). Experiment A, with g = 220, is different from the regular face-centered-cubic pattern of experiment B. The symbols 1, 2, 3, etc., indicate the peaks in the theoretical diffraction pattern of icosahedral structure. The Miller indices for fcc structure are indicated in experiment B.

where n is the number of atoms in a unit cell, $n = 4$ for face-centered cubic (fcc) structure, and $\ell = d$ from equation (4).

As an example, consider the diffraction data for Ar clusters shown in Figure 5. Diffraction pattern B, with size $g = 900$, exhibits features of the bulk fcc structure, while pattern A with $g = 200$ is clearly not the same structure.

The variation of unit cell size, a, with cluster size g reveals at most a few percent decrease as g decreases, in agreement qualitatively with theoretical predictions (Scherrer, 1918) and observations of metal clusters (Vergand, 1975; Yokozeki and Stein, 1978). Krypton shows the least variation in unit cell dimension. Changes in a, however, cannot be attributed solely to size variation, since cluster temperature changes over an estimated range of 40°C as beam parameters p_o, T_o, and χ_o are varied.

Estimates of cluster diameter d using equation (4) and the number of atoms per cluster g using equation (5) are shown in Figure 6 as a function of starting pressure p_o. Results for nozzles 11 and 12 show that nozzle 11 is a more efficient source since it produces a given size at lower p_o and thus at lower source mass flow rate. These two nozzles have the same entrance diameter but nozzle 11 diverges at a greater angle and is shorter than nozzle 12 (see Table 1 and Kim and Stein (1982)). All gases show an increase in size as p_o increases. The clusters appear in the order Xe, Kr, and Ar with increasing p_o as expected from their respective vapor-solid equilibrium curves.

Cluster temperature T_c is seen in Figure 7 to

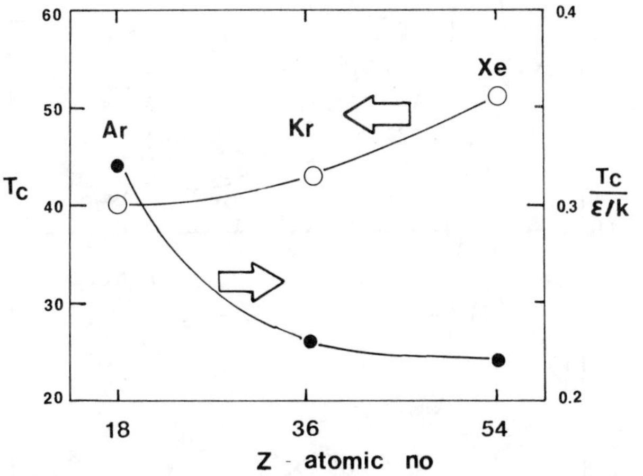

Fig. 7. The cluster temperature T_c and $T_c/(\varepsilon/k)$ nondimensionalized by the respective van der Waal's potential well depth are shown as a function of atomic number Z. The temperature is one of the adjustable parameters in the theoretical model described in the text which is compared peak by peak to a particular diffraction experiment. These temperatures have been averaged over all of the diffraction patterns analyzed (Kim and Stein, 1981), i.e., about 30.

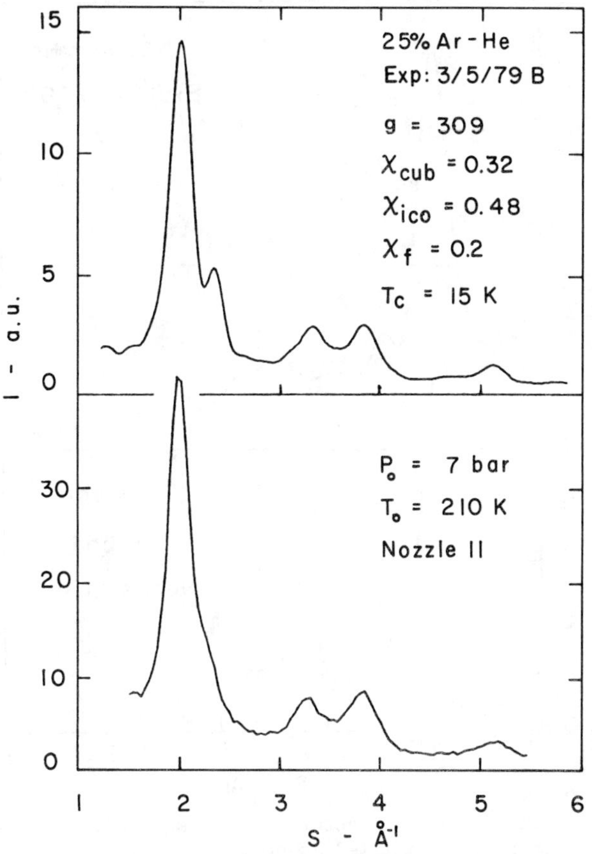

Fig. 8. The theoretical and experimental diffraction intensities are shown above and below, respectively, as a function of the scattering parameter s for Ar with $g = 309$. The theoretical structure is best fitted to the data having 34 percent cubic structure.

increase with atomic number (really with potential well depth ε) as one might expect. The deeper the potential well, the warmer the cluster may become without "breaking" the van der Waal's bonds. However, the nondimensional cluster temperature $T_c/(\varepsilon/k)$, k being Boltzmann's constant, decreases with increasing Z. Thus the clusters are more stable in the order Xe, Kr, and Ar.

Cluster Structure

Theoretical predictions for the structure of van der Waal's (i.e., noble gas) clusters have been made by several investigators using static calculations, molecular dynamics, and Monte Carlo techniques (Hoare and Pal, 1971, 1972; Barker, 1977; Abraham and Dave, 1971a, b). They have predicted that as g decreases the minimum energy configuration changes from fcc to icosahedral structure. Since a structure change has been seen previously in Ar clusters (Farges et al., 1977) as well as in this investigation, a mixture con-

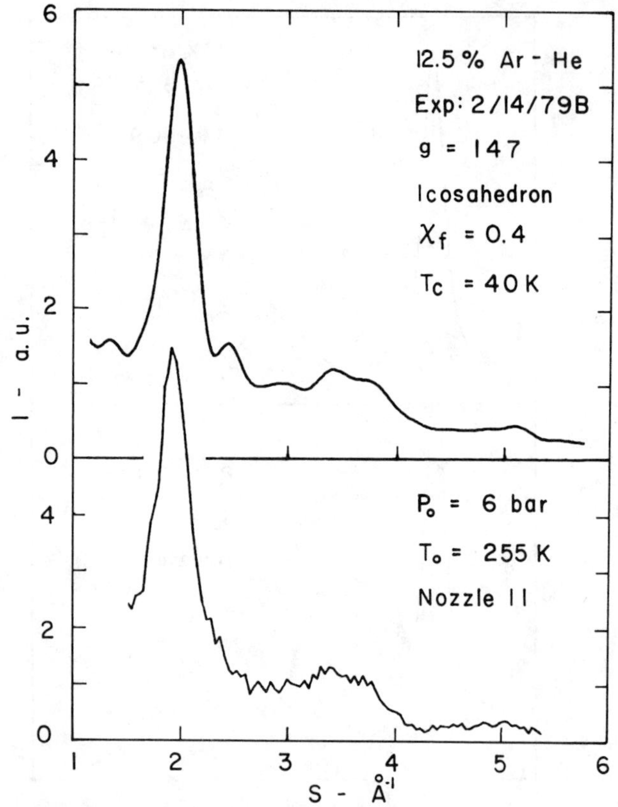

Fig. 9. The structure for this Ar cluster beam is best modeled as completely icosahedral with g = 147. (Theoretical and experimental are shown above and below, respectively.)

sisting of fcc, icosahedral, and free, uncondensed atoms having mole fractions χ_c, χ_{ic}, and χ_f, respectively, is used to approximate the cluster beam as it crosses the electron beam in the scattering chamber:

$$\chi_c + \chi_{ic} + \chi_f = 1 \qquad (6)$$

The contributions to the cluster structure from fcc cuboctahedra and icosahedra are calculated using their known structure with equation (1). The uncondensed gas in the cluster beam is a monotonically decreasing function of scattering angle. The free atoms in the beam are assumed to all be that of the condensable species, since He will preferentially diffuse or be scattered out of the beam by the skimmer interaction with the beam. The Debye-Waller cluster temperature, unit cell dimension, and the mole fractions in equation (6) are varied until a best fit to the data is obtained. The effects of possible multiple electron scattering within the cluster, using both the Blackman two-beam theory (Blackman, 1939) and a formulation using a modified Glauber theory (Bartell et al., 1977), are estimated to be negligible.

The fit to data is virtually all within 10 percent and a majority within 1 to 4 percent (Kim and Stein, 1982). Several examples are shown in Figures 8 to 11. The bulk fcc structure can be seen in the data of Figure 8 for 25 percent Ar expanded at a cold T_o = 210 K. The second peak shows up as a shoulder in the data (lower curve) but the next three peaks agree quite well. For Ar at smaller size g ≅ 147, shown in Figure 9, the data are fit best by an icosahedral calculation plus atomic (monomeric) Ar. There is also a hint of the small peak at s = 3. The discrete theoretical sizes correspond to closed or completely filled shells for both the icosahedra and cuboctahedra. Krypton data and calculation are shown in the lower and upper parts of Figure 10, respectively, for g = 147 and T_o at room temperature. Here again the second peak is unresolved with only a suggestion of the small third peak. The data are consistent with the icosahedral model and cannot be made to fit a cubic structure. Data and calculation for Xe are presented in Figure 11. A complete set of data and calculations is given by Kim and Stein (1982).

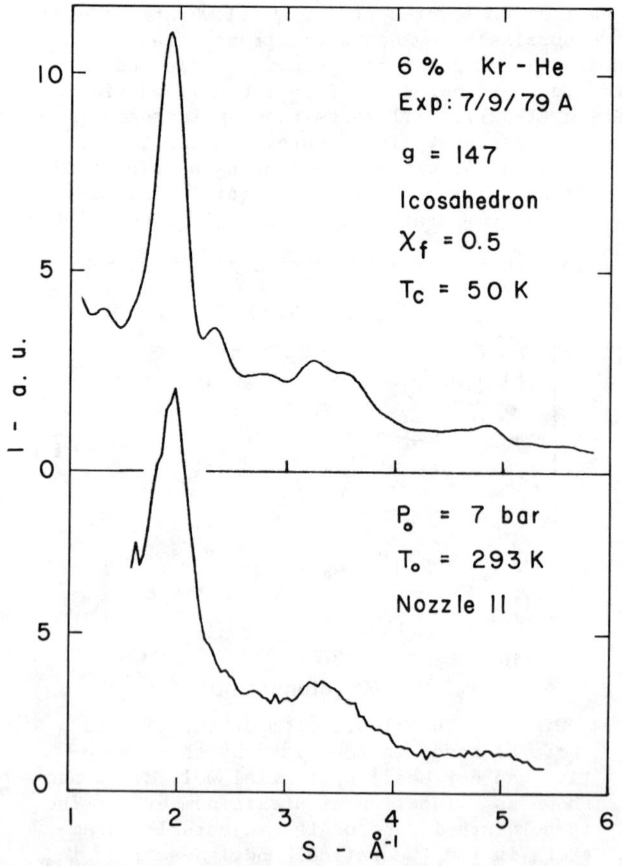

Fig. 10. An example of Kr clusters is presented with a computed structure that is all icosahedral and g = 147. (Theoretical and experimental are shown above and below, respectively.)

Conclusions

As cluster size g is reduced from 1000 to 50 the noble gas structure undergoes a transformation from bulk fcc to a form which is consistent with an icosahedral arrangement. It is certain that there has been a change in structure from that of the bulk fcc (cube octahedral) as size is reduced. The multishell icosahedral structure is one of the theoretically proposed derivations from that of the bulk for small clusters. Other configurations have been proposed involving compact noncrystallographic packing such as tetrahedral, pentagonal, and icosahedral symmetries, all of which are calculated to be more stable than fcc microcrystals having the same number of atoms (Hoare and Pal, 1971, 1975; Hoare and McInnes, 1976). Although the fit to data is good, other possible models may result in still better agreement with the data. This modeling effort is an ongoing activity in our laboratory.

There have been numerous attempts to describe noncrystalline structure in solids (Paul and Connell, 1976; Chaudhari et al., 1974; Cargill, 1976) using models such as continuous random network (Bell and Dean, 1966, 1972; Polk, 1971; Polk and Boudreaus, 1973), and dense random packed (Bernal, 1960; Leung and Wright, 1974a, b) and amorphous structures (Tilton, 1957; Grigorovici and Nanaila, 1968). Several authors have predicted a transition from solid noncrystalline or amorphous phase to a liquid phase as cluster temperature is increased (Barker, 1977; Briant and Burton, 1975). For Ar clusters a transition to liquid is predicted at temperatures in the range of 40 to 60 K, depending on cluster size. This is well below the bulk melting temperature of 84 K and the pressures external to the cluster are well below that of the triple point. A model structure composed of a combination of bulk crystalline plus liquid has been proposed (Yokozeki, 1978) for Pb and Ar while others interpret Ar as icosahedral structures in a polytetrahedral configuration (Farges et al., 1975).

A source that could produce colder clusters and a narrower size distribution would help elucidate more clearly the precise nature of the structure transformation. Prior studies of gas-phase nucleation in continuous supersonic Laval nozzles, using a noncondensable carrier gas and a small mole fraction of the condensable, produced the desirable effects mentioned above (Wegener and Pouring, 1964; Wegener and Stein, 1969). This experienced was applied to the small-nozzle molecular beam work reported here (Kim et al., 1981; Wu et al., 1978; Abraham et al., 1981a, b). Very striking cooling effects were observed in spectroscopic studies of small free-jet expansions of unclustered or dimeric species of low molecular mole fractions in He (Smalley et al., 1974, 1977; Wharton and Levy, 1979).

The results presented here for noble gas clusters show that structure changes have occurred. Structure changes have been observed in metals as well. The effects of these transformations are predicted to produce changes in other properties, among which are differences in the electronic structure. Small particle size has been demonstrated to enhance catalysis of some reactions. Thus there is no reason to think that this phenomenon can be excluded from atmospheric processes since aerosol production, for example by solar photochemistry, can result in particulates in this interesting transition size regime.

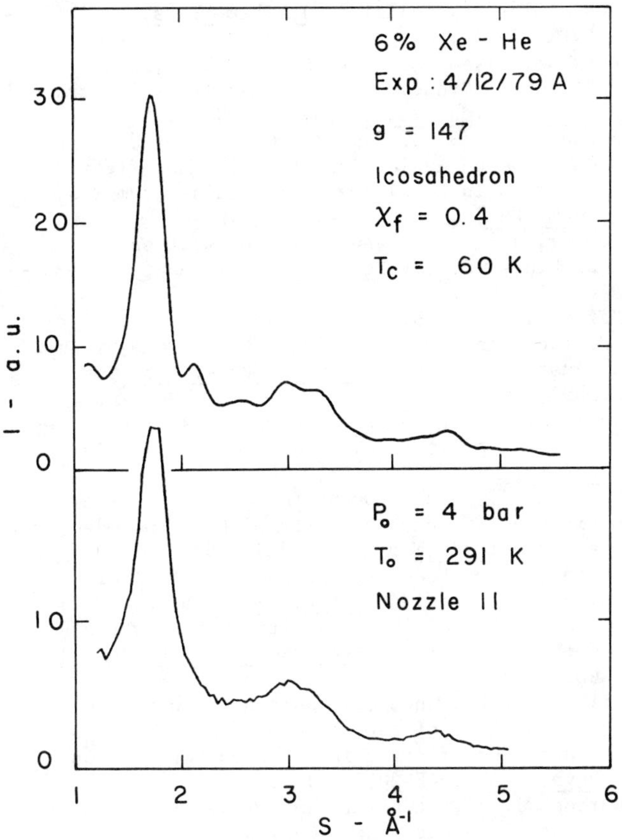

Fig. 11. Xenon cluster diffraction patterns shown here are similar to those in Figure 9 for Kr. (Theoretical and experimental are shown above and below, respectively.)

Acknowledgements. The authors are greatly indebted to the Engineering Energetics Program of the National Science Foundation and the Chemistry Division of the Office of Naval Research for partial financial support of this research.

References

Abraham, F. F., and J. V. Dave, Thermodynamics of microcrystallites and its relation to nucleation theory, J. Chem. Phys., 55, 1587-1597, 1971a.

Abraham, F. F., and J. V. Dave, On a generalized theory for the thermodynamics of planar surfaces and microcrystallites, J. Chem. Phys., 55, 4817-4821, 1971b.

Abraham, O., J. H. Binn, B. G. DeBoer, and G. D. Stein, Gasdynamics of very small Laval nozzles, Phys. Fluids, 24, 1017-1031, 1981a.

Abraham, O., S. S. Kim, and G. D. Stein, Homogeneous nucleation of sulfur hexafluoride clusters in Laval nozzle molecular beams, J. Chem. Phys., 75, 402-411, 1981b.

Barker, J. A., The geometries of soft-sphere packings, J. Physique, 38, Colloque C2, C2-37 - C2-45, 1977.

Bartell, L. S., B. Raoult, and G. Torchet, Dynamic effects in the scattering of electrons by small clusters of atoms, J. Chem. Phys., 66, 5387-5392, 1977.

Bell, R. J., and P. Dean, Properties of vitreous silica: Analysis of random network models, Nature, 212, 1354-1356, 1966.

Bell, R. J., and P. Dean, The structure of vitreous silica: Validity of the random network theory, Phil. Mag., 25, 1381-1398, 1972.

Bernal, J. D., Geometry of the structure of monatomic liquids, Nature, 185, 68-70, 1960.

Blackman, M., On the intensities of electron diffraction rings, Proc. Roy. Soc. London, 173, 68, 1939.

Briant, C. L., and J. J. Burton, Molecular dynamics study of the structure and thermodynamic properties of argon in microclusters, J. Chem. Phys., 63, 2045-2058, 1975.

Burton, J. J., Structure and properties of microcrystalline catalysts, Cat. Rev. Sci. Eng., 9, 209-222, 1974.

Cargill, G. S. III, Structure of amorphous solids. Recurring themes in the structure of glassy solids, Ann. N.Y. Acad. Sci., 279, 208-222, 1976.

Chaudhari, P., J. F. Graczyk, and S. R. Herd, Physics of Structurally Disordered Solids, edited by S. S. Mitra, pp. 45-91, Plenum Press, New York/London, 1976.

DeBoer, B. G., S. S. Kim, and G. D. Stein, Molecular beam studies of sulfur hexafluoride clustering in an argon carrier gas from both free jet and Laval nozzle sources, Rarefied Gas Dynamics, edited by R. Camparague, Commissariat à L'Energie Atomique, Paris, 1151-1160, 1979.

Farges, J., M. F. de Feraudy, B. Raoult, and G. Torchet, Dense random packing structure and polytetrahedral models of molecular aggregates, J. Phys. (Paris), 36, C2-13 - C2-17, 1975.

Farges, J., M. F. de Feraudy, B. Raoult, and G. Torchet, Transition in local order of aggregates of some tens of atoms, J. Phys. (Paris), 38 (7), C2-47, 1977.

Grigorovici, R., and R. Manaila, Structural model for amorphous germanium layers, Thin Solid Films, 1, 343-352, 1968.

Hoare, M. R., and J. McInnes, Statistical mechanics and morphology of very small atomic clusters, Disc. Faraday Soc., 61, 12-24, 1976.

Hoare, M. R., and P. Pal, Statics and stability of small cluster nuclei, Nature Phys. Sci., 230, 5-8, 1971.

Hoare, M. R., and P. Pal, Geometry and stability of "spherical" f.c.c. microcrystals, Nature Phys. Sci., 236, 35-37, 1972.

Hoare, M. R., and P. Pal, Physical cluster mechanics: Statistical thermodynamics and nucleation theory for monatomic systems, Adv. Phys., 24, 645-678, 1975.

Ibers, J. A., and W. C. Hamilton (Eds.), International Tables for X-Ray Crystallography, Vol. IV, Kynoch Press, Birmingham, England, 1973.

Kim, S. S., D. C. Shi, and G. D. Stein, Noble gas condensation in controlled-expansion beam sources, in Rarefied Gas Dynamics, edited by S. S. Fisher, Amer. Inst. Aeronaut. and Astronaut., New York, 1211-1224, 1981.

Kim, S. S., and G. D. Stein, Creation and structure study of vacuum-isolated clusters of argon, krypton, and xenon, J. Coll. Inter. Sci., in press, 1982.

Leung, P. K., and J. G. Wright, Structural investigations of amorphous transition element films I. Scanning electron diffraction study of cobalt, Phil. Mag., 30, 185-194, 1974a.

Leung, P. K., and J. G. Wright, Structural investigations of amorphous transition element films II. Chromium, iron, manganese, and nickel, Phil. Mag., 30, 995-1008, 1974b.

MacGillary, C. H., and G. D. Reick (Eds.), International Tables for X-Ray Crystallography, Vol. III, Kynoch Press, Birmingham, England, 1962.

Paul, W., and G. A. N. Connell, Physics of Structurally Disordered Solids, edited by S. S. Mitra, pp. 45-91, Plenum Press, New York/London, 1976.

Pirenne, M. H., The Diffraction of X-Ray and Electrons by Free Molecules, Cambridge Univ. Press, London, 1946.

Polk, D. E., Structural model for amorphous silicon and germanium, J. Noncrystal Solids, 5, 365-376, 1971.

Polk, D. E., and D. S. Boudreaux, Tetrahedrally coordinated random-network structure, Phys. Rev. Lett., 31, 92-98, 1973.

Scherrer, P., The space lattice of aluminum, Z. Physik, 19, 23, 1918.

Smalley, R. E., B. L. Ramakrishna, D. H. Levy, and L. Wharton, Laser spectroscopy of supersonic molecular beams: Application to the NO_2 spectrum, J. Chem. Phys., 61, 4363-4364, 1974.

Smalley, R. E., L. Wharton, and D. H. Levy, Molecular optical spectroscopy with supersonic beams and jets, Acc. Chem. Res., 10, 139-145, 1977.

Stein, G. D., Atoms and molecules in small aggregates. The fifth state of matter, Phys. Teacher, 17, 503-512, 1979.

Tilton, L. W., Noncrystal ionic model for silica glass, J. Res. Nat. Bur. Stand., 59, 139-154, 1957.

Vergand, P. F., Influence de la taille des grains sur la distance interatomique des terres rares lourdes, Phil. Mag., 31, 537, 1975.

Wegener, P. P., and A. A. Pouring, Experiments on

condensation of water vapor by homogeneous nucleation in nozzles, Phys. Fluids, 7, 352-361, 1964.

Wegener, P. P., and G. D. Stein, Twelfth Symposium (International) on Combustion, The Combustion Institute, Pittsburgh, 1183, 1969.

Wharton, L., and D. H. Levy, Jet supercooling and molecular jet spectroscopy, in Rarefied Gas Dynamics, edited by R. Camparague, Commissariat à L'Energie Atomique, Paris, 1009-1028, 1979.

Wu, B. J. C., P. P. Wegener, and G. D. Stein, Homogeneous nucleation of argon carried in helium in supercooled nozzle flow, J. Chem. Phys., 69, 1776-1777, 1978.

Yokozeki, A., Lead microclusters in the vapor phase as studied by molecular beam electron diffraction: Vestige of amorphous structure, J. Chem. Phys., 68, 3766-3773, 1978.

Yokozeki, A., and G. D. Stein, A metal cluster generator for gas-phase electron diffraction and its application to bismuth, lead, and indium: Variation in microcrystal structure with size, J. Appl. Phys., 49, 2224-2232, 1978.

CHEMICAL REACTIONS WITH AEROSOLS

Egon Matijević

Department of Chemistry and Institute of Colloid and Surface Science,
Clarkson College of Technology, Potsdam, New York 13676

Abstract. Chemical reactions of aerosol droplets with vapors are discussed. Examples are given in which liquid aerosols of 1-octadecene of narrow size distribution are converted to 1,2-dibromooctadecane with bromine vapor. It was shown that the chemical reaction in the droplet controls the kinetics of this process.

The application of chemical reactions with aerosols to the formation of pure, uniform spherical particles of metal oxides is also described. Droplets of metal alkoxides rapidly react with water vapor to yield well-defined powders. The technique was used to prepare titanium dioxide, aluminum oxide, and particles consisting of both metal oxides. This procedure allows generation of powders of predetermined size and composition.

Introduction

Colloidal dispersions in gaseous media (aerosols) have been studied much less extensively than dispersions in liquids (sols or emulsions). The reasons may be due in part to difficulties encountered in aerosol research; the latter requires unconventional techniques and sophisticated analytical procedures.

Most of the investigations with aerosols have dealt with physical aspects of these systems, such as nucleation, particle or droplet growth, evaporation, coagulation, capture, etc. The amount of work on chemical reactions in these systems is relatively small. Standard texts on aerosols do not even discuss chemical aspects of such dispersions, although these are significant from two rather diverse points of view: (1) aerosols may form by chemical reactions of different gases or vapors, and (2) aerosols (liquid or solid) may interact with gases or vapors.

It is the latter case which has been studied least. A certain amount of work is available on chemisorption of gases on solid aerosol particles. For example, it was noted that irritant gases (or vapors) show enhanced effects on respiratory systems if they are introduced in adsorbed form. Even these investigations had little to do with the actual chemistry of the particle/vapor interfaces.

This article will review some work carried out in the author's laboratory which deals with two types of chemical reactions in aerosol systems: (1) chemical reactions of liquid aerosols of narrow size distribution with vapors, and (2) preparation of well-defined solids by reaction of liquid droplets with vapors.

The second part offers a new technological approach for generation of materials of great purity of controlled particle size and shape. One of the advantages of this technique is that each droplet acts as a separate "reaction container".

Aerosol Preparation

The first condition for carrying out quantitative work on chemical reactions with aerosols is to have particles (or droplets) uniform in size. Since, in many cases, the dispersed phase consists of spheres, light scattering can be used for the evaluation of particle size distribution and of the optical properties of these systems. Such analysis offers significant advantages over any other technique since no denaturation of the materials takes place during sampling. As a result, one can follow particle growth (or decrease) or the change in composition **in situ**.

Spherical particles of various solids were obtained by an evaporation/condensation technique with carrier gases laden with appropriate nuclei. Thus, "monodispersed" aerosols of sodium chloride (Matijević et al., 1963; Espenscheid et al., 1964), silver chloride (Matijević et al., 1960; Espenscheid et al., 1965), and vanadium pentoxide (Jacobsen et al., 1967) were produced over a broad range of modal diameters.

To prepare liquid aerosols of narrow size distributions in a reproducible manner, a falling film generator was designed (Nicolaon et al., 1970, 1971). This equipment has been used to generate "monodispersed" droplets which were then reacted with vapors as described in the following section.

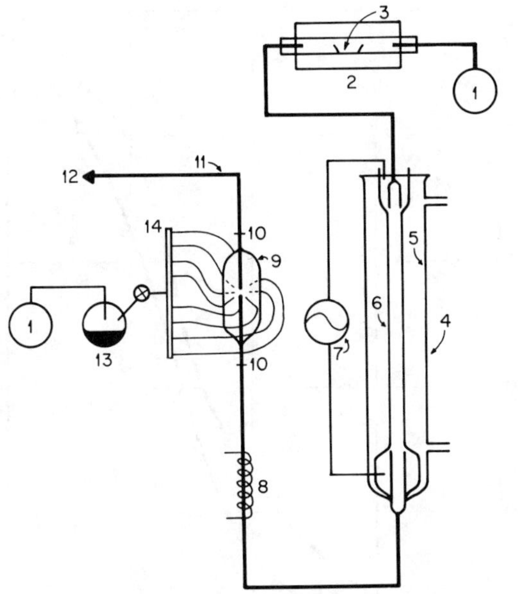

Fig. 1. Schematic diagram of the apparatus used to study the bromination of aerosol droplets: (1) helium tanks, (2) nucleating furnace, (3) ceramic boat containing nuclei material (AgCl), (4) falling film aerosol generator, (5) constant-temperature oil bath, (6) boiler tube, (7) pump for recycling aerosol liquid, (8) heat tape, (9) bromine feed chamber, (10) joints for changing reaction tubes, (11) reaction tube, (12) exit to light-scattering photometer or thermal precipitator, (13) flask containing bromine, (14) manifold to distribute bromine vapor to feed chamber (McRae et al., 1975).

Examples of Chemical Reactions with Aerosols

Bromination of 1-Octadecene

As an example of a chemical reaction in an aerosol droplet a comprehensive study of the bromination of 1-octadecene to 1,2-dibromooctadecane was carried out (McRae et al., 1975). Liquid aerosols of octadecene were exposed to bromine vapor and the degree of bromination was then determined spectrophotometrically as a function of various parameters (reaction time, bromine vapor pressure, temperature, droplet size, etc.) (McRae et al., 1978).

This type of research is fraught with experimental difficulties. However, the chosen system was well suited for the study because, depending on conditions, partial to complete bromination took place between 5 and 30 seconds. These rates allowed the investigation of the reaction in a continuously flowing system in equipment of reasonable dimensions. Because of rather low aerosol concentrations (necessary to avoid coagulation and to eliminate secondary scattering) up to 800 collections were necessary to obtain analytically meaningful samples. To account for the bromination that occurred during or after aerosol collection, only the percent bromination obtained by extrapolation to zero collection time has been considered in the interpretation of the data.

The schematic diagram of the equipment used in these studies is given in Figure 1. Octadecene droplets were generated in the falling film generator and then reacted with gaseous bromine vapor of known concentration in a specially designed chamber in which the mixing of the reacting components was rapid yet without turbulence. The time of reaction was adjusted by altering the length of the reaction tube. The reacted aerosol droplets were separated from the carrier gas and excess bromine vapor by a precipitator, which uses thermal gradient, and were collected on aluminum foil. The samples were dissolved in chloroform, the solvent was stripped off using an aspirator and a steam bath, and the

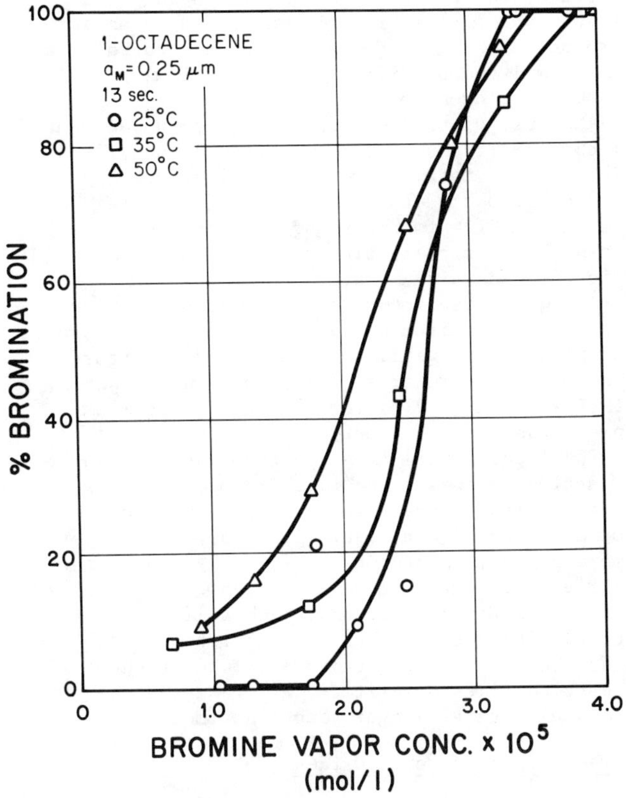

Fig. 2. Percentage bromination of 1-octadecene in the 90-cm reaction tube (average residence time 13 sec) as a function of the bromine vapor concentration at reaction temperatures of 25° (○), 35° (□), and 50°C (△). Aerosol generation conditions: generator temperature 90°C, flow rate 2.0 l/min, AgCl nuclei furnace temperature 596°C, and modal droplet radius 0.25 μm (McRae et al., 1978).

degree of bromination was determined by infrared spectroscopy at 1641.5 cm^{-1}.

The degree of bromination of 1-octadecene as a function of the bromine vapor concentration after 13 seconds of contact at three different temperatures is shown in Figure 2. At the flowrate used the reaction tube was 90 cm long. Obviously, complete bromination can be achieved under the conditions of these experiments and the change in the percentage of bromination rises rather steeply over a small increase in bromine content in the vapor phase. The effect of temperature on the studied process is rather small.

The percentage of bromination of 1-octadecene aerosols of two different droplet sizes is given in Figure 3. Apparently, the larger particles are somewhat more brominated under otherwise identical conditions.

These results were analyzed in terms of the rate-controlling processes for the droplet-vapor interaction. Three possible mechanisms were considered: (1) bromine diffusion to the droplet surface from the bulk gas-phase and internal liquid-phase diffusion of the reactants and/or products, (2) liquid-phase reaction control, and (3) simultaneous diffusion and chemical reaction within the droplet.

Analysis of the reaction kinetics showed that mechanism (1) was unlikely (McRae et al., 1975). The time of total conversion of 1-octadecene to 1,2-dibromooctadecane was approximately 10 seconds, which is orders of magnitude greater than the characteristic times for gas-phase and liquid-phase diffusion.

A liquid reaction model was developed which explained the bromination of 1-octadecene aerosol droplets quite well, as illustrated in Figure 4. The reader is referred to the original paper (McRae et al., 1978) for the mathematical expressions used in drawing the theoretical lines.

The experimental data fit two liquid-phase reaction schemes: one is third order with respect to the concentration of bromine and the other is an autocatalytic process. The latter is more probable since it was found by others that HBr acts in small amounts as a powerful catalyst for bromination reactions. The validity of those conclusions was further substantiated by the fact that the kinetic constants obtained at shorter reaction times quantitatively predicted the experimental data at longer reaction times.

Preparation of Metal Oxides

Chemical reactions in aerosols lend themselves well to the preparation of well-defined and very pure metal (hydrous) oxides consisting of spherical particles of narrow size distribution. The method consists of the formation of a "monodispersed" liquid aerosol of a reactive metal compound which is then hydrolyzed with water vapor. Since the reaction takes place in the droplet, the resulting particles are spherical. Furthermore, it is possible to predict the size

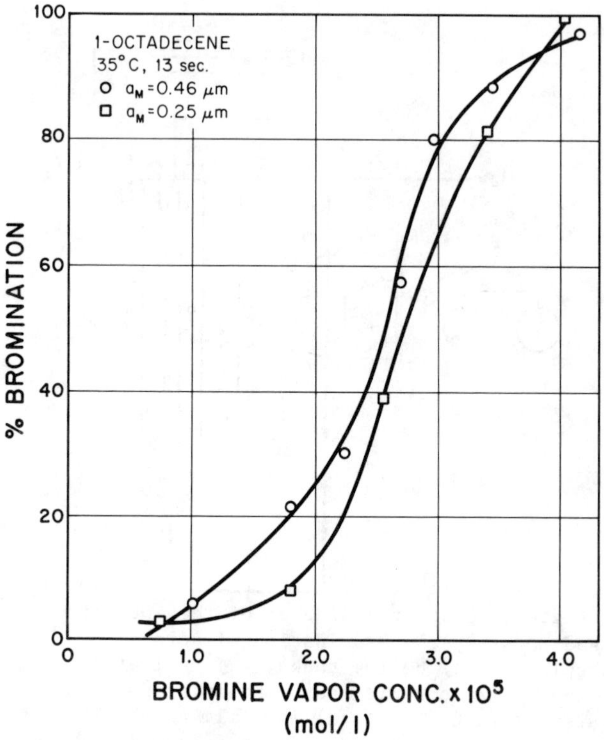

Fig. 3. Percentage bromination of 1-octadecene at 35°C in the 90-cm reaction tube (average residence time 13 sec) as a function of the bromine vapor concentration for droplets with a modal radius of 0.25 μm (□) and 0.46 μm (○). Aerosol generation conditions as in Figure 2 (McRae et al., 1978).

of the final product from the size of the liquid aerosol.

Titanium dioxide. Spherical titanium dioxide particles were obtained by hydrolysis of aerosols consisting of either $TiCl_4$ or Ti(IV) alkoxides (Visca and Matijević, 1979; Matijević and Visca, 1979). In the latter case either titanium(IV) ethoxide or titanium(IV) isopropoxide has been used. On reacting with water vapor $TiCl_4$ yields titanium dioxide and HCl, whereas the alkoxides yield the corresponding alcohols in addition to TiO_2.

The liquid aerosols were prepared in the falling film generator and then reacted in the chamber as illustrated in Figure 1. Other designs for the hydrolysis vessel have also been developed.

The particle size could be varied depending on the conditions of the aerosol generator, i.e., the flow rate, temperature, and the nuclei concentration. Thus, the modal diameters of the resulting titanium dioxide spheres varied between 0.1 and 1 μm, although in principle larger and smaller particles could be prepared by the same procedure.

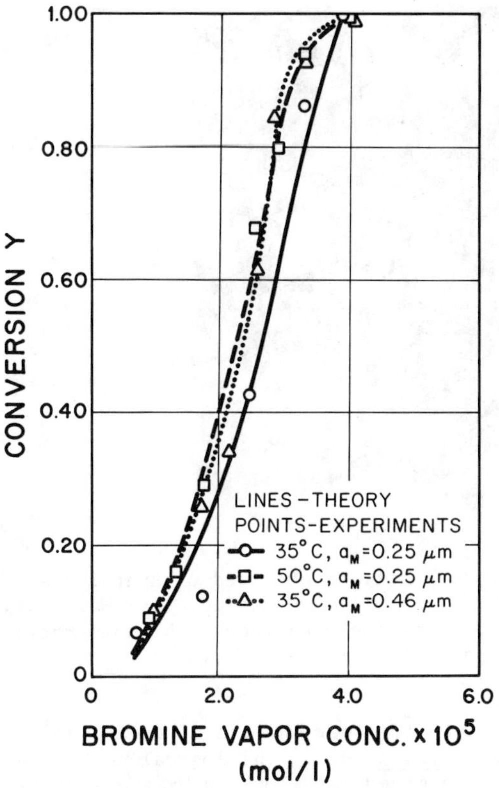

Fig. 4. Comparison of the theoretical curves (lines) based on model which assumes a liquid-phase chemical reaction control of the bromination with a reaction order of $m = 3$ with respect to bromine. The experimental points refer to droplets of a modal radius of 0.25 μm at 35° (○) and 50°C (□) and for droplets of a modal radius of 0.46 μm at 35°C (△) (McRae, 1978).

An example of titanium dioxide solids obtained from Ti(IV) ethoxide is illustrated in the transmission electron micrograph in Figure 5. The powder so obtained is amorphous, but on heating to 450°C the particles crystallize into essentially pure anatase form. On heating the same material at temperatures ≥900°C recrystallization from anatase to rutile takes place. The percentage in the rutile form depends on the duration of the heat treatment. The important aspect of this procedure is that the particles retain their spherical shape even when suspended in water.

Dispersibility of the particles in aqueous solutions is excellent; when necessary it could be aided by treatment in an ultrasonic bath. The stability of the hydrosols so prepared was dependent on pH. The isoelectric point was found at pH ~5.5 which is very close to the "best" value (Furlong and Parfitt, 1978).

One of the important advantages of the aerosol procedure for the preparation of titanium dioxide from Ti(IV) alkoxides is the absence of anionic impurities in the resulting particles. This result is of particular importance in the possible use of this material as catalyst, in which case the absence of impurities may be essential to the performance. Another application of such titanium dioxide powders is for use in areas in which their optical properties are important. Since the latter are fully understood for spherical geometry, solids of exact particle size necessary to achieve a given optical effect can now be prepared.

The amount of material produced in a falling film generator is exceedingly small. However, the liquid aerosols can be obtained by various other procedures (such as different nebulizers) and then reacted with water vapor to yield titanium dioxide in larger quantities.

Aluminum (hydrous) oxide. The procedure described previously was also used to prepare spherical particles of aluminum (hydrous) oxide (Ingebrethson and Matijević, 1980). The most convenient starting material was found to be aluminum *sec*-butoxide, which is readily hydrolyzed. The obtained particle size range was similar to that of titanium dioxide.

Spheres so produced were amorphous and had a chemical analysis consistent with a hydrous oxide. Heating the collected aerosol powder in an oven to 700°C for over 8 hours resulted in a slight decrease (~10 percent) in particle size, yet the spherical shape was retained. The X-ray analysis of the calcined particles gave a pattern characteristic of γ-Al_2O_3.

The powder was most readily dispersible in water. The aqueous sol had an isoelectric point (i.e.p.) of pH 9.3, which is another indication of the absence of contaminations. As is well known, anionic inclusions into aluminum (hydrous)

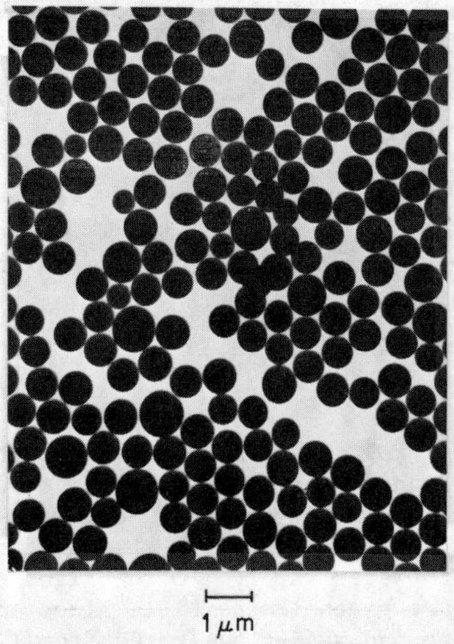

Fig. 5. Transmission electron micrograph of titanium dioxide particles obtained by hydrolysis of Ti(IV) ethoxide aerosol droplets with water vapor at 99°C.

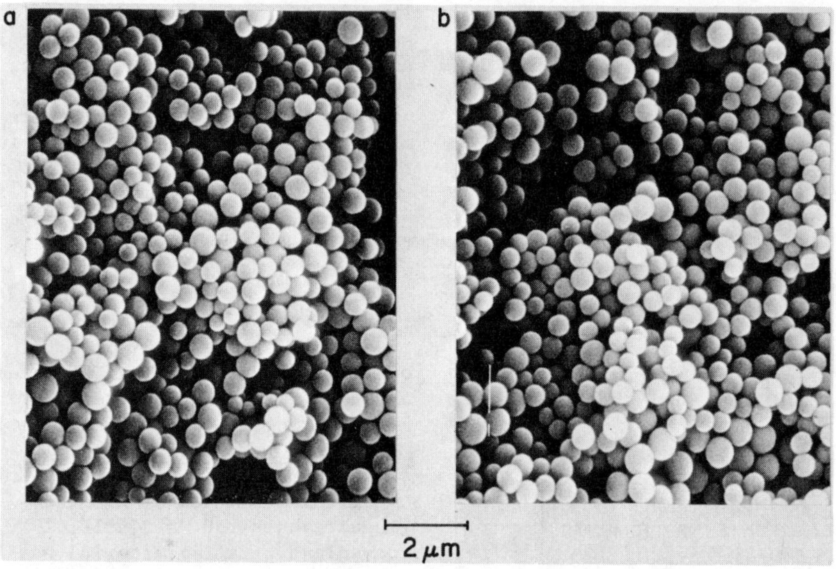

Fig. 6. Scanning electron micrographs of Al-Ti(IV) oxide particles obtained by hydrolysis of aerosol droplets consisting of a mixture of Al *sec*-butoxide and Ti(IV) ethoxide. (a) Mole ratio of Ti:Al = 1.4:1.0 in droplet; mole ratio of Ti:Al = 13:1 on particle surface. (b) Mole ratio of Ti:Al = 0.03:1 in droplet; mole ratio of Ti:Al = 1:2 on particle surface.

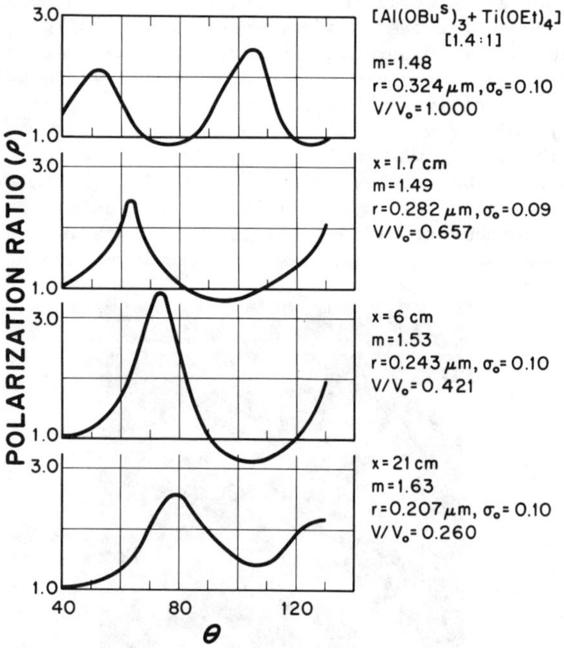

Fig. 7. Polarization ratios of scattered light for an aerosol, the droplets of which contained a mixture of Al *sec*-butoxide and Ti(IV) ethoxide, at various stages of hydrolysis with water vapor. To the right of each diagram are given the particle sizes and refractive indexes as evaluated from light-scattering. The progress of the process was followed along a reaction tube. The shift of 2 cm along the tube corresponds to a reaction time of ~13 millisec.

oxides tend to lower the i.e.p. (below 9.3).

Mixed metal oxides. The logical next step is to prepare mixed metal oxides by the aerosol technique. It was indeed shown that liquid aerosols consisting of a mixture of aluminum *sec*-butoxide and titanium(IV) ethoxide can react with water vapor to give spherical particles consisting of titanium and aluminum oxides (Ingebrethsen, 1982). It is noteworthy that the surface composition of the solids as determined by ESCA shows a different ratio of Al/Ti than in the liquid droplet; in all cases there is an enrichment of Ti in the interface as compared to the bulk.

Scanning electron micrographs in Figure 6 illustrate two Al-Ti(IV) oxide powders as obtained by the aerosol technique under the conditions described in the legends. It is quite evident that the particles are spherical and of narrow size distribution.

The process of solidification of mixed alkoxide droplets in the course of reaction with water vapor was followed by light-scattering along the reaction tube. The modal particle size, the width of the distribution, and the refractive index were determined using the polarization ratio method (Kerker et al., 1964). In Figure 7 a series of curves are shown which represent the polarization ratios as a function of angle for the same system which originally consisted of droplets containing Al *sec*-butoxide and Ti(IV) ethoxide in the molar ratio 1.4:1 and was subsequently exposed to water vapor. Calculated size parameters and the volume ratios (relative to the original droplet volume) are given beside each curve; the listed refractive indexes are those which best fit the data. From the average velocity of the flowing aerosol it was calculated that a 2-cm distance corresponded to a reaction time of ~13 milliseconds. There is a significant change in the light-scattering pattern which is caused by the chemical process taking place within the droplets. Despite the very rapid reaction the systems are quite reproducible, which allows for a quantitative interpretation of the data.

The given example indicates that the aerosol procedure offers significant opportunities for the preparation of mixed metal oxides. Now it is possible to obtain such materials in the form of spherical particles of well-defined (and predictable) sizes and of a desired composition. The possibility of making a metal oxide doped with another metal is of particular interest in heterogeneous catalysis.

Acknowledgment. The author is indebted to his associates Dr. Douglas McRae, Dr. Mario Visca, and Mr. Bradley J. Ingebrethsen, on whose work this presentation is based. The collaboration of Professor E. James Davis in the aerosol bromination project is gratefully acknowledged. This work was supported by the National Science Foundation under Grant No. CHE-80 13684.

References

Espenscheid, W. F., E. Matijević, and M. Kerker, Aerosol studies by light scattering. III. Preparation and particle size analysis of sodium chloride aerosols of narrow size distribution, J. Phys. Chem., 68, 2831-2842, 1964.

Espenscheid, W. F., E. Willis, E. Matijević, and M. Kerker, Aerosol studies by light scattering. IV. Preparation and particle-size distribution of aerosols consisting of concentric spheres, J. Coll. Inter. Sci., 20, 501-521, 1965.

Furlong, D. N., and G. D. Parfitt, Electrokinetics of titanium dioxide, J. Coll. Inter. Sci., 65, 548-554, 1978.

Ingebrethsen, B. J., Formation of mixed metal oxides by chemical reactions in aerosols, Ph.D. Thesis, Clarkson College, Potsdam, N.Y., 1982.

Ingebrethsen, B. J., and E. Matijević, Preparation of uniform colloidal dispersions by chemical reactions in aerosols - 2. Spherical particles of aluminum hydrous oxide, J. Aerosol Sci., 11, 271-280, 1980.

Jacobsen, R. T., M. Kerker, and E. Matijević, Aerosol studies by light scattering. V. Preparation and particle size distribution of aerosols consisting of particles exhibiting high optical absorption, J. Phys. Chem., 71, 514-520, 1967.

Kerker, M., E. Matijević, W. Espenscheid, W. A. Farone, and S. Kitani, Aerosol studies by light scattering. I. Particle-size distribution by polarization-ratio method, J. Coll. Inter. Sci., 19, 213-222, 1964.

Matijević, E., W. F. Espenscheid, and M. Kerker, Aerosols consisting of spherical particles of sodium chloride, J. Coll. Inter. Sci., 18, 91-94, 1963.

Matijević, E., M. Kerker, and K. F. Schulz, Light scattering of coated aerosols. Part 1.- Scattering by the AgCl cores, Disc. Faraday Soc., 30, 178-184, 1960.

Matijević, E., and M. Visca, Titanium dioxide, Ger. Offen. 2,924,072 (Cl. C01G23/06), 20 December 1979.

McRae, D., E. Matijević, and E. J. Davis, Chemical reactions in aerosols. I. Bromination of octadecene droplets, J. Coll. Inter. Sci., 53, 411-421, 1975.

McRae, D., E. Matijević, and E. J. Davis, Chemical reactions in aerosols. II. The effects of various parameters on the bromination of 1-octadecene droplets, J. Coll. Inter. Sci., 67, 526-537, 1978.

Nicolaon, G., D. D. Cooke, M. Kerker, and E. Matijević, New liquid aerosol generator, J. Coll. Inter. Sci., 34, 534-544, 1970.

Nicolaon, G., D. D. Cooke, E. J. Davis, M. Kerker, and E. Matijević, New liquid aerosol generator. II. Effect of reheating and studies on the condensation zone, J. Coll. Inter. Sci., 35, 490-501, 1971.

Visca, M., and E. Matijević, Preparation of uniform colloidal dispersions by chemical reactions in aerosols. I. Spherical particles of titanium dioxide, J. Coll. Inter. Sci., 68, 308-319, 1979.

PHOTOPHORETIC SPECTROSCOPY: A SEARCH FOR THE COMPOSITION OF A SINGLE AEROSOL PARTICLE

S. Arnold and M. Lewittes

Department of Physics, Polytechnic Institute of New York, Brooklyn, New York, 11201

Abstract. Photophoretic spectroscopy is discussed as a means for obtaining the composition of a single particle in situ, in real time, and in a controlled environment. Theoretical relationships relating the force spectrum to the particle absorption spectrum are presented along with experimental results.

What is Photophoretic Spectroscopy

Photophoretic spectroscopy (PPS) (Pope et al., 1979; Arnold and Amani, 1980; Arnold et al., 1980) is a possible means for determining the composition of a single fixed aerosol particle in real time in a nearly in situ environment. This spectroscopy measures the optical absorptive properties of the particle, and therefore, processes such as the heterogeneous catalysis of $SO_3^{=}$ to $SO_4^{=}$ may be monitored by observing the photophoretic response in the infrared. Since only one particle is involved, basic information concerning catalysis such as the role of size and dielectric response may be obtained without the statistical interference which is inherent in a polydispersed system.

Photophoresis is a term first used in 1917 by Ehrenhaft to describe the motion induced by light on a small particle in a gaseous medium. A great deal of work has established that this force is principally radiometric in origin (Preining, 1966). Although more will be said in what follows about photophoresis, its radiometric nature suggests that a measure of the spectral dependence of this phenomenon could be used to investigate the character of picogram quantities of material. Initial experiments by Pope et al. (1979) on several irregular shaped particles of perylene at three Ar^+ laser wavelengths revealed a correspondence between the photophoretic force and optical absorption in bulk. Further experiments by Arnold and Amani (1980) on several irregular CdS particles (diameter ~5 μ) using a broadband tunable source (Xe-arc/monochromator) gave a full spectrum (see Figure 4) from which the energy band gap of CdS was easily identified.

From the standpoint of heterogeneous catalysis the most important spectral region is the near and mid-infrared where fingerprint spectra for most molecular pollutants are to be found. For instance, the main IR active vibrational line of $SO_4^{=}$ is at 1100 cm^{-1}; a wavelength $w \simeq 9$ μ. Since many aerosols having reasonable lifetimes in the atmosphere are of a radius $R < 2$ μ (Lodge et al., 1981), investigations of heterogeneous catalysis require knowledge of dielectric response for $R/\lambda << 1$. We will call this the small particle region. This is not to say that the visible region is not also of interest. The detection of aromatic molecules (i.e., PAH) is certainly possible (as demonstrated for perylene (Pope et al., 1979)) by photophoretic spectroscopy; however, the technique should play its major role in the IR where fluorescence detection is unavailable and thermal techniques are traditional.

In addition to the analytical possibilities which photophoretic spectroscopy opens, the phenomena provides a test for theory involving the manner in which heat sources are distributed within, as well as on, the surface of spheres. This distribution of sources is known as the source function. In what follows we will show that there are many misconceptions concerning the appropriate source function to be used for photophoresis.

The Physics of Small Particle Photophoresis

The origin of photophoresis is best understood in the free molecule regime, in which the mean free path in the gas λ is considerably larger than the particle size R; Knudsen number $K_n = \lambda/R >> 1$. In this limit, each of the colliding molecules may be thought of as acting independently from a gas at infinity with a constant temperature T_∞. If we assume energy and momentum accommodation at the surface to be 1, then it is straightforward to show that the pressure P_s on a surface at temperature $T_s (T_s \gtrless T_\infty)$ due to momentum transfer from the gas at pressure P_∞ is $P_s = 1/2\, P_\infty[(T_s/T_\infty)^{\frac{1}{2}} + 1]$. For a sphere such as that shown in Figure 1, the net force is

$$F_z = -\int P_s \hat{n}\, dA = -P_\infty \int_0^\pi \frac{1}{2}\left[\left(\frac{T_s(\theta)}{T_\infty}\right)^{\frac{1}{2}} + 1\right] \times \cos\theta\; 2\pi R^2 \sin\theta\; d\theta \qquad (1)$$

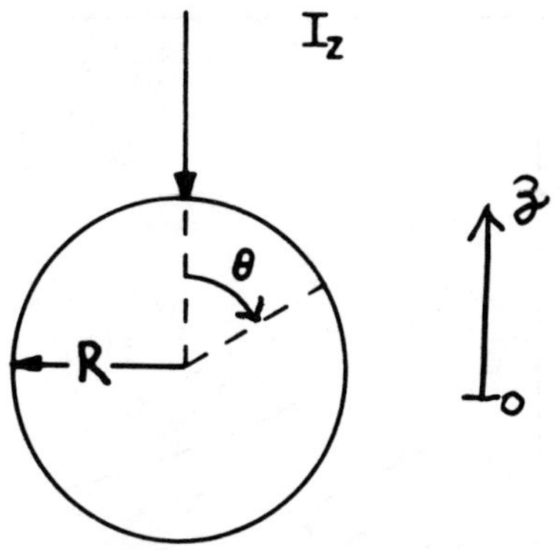

Fig. 1. The coordinate system used in evaluating equation (1).

where n is the unit normal to the sphere. The constant pressure background P_∞ in the presence of an isotropic surface temperature contributes nothing to the net force so that

$$F_z = -\frac{P_\infty \pi R^2}{2} \int_0^\pi \left[\frac{T_s(\theta)}{T_\infty}\right]^{\frac{1}{2}} \sin 2\theta \, d\theta \qquad (2)$$

The integrand in equation (2) clearly gives the sign of the force. If the average value of the integrand is larger over the front surface than the back (i.e., if the front surface is warmer than the back), the force will be in the minus z-direction (i.e., in the direction of the light). If the opposite is true, the force will be in the plus z-direction (i.e., toward the light). In order to determine the sign and magnitude of the force correctly, one must find $T_s(\theta)$. This is the most difficult part of the problem, and accordingly is the one containing the greatest number of misconceptions in the literature. To establish $T_s(\theta)$, one must solve the Fourier heat transfer equation

$$K_p \nabla^2 T(r,\theta) = -S(r,\theta) \qquad (3)$$

where K_p is the thermal conductivity of the particle, $T(r,\theta)$ is the internal temperature at position (r,θ), and S is the electromagnetic source function. The thermal boundary conditions are well known; the surface of the particle is considered to be impenetrable to gas molecules, and the heat transferred to the surface from within the particle per unit area is equal to the gain in kinetic energy of the colliding molecules per unit area. $T_s(\theta)$ has been determined in innumerable papers for a so-called "opaque" particle (Reed, 1977; Hidy and Brock, 1967). An "opaque" particle is one in which the incident light of intensity I_z is absorbed only on the front surface. The resultant force to 1st order is (Hidy and Brock, 1967)

$$F_z = \frac{-\pi R^3 I_z P_\infty}{6\left(\frac{1}{2} P_\infty v R + K_p T_\infty\right)} \qquad (4)$$

where v is the gas molecular velocity. Clearly such an approximation eliminates much of the labor in solving equation (3), especially if all one wants to do is determine the maximum possible force. However, the "opaque" particle source function is unrealistic in visible light for sizes below 2 μ although it has been applied in several recent papers down to 0.1 μ (Reed, 1977; Hidy and Brock, 1967). Simply stated, particles comparable to the wavelength cannot possibly form sharply defined shadows. Light hitting such small particles washes (by diffraction) around the sides. This is a consequence of a self-consistent solution of Maxwell's equations. The solution for the internal electromagnetic field is given by Mie (1908). This solution is in terms of a polynomial series which gives a corresponding source function $S(r,\theta)$ as derived by Kerker and Cooke (1973). Visualizations of these source functions can only be had at present through computer calculations. The most extensive calculations in this area are those of Dusel et al. (1979). Fortunately, for wavelengths $w \gg R$, a considerable simplification can be made. A solution in this limit has been arrived at by Yalamov et al. (1976), and a variation of this solution is presented in what follows.

Due to the virtual absence of interference within the particle, the square of the electric field $E^*(r,\theta)E(r,\theta)$ is given by (Yalamov et al., 1976)

$$|E(r,\theta|^2 = \frac{3E_o^2|3 + 2N^2|^{-2}}{|N^2 + 2|^2} \{3|3 + 2N^2|^2 + 2n\kappa k_o r \cos\theta(|3 + 2N^2|^2 - 5)\} \qquad (5)$$

where E_o is the electric field intensity of the incident light, n and κ are the real and imaginary parts of the index of refraction N (i.e., $N = n - i\kappa$), and k_o is the incident wave vector (i.e., $k_o = 2\pi/w_o$). It is assumed in equation (5) that not only is the particle small in comparison to the wavelength of light (i.e., $R/w \ll 1$), but also small in comparison to the absorption depth of the light in bulk material (i.e., $\kappa R/w \ll 1$). We note that equation (5) has a very simple physical form. The energy density available for absorption diminishes linearly as we traverse the particle from its irradiated pole to its opposite pole. The actual source function or heat power generated per unit volume must be found from

$$S = \frac{1}{2} \sigma |E(r,\theta)|^2 \qquad (6)$$

where σ is the effective conductivity of the medium given by

$$\sigma = \frac{2n\kappa k_o}{\mu c} \tag{7}$$

where μ is the magnetic permeability and c is the speed of light. The part of the source function which depends on position within the particle $S_v(r,\theta)$ is the only component which can give a photophoretic force. From equations (6), (7), and (5) this contribution is

$$S_v(r,\theta) = \frac{2n\kappa k_o}{\mu c|3+2N^2|^2|N^2+2|^2}[3|E_o|^2(2n\kappa k_o r \cos\theta) \times (|3+2N^2|^2 - 5)] \tag{8}$$

Surprisingly we see that S_v is proportional not to κ, a quantity proportional to the density of absorbers, but κ^2. The factor $24n\kappa\rho/|N^2 + 2|^2$ is just k_a, the absorption factor for a small sphere (i.e., power absorbed is $\pi r^2 k_a I$). In addition, for MKS units $|E_o|^2 = 2I\left(\frac{\mu_o}{\varepsilon_o}\right)^{\frac{1}{2}}$ so that $S_v(r,\theta)$ may be rewritten as

$$S_v = \frac{k_a}{R}\left[In\kappa k_o r \cos\theta\left(1 - \frac{5}{|3 + 2N^2|^2}\right)\right] \tag{9}$$

A physical way to express the distribution of sources is through the asymmetry factor

$$\vec{A} = \frac{\int_s |E(R,\theta)|^2 \hat{n}\, dA}{\int_s |E(R,\theta)|^2\, dA} \tag{10}$$

This factor represents the asymmetry of the sources at the surface. The derived factor using equations (10) and (5) is

$$\vec{A} = +\frac{2}{9}n\kappa\rho\left(1 - \frac{5}{|3 + 2N^2|^2}\right)\hat{k} \tag{11}$$

where ρ is the size parameter k_oR and $\hat{k}$ is a unit vector in the z-direction. Figure 2 shows the influence of the refractive index on $\vec{A}$ for a particle having a $\rho = 0.5$ (i.e., R/w = 0.08). We note that in the region of interest ($1 < n < 2$, $0 < \kappa < 1.8$), $\vec{A}$ is in the positive z-direction (this corresponds to a photophoretic force in the negative z-direction). The solid lines are taken from our asymptotic equation (equation (13)). In addition, Figure 2 shows numerical values of $\vec{A}$ calculated using the full Mie description for the internal fields (M. S. Kerker, unpublished manuscript, 1981). We see that equation (11) appears to underestimate the value of $\vec{A}$ by no more than 30 percent. This error is small in comparison with the "opaque" particle picture. For example, a particle of polycrystalline graphite has an index N = 1.95 − i(0.66). From Figure 2 this index corresponds to an $\vec{A}$ of +0.13; however, for an "opaque" particle, $\vec{A}$ is +0.5, which is an overestimate of ~300 percent.

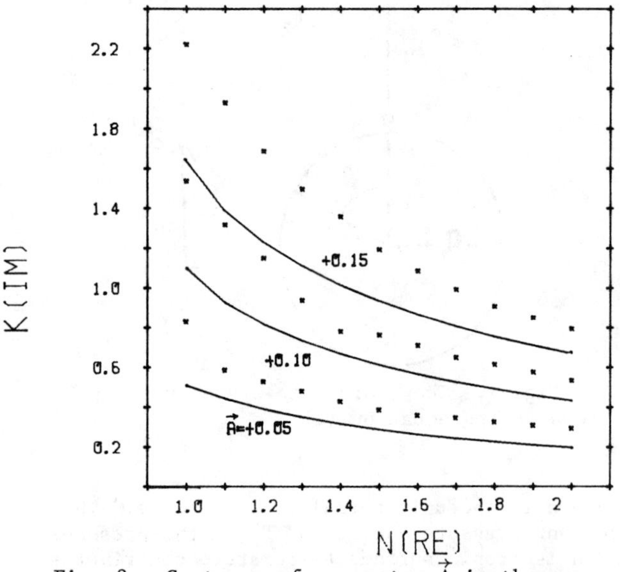

Fig. 2. Contours of asymmetry $\vec{A}$ in the refractive index domain, κ (imaginary part of index) vs. n (real part of index), for ρ = 0.5. The solid lines are a contour plot of the asymptotic solution for $\vec{A}$ in the limit $\rho \ll 1$. The points are exact computer contours generated by M. S. Kerker (unpublished manuscript, 1981).

The solution to equation (3) to first order using the source function given in equation (9) is

$$T_v(R,\theta) = +\frac{9}{20}\frac{A_z k_a IR \cos\theta}{\left(K_p + Rn_\infty\sqrt{\frac{2kT_\infty}{\pi m}}\right)} \tag{12}$$

where A_z is obtained from equation (11), n_∞ is the molecular density, and m is the mass of a molecule. The second term in the parentheses in the denominator of equation (12) is the effective thermal conductivity of the gas. In practice, the thermal conductivity of a particle K_p is at least an order of magnitude greater than that of the gas so that the second term in the denominator of equation (12) can be neglected. Finally, the substitution of equation (12) into equation (2) gives the small particle force to be

$$F_z = -\frac{3}{20}\frac{\pi R^3 p_\infty I k_a A_z}{K_p T_\infty} \tag{13}$$

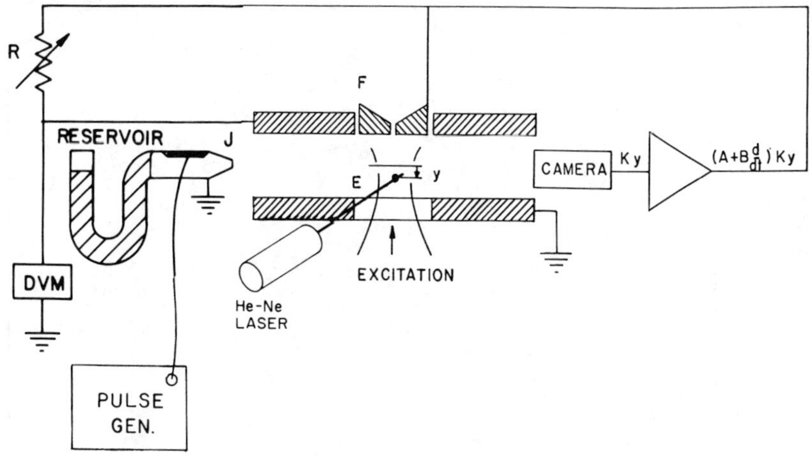

Fig. 3. Photophoretic spectrometer (Arnold et al., 1980): The excitation sources are described in the text. J is an impulse jet for injecting liquid particles. The entire chamber is enclosed in an evacuable cell.

Equation (13) provides a basis for the spectroscopy of small particles. From the definitions of A_z (equation (11)) and k_a we see that F_z is proportional to κ^2. In other words, if the concentration of absorbers is doubled, the force should change by a factor of 4 at a given wavelength. Assuming that n is relatively independent of wavelength, the square root of the force spectrum is therefore the spectrum of κ (i.e., the absorption spectrum).

Intermediate and Large Particle Photophoresis

As particles become larger than the wavelength of the exciting light, diffraction effects begin to fade and geometrical optics begins to take over. Thus, the light can be focused strongly on the back side of the particle leading to reversed photophoresis (i.e., a force into the light). The transition from small to large particles has never been observed in a continuous fashion, and there are no closed-form solutions available for this region. At even larger sizes some asymptotic solutions (Rubinowitz, 1926) are available, and these solutions predict that the reversed force will be proportional to the complex index of refraction κ as long as the absorption depth of the light is considerably larger than the particle size. This description may be applied to the photophoretic spectroscopy of weak absorbing aerosols in the visible so long as the particles have sizes $R >> 0.5\ \mu$. As the particle enlarges further, the absorption depth of the light will eventually become small in comparison to its size so that it becomes difficult to deliver energy to the back side; the reversed effect will now turn into a forward effect. An example of photophoretic spectroscopy in this regime is exemplified by our work on CdS (Arnold and Amani, 1980). The size dependence of photophoresis should, when coupled with theory for the internal field, provide an absolute measurement for the complex refractive index κ, a parameter which is difficult to obtain from light scattering data.

Photophoretic Spectrometer

The photophoretic spectrometer shown schematically in Figure 3 is described by Arnold et al. (1980). A short summary is presented here for completeness. Excitation is provided by, (A) a 1000 Xe lamp - followed by a 0.125M monochromator, or (B) a tunable CO_2 laser. This excitation enters a highly modified Millikan capacitor (Arnold, 1979) (known as the Millikan-Pope-Arnold chamber (Pope et al., 1979)) either through a transparent conducting electrode (for (A) excitation see E in Figure 3) or a thru hole (1.5 mm diameter for (B) excitation). A horizontal potential well is created for the particle by the use of an insulated electrode F in the top plate (Fletcher, 1914). This electrode also acts as a reservoir for solid particles that are injected through it. Liquid particles are injected horizontally via an impulse jet J. Vertical stability is maintained by use of an electro-optic servo-loop. A camera detects laser light scattered from the particle at 6328 Å and provides a signal proportional to the vertical offset y from the center of the chamber. The camera signal Ky is conditioned by the operator A + B(d/dt) and fed back to the top plate of the capacitor. The use of a voltage divider R and DVM (the resistance of our digital voltmeter) assures that the reading on the DVM is proportional to the electric field at the particle. In this way the change in the holding voltage V upon turning on the laser divided by the voltage V in the absence of excitation is just equal to the photophoretic force F_p divided by the weight (mg).

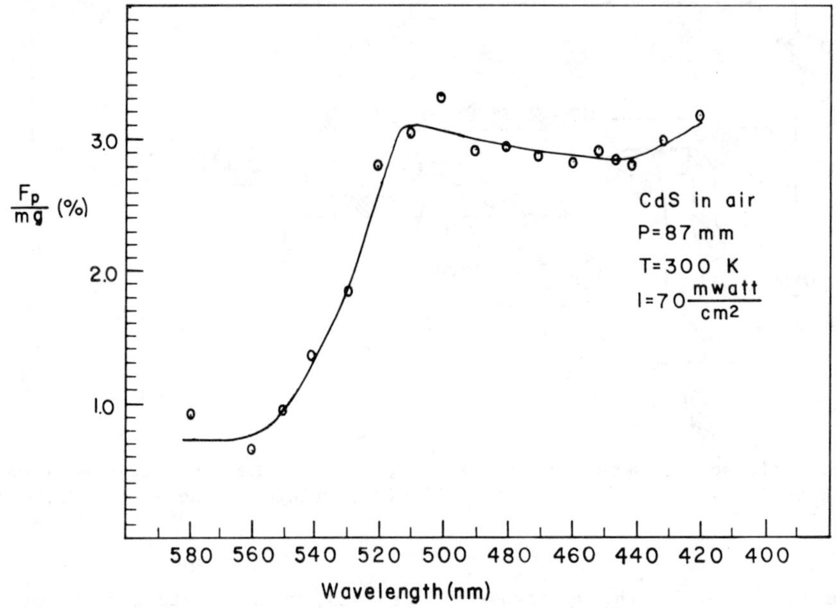

Fig. 4. Spectrum of a CdS particle taken in air. The particle has a size between 5 and 10 microns.

Experiments

Figure 4 shows a typical photophoretic spectrum of CdS (size between 5 and 10 μ) (Arnold and Amani, 1980; Arnold et al., 1980). The force measured was in the direction of the incident light. The spectrum was taken at high pressure P_H (70 mm Hg in air) such that $\lambda << R$; however, the shape of the spectrum is not expected to be sensitive to pressure since the proportionality factor connecting the high (i.e., $R >> \lambda$) and low pressure P_L (i.e., $R << \lambda$) photophoretic forces F_H and F_L to first order is (Rosen and Orr, 1964)

$$\frac{F_H}{F_L} = \frac{3\mu^2 CT}{\alpha P_H P_L M R^2} \quad (14)$$

where μ is the gas viscosity, C is the universal gas constant, T is the ambient temperature, α is the accommodation coefficient, and M is the gas molecular weight. The spectrum in Figure 4 is surprisingly similar to that obtained by photoacoustic spectroscopy (Rosencwaig, 1977). Both reveal a plateau at 2.1 ± 0.03eV, the band gap of CdS; however, both are considerably broader at the band edge (after correcting for source bandwidth) than the data for the single crystal absorption (Dutton, 1958). In the case of the photophoretic spectrum, the reason for this disparity is difficult to uncover since the particle is irregular in shape; however, spectra taken on other particles were similar to the ones shown in Figure 2. The reason for their similarity is certainly connected with the rotational averaging which takes place in the experiment. Due to radiometric torques the particle tumbles in the light beam. This is revealed by increased noise in our detection electronics in the presence of light and lateral oscillations of the particle's image. However, such rotational motion should not reduce the photophoretic force since the relaxation time associated with the force is expected to be less than $\sim 10^{-4}$ sec for size R below 10 μ (Fuchs, 1964) which is considerably smaller than the time needed for a few degrees of rotation. Although the photophoretic force is expected to reverse for weak absorption by a sphere, no such effect is seen on these crystallites even at 5250 Å where the absorption depth in bulk is ~200× the particle size. The lack of reversal and the large photophoretic force in this region are most probably associated with the irregular morphology of the crystallites; irregular morphology will prevent focusing. Clearly, elucidation of the internal fields requires experiments on well-characterized particles. A liquid sphere is an apparent choice.

The distribution of the internal field only depends upon the index of refraction N and the size parameter ρ. Since particles below 0.3 μ are difficult to contain in the present apparatus, experiments in the visible could not possibly cover all regions in size parameter. In particular, the transition from large (optical) to small size near $\rho = 1$ (i.e., $2\pi R/w_o = 1$) would be exceedingly difficult to measure. It was chosen therefore to work at a wavelength of ~10 μ. In order to cover a large range in size parameter in a precise manner, a liquid droplet was allowed to evaporate in place at a fixed pressure. Glycerol

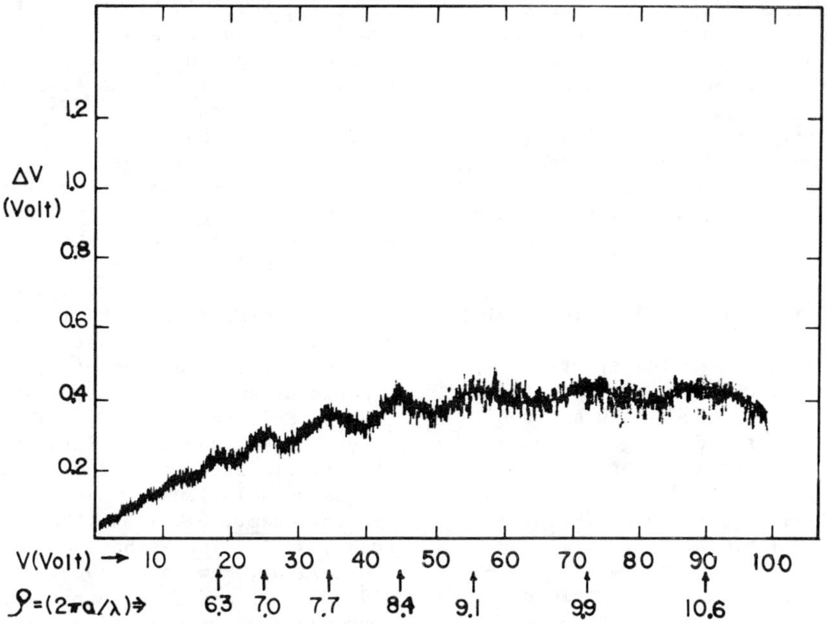

Fig. 5. The dependence of the photophoretic difference voltage ΔV on the holding voltage V for a pure glycerol droplet evaporating in CO_2 laser radiation at 10.66 μ and 250 m Watts/cm^2. The fast fluctuations with voltage are system noise. Positions in ρ of the peaks in the periodic structure are shown on the lower scale. The index of refraction for glycerol at 10.66 μ is 1.57 - 0.038i.

was found to be a good candidate. It has a good deal of absorption near 10 μ, and, since it is a pure substance, its refractive index is independent of size, so that in principle, one can examine the size dependence independent of changes in N.

In practice, the glycerol was jetted into the chamber (see Figure 4) horizontally through a small hole in the side. The velocity of the particle was adjusted so that drag slowed it to a near stop precisely at the center of the chamber. In this way it was found that one could start with a particle having a diameter of ≃40 μ (measured by back lighting and projection). Since evaporation is diffusion-limited near atmospheric pressure for such a particle, the rate of evaporation was easily controlled by lowering the pressure. In the normal course of an experiment, measurements were made of the difference voltage ΔV (excitation on minus excitation off) vs. the holding voltage (excitation off).

Figure 5 shows a typical run made on glycerol at 10.66 μ (excitation via CO_2 laser) in air at 22 torr. Transmission and reflection measurements on flat films gave a value for the refractive index of 1.57 - i(0.038). The photophoretic force was found to be reversed (toward the laser) over the entire size range $2 < \rho < 11$.

The reversed photophoresis is understood by the fact that the light is weakly absorbed (i.e., $\kappa\rho$ is less than 1 over the whole range of size in Fig. 5). Although the fast fluctuations are system noise, the modulating ripple is clearly an optical effect as indicated by its periodicity in size parameter (period = 0.72). What is unclear at present is how much of this effect is due to change in absorption factor (see section on small particle photophoresis) and how much is due to the redistribution of internal sources. The smooth curve which the ripples appear to modulate can be shown to have an approximate functionality, $\Delta V \propto V^{2/3} \propto R^2$. The present experiment was carried out at high pressure (i.e., $\lambda << R$); however, from equation (14) we anticipate that at low pressure, with all other things remaining equal, $\Delta V_1 \propto (\Delta V)R^2 \propto R^4$. Although no current analytical solutions are available for photophoresis by a weakly absorbing sphere at high pressure, the low pressure functionality which we conjecture is precisely that given by the asymptotic solution of Rubinowitz (1926) for $\rho >> 1$. Clearly, the current experiments show the capability for investigating photophoresis in a continuous manner from large to small size. The results of such an investigation by S. Arnold and M. Lewittes are forthcoming.

Future

The question of particle composition as it pertains to heterogeneous catalysis is the area to which our own interest in photophoretic spectroscopy is directed. This once again means investigations in the infrared. Fortunately, the beginning work can be done with a tunable CO_2 laser by taking advantage of coincidences between its spectrum and the spectrum of surface impurities. Eventually, broadband photophoretic spectra of surface species may be accomplished. However, since broadband sources of infrared are relatively low in intensity compared with visible sources, one must multiplex the input light. This is accomplished by using a source consisting of a glow bar followed by an interferometer. As one of the interferometer mirrors moves by a distance x, the output voltage $\Delta V(x)$ in our chamber will be

$$\Delta V(x) \propto \int I_s(K) R_p(K)(1 + \cos K x) dK \quad (15)$$

where $K = 2\pi/\lambda$, $I_s(K)$ is the source spectrum, and $R_p(K)$ is the photophoretic response. The inverse transform gives $I_s(K)R_p(K)$ which in turn gives $R_p(K)$ since $I_s(K)$ can be found using an independent detector. This Fourier transform IR photophoretic spectroscopy has already produced interferograms for polystyrene in the near IR (M. Lewittes et al., unpublished data, 1981). The future will tell whether it can be used in following catalytic processes on the particle surface.

At this point it should be emphasized that photon stimulated catalysis such as the photocatalytic conversion of $SO_3^{=}$ to $SO_4^{=}$ on ZnO (Frank and Bard, 1977) will depend for its yield on the dielectric response of ZnO in the presence of radiation which can produce electron-hole pairs. In the past few years we have come to realize from work on the surface enhanced Raman effect that the field within a small particle can exhibit resonant effects which enhance the field at the surface. Therefore, it is not sufficient to measure the yield in photocatalytic conversion on a flat surface. One must deal with the collective dielectric response of the small particle + surface molecule (or surface layer) system. Photophoretic spectroscopy is directed toward this end.

Acknowledgements. We would like to thank M. Kerker of Clarkson College for discussions of the electromagnetic source function, and M. J. D. Low of New York University for suggesting the Fourier transform approach to photophoretic spectroscopy. This work is supported by the Atmospheric Chemistry Division of the National Science Foundation (Grant ATM 8006580).

References

Arnold, S., Determination of particle mass and charge by one electron differentials, J. Aerosol Sci., 10, 49-53, 1979.

Arnold, S., and Y. Amani, Broadband photophoretic spectroscopy, Opt. Lett., 5, 242-244, 1980.

Arnold S., Y. Amani, and A. Orenstein, Photophoretic spectrometer, Rev. Sci. Inst., 51, 1202-1204, 1980.

Dusel, P. W., M. Kerker, and D. Cooke, Distribution of absorption centers within irradiated spheres, J. Opt. Soc. Am., 69, 55-59, 1979.

Dutton, D., Fundamental absorption edge in cadmium sulfide, Phys. Rev., 112, 785-792, 1958.

Ehrenhaft, E., Towards a physics of millionths of centimeters, Z. Physik, 18, 352-370, 1917.

Fletcher, H., A determination of Avogadro's constant N from measurements of the Brownian movements of small oil drops suspended in air, Phys. Rev., 4, 440-453, 1914.

Frank, S. N., and A. J. Bard, Heterogeneous photocatalytic oxidation of cyanide and sulfite in aqueous solutions at semiconductor powders, J. Phys. Chem., 81, 1484-1488, 1977.

Fuchs, N. A., The Mechanics of Aerosols, Chap. II, Macmillan, New York, 1964.

Hidy, G. M., and J. R. Brock, Photophoresis and the descent of particles into the lower stratosphere, J. Geophys. Res., 72, 455-460, 1967.

Kerker, M., and D. Cooke, Radiation pressure on absorbing spheres and photophoresis, Appl. Opt., 12, 1378-1379, 1973.

Lodge, J. P., Jr., A. P. Waggoner, D. T. Klodt, and C. N. Crain, Non-health effects of airborne particulate matter, Atmos. Environ., 15, 431-482, 1981.

Mie, G., Contributions to the optics of turbid media, especially colloidal metal solutions, Ann. Physik, 25, 377-445, 1908.

Pope, M., S. Arnold, and L. Rozenshtein, Photophoretic spectroscopy, Chem. Phys. Lett., 62, 589-591, 1979.

Preining, O., Photophoresis, in Aerosol Science, edited by C. N. Davies, pp. 111-135, Academic Press, New York, 1966.

Reed, L. D., Low Knudsen number photophoresis, J. Aerosol Sci., 8, 123-131, 1977.

Rosen, M., and C. Orr, Jr., The photophoretic force, J. Colloid Sci., 19, 50-60, 1964.

Rosencwaig, A., Solid state photoacoustic spectroscopy, in Optoacoustic Spectroscopy and Detection, edited by Y. -H. Pao, pp. 192-234, Academic Press, New York, 1977.

Rubinowitz, A., Radiometric forces and Ehrenhaft's photophoresis!, Ann. Physik, 62, 691-737, 1926.

Yalamov, U. I., V. B. Kutukov, and E. R. Shchukin, Motion of small aerosol particle in a light field, J. Eng. Phys., 30, 648-652, 1976.

ELECTRON BEAM STUDIES OF INDIVIDUAL NATURAL AND ANTHROPOGENIC MICROPARTICLES: COMPOSITIONS, STRUCTURES, AND SURFACE REACTIONS

Peter R. Buseck and John P. Bradley

Departments of Chemistry and Geology, Arizona State University, Tempe, Arizona 85287

Abstract. Quantitative chemical data can be obtained from individual submicron particles from aerosols by using an electron microprobe, analytical scanning electron microscope (SEM), or transmission electron microscope (TEM). Such analytical data, in combination with high-resolution TEM images, can reflect both the origin and events in the evolution of microparticles. Surface deposits on microparticles appear to be widespread. Such deposits are prominent features of fly ash from copper smelters, where we have detected species of Zn, S, and C. Complex reaction sequences have also been recognized on aerosol particles; carbonaceous pseudomorphs after ZnO reflect an evolutionary sequence, including a change from oxidizing to reducing conditions. Long carbon filaments containing CuZn particles indicate catalytic growth and are distinctive indicators of their sources. All of these unique particles can be used for source attribution.

Introduction

Fine-grained airborne particles are ubiquitous and represent an important fraction of both urban and nonurban aerosols. A large number enter the respiratory system with each breath. Their compositions are extremely varied and their small sizes and large surface areas make them excellent substrates for heterogeneous catalytic reactions. However, those same small sizes often make the total chemistry of microparticles difficult to study and their surface chemistry even more difficult to determine.

Recent research on airborne particulate matter has been directed towards developing ever more sensitive methods for determining its bulk chemistry, i.e. analyzing large masses (and thus numbers) of particles. However, such analyses provide highly limited data regarding the characteristics of the individual particles, and it is the single particles that participate in chemical reactions. For this reason, we have been emphasizing the study of single particles, their morphology, and especially their chemistry.

Current techniques and types of instruments that are capable of supplying chemical and structural information about individual microparticles are limited. In order to study submicron-sized particles, an incident radiation is required which has a wavelength less than 1 μm and can be focused into a beam of at least comparable dimensions. Electron beams fulfill these requirements; 100-keV electrons, for example, have a wavelength of 0.037 Å and can be focused into extremely narrow probes, in some cases to less than 10 Å diameter. For these reasons, we as well as other groups have chosen electron beam techniques to explore the realm of microparticles.

Electron beam instruments can be used to furnish a variety of types of data. For example, secondary electrons are scattered from points at or near the sample surface and thus can provide detailed topographical information, as in the familiar scanning electron microscope (SEM) images. Electrons transmitted through the specimen can provide both structural and chemical information from even smaller regions than the SEM. Elastically scattered transmitted electrons provide information about the crystalline structures of particles, and inelastically scattered transmitted electrons yield chemical data. The presence of light elements ($Z \lesssim 30$) can be determined by measuring the loss in energy of the electrons after they pass through the sample, and elements with $Z \gtrsim 6$ can be determined by examining the X-ray emission spectrum produced when electrons are scattered inelastically. These techniques are most useful for inorganic materials; therefore this discussion is limited to the inorganic fraction of the aerosol.

In this paper we review the instrumental and analytical procedures used to analyze individual aerosol particles with electron beams. Following a brief discussion of methods of surface analysis, we consider several examples in which the study of fine-grained particles has provided data regarding their origin and subsequent chemical reactions. In some of these examples the particles exhibit evidence of having participated in heterogeneous catalysis reactions.

Analysis of Individual Microparticles

There are basically three types of information that we wish to obtain from microparticles: size

TABLE 1. Electron Beam Instruments (and Acronyms) Used for Studying Individual Microparticles

Instruments		Used For:
SEM	Scanning electron microscope	Morphology and chemistry
EPMA	Electron probe microanalyzer	Chemistry
TEM	Transmission electron microscope	Structure
STEM	Scanning transmission electron microscope	Structure and chemistry
Chemical Measurements (Inelastically Scattered Electrons)		
Electron-induced X-ray emission		
EDS	Energy-dispersive spectrometry	Rapid data collection; elements with $Z \gtrsim 10$
WDS	Wavelength-dispersive spectrometry	Light elements ($Z \gtrsim 6$) and high accuracy
Electron spectroscopy		
EELS	Electron energy loss spectrometry	Light elements ($Z \lesssim 30$)
Structural Measurements (Elastically Scattered Electrons)		
Diffraction		
SAED	Selected area electron diffraction	Areas $\gtrsim 0.5$ µm
CBED	Convergent beam electron diffraction	Areas down to 10 Å
Imaging		
HRTEM	High-resolution transmission electron microscopy	Imaging of crystal structures

and shape, chemical composition, and crystalline structure. Particle sizes can be determined by any of several methods, but those that concern us here are direct imaging techniques. For example, the SEM can provide information about morphology (and surface topography, including coatings) from particles as small as 0.05 µm. For smaller particles the instrument of preference is the transmission electron microscope (TEM). With the TEM it is possible to image particle outlines and to observe coatings that project from the particle surface in directions approximately perpendicular to the electron beam. The procedures employed for determining composition and structure are more complex and therefore require more detailed discussion. A number of instruments and techniques are commonly used; their names (and acronyms) are summarized in Table 1.

Chemical Composition

Despite the obvious application that electron beam instruments have for performing quantitative analyses of individual particles, only relatively few reliable quantitative analyses have appeared in the literature to date. This is largely the result of difficulties inherent in correcting observed X-ray emission data from materials with irregular surfaces and thicknesses less than the incident electron range. For particles larger than a few microns in diameter, major problems arise from the effects of absorption resulting from particle geometry, whereas for smaller particles, corrections for atomic number and thickness effects become severe. However, there are now procedures for obtaining reliable quantitative analyses of particles. For relatively large particles we use the theoretical corrections of Armstrong and Buseck (1975) and for smaller particles a combination of empirical and theoretical corrections (Aden and Buseck, 1979). Although we do not generally use them, other procedures for correcting particle analyses have been proposed and are discussed by Heinrich (1980) and Small et al. (1980).

We can now routinely perform quantitative analyses for major elements in individual particles >0.5 µm in diameter to an accuracy of 5 to 10 percent relative. For particles 2 µm or larger, we are able to analyze for elements (Na to U) present in amounts greater than 0.1 weight percent. The minimum weight percent for accurate analysis increases to about 0.5 in a 1-µm particle.

There are two major methods of collecting X-ray emission data, wavelength-dispersive spectrometry (WDS) and energy-dispersive spectrometry (EDS). Both provide similar types of information; EDS is far more rapid, while WDS is generally more accurate and sensitive for low concentrations. WDS is also useful for the lighter elements ($Z \lesssim 11$), for which EDS is not currently very reliable. A limitation, however, in using WDS for the study of small particles is that some of them are unstable at the higher beam currents required.

Wavelength-dispersive spectrometers are used routinely in electron microprobes, are rather unusual on SEMs, and are almost never used on TEMs, primarily because of the low count rates produced on TEMs and the far superior counting efficiencies (as opposed to energy resolutions) of EDS units. Energy-dispersive spectrometers, on the other hand, are used on all three of these instruments. However, a word of caution is necessary. Most energy-dispersive units are now manufactured with built-in microprocessors or minicomputers so that they yield numerical output. Great care must be used in interpreting these numbers; their apparent accuracy is commonly far greater than their true accuracies, especially when used for irregularly shaped materials such as individual particles and complex compounds like silicates. Unless special care has been given to the use of good standards and properly tested data reduction procedures, it is probable that most SEMs and TEMs equipped with energy-dispersive spectrometers currently yield only qualitative or perhaps semiquantitative data. Nonetheless, with proper attention X-ray emission analysis has the capability of providing accurate analyses of small particles.

In the TEM it is also possible to monitor the energies of the electrons transmitted through the specimen. The losses in their energies are proportional to the atomic numbers of the elements in the sample; thus electron energy loss spectrometry (EELS) is another source of compositional data. EELS holds much promise for the analysis of elements with $Z \lesssim 30$ and especially for the light elements, including those not readily detectable by X-ray emission analysis (Egerton, 1978; Joy et al., 1979). Furthermore, there is minimal lateral electron scatter and so EELS has the added advantage of spatial resolutions approaching the diameter of the electron beam. Except for samples that are only tens of angstroms thick, X-rays are produced from an excited volume that limits accurate analyses to regions ~200 Å in diameter, whereas EELS data theoretically can be produced from regions a few tens of angstroms in diameter.

Structural Data

For the study of individual particles, the TEM can provide data of a unique sort. Both electron diffraction patterns and high-resolution images can be obtained from and reveal aspects of the crystalline structures of single particles. The two types of data bear a direct relationship to one another (one is the Fourier transform of the other, properly corrected for imaging effects produced by the optical system of the TEM), but they have different appearances and yield somewhat different types of information (Buseck and Iijima, 1974). It is generally easier to obtain diffraction data than high-resolution images.

Electron diffraction patterns provide data similar to those obtained by X-ray diffraction, except that smaller grains can be studied. The electron beam can be limited by an aperture to produce selected-area electron diffraction (SAED) from areas down to ~0.5 μm in diameter. Alternatively, the incident electron beam can be focused to a small probe, in special cases as small as 5 or 10 Å. Such strongly focused beams produce convergent beam electron diffraction (CBED). As distinct from SAED patterns, those resulting from CBED can produce interpretable information from higher-order Laue zones and thus can yield three-dimensional structural data, and from considerably smaller volumes than with SAED. Both methods have potential applicability for particle studies. In order to produce CBED patterns, a TEM with a rastered electron beam is used. Such a scanning transmission electron microscope (STEM) is also used for obtaining the EDS and EELS chemical measurements.

Using high-resolution transmission electron microscopy (HRTEM) it is possible to obtain images

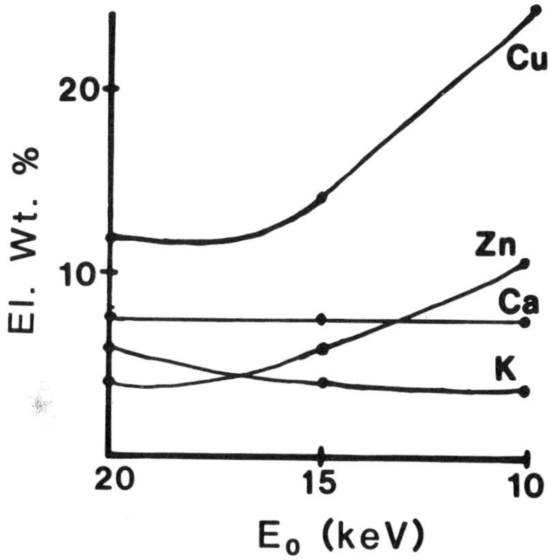

Fig. 1. Measured concentrations of copper, zinc, calcium, and potassium (elemental weight percents) in an individual particle as a function of electron beam accelerating potential (Eo). At the lower accelerating potentials, only surface material is analyzed. Thus copper and zinc are concentrated on the particle surface while calcium and potassium are evenly distributed throughout this clay. (After Armstrong and Buseck, 1977.)

TABLE 2. Components of the Aerosol in the Salt River Canyon Area, Arizona on Two Days in 1980 (after Thomas and Buseck, 1982).

Type of Particle	Abundance by Number	Estimated Relative Weight	Elements in X-Ray EDS Spectra	Size Range (μm)
Sulfates, probably primarily ammonium sulfate; minor H_2SO_4 droplets	>95%	May: 20.0% Oct.: 14.2%	Usually only S; rarely with minor Fe, Cu, Zn, or Cd	0.05 to 1.0
Silicate and carbonate mineral fragments	<5%	May: 79.2% Oct.: 82.0%	Na, Mg, Al, Si, K, Ca, T, Fe; in some case minor S, Cl	0.3 to ±10
Mixed salts, sulfates and chlorides; possibly nitrates and/or carbonates	<0.07%	May: 0.8% Oct.: 3.8%	Na, S, Cl, with lesser Mg, K, and Ca	0.4 to 6.0
Fly ash	Rare	-	Al and Si, with lesser Na, Mg, K, Ca, Ti, Ba, and Fe; minor S, Cl	0.5 to 4.5
Metal sulfides, sulfates, and oxides	Rare	-	Cu, Zn, Fe, Cd with S in stoichiometric proportions, or only metal peak	0.4 to 2.7
Pollen and spores	Rare	-	Only background	$\lesssim$5

with point-to-point resolutions of 3 to 4 Å. It is thus possible to observe structural details within unit cells, making observations at close to the atomic level. Such images are of special interest where materials are nonperiodic, i.e. contain "defects", and where crystals are too small or too thin to produce good diffraction data. The latter is the case for some of the surface coatings, described in the following sections, that we have observed on fly-ash spheres.

Particle Surfaces

It is, in general, more difficult to analyze the surface of a microparticle than to obtain an analysis of the total particle. There are, however, several methods that can be used, including Auger electron spectroscopy (AES), secondary ion mass spectroscopy (SIMS), X-ray depth profile microanalysis, and direct observation by high-resolution transmission electron microscopy.

AES is the method best suited for surface analysis. Auger electrons have very low energies (typically <1 to 2 keV) and so are absorbed or perturbed if they are produced at depths greater than roughly 20 Å. A consequence is that the outer few atom layers of a surface can be sampled and analyzed quite precisely with sensitivities down to 0.1 to 1 atomic percent (Keyser et al., 1978). A present limitation of Auger spectroscopy is that the spatial resolution is insufficient for the analysis of small single particles, but it is probably just a matter of time before such analyses are feasible.

In SIMS, material is physically removed from the surface of a sample by bombardment with a focused beam of ions, usually negative oxygen ions (Keyser et al., 1978). Secondary ions that are sputtered from the surface are analyzed by conventional mass spectrometry. This technique offers very high sensitivity, but its application to individual microparticles is limited since the primary ion beam generally can only be focused to a diameter of a few microns.

In our initial studies of surface deposits we used X-ray depth-profile microanalysis (Armstrong and Buseck, 1977; Armstrong, 1978). This is simply a method of using variable accelerating potentials (Eo) for the electron beam of an electron microprobe or an analytical SEM. The effect is to vary the depths of electron penetration and thus X-ray emission. We were able to detect surface coatings of Cu and Zn on clay particles collected downwind from a copper smelter (Figure 1). This method is useful for analyses of individual particles, but offers relatively poor spatial resolution. Since the depth resolution is also very much greater than the thickness of the surface layer that is normally of interest, and is indeed greater than the diameters of small particles, the surface analysis capabilities by this technique are rather limited.

The transmission electron microscope, because of its great spatial resolution, has opened up a new realm of particle surface analysis. As described in the following section, we have been able to obtain various types of information, including direct crystal structure images at the unit cell scale, of a range of coatings on copper smelter fly-ash spheres. We now quite regularly use a TEM or STEM in conjunction with the analytical SEM to study microparticle surfaces and have found this to be a powerful combination.

Types of Airborne Microparticles (Arizona Examples)

The range of particle sizes and compositions in aerosols is extremely wide, as is the number of possible mineral types and synthetic compounds. Fortunately, although many phases have been identified, only relatively few species are abundant.

The emphasis of our research to date has been on the aerosol of arid, urban areas such as occur in the southwestern part of the United States. Although the majority of the airborne particles are minerals, a significant fraction in urban areas is anthropogenic. Furthermore, in these urban regions even the natural particles carry surficial deposits of an anthropogenic character, clearly the result of the efficient scavenging ability of fine-grained dusts. Indeed, in arid regions such scavenging may well be one of the more effective means of cleaning up the atmosphere of gaseous emissions.

We are currently completing a survey of the ambient particle constituents of the aerosol in and around Phoenix (Post and Buseck, 1982). These results are in agreement with earlier data of Armstrong (1978) that showed quartz, feldspar, illite and other clays, and mica to be the major mineral constituents. It is too early to summarize the range of anthropogenic particles found in the Phoenix urban area, except to say that it is very wide in both composition and structure. We have observed alloys, oxides, chlorides, sulfides, sulfates, and so forth. Among the more "exotic" elements, relative to common minerals, that we have observed as important constituents of anthropogenic particles are Ti, V, Cr, As, Br, Sr, Zr, Te, Hg, and Pb. Some, such as As, have also been observed as surface deposits (Armstrong and Buseck, 1977), and these can produce an obvious health hazard.

We have also completed a study of the aerosol in a rural area in eastern Arizona (Thomas and Buseck, 1982). This study was undertaken to be able to compare urban and rural aerosols (Table 2). The predominant aerosol constituents greater than 1 μm in diameter are common minerals such as clays, calcite, and the rock-forming silicates. Other important constituents, and especially for sizes less than 1 μm, include sulfate compounds and alkali and alkaline-earth salts (chlorides and sulfates). The alkali and alkaline-earth salts are probably derived from playas. These desert lakes are dry for a large part of the year, so extensive salt flats are exposed to wind erosion. A fraction of the salt particles is probably sea spray from the Pacific Ocean.

In both urban and rural areas the most abundant particles by number are submicron sulfates. Under Arizona conditions sulfuric acid droplets are probably quite rare, and the majority of the sulfates appear to be ammonium sulfates. However, large numbers of sodium sulfate particles were observed in a submicron aerosol sample from the northern part of the state (Aden and Buseck, 1980). Although less abundant, they have also been observed in Phoenix and appear to be sparse at the eastern Arizona site. The sulfates are derived from both natural and anthropogenic sources, and long-term transport may be important. The copper smelters and coal-fired power plants in Arizona release large quantities of sulfur-bearing gases and are probably the source of an important fraction of the particles of secondary sulfate aerosol in both urban and rural areas of Arizona (Pueschel and Van Valin, 1978; Arizona Dept. of Health Services, 1980).

The copper smelters are a major industry in Arizona and a large source of high-temperature anthropogenic emissions. Because of their importance and the variety of processes and thus particles produced, we have done considerable work on smelter emissions. In the remainder of this paper we present a number of examples of complex particle histories and reactions; they are all selected from among the smelter outputs. Presumably, similarly detailed genetic information can and will be determined from other particle types as more work is done.

Analysis of Surfaces of Copper Smelter Fly Ash by Electron Microscopy

Fly ash, which is produced by a wide variety of high-temperature industrial processes, is one of the major solid waste products generated within the United States. It consists largely of impure alumino-silicate glass together with small amounts of several crystalline phases (Hulett et al., 1980; Henry and Knapp, 1980). Fly-ash spheres also carry minor amounts of potentially toxic volatile elements such as Pb, As, Se, Zn, Cd, and S (Campbell et al., 1978; Natusch et al., 1977). A variety of analytical procedures has been employed to determine both the location and chemical state of the volatile species within fly ash, since these largely determine the environmental impact of fly-ash emissions.

Several studies have indicated that the volatile elements are concentrated on the surfaces of fly-ash spheres (Linton et al., 1976; Pueschel, 1976), although the chemical nature of these surficial enrichments is unknown. Other studies suggest

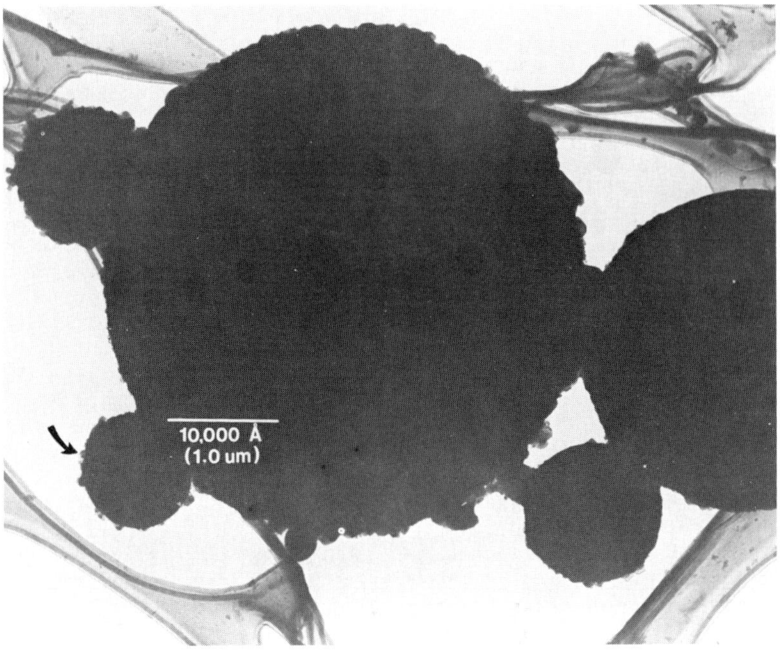

Plate 1(a)

Plate 1. Transmission electron micrograph (increasing in magnification, (a) through (d)) of a copper smelter fly ash agglomerate mounted on a holey-carbon substrate. One of the spheres carries a surficial microcrystalline deposit (arrowed in micrographs (a) through (c)). High-resolution images of such features (micrograph (d)) provide a means of directly observing structural features (spacings of atomic planes, in this instance) of such deposits.

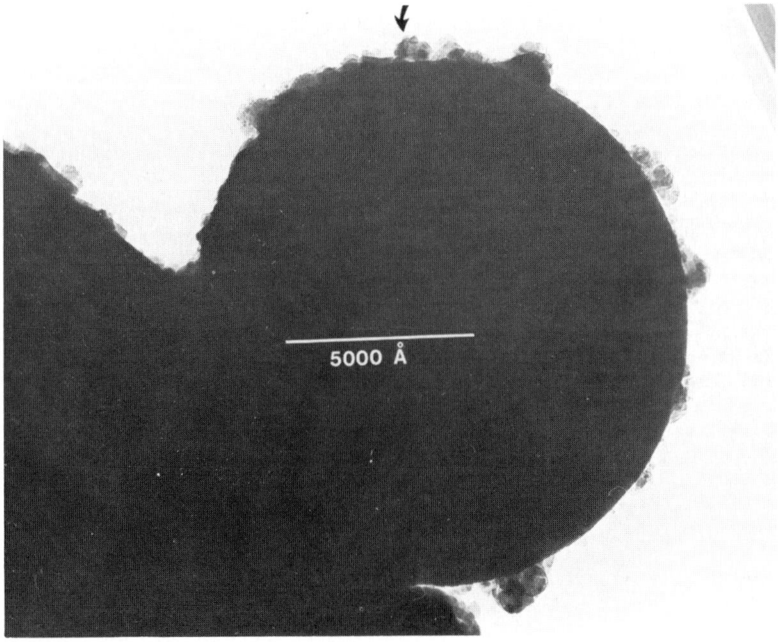

Plate 1(b)

Plate 1(c)

Plate 1(d)

that the volatile species are dissolved within the fly-ash matrices (Hulett et al., 1979, 1980; Hulett and Weinberger, 1980). Clearly a distinction between these hypotheses is important; if volatile species are indeed concentrated on the particle surfaces then they are directly accessible to leaching (or other types of chemical reaction) by the natural environment, including biological systems. On the other hand, if these elements are dissolved within the fly-ash matrices, then they are contained within relatively stable phases and hence are less susceptible to chemical attack.

We have recently been studying individual spherical particles formed within and emitted from Arizona copper smelters. These particles are similar in both morphology and gross chemical composition to the fly ash that is generated during coal combustion. Therefore, we refer to this material as copper smelter fly ash. However, it invariably contains higher concentrations of volatile elements. We have been able to observe directly a variety of crystalline compounds, among them condensates of volatile elements, that are located on the outer surfaces of these fly-ash spheres.

Two types of fly ash samples were recovered from an Arizona copper smelter. Bulk samples were retrieved from electrostatic precipitators that remove particles from reverberatory furnace waste gases, while dispersed individual particle samples were collected by low-volume samplers at ground-level sampling stations located several kilometers from the smelter. Individual particles were characterized by using the analytical SEM; their major constituents are Al, Si, Ca, Ti, Fe, Na, and Mg, together with minor Zn and S. Bulk chemical analyses have also shown these elements to be highly enriched in particulate emissions from copper smelters (Germani et al., 1981; Small et al., 1981).

The aerosol samples are collected directly onto TEM grids containing holey-carbon support films. The grids are mounted on Nuclepore substrates which are placed within sampling devices downwind of the smelter. Surface features are examined by positioning the electron beam of the STEM at a tangent to the curved surface of the selected sphere (Plates 1 and 2). By tilting the specimen with respect to the electron beam it is possible to examine large areas of the particle surface.

The fly ash retrieved from electrostatic precipitators within the smelter is prepared by crushing it to a fine powder and dispersing the resulting fragments onto holey carbon TEM grids. Among these fragments are shards of aluminosilicate glass from the surfaces of fly-ash spheres, some of which are suitably thin ($\lesssim$500 Å) for STEM imaging and analyses. The selected fragment is positioned normal to the electron beam and examined by the usual means of imaging thin crystals. There is a potential problem with interpretation of such images (cf. Plate 3) since we cannot be certain that the microdeposits are indeed on the surfaces. However, such features consist of volatile phases, while the matrices consist of relatively refractory materials. Thus we presume that we are observing surficial condensates on flakes from cenospheres.

We have observed fine-grained deposits on the surfaces of many individual spheres. A variety of microcrystalline deposits, including zinc and sulfur compounds (Plate 3) and graphitized carbon (Plate 2) can be observed, and it is usually possible to orient the specimen to obtain high-resolution images of the structures of the crystals deposited on the fly-ash surfaces. The measured lattice fringes in the images, coupled with data from electron diffraction, EDS, and EELS, provide a powerful means of characterizing compounds on the surfaces of copper smelter fly ash.

The furnaces used for copper smelting are fired by coal, light fuel oil, or natural gas; as with many such high-temperature processes, large amounts of amorphous carbon (soot) are released. It is of interest that the carbon deposited on some of the fly-ash surfaces is crystalline (Plate 2), suggesting that these surfaces may have promoted graphitization of the emitted carbon.

At the present time we are examining fly ash from other sources and climates to determine the abundance and character of surficial deposits. Transmission electron microscopy has shown surface deposits on all fly-ash samples that we have examined. Such observations are consistent with the view that particles formed at high temperatures act as substrates for the condensation of other materials during cooling. It is clear that fly ash and other airborne particles contain a range of surficial deposits. We believe that their study will reveal much data of interest and importance regarding both the history and travels of the particles as well as their chemical reactivity and thus health effects.

Examples of Complex Microparticle Reaction Sequences

Zinc Oxide Fourlings

Highly branched particles (Plate 4), on occasion agglomerated into clusters resembling a crown of thorns (Plate 5), are among the more unusual and fascinating particle types observed in aerosol samples recovered downwind of an Arizona copper smelter (Bradley et al., 1981). These particles consist of zinc oxide crystals, twinned to produce approximately tetrahedrally shaped units (fourlings) that are typically coated with carbon. The occurrence of these fourlings is of special significance because they carry with them detailed information about their origin and the nature of their parent environment.

In order to understand the structure and chemistry of the particles it was necessary to employ

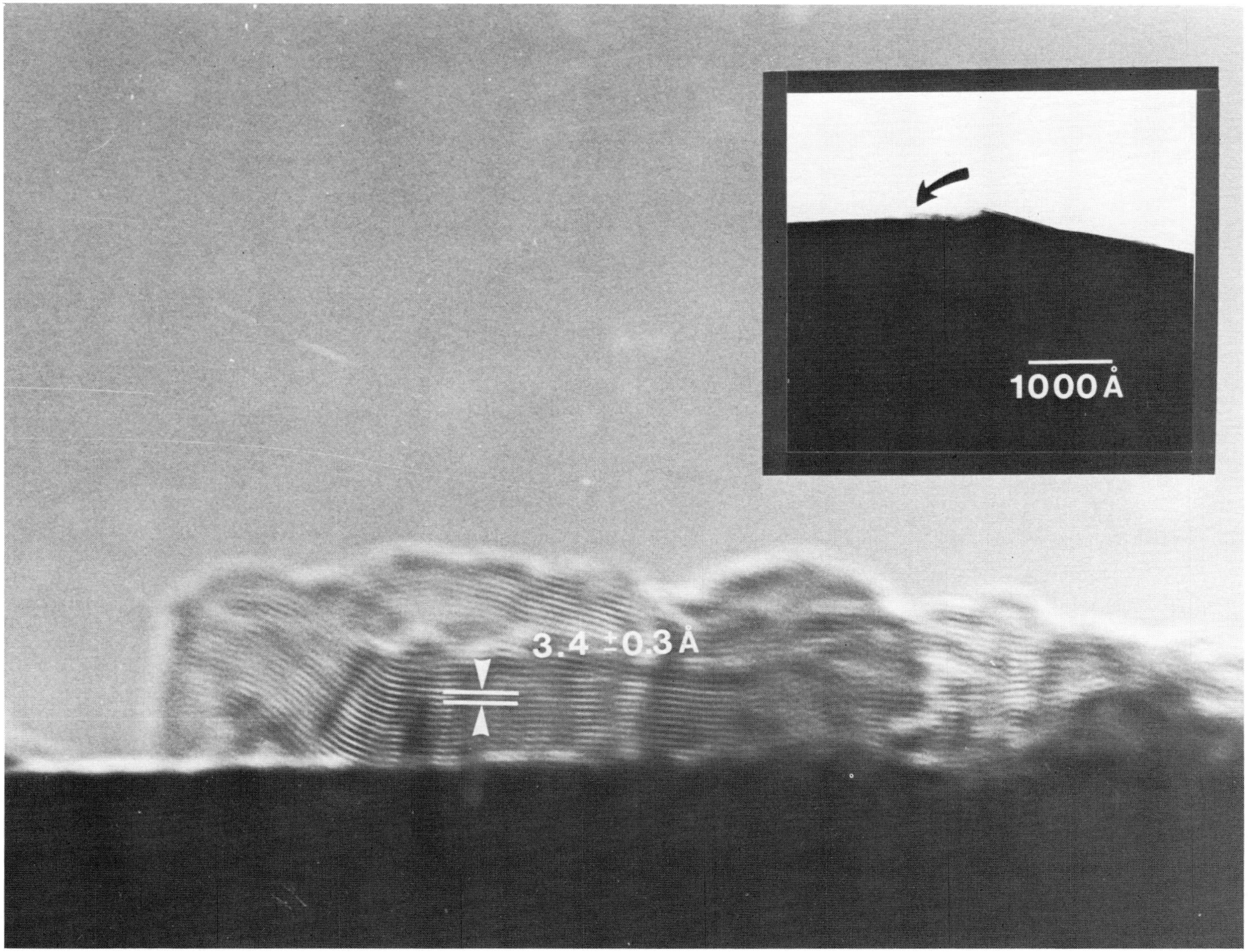

Plate 2. Transmission electron micrograph of a surface feature on a fly ash particle. The periodicity of the observed lattice fringes (3.4 Å) is consistent with the (002) interplanar spacings for graphitized carbon. Inset (upper right) shows the alumino-silicate sphere upon which this surface material (arrowed) is deposited.

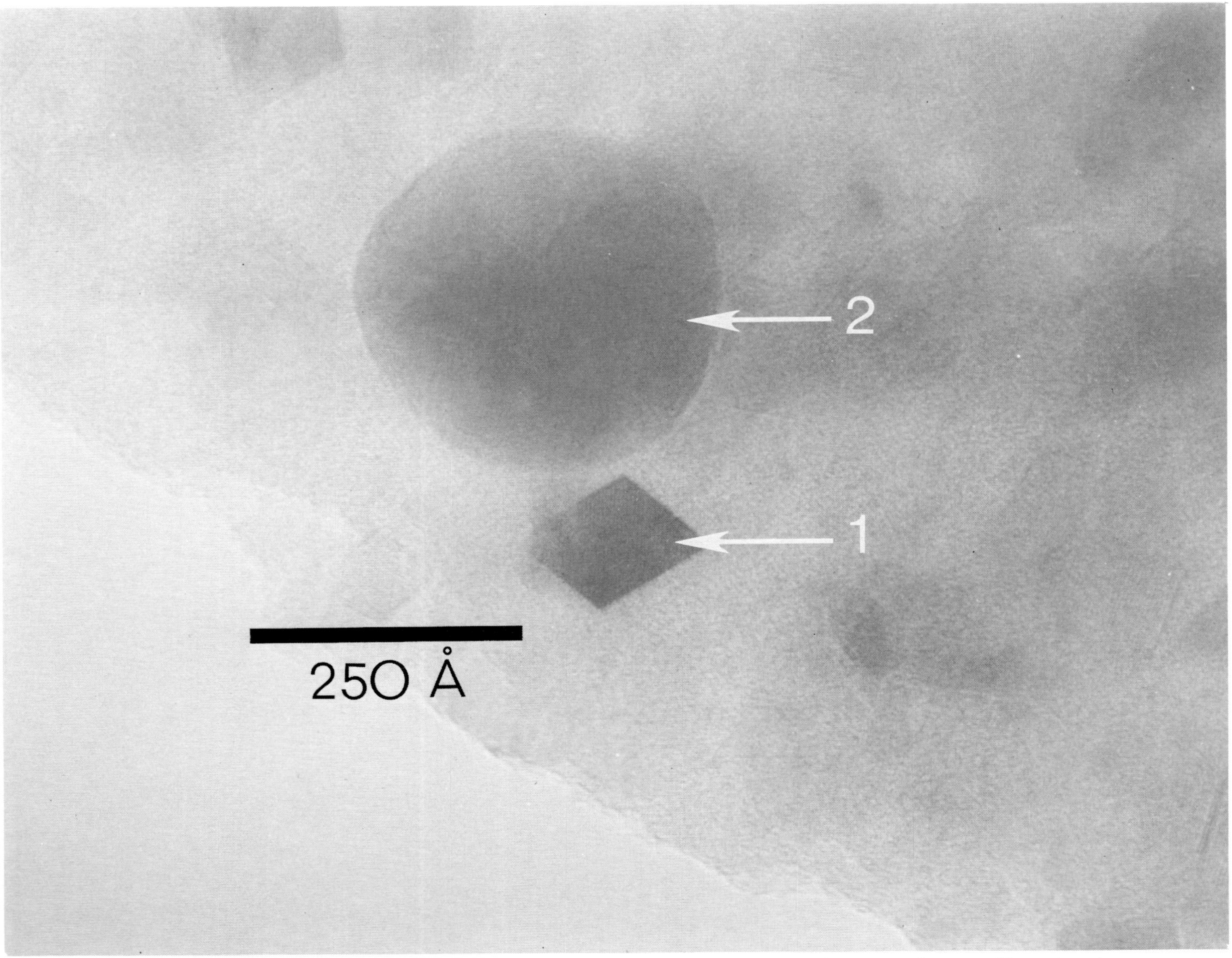

Plate 3. Transmission electron micrograph of an alumino-silicate fragment from a fly ash sphere. The crystalline phase is a zinc compound (arrow 1), while the dark circular area (arrow 2) corresponds to a high concentration of a sulfur compound; evidence suggests these are deposits on the fly ash surface.

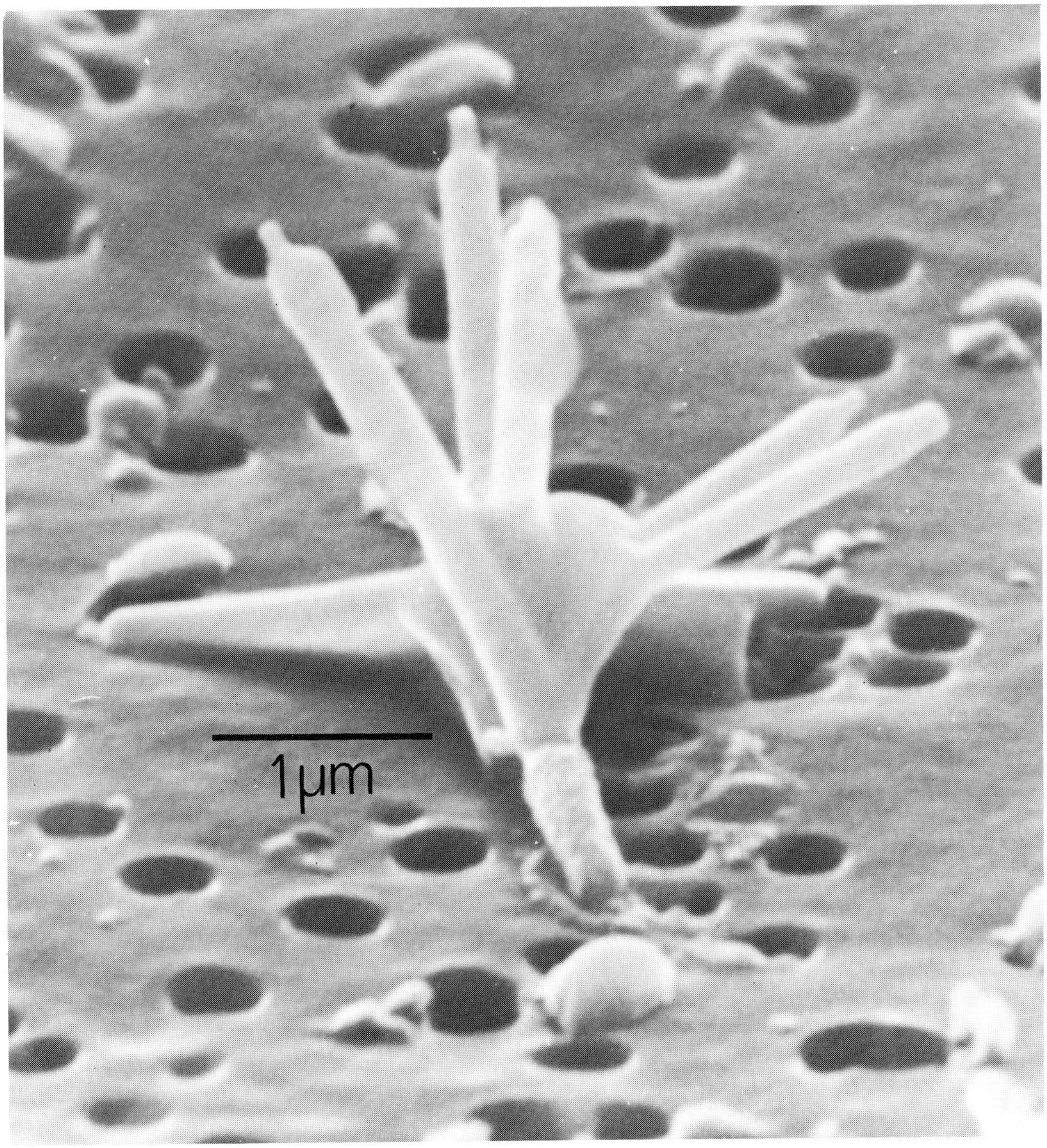

Plate 4. Scanning electron micrograph of a carbon-coated zinc oxide particle (Nuclepore substrate) collected from fugitive emissions from a copper smelter. (After Bradley et al., 1981.)

the entire range of techniques and instruments discussed in the section on analysis of individual microparticles. The characteristics of these particles are summarized in Plate 6. They exhibit a progressive chemical and morphological variation that we arbitrarily subdivided into four categories: type a - sharply angular zinc oxide crystals twinned into fourlings; type b - rounded carbon-coated fourlings; type c - rounded carbon-coated fourlings in which the zinc oxide core crystals have suffered substantial decomposition; type d - well-rounded carbonaceous pseudomorphs containing little or no zinc oxide. These are the least abundant of the particle types and commonly are fragmental.

The chemical variations (obtained from the analytical SEM) are shown in columns ii and iii of Plate 6. There is a progressive decrease in zinc content and a corresponding increase in carbon content from types a through d. All particle types contain minor amounts of sulfur; Novakov et al. (1974) and Tartarelli et al. (1978) have suggested that carbon typically adsorbs SO_2 to produce surficial sulfate compounds, and that may also be the case with these particles.

Structural variations, observed by electron diffraction, correspond to the chemical variations. Individual arms of type a fourlings yield sharp SAED patterns of single crystals of zinc oxide. On the other hand, SAED patterns

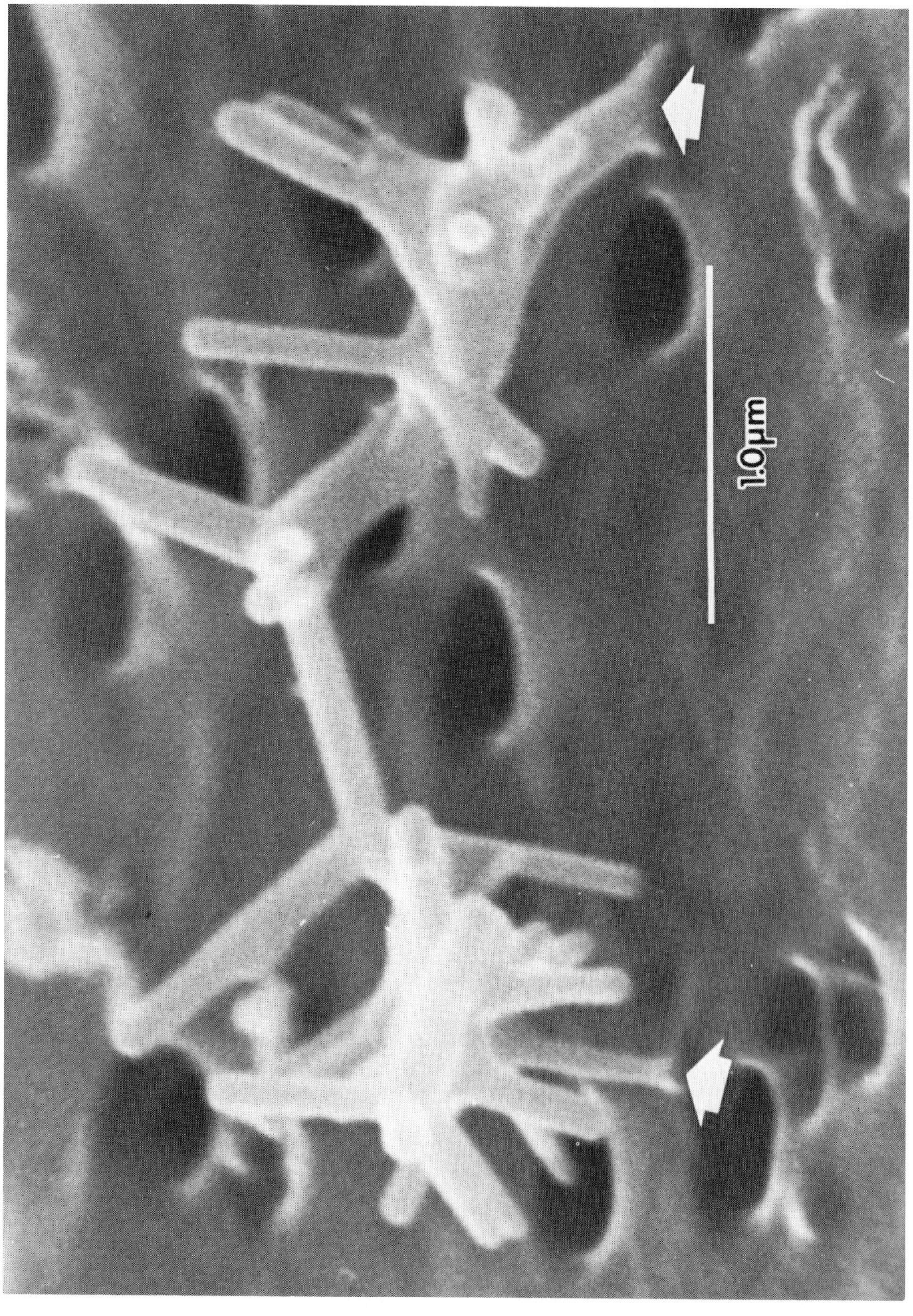

Plate 5. Scanning electron micrograph of a "crown of thorns" aggregate of carbon-coated zinc oxide particle. The arrows indicate sites of reaction with the Nuclepore substrate, which may result from sulfuric acid adsorbed onto the particle. (After Bradley et al., 1981.)

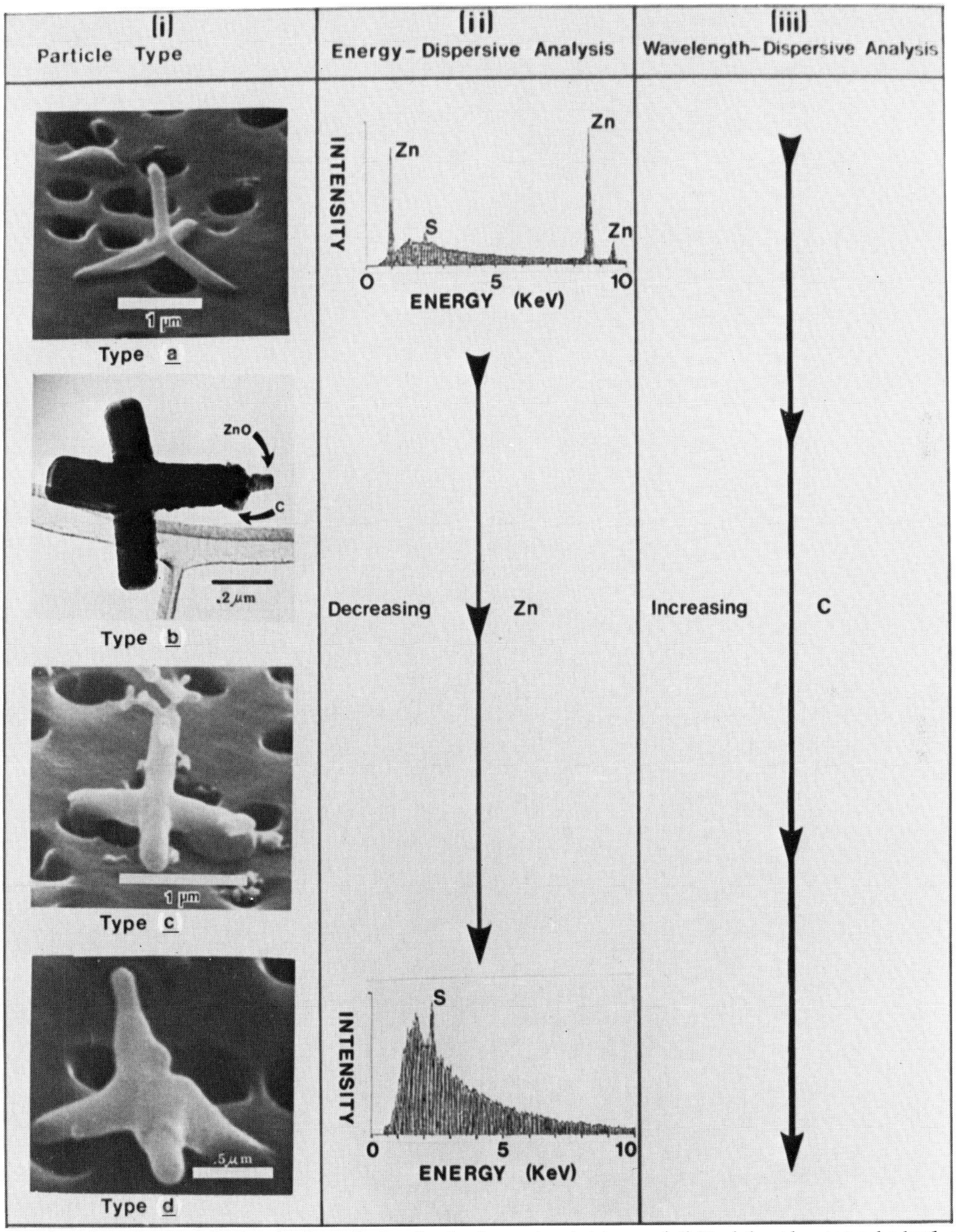

Plate 6. Continuum of physical and chemical characteristics exhibited by the tetrahedral particles. Column i - electron micrographs of the four particle types, whose individual arms range from 0.1 to 1.0 µm in length: type a, zinc oxide fourling (SEM image, Nuclepore substrate); type b, fourling with zinc oxide core within amorphous carbon shell (TEM image, holey-carbon substrate); type c, partially decomposed zinc oxide fourling within amorphous carbon shell (SEM image, Nuclepore substrate); type d, carbonaceous pseudomorph that has suffered complete loss of its zinc oxide core (SEM image, Nuclepore substrate). Columns ii and iii - corresponding chemical variations as indicated by X-ray spectroscopy, showing correlated decrease in zinc and increase in carbon with progress from type a to type d. (After Bradley et al., 1981.)

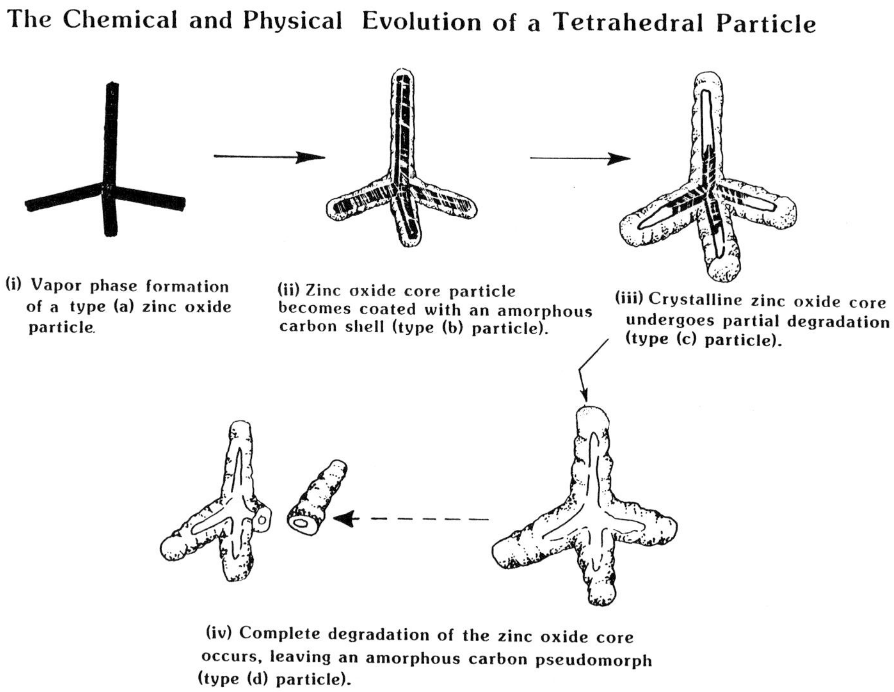

The Chemical and Physical Evolution of a Tetrahedral Particle
(i) Vapor phase formation of a type (a) zinc oxide particle.
(ii) Zinc oxide core particle becomes coated with an amorphous carbon shell (type (b) particle).
(iii) Crystalline zinc oxide core undergoes partial degradation (type (c) particle).
(iv) Complete degradation of the zinc oxide core occurs, leaving an amorphous carbon pseudomorph (type (d) particle).

Fig. 2. Chemical and physical evolution of a zinc oxide fourling particle. (Modified from Bradley et al., 1981.)

obtained from type d particles yield only the characteristic rings of amorphous carbon. Particle types b and c yield diffraction data intermediate to types a and d, indicating they are composites of zinc oxide and carbon. It is apparent that there is a gradational structural sequence between types a and d.

The continuum of particle characteristics provides detailed information about the history of the particles. For example, the distinctive morphology of type a zinc oxide particles is indicative of the reaction conditions leading to their formation. When zinc metal is heated in air the emerging vapor condenses to form zinc oxide fourlings. Each fourling consists of four single zinc oxide crystals related to one another by twinning on $(11\bar{2}2)$ planes (Cowley et al., 1951; Fuller, 1944). This distinctive morphology thus provides a clue to the origin of the particles under investigation; their probable source is condensation of zinc oxide from a vapor. In the copper smelting process this can occur wherever zinc vapor comes into contact with the atmosphere.

The intimate association of carbon and zinc oxide (particle types b and c) is strongly suggestive of the reaction conditions responsible for their formation. Deposition of carbon on zinc oxide occurs during nonferrous smelting processes when a hydrogen/hydrocarbon gas mixture is used to reduce zinc oxide to metallic zinc (Doerner, 1936; Fulton, 1919). This reaction proceeds rapidly under the conditions prevailing during copper smelting, and large amounts of carbon are generated by thermal decomposition of both the hydrocarbon gas and a reaction byproduct, carbon monoxide. Therefore, type a zinc oxide fourlings become coated with carbon, together with partial or complete reduction of zinc oxide (and subsequent vaporization of zinc), to yield particle types b through d.

Within the copper smelting process there are several high-temperature processes, each of which yields particles containing zinc (Germani et al., 1981). However, only the fire refinery fulfills the criteria necessary to explain adequately the observed characteristics of the fourling particles. The product that is treated in the fire refinery is blister copper, a molten liquid (at ~1150°C) containing approximately 97.5 percent (by weight) copper and 2.5 percent impurities, including zinc.

The bulk of the impurities is removed in the fire refinery during a two-stage redox procedure. Initially, during the oxidation step, compressed air is introduced at the base of the refining vessel and is blasted through the molten blister copper. Residual sulfur is liberated as SO_2, but a fraction of the metals is also oxidized. Consequently an oxide slag (including zinc oxide) accumulates on the surface of the blister copper; some zinc oxide fourlings are probably blown into the air above the slag. The second stage of fire refining consists of a reduction process in which the compressed air is replaced by a mixture of hydrogen and propane. Under such strongly reducing conditions, the oxides are reduced to their metals. However, thermal decomposition of the carbon-bearing gases occurs, which coats carbon onto many of the type a zinc oxide fourlings that have accumulated on the surface of the blister copper. Thus, during the reduction stage of fire refining particle types b through d are generated.

By studying the individual particles of an aerosol we have been able to determine their origin and evolution, from formation in the fire refinery to carbon-coating in the refinery atmosphere and eventual release in the gaseous emission from the smelter (Figure 2). These airborne particles thus carry the traces not only of their source region, but also of the specific stage of the smelting process where they were generated.

Filamentous Carbon Particles With Attached Metal Catalysts

The Arizona copper smelters are a source of a particle type with a fibrous morphology that is also generated from and characteristic of a number of other high-temperature processes. Elongated fibers of carbon are produced by the catalytic decomposition of carbon-bearing gases (usually hydrocarbons or carbon monoxide) between 600 and 1300 K (Baker and Harris, 1978). Such gases are utilized during many industrial activities, and thus the resulting fibers may be rather widespread. For example, they are known to be produced in the petroleum, petrochemical, and nuclear industries, where deposition of carbon filaments adversely affects heat transfer efficiencies and is associated with corrosion of metallic components (Baker and Harris, 1978; Everett et al., 1968). Their occurrence was first reported over 20 years ago by Radushkevich and Luk'Yanovich (1952). The fibrous morphology gives this form of carbon a high probability of retention in the lung, and its unusual shape also makes it potentially useful for source identification. For these reasons we were interested in using our single-particle techniques for characterizing these particles (Bradley and Buseck, 1982).

A typical filamentous carbon particle that was collected in an aerosol sample emitted from the fire refinery of a smelter is shown in Plate 7. Such particles consist of a large number of individual filaments often branching and radiating outwards from a common base. The filaments exhibit a wide variation in dimensions, with diameters between $\sim 10^2$ and 2×10^4 Å and lengths between $\sim 10^3$ and 10^7 Å (1 mm). Aspect ratios greater than 20 are typical. Metal particles are typically embedded within carbon at the ends of many of the fibers (Plate 8).

TEM images of individual filaments reveal a thick sheath surrounding a hollow core (Plates 8 and 9). The measured lattice fringes within the sheaths (3.4 Å), together with EELS data, confirm that the filaments are composed of graphitic carbon. Backscattered electron images, SAED patterns, and EDS analyses on both the SEM and STEM indicate an association of a copper-zinc alloy with the filaments. The alloy typically forms the bases of the filamentous clusters, in places it is dispersed within the hollow cores of individual filaments, and elsewhere it occurs in particles at the tips of the fibers (inset, Plate 8).

The mechanism of formation of carbon filaments is of interest. Baker et al. (1972), in a study of the decomposition of acetylene on isolated nickel particles, showed that the formation of filaments occurs by the buildup of carbon on one side of the metal particle (Figure 3). Initially, a carbon-bearing gas decomposes on the exposed surface of a metal catalyst particle. Decomposition releases hydrogen and carbon, which dissolve in the particle. The dissolved carbon diffuses through the particle and is precipitated on the rear face to form the body of the growing filament. Propagation of filament growth continues in this manner so long as the supply of the carbon-bearing gas remains constant and the active face of the catalyst is exposed.

The fire refinery of a copper smelter provides an ideal environment for the generation of carbon filaments. At regular intervals during smelting the refining vessel is charged with large quantities of molten metal. The surface of this liquid acts as a support substrate for the growth of carbon filaments in the steep thermal gradient above the refining vessel; such growth occurs when a hydrogen/hydrocarbon gas mixture is blown through the metal melt. Under these conditions large amounts of soot and filamentous carbon are generated, which are then discharged directly into the atmosphere.

These filamentous carbons are another example of an individual airborne particle type that carries with it a detailed record of both its source environment and the reactions that determined its formation. Detailed studies of the individual particles have led to several useful observations. (1) The shapes of the carbon filaments and their characteristic association with small metal grains indicate that they are the products of a catalysis reaction. (2) Microanalysis of the associated metal grains allows identification of the catalyst as a CuZn alloy. (3) The observations described above, combined with data in the literature, indicate (a) the catalytic growth mechanism by which the fibers were formed and (b) their gaseous precursors. (4) Finally, the occurrence of the fibers in the smelter plume indicates their source, and the details of their morphology and composition limit them to a specific stage, fire refining, within the smelting process.

Although carbon filaments have many potential sources and origins, the association of the

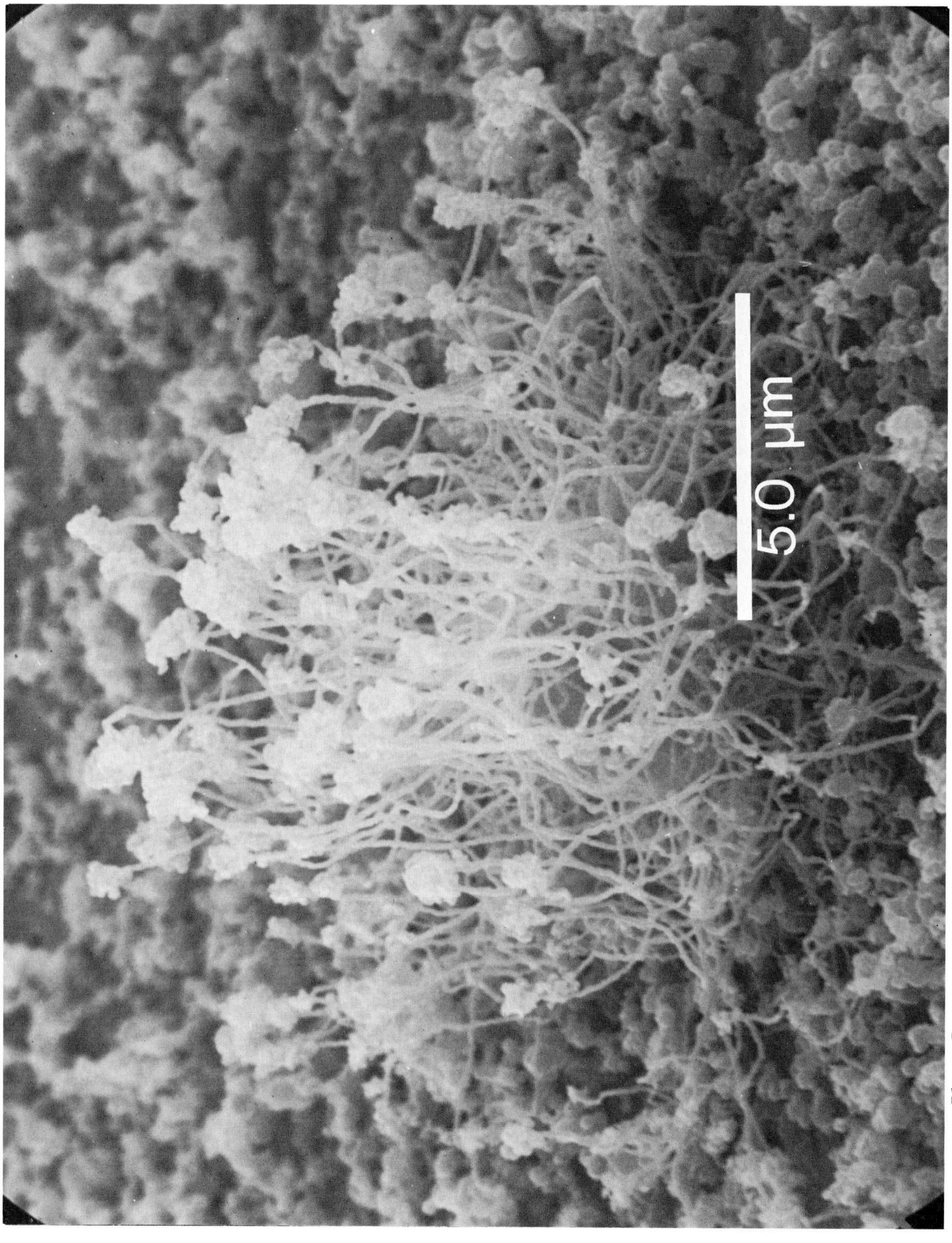

Plate 7. Scanning electron micrograph of filamentous carbon cluster (soot-coated Nuclepore substrate). This particle consists of a large number of individual carbon filaments radiating outwards from a CuZn alloy base. The tips of each filament contain a CuZn particle embedded within carbon. (After Bradley and Buseck, 1982).

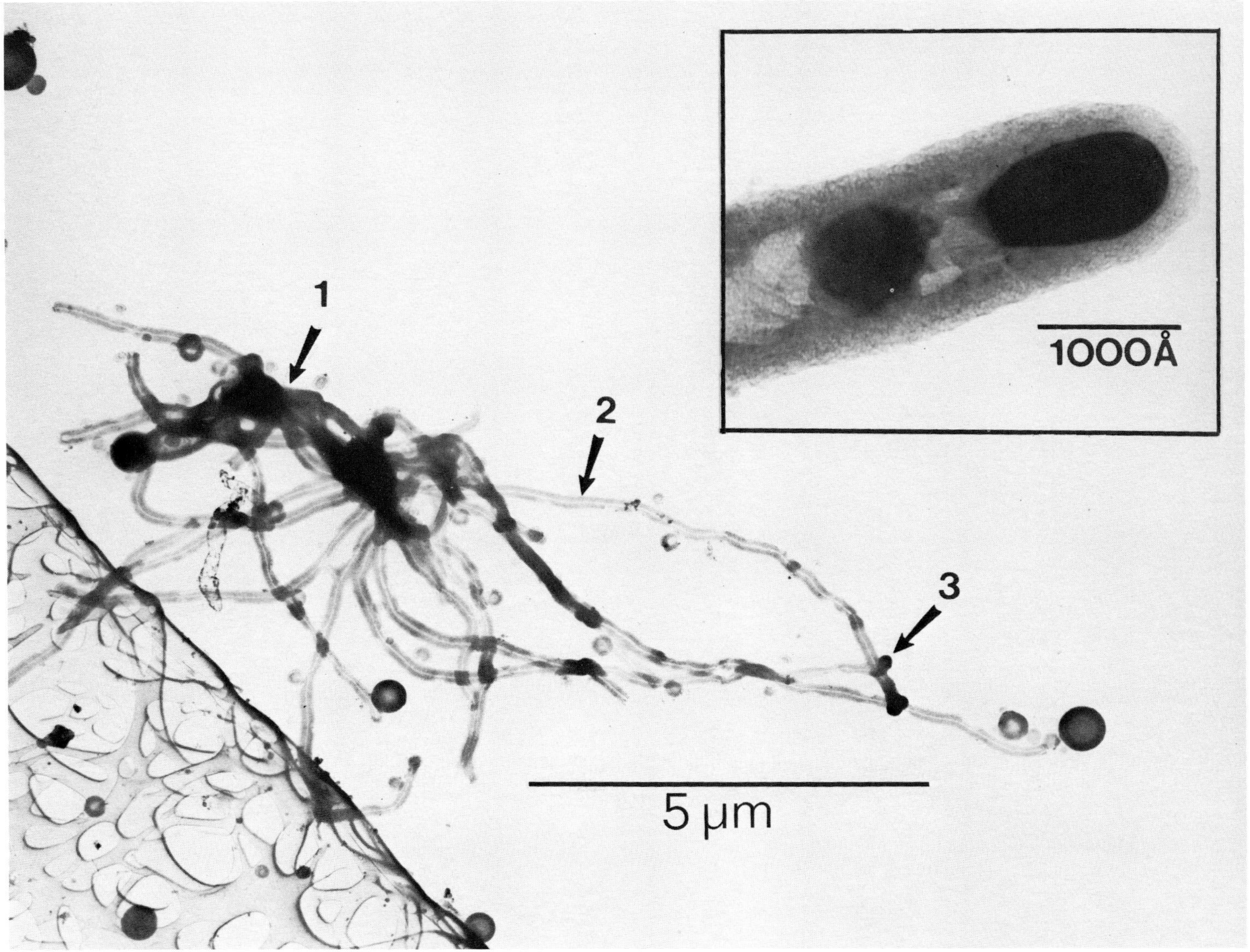

Plate 8. Transmission electron micrograph of filamentous carbon cluster. Arrow 1 indicates CuZn alloy base of cluster. Arrow 2 points to hollow core structure of individual filaments. Arrow 3 indicates an example of a CuZn alloy particle embedded at the tip of an individual filament (mesh structure, lower left, is part of holey carbon supporting film). (After Bradley and Buseck, 1982.) Inset (upper right) is a high-magnification image of the tip of an individual carbon filament, showing a CuZn particle embedded within carbon and the hollow core that runs the length of each carbon fiber.

Plate 9. High-resolution transmission electron micrograph of individual filament. This shows the 3.4-Å spacings which are characteristic of the (002) interplanar spacing for graphite, as well as the hollow core (free of fringes) typical of these fibers.

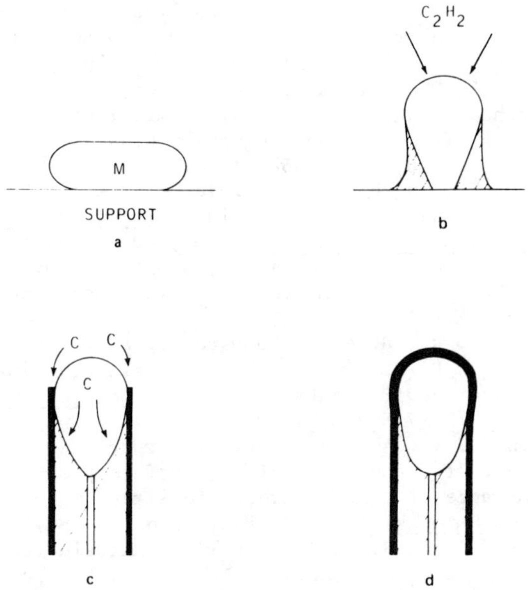

Fig. 3. Stages in the catalytic growth of carbon filaments, where M represents the metal alloy which promotes decomposition of the hydrocarbon gas (symbolized as C_2H_2) and deposition of a carbon fiber behind the metal particle.

Arizona fibers with CuZn alloys greatly limits the possible places from which they were derived. Too little work has been done yet to conclude that the details of the association restrict the particles to one particular smelter among the several that operate in the region. However, further work could determine the range of catalytic alloy compositions resulting from a given smelter and the differences among smelters. If indeed there are significant compositional differences among alloys from several smelters, the study of individual carbon filaments with their attached alloy grains can be used for source attribution and for gaining insight into the atmospheric reactions they have undergone.

Conclusions

The combined capabilities of modern electron beam instruments provide a means of characterizing virtually the entire size range of individual particles occurring in aerosols. Despite the small sizes of these particles, highly specific structural, compositional, and morphological information can be obtained. Such information provides useful insight into the details of particle formation and evolution.

The particles discussed in this paper illustrate types of data that can be obtained from analyses of single particles. Each of the particle types that is discussed carried decipherable information about its origin and subsequent history. For example, examination of fly ash from copper smelter emissions has confirmed the important role played by particle surfaces. These surfaces provide a substrate for nucleation and growth of volatile condensates; they may also provide a catalytically active substrate, as suggested by the presence of partially and completely graphitized carbon.

Carbon-coated zinc oxide fourlings exhibit evidence of both their chemical and physical evolution, as well as the exact stage of copper smelting from which they originated. Likewise, the filamentous carbon particles carry evidence of their source environment. These fibrous particles also illustrate the role of heterogeneous catalysis in the formation of certain anthropogenic particle types.

Although the range of compositions of most of the particles within aerosols is limited, aerosols do contain "exotic" species such as some of the particles discussed in this paper. Exotic particles are commonly sparse, but they are very important because they can be diagnostic of their sources and can provide information about the mechanisms of particle formation.

Acknowledgments. We acknowledge with appreciation the helpful discussions and reviews of Gary Aden, Mark Germani, Jeff Post, and Ellen Thomas, and the assistance of Melodie Carr, Donna Colletta, Greg Goldman, and Jennie Needham. Financial support was provided by grant ATM-8022849 from the Atmospheric Sciences Division of the National Science Foundation. Transmission electron microscopy was performed at the electron microscopy facility established with support from the National Science Foundation Regional Instrumentation Facilities Program (grant CHE-7916098).

References

Aden, G. D., and P. R. Buseck, Rapid quantitative analysis of individual particles by energy-dispersive spectrometry, in Proc. 15th Ann. Conf. Microbeam Anal. Soc., edited by D. E. Newbury, San Francisco Press, 254-258, 1979.

Aden, G. D., and P. R. Buseck, The chemical composition of individual submicron sulfate particles from an aerosol near a power plant, Proc. 15th Ann. Conf. Microbeam Anal. Soc., edited by D. B. Wittry, San Francisco Press, 203-207, 1980.

Arizona Dept. of Health Services, 1979 Air Quality Data for Arizona, Arizona Dept. of Health Services, 1980.

Armstrong, J. T., Methods of quantitative analysis of individual microparticles with electron beam instruments, SEM/1978, 1, Scanning Electron Microscopy, Inc., Illinois, 455-468, 1978.

Armstrong, J. T., and P. R. Buseck, Quantitative chemical analysis of individual microparticles using the electron microprobe: Theoretical, Anal. Chem., 47, 2178-2192, 1975.

Armstrong, J. T., and P. R. Buseck, The chemical

composition, morphology, and surface properties of individual airborne microparticles, in Proc. 4th Int. Clean Air Congress, edited by S. Kasuga, The Japanese Union of Air Pollution Prevention Association, 617-620, 1977.

Baker, R. T. K., M. A. Barber, P. S. Harris, F. S. Feates, and R. J. Waite, Nucleation and growth of carbon deposits from the nickel catalyzed decomposition of acetylene, J. Catal., 26, 51-62, 1972.

Baker, R. T. K., and P. S. Harris, The formation of filamentous carbon, Chem. Phys. Carbon, 14, 83-165, 1978.

Bradley, J. P., and P. R. Buseck, Airborne filamentous carbon, submitted to Science, 1982.

Bradley, J. P., P. Goodman, I. Y. T. Chan, and P. R. Buseck, Structure and evolution of fugitive particles from a copper smelter, Environ. Sci. Tech., 15, 1208-1212, 1981.

Buseck, P. R., and S. Iijima, High resolution electron microscopy of silicates, Am. Min., 59, 1-21, 1974.

Campbell, J. A., J. C. Laul, K. K. Nielson, and R. D. Smith, Separation and chemical characterization of finely-sized fly-ash particles, Anal. Chem., 50, 1032-1040, 1978.

Cowley, J. M., A. L. G. Rees, and J. A. Spink, The morphology of zinc oxide smoke particles, Proc. Phys. Soc. London, 64, 638-644, 1951.

Doerner, H. A., Reduction of zinc ores by natural gas, Trans. Am. Inst. Min. Metall. Eng., 121, 636-677, 1936.

Egerton, R. F., Quantitative energy-loss spectroscopy, Proc. 11th Ann. SEM Symposium, 1, Scanning Electron Microscopy, Inc., Illinois, 13-23, 1978.

Everett, M. R., D. V. Kinsey, and E. Roemberg, Carbon transport studies for helium-cooled high-temperature nuclear reactions, Chem. Phys. Carbon, 3, 289-436, 1968.

Fuller, M. L., Twinning in zinc oxide, J. App. Phys., 15, 164-170, 1944.

Fulton, C. H., Condensation of zinc from its vapor, Trans. Am. Inst. Min. Metall. Eng., 60, 280-302, 1919.

Germani, M. S., M. Small, W. H. Zoller, and J. L. Moyers, Fractionation of elements during copper smelting, Environ. Sci. Tech., 15, 299-305, 1981.

Heinrich, K. F. J., Electron Beam X-Ray Microanalysis, Van Nostrand Reinhold Company, New York, 450-455, 1980.

Henry, W. M., and K. T. Knapp, Compound forms of fossil fuel fly ash emissions, Environ. Sci. Tech., 14, 450-456, 1980.

Hulett, L. D., Jr., J. F. Emery, J. M. Dale, A. J. Weinberger, H. W. Dunn, C. Feldman, E. Ricc, and J. O. Thomson, Proceedings: Advances in Particle Sampling and Measurement, EPA-600/7-79-065, Environmental Protection Agency, Washington, D. C., 1979.

Hulett, L. D., Jr., and A. J. Weinberger, Some etching studies of the microstructure and composition of large aluminosilicate particles in fly ash from coal-burning power plants, Environ. Sci. Tech., 14, 965-970, 1980.

Hulett, L. D., Jr., A. J. Weinberger, K. J. Northcutt, and M. Ferguson, Chemical species of fly ash from coal-burning power plants, Science, 210, 1356-1358, 1980.

Joy, D. C., The basic principles of electron energy loss spectroscopy, in Introduction to Analytical Electron Microscopy, edited by J. J. Hren, J. I. Goldstein, and D. C. Joy, pp. 223-244, Plenum Press, New York and London, 1979.

Keyser, T. R., D. F. S. Natusch, C. A. Evans, Jr., and R. W. Linton, Characterizing the surfaces of environmental particles, Environ. Sci. Tech., 12, 768-773, 1978.

Linton, R. W., A. Loh, D. F. S. Natusch, C. A. Evans, Jr., and P. Williams, Surface predominance of trace elements in airborne particles, Science, 191, 852-854, 1976.

Natusch, D. F. S., C. F. Bauer, H. Matusiewicz, C. A. Evans, J. Baker, A. Loh, R. W. Linton, and P. K. Hopke, Characterization of trace elements in fly ash, Institute for Environmental Studies Report, UILU-IES 77 0003, Univ. of Illinois at Urbana-Champaign, 1977.

Novakov, T., S. G. Chang, and A. B. Harker, Sulfates as pollution particulates: Catalytic formation on carbon (soot) particles, Science, 186, 259-261, 1974.

Post, J. E., and P. R. Buseck, Characterization of the Phoenix urban aerosol using electron beam instruments, submitted to Atmos. Environ., 1982.

Pueschel, R. F., Aerosol formation during coal combustion: Condensation of sulfates and chlorides on fly ash, Geophys. Res. Lett., 3, 651-653, 1976.

Pueschel, R. F., and C. C. Van Valin, Cloud nucleus formation in a power plant plume, Atmos. Environ., 12, 307-312, 1978.

Radushkevich, L. V., and V. M. Luk'Yanovich, Zh. Fiz. Khim., 26, p. 88, 1952 (taken from Baker and Harris, 1978).

Small, J. A., K. F. J. Heinrich, D. E. Newbury, R. L. Myklebust, and C. E. Fiori, Procedure for the quantitative analyses of single particles with the electron probe, in Characterization of Particles, edited by K. F. J. Heinrich, NBS Special Publication 533, U.S. Govt. Printing Office, Washington, D.C., 29-38, 1980.

Small, M., M. S. Germani, A. M. Small, W. H. Zoller, and J. L. Moyers, Airborne plume study of emissions from the processing of copper ores in southeastern Arizona, Environ. Sci. Tech., 15, 293-299, 1981.

Tartarelli, R., P. Davini, F. Morelli, and P. Corsi, Interactions between SO_2 and carbonaceous particulates, Atmos. Environ., 12, 289-293, 1978.

Thomas, E., and P. R. Buseck, Characterization of the aerosol at a rural location in eastern Arizona, submitted to Atmos. Environ., 1982.

ELECTRONIC STRUCTURE THEORY FOR SMALL METALLIC PARTICLES

R. P. Messmer

General Electric Co., Corporate Research and Development, Schenectady, New York 12301

Abstract. The electronic structure of small metallic particles or clusters determines their chemical and physical properties. Thus such diverse properties as the "molecular structure" - i.e., the geometric disposition of the atoms in a cluster, the magnetic properties, the catalytic properties, and the chemical interactions with molecules in the environment - are a consequence of the electronic structure of the metal clusters. Unlike the situation in molecular physics, where much theoretical and spectroscopic information exists on a very large number of molecules and provides detailed information on electronic structure, the situation for metal clusters is rather primitive. Indeed, the science of metal clusters is still in its infancy.

With this perspective in mind, the present paper reviews the recent theoretical work which has been carried out in an effort to characterize the electronic structure of metal clusters.

Introduction

Unlike the situation for bulk crystalline metals, where a vast literature exists on electronic structure, there is a relatively modest literature concerned with the electronic structure of metal clusters. From the theoretical point of view this is quite understandable. First, the crystal structures of metals are well known, and second, the periodicity characteristic of crystalline materials results in great mathematical simplifications in a quantum mechanical treatment. For example, such simplifications allow the calculation of electronic structure via energy band theory (Slater, 1965). By mathematical construction, the band theoretic treatment of bulk crystalline metals cannot address the properties of the surface of these metals. However, in recent years extensions of band theory have been made to treat slabs of a metal (e.g., Feibelman et al., 1979). In this case periodicity is retained in two dimensions, parallel to the slab. In the third dimension, periodicity is broken, creating two free surfaces. Energy band calculations have been performed for such slabs of variable thickness, i.e., different numbers of atom layers. Such calculations are more difficult and expensive than bulk calculations, and the complexity increases with the slab thickness.

If one eliminates the periodicity in the other two dimensions, one arrives at a cluster of atoms. Under these circumstances the energy band theory reduces to the molecular orbital theory - a theory long familiar in the treatment of discrete molecules. Thus the use of molecular orbital theory to treat the electronic structure of metal clusters is seen to be a natural consequence of starting with band theory for the bulk metal and then eliminating the assumption of periodicity in space.

This all sounds very simple and natural; however, at this stage one is faced with a number of nontrivial problems. That is, if one wishes to do a molecular orbital theory calculation for a metal cluster, at least three major issues have to be addressed, as follows. What is the arrangement of atoms in the cluster? The atoms may not have the same spatial arrangement as a small chunk of the bulk. What form of molecular orbital theory should one use? There is a variety of approximate methods available - which ones are most appropriate? How many atoms do we wish to consider? If one wants to retain any degree of rigor in the molecular orbital calculations, a dozen atoms already constitute a very significant computational task. Thus there must be a trade-off between rigor and cluster size.

Before taking up these issues it is useful to make a few general observations about metal clusters. The most obvious characteristic of small metal particles is that a large percentage of the atoms in the cluster are surface atoms. For example, if we assume that a particle has a simple cubic geometry, then any particle containing fewer than roughly 1000 atoms will contain over 50 percent surface atoms. Most of the clusters to be discussed in this paper contain far fewer than 1000 atoms, and hence the question of surface effects and their contribution to observed properties is a very important one. As many of the atoms in a cluster are surface atoms and these surface atoms can have different electronic behavior than interior atoms, it is important in a theoretical calculation to allow the electrons to determine the potential (in which they find themselves) in a self-consistent way. That is, employing self-consistent field (SCF) calculations is of the utmost importance. Likewise, the effects of the cluster environment (i.e., the substrate or support with which it interacts or the atmospheric gases with which it is in contact) on the electronic structure of a cluster can be appreciable if contact is over a reasonably large fraction of the cluster's surface area. Thus one must be careful in comparing calculations which deal with intrinsic electronic properties of isolated metal clusters with experiments on metal clusters interacting with unspecified or poorly characterized environments.

A question which is often raised with respect to metal clusters concerns how many atoms a cluster should contain in order to exhibit bulk-like properties. Again using a cluster of simple cubic geometry, if one considers the fraction of atoms on the surface as a criterion for convergence to the bulk, then even with a million atoms (corresponding to a particle diameter of a few hundred angstroms) 6 percent of the atoms are on the surface. Thus one might suppose that clusters of this size are necessary to attain reasonable convergence to the bulk. However, it should be clear that one cannot discuss convergence of cluster properties to the bulk values in such simple terms. There will be some properties, arising from local electronic structure effects, which may require relatively few atoms to describe. On the other hand, to obtain the properties of the bulk Fermi surface may require very many atoms. Thus, in general, no simple statement regarding the convergence of cluster properties can be made. The rate of convergence to the bulk values of a property will depend upon the property.

Intrinsic Properties of Metal Clusters

Before discussing what has been learned recently regarding the intrinsic electronic structure of metal clusters, we must first make an attempt to address the issues raised in the introduction.

The first issue concerns the structure of metal particles. A primary problem which plagues all electronic structure calculations of clusters is the lack of experimental information on the geometrical structure (i.e., the positions of the atoms) of the clusters. For example, consider Figure 1, which shows two high-symmetry arrangements for 13 metal atoms. One polyhedron is a cubooctahedron; the other is an icosahedron. Which of these structures, if either, is the appropriate geometry for a cluster of 13 nickel atoms, for example? In most situations in molecular and solid state physics, this structural information is known at the outset. However, for the case of clusters one is forced to assume "reasonable structures" in order to proceed. The usual wisdom is to assume highly symmetrical clusters (e.g., those of Figure 1) most often reflecting the bulk structure (e.g., the cluster of Figure 1(a) for the case of a face-centered cubic metal) and to assume internuclear distances close to the bulk values. It is unlikely that these assumptions will be true for extremely small clusters; that is, on the order of 10 atoms or less. Experimental information on structure is badly needed.

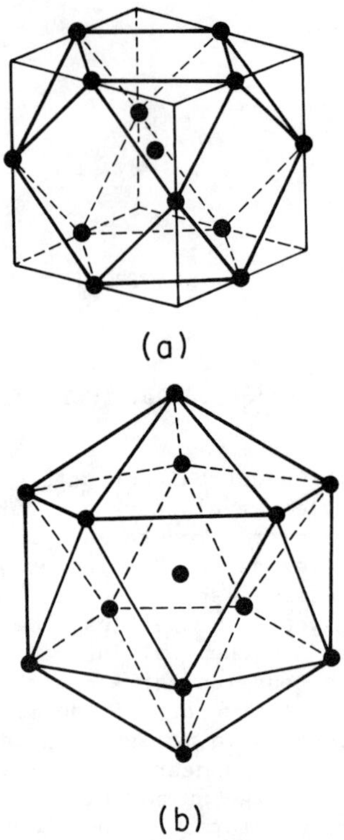

Fig. 1. Two 13-atom clusters: (a) Cubo-octahedral geometry, (b) Icosahedral geometry.

One might wonder why the electronic structure calculations are not used to determine the geometry of clusters. Unfortunately, at present this appears to be an almost impossible task. The simple models which can treat reasonably large clusters (>25 atoms) are not accurate enough to reliably determine the small energy differences between various geometric configurations. On the other hand, the sophisticated models are so computationally complex that only a few atoms can be treated.

A recent breakthrough in the experimental production of metal particles which have been mass analyzed from single atoms to clusters of over 100 atoms (Sattler et al., 1980) is of fundamental importance. It promises to offer a source of metal clusters to experimentalists interested in the study of the intrinsic structure and properties characterization of such clusters. Such work will be essential to future progress in this field.

The second issue to be addressed is the form of molecular orbital theory one should use to calculate the electronic structure of metal clusters. It is this author's opinion that the best "first principles" calculations possible should be carried out on clusters of 10 to 20 atoms to explore the properties of such clusters. With these results in hand, the simple semiempirical methods can then be properly parameterized to reproduce, as well as possible, the more accurate calculations for small clusters. These methods then could be logically extended to calculate the properties of much larger clusters.

The term "first principles" methods means that such methods avoid, to the maximum extent possible, the introduction of approximations and parameters beyond the basic premises of the methods. There are two such types of molecular orbital methods in common use: 1) the Hartree-Fock method, and 2) the density functional and Xα methods. The former method has been the workhorse of quantum chemical applications to molecules, but is not applied to the treatment of bulk metals via band theory because of its well-documented problems in the description of metals (Kittel, 1963; Monkhorst, 1979). The Xα method (Slater, 1972) (which is a particular form of density functional theory), on the other hand, has been widely used to treat the electronic structure of bulk metals by band theory and over the last 10 years has been extended to the investigation of molecules.

Because of its proven ability to treat the electronic structure of bulk metals, the author believes that the Xα method is the most reasonably suited first principles technique for initial investigations of metal clusters. Thus far it has been the most widely applied method for the investigation of metal clusters, and it is the results of this method which will dominate the discussion to follow.

The third issue of how many atoms to consider is largely resolved de facto by the complexity of the computational scheme chosen. For the Xα method implemented by the scattered wave formalism (Johnson, 1973), one is restricted at present (due to computational considerations) to metal clusters of 50 atoms or less.

Cluster Density of States

One of the most readily accessible theoretical characteristics of a metal cluster is its density of states (DOS). The DOS is a function which gives the number of energy levels per unit of energy as a function of energy. It is usually calculated for bulk metals from band theory computations. The ability to obtain such information from cluster calculations allows a comparison to be made between the DOS of clusters and that of the bulk metal.

The calculated energy level diagram for a Ni_{13} cluster with the geometry of Figure 1(a) and a metal-metal distance appropriate to bulk Ni is shown in Figure 2. The method of calculation is the self-consistent-field Xα scattered-wave (SCF-Xα-SW) method (Slater, 1972; Johnson, 1973). Except for the change of boundary conditions to

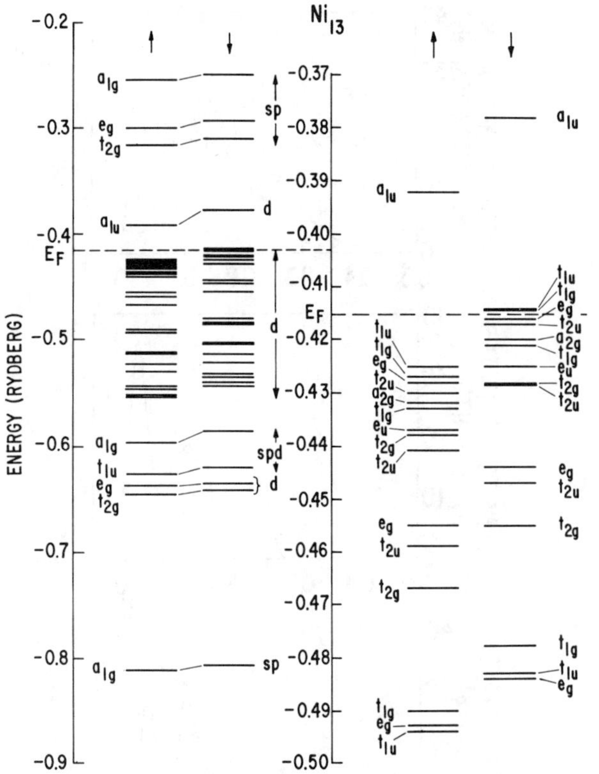

Fig. 2. Spin-polarized SCF-Xα-SW electronic energy levels for a cubooctahedral Ni_{13} cluster with nearest neighbor internuclear distance equal to that of bulk crystalline nickel. The levels are labeled according to the irreducible representations of the O_h point group.

those appropriate to a cluster, the method is the same as the well-known Korringa-Kohn-Rostoker (KKR) method (Korringa, 1947; Kohn and Rostoker, 1954) of band theory. The calculations were carried out in a spin-polarized fashion, allowing different molecular orbitals for different spins. The spin-polarized energy levels of Figure 2 are displayed on two energy scales, one encompassing the entire d band and overlapping sp levels and the other to the right resolving the upper part of the d band on a finer energy mesh. The Fermi level, E_F, separates the unoccupied spin orbitals from the occupied ones. The Ni_{13} cluster has a net magnetic moment and the moments of the central and peripheral atoms are aligned; i.e., the cluster is ferromagnetic. This is particularly interesting as the bulk metal is known to be ferromagnetic. Even more striking is the fact that similar calculations for Pd_{13} and Pt_{13} clusters are not magnetic, which again is consistent with the bulk situation.

One can characterize the complete set of levels in Figure 2 as a dense band of d levels through which the Fermi level passes at the upper end, bounded above and below with sp-like levels. Thus

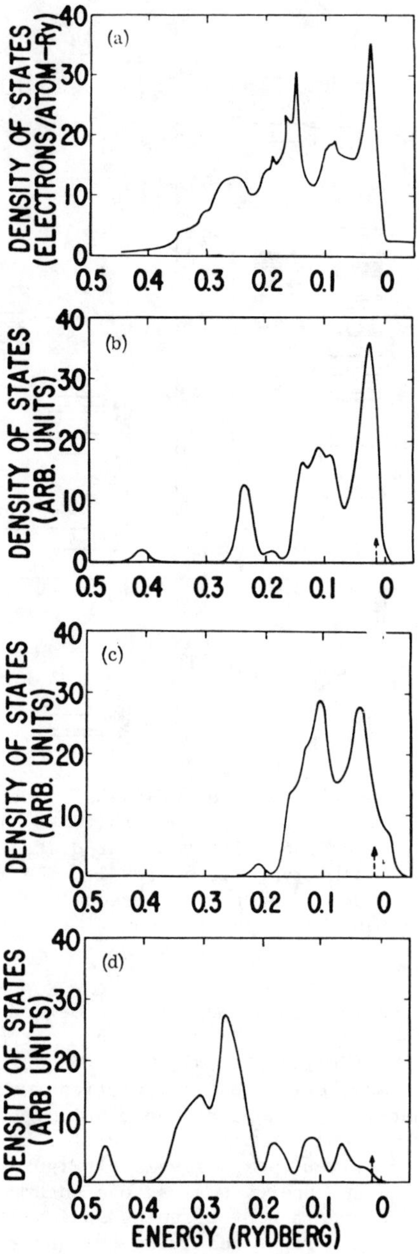

Fig. 3, Comparison of the electronic density of states of nickel as determined from: (a) bulk band structure; Ni_{13} clusters using (b) SCF-Xα-SW method, (c) extended Hückel method, and (d) CNDO method.

one finds that even for 13 atoms (actually, even for 2 to 4 atoms) the d band is totally overlapped by the sp band as in the solid. Lest one think that such a cluster mirrors all the properties of the bulk, we hasten to point out that there are also considerable differences between the cluster and the bulk as discussed previously (Messmer et al., 1976). For example, the magnetic moments of the central and surface atoms are different (which is not surprising) and the surface atoms have a deficit of electron charge whereas the central atom has a surplus. Nonetheless, the number of qualitative similarities is rather striking.

One may generate a DOS from an energy level diagram such as that shown in Figure 2 by centering a Gaussian of finite width at the position of each energy level. This has been done for a spin-restricted Ni_{13} calculation and is compared in Figure 3 with the bulk DOS of a band structure calculation as well as with two cluster calculations based on semiempirical molecular orbital methods (Messmer et al., 1976). The band theory results are from an Xα-LCAO (linear combination of atomic orbitals) calculation (Callaway and Wang, 1973); the results labeled EH are extended Hueckel cluster results (Anderson and Hoffmann, 1974), and the results labeled CNDO are "complete neglect of differential overlap" approximate MO (molecular orbital) cluster results (Blyholder, 1975). In Figure 3 the arrows which appear in the panels of the cluster results designate the energy position of the highest occupied MO (the Fermi level). From Figure 3 one can say that the CNDO and EH methods, with the parameterizations which have appeared in the literature, do not give DOS curves which resemble the main features of the bulk metal DOS. On the other hand the Xα-SW cluster results, while clearly different in detail, give an overall qualitative resemblance to the bulk DOS. This situation obtains for other transition metals as well. It is the author's view that these differences in the cluster results are due to inappropriate and rather arbitrary parameterizations of the semiempirical methods (CNDO and EH), as pointed out previously (Messmer et al, 1976). This view is supported by two calculations which have been published using the EH method for Cu clusters (Baetzold and Mack, 1975; Baetzold, 1978). The calculated d-band widths differ by an order of magnitude in the two calculations, which use different parameterizations. Thus, it would appear that until the semiempirical methods have found a consistent and reliable approach to parameterization, one must rely upon a nonempirical first principles theory such as the Xα method to discuss the electronic structure of metal clusters.

The other type of nonempirical MO method besides the density functional/Xα methods is the Hartree-Fock approach mentioned previously. This method has been applied to transition metal clusters only rather recently (Basch et al., 1980; Bachmann et al., 1980). Both Ni (Basch et al., 1980) and Cu (Bachmann et al., 1980) clusters have been considered, in the former case Ni_2 to Ni_6 and in the latter case clusters up to Cu_{13}. One striking difference between the Hartree-Fock (HF) and Xα methods for such clusters is that, in contrast to the Xα method, the HF method produces s-like levels which are always above the d bands. That is, s-like character does not overlap the d bands as in the bulk or as in the Xα results for Ni and

Cu clusters. Thus the studies based on the HF method conclude that such small clusters are fundamentally different from the bulk even in their qualitative characteristics. This raises a substantial conflict between the conclusions of the HF and $X\alpha$ methods for metal clusters.

Although it is too early to say that there is a generally accepted resolution of this conflict, some insight into how this situation might be resolved can be obtained by looking at some previous conflicts between the conclusions of the two methods. Only three cases will be cited here. The first involves calculations on the model system $Cr_2(O_2CH)_4$. It is known experimentally that such molecules have substantial metal-metal bonding, yet a Hartree-Fock calculation (Garner et al., 1976) concluded that there was no net bonding between the metal atoms. $X\alpha$-SW calculations for the same system showed that there was metal-metal bonding in agreement with experiment (Cotton and Stanley, 1977). Subsequent work (Benard and Veillard, 1977; Benard, 1979) has shown that the usual Hartree-Fock method is inappropriate because it can not take into account some important electronic correlation effects. When these effects are taken into account by a configuration interaction calculation, which introduces much additional computational complexity, a "completely different description of the bonding is achieved" (Benard and Veillard, 1977; Benard, 1979). That is, a Cr-Cr bond is found to exist, in agreement with the $X\alpha$ results and experiment.

A second example is the ozone molecule, which is perhaps of more interest to atmospheric science. In this case the Hartree-Fock method gives the incorrect ground state and the wrong ordering of several excited states (Hay et al., 1975). Again, if electronic correlation effects are taken into account (i.e., going beyond the HF molecular orbital approximation) by generalized valence bond (GVB) and configuration interaction (CI) approaches, then a description consistent with experiment is achieved. However, it has been shown (Messmer and Salahub, 1976) that at the $X\alpha$ molecular orbital level one already obtains the correct ground state and a reasonable description of a number of excited states. Thus in these cases, as well as others, the $X\alpha$ MO method appears to give a much better first-order description of the systems than is achieved by the Hartree-Fock method. It should be clear that GVB and CI methods which incorporate correlation effects, if carried far enough, can provide very accurate descriptions of any system, descriptions which are far superior to those obtained with the $X\alpha$ method. However, that is not the point. The issue is whether the density functional/$X\alpha$ or the Hartree-Fock description is more appropriate at the molecular orbital level of approximation. This is particularly important for metal clusters, as the GVB and CI methods are so computationally demanding in their present forms that they are unlikely to be applied to a cluster such as Ni_{13} for some time to come. Thus one will be confined to results at the MO level of approximation for the time being, and it is thus essential to know which approach bears the greater resemblance to the actual situation in metal clusters.

The last example to be discussed is the most relevant to the metal cluster problem, namely Ni_2. For the X_α results (Roesch and Rhodin, 1974) one arrives at a situation qualitatively similar to that found for $X\alpha$ calculations on larger clusters (see, e.g., Figure 2). That is, there is an overlap of s- and d-like excitations for Ni_2, although the d levels are not bounded below by sp-like levels as for the larger clusters. For the Hartree-Fock case (Basch et al., 1980), on the other hand, the sp-like levels are all above the d levels in energy. That is, there is no overlap of d- and sp-like levels in the HF calculations. However, for the Ni_2 molecule extensive GVB and GVB-CI calculations have been performed, in addition to HF calculations, by Upton and Goddard (1978). They conclude that the states of Ni_2 and the ion states of Ni_2^+ cannot be approximately described by Hartree-Fock calculations and that one must go to a more sophisticated treatment such as the GVB method to obtain meaningful results. It is important to note that in Figure 6 of their paper (Upton and Goddard, 1978) they show for the ion states of Ni_2^+ that there is an overlap of the s- and d-type excitations in qualitative agreement with band theory and $X\alpha$ results. It should be kept in mind that when angle-resolved photoelectron spectroscopy is used to determine energy-band dispersions (Dietz and Eastman, 1978) it is the ion states which are of relevance.

It is probably worthwhile to reiterate a few points. First, when accurate calculations of the GVB or CI type can be carried out, they will always be more reliable than either $X\alpha$ or HF results and contain more detailed information than either MO method is capable of giving. Second, as such accurate calculations will not be routine for large transition metal clusters for some time, one will probably have to be content temporarily with results at the MO level of approximation. Third, on the basis of known examples, three of which have been discussed, the density functional/$X\alpha$ results are in better qualitative agreement with the more accurate calculations than are the Hartree-Fock results. Thus, one is led to conclude, on the basis of the information presently available, that the density functional/$X\alpha$ methods are likely to give the most reliable information on the electronic structure of metal clusters of any of the presently available molecular orbital theories.

Cluster Magnetism

One of the fascinating aspects of investigating transition metal clusters is the prospect of finding unusual magnetic effects, again due to the large surface-to-volume ratio of such clusters. Of the cluster calculations carried out to date using various methods, only the Xα-SW method has been used to investigate the magnetic properties of various metal clusters.

The first Xα-SW work on transition metal clusters (Messmer et al., 1976) was concerned with Ni_8, Ni_{13}, Cu_{13}, Pd_{13}, and Pt_{13}. It was found that Ni_8 and Ni_{13} clusters were magnetic while Pd_{13} and Pt_{13} were not. For the Ni_8 cluster the calculated average spin magneton number per atom was 0.25, whereas for the Ni_{13} cluster the value was 0.46. These may be compared with the bulk value of 0.54. As pointed out above, the spin density on the central atom is quite different from that found on the 12 surface atoms.

Although it is rather remarkable that the 13-atom clusters of Ni, Pd, and Pt, which all have the same number of valence electrons, turn out to be magnetic for Ni and nonmagnetic for Pd and Pt, thus mimicing the behavior found in the bulk metals, one cannot expect that this will be a general phenomen. In fact, recent calculations (Salahub and Messmer, 1981) on a 15-atom Cr cluster have shown that such clusters carry a net magnetic moment, while it is known that the bulk metal is antiferromagnetic. Nontheless, the Cr_{15} cluster results try to mimic the antiferromagnetic behavior to a large degree, but due to surface effects there is not an exact cancellation of moments of opposite spin orientation. It is interesting to consider the Cr_{15} results in a broader context, however; i.e., to study the magnetic properties of clusters of three different elements which have the same crystal structure in the bulk (body-centered cubic) and exhibit three different types of magnetic ordering. The metals are vanadium (nonmagnetic), chromium (antiferromagnetic), and iron (ferromagnetic).

Figure 4 shows the orbital energies from spin-polarized Xα-SW calculations for V_{15}, Cr_{15}, and Fe_{15}. The cluster geometry is shown at the bottom of the figure and consists of 15 atoms representing the body-centered cubic (bcc) structure, with a central atom, the eight nearest neighbors at the corners of a cube of side a (appropriate to the observed bulk internuclear distance) and six second neighbors in the faces of a cube of side 2a. The lines connecting up-spin and down-spin energy levels of Figure 4 merely relate the same symmetry types in sequential order.

The calculations were carried out fully self-consistently, allowing a flow of electrons between the two spin manifolds such that, at convergence, Fermi statistics are satisfied. In the case of V_{15}, the added degree of freedom provided by spin-polarization has a negligible effect. The converged result has a single extra majority spin electron and the exchange splitting is very small. Thus V_{15} mimics, as far as possible for a system with an odd number of electrons, the nonmagnetic situation of bulk vanadium.

For Cr_{15}, an approximate calculation of the magnetic moment (Salahub and Messmer, 1981) for the various atoms yields $-0.7\ \mu_B$ for the central atom, $+4.1\ \mu_B$ for the first neighbors, and $-3.4\ \mu_B$ for the second neighbors, which may be compared to $\pm 0.7\ \mu_B$ for antiferromagentic chromium as determined from neutron diffraction data (Slater, 1974). The near-perfect agreement of the central atom value with experiment must be considered largely fortuitous. Thus, while the alternation in sign characteristic of bulk chromium is well represented by the results, they also indicate that the 15-atom cluster in the bulk geometry should have a permanent magnetic moment and also that the moments for the surface atoms may be quite different from those of atoms in the interior. Adding further shells of atoms will undoubtedly lead to equivalent spin-up and spin-down potentials in the interior, but the overall magnetic ordering will remain, as it is largely determined by local short-range interactions.

Turning finally to the case of iron, it can be seen from Figure 4 that Fe_{15} shows a very large spin polarization and has an excess of 40 majority spin electrons. Neglecting the contribution to the magnetization from orbital angular momentum

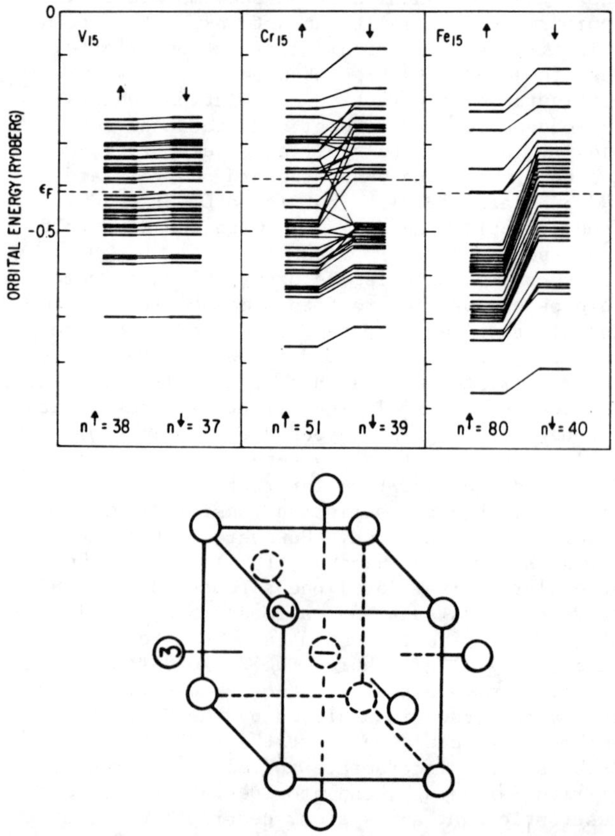

Fig. 4. Orbital eigenvalues from spin-polarized SCF-Xα-SW calculations on 15-atom clusters of vanadium, chromium, and iron with geometry shown at bottom and where the internuclear distances are equal to those of the respective bulk metals. The dashed lines mark the Fermi levels. The numbers of spin-up and spin-down valence electrons are given at the bottom of each panel.

(which is small for bulk iron (Shull and Mook, 1966)), this yields an average magnetic moment per iron atom of 2.7 μ_B which compares to the experimental value of 2.2 μ_B for bulk iron. However, the three groups of symmetry-equivalent iron atoms in the Fe_{15} cluster are in different environments, and one would not expect that the charge and spin densities for them would be the same. An approximate calculation (Salahub and Messmer, 1981) of magnetic moments yields 1.1 μ_B for the central atom, 2.7 μ_B for the first neighbors, and 2.8 μ_B for the second neighbors; thus the cluster is ferromagentic. A detailed discussion of the Fe cluster results can be found in Yang et al. (1981). Thus it can be seen again that while the qualitative features of magentism of bulk iron are reasonably well interpreted with the results of an Fe_{15} cluster, there are definite effects which arise from the finite size of the cluster (surface effects).

Summarizing cluster magentism for 15-atom clusters of three bcc transition metals, one finds that the clusters appear to be sufficiently large to exhibit magnetic ordering which is qualitatively similar to that shown by the bulk metals, and hence insight into certain aspects of bulk properties may be obtained. However, the clusters also exhibit characteristic differences as compared to the bulk, and these intrinsic differences are important and of interest in their own right.

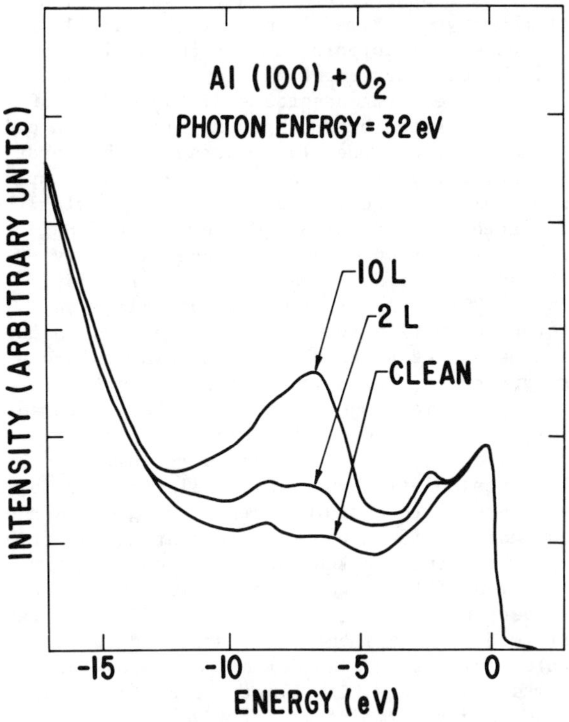

Fig. 5. Photoelectron spectra for oxygen chemisorption on Al(100) surface. Spectra are for the clean Al surface and for exposures of 2 and 10 langmuirs of oxygen (Eberhardt and Kunz, 1978).

The Interaction of Metal Clusters With Atoms and Molecules

As stressed in the last section, clusters appear to exhibit qualitative aspects of the bulk metal together with definite surface characteristics. It is thus obvious to ask how well such clusters interacting with atoms or molecules might model the behavior of these species interacting with bulk metal surfaces. A number of such studies have been undertaken and two will be discussed in this section. An important question is how many atoms the cluster must contain in order to give a reasonably converged description of the electronic structure of the actual surface-adsorbate interaction under consideration.

The first example deals with the chemisorption of oxygen atoms on an Al(100) surface, as modeled by clusters of Al atoms (Messmer and Salahub, 1977). In the calculations it was assumed that an oxygen atom would chemisorb in a fourfold site on the surface and clusters of 5, 9, and 25 Al atoms were used to model the (100) surface. Xα-SW calculations were performed for the oxygen atom at 0.0, 2.0, and 4.0 bohr (1 bohr = 0.5292 Å) above the surface plane of aluminum atoms. The calculations at 0.0 bohr were used to simulate the incorporation of the oxygen atom into the Al lattice. Various DOS curves were generated and compared with photoelectron spectra available at the time. It was found that only the 25-atom cluster of Al interacting with oxygen provided a reasonably converged description of the metal-adsorbate interaction. This was consistent with a study of the convergence of the DOS for pure Al clusters containing up to 43 atoms (Salahub and Messmer, 1977).

Experimental results from techniques such as photoelectron spectroscopy give information about electronic structure, but not without some theoretical input. They also cannot directly give the chemisorption site or the adsorbate-surface distance. Theory plays an important role in helping to extract such information from experiments.

As the Xα-SW method, in its present form, calculates only an approximate total energy, it cannot be used reliably to compute the equilibrium metal-adsorbate distance. Thus, such information must be arrived at by indirect means. That is, one must calculate some properties of the system as a function of metal adsorbate separation and compare these to the experimentally observed properties in order to deduce the most likely separation. For oxygen atoms chemisorbed on aluminum two types of information were available with which to make a comparison with theory. These were photoelectron spectra of the valence electron region and the Al 2p-level shifts due to chemisorption of oxygen as determined by photoelectron spectroscopy.

From a comparison of the theoretical results

with experimental information, it was concluded that the oxygen atoms did not sit on the Al surface but were incorporated into the Al lattice; i.e., the best agreement with experiment was obtained at the metal-adsorbate distance d = 0.0 bohr. The photoelectron spectra for clean aluminum and for two rather low coverages (Eberhardt and Kunz, 1978) are shown in Figure 5. A comparison of the experimental data of Figure 5 with the calculated results for $Al_{25}O$ (Messmer and Salahub, 1977), which produces peaks at approximately -3, -7.5 and -10 eV, shows rather good agreement.

The Al 2p-level shifts which take place on interaction with an oxygen atom were also calculated using the $Al_{25}O$ cluster. Again, the results for lattice incorporation of the oxygen (i.e., d = 0.0) are more consistent with the experimental finding (Flodstroem, 1976) than are those for oxygen above the surface (i.e, d = 2.0).

These calculations were also able to describe the nature of the bonding between oxygen and aluminum and provide a detailed account of the origin of the structure in the photoelectron spectrum. A prediction of the light polarization dependence of the spectrum (Messmer and Salahub, 1977) was later confirmed by experiment (Eberhardt and Kunz, 1978). This study also demonstrated how such results can change with cluster size, therefore emphasizing the importance of investigating the convergence of cluster results. Subsequent work has also investigated oxygen chemisorption on the (111) surface of Al (Salahub et al., 1978).

A second example involves the interpretation of satellite structure in the X-ray photoelectron spectra of CO adsorbed on a Cu(100) surface. Xα-SW calculations for Cu_5CO and Cu_9CO clusters were used to model the problem (Messmer et al., 1980, 1982). The qualitative results of the two cluster models were very similar, but the quantitative differences suggested that the Cu_9CO cluster was the better model.

Experimentally a three-peak structure is observed in both the O1s and C1s regions of the X-ray photoelectron spectrum. The spectrum for the C1s region (Fuggle et al., 1978) is shown in the upper panel of Figure 6. A simple theoretical model based on the results of the calculations is shown in the schematic energy level diagram of the lower panel of Figure 6. The experimental spectrum shows that the first peak at lowest binding energy is followed by a second peak at 2 to 3 eV higher binding energy and a third peak at 7 to 8 eV higher energy with respect to the first peak. The theoretical interpretation is as follows. At the right of the lower panel the 1π and 2π energy levels of the isolated CO molecule are shown (the next column to the left shows the levels of CO when a C1s electron is removed by photoionization). The interaction of the 2π level of CO with the Cu surface results in a mixing between metal and the 2π orbital, producing two levels, $2\tilde{\pi}_a$ and $2\tilde{\pi}_b$. The $2\tilde{\pi}_a$ level, which is higher in energy than the $2\tilde{\pi}_b$, has far more CO 2π character than the $2\tilde{\pi}_b$ level. However, when a C1s hole is produced by photoionization of the chemisorbed CO, resulting in the $2\tilde{\pi}'_a$ and $2\tilde{\pi}'_b$ levels shown in Figure 6, it was found that the character of these orbitals is considerably changed from that of the $2\tilde{\pi}_a$ and $2\tilde{\pi}_b$ orbitals. In fact, the $2\tilde{\pi}'_b$ becomes more strongly CO 2π-like and the $2\tilde{\pi}'_a$ becomes more strongly Cu sp-like. The details have been discussed (Messmer et al., 1980, 1982). The $2\tilde{\pi}'_b$ level which is strongly CO 2π in character is partially occupied with one electron. Thus the first peak in the experimental spectrum can be attributed to a transition between the ground state of the neutral chemisorbed system and a final state in which a core hole on CO is produced, together with a transfer of an electron from Cu to the $2\tilde{\pi}'_b$ orbital (arrow labeled 1 in Figure 6). This $2\tilde{\pi}'_b$ orbital, containing very significant CO 2π character, contributes to the screening of the core

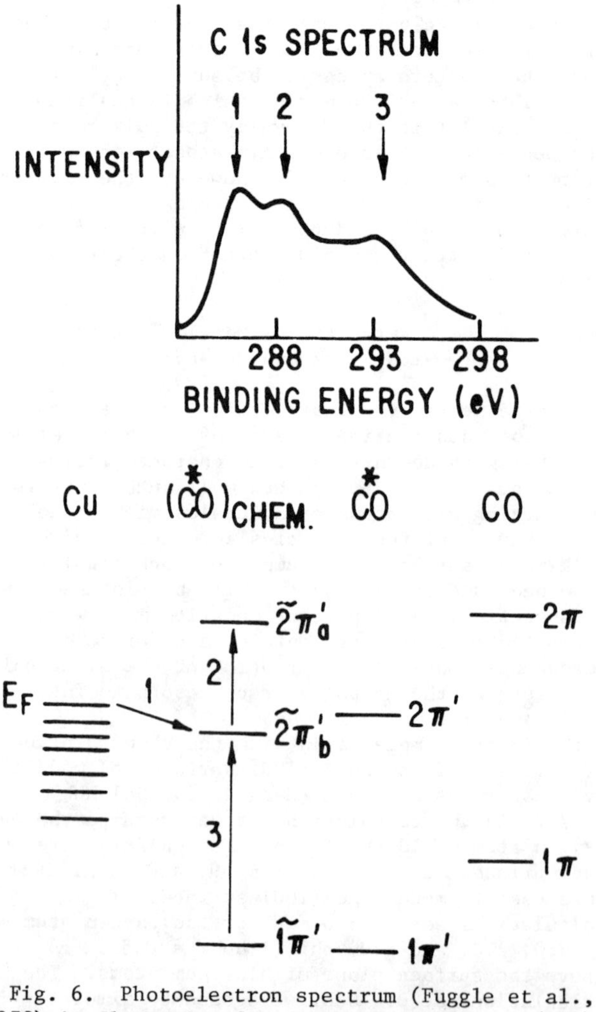

Fig. 6. Photoelectron spectrum (Fuggle et al., 1978) in C1s region for CO chemisorbed on Cu(100), together with a schematic representation of the nature of the model (Messmer et al., 1980, 1982) used to explain the three-peak structure of the spectrum.

hole. This final state, i.e., the final state associated with peak 1, is the calculated ground state of the chemisorbed core hole ion system.

If one chooses this ion state as the zero of energy for discussing the experimental spectrum of Figure 6, then the other two peaks represent shakeup states because they result from excitations from this core hole ion ground state. That is, the transitions labeled 2 and 3 in Figure 6 must be considered to take place together with transition 1. Thus peak 2 can be viewed as a transition from this ground ion state to an excited ion state by virtue of excitation of an electron from the $2\tilde{\pi}_b'$ orbital to the $2\tilde{\pi}_a'$ orbital. Considering the character of these orbitals as discussed above, the transition may be described rather well as a charge transfer from the CO 2π level to the Cu. Peak 3 is described as a one-electron excitation from the $1\tilde{\pi}'$ level to the $2\tilde{\pi}_b'$ level. This is the analog of the $1\pi \rightarrow 2\pi$ shake-up in isolated CO, which is found at ~8 eV above the main peak in the molecular core hole spectrum (Carlson et al., 1971).

Conclusions

Although the theory of electronic structure of metal clusters is at a rather primitive stage, a number of interesting features have already been found. These have been discussed in the text of this paper. There still exist conflicts between results of different methods, however, and it would be very beneficial to have these resolved as soon as possible.

Experiments on isolated metal clusters will be crucial to provide guidance to theory, but as yet no data exist on the properties of isolated transition metal clusters.

A number of studies have been reported which use clusters of atoms to model the interaction of adsorbates with metal surfaces. The results have been quite encouraging and further progress in this area can be expected.

References

Anderson, A. B., and R. Hoffman, Molecular orbital studies of dissociative chemisorption of first period diatomic molecules and ethylene on (100) W and Ni surfaces, J. Chem. Phys., 61, 4545-4559, 1974.

Bachmann, C., J. Demuynck, and A. Veillard, Structure and electronic properties of copper clusters. An ab initio LCAO-MO-SCF study, Faraday Symp., 14, 170-179, 1980.

Baetzold, R. C., Size and geometric effects in copper and palladium metal clusters, J. Phys. Chem., 82, 738-744, 1978.

Baetzold, R. C., and R. E. Mack, Electronic properties of metal clusters, J. Chem. Phys., 62, 1513-1520, 1975.

Basch, H., M. D. Newton, and J. W. Moskowitz, The electronic structure of small nickel atom clusters, J. Chem. Phys., 73, 4492-4510, 1980.

Benard, M., A study of Hartree-Fock instabilities in $Cr_2(O_2CH)_4$ and $Mo_2(O_2CH)_4$, J. Chem. Phys., 71, 2546-2556, 1979.

Benard, M., and A. Veillard, Nature of metal-metal interaction in binuclear complexes of chromium and molybdenum, Nouv. J. Chim., 1, 97-99, 1977.

Blyholder, G., CNDO MO calculation for hydrogen atom adsorption on nickel atom clusters, J. Chem. Phys., 62, 3193-3197, 1975.

Callaway, J., and C. S. Wang, Self-consistent calculation of energy bands on ferromagnetic nickel, Phys. Rev. B, 7, 1096-1103, 1973.

Carlson, T. A., M. O. Krause, and W. E. Moddeman, Excitation accompanying photoionization in atoms and molecules and its relationship to electron correlation, J. Phys. (Paris), 32, C4-76, 1971.

Cotton, F. A., and G. G. Stanley, Existence of direct metal-to-metal bonds in dichromium tetracarboxylates, Inorg. Chem., 16, 2668-2671, 1977.

Dietz, E., and E. E. Eastman, Symmetry method for the absolute determination of energy-band dispersions E(k) using angle-resolved photoelectron spectroscopy, Phys. Rev. Lett., 41, 1674-1677, 1978.

Eberhardt, W., and C. Kunz, Oxidation of Al single crystal surfaces by exposure to O_2 and H_2O, Surface Sci., 75, 709-720, 1978.

Feibelman, P. J., J. A. Appelbaum, and D. R. Hamann, Electronic structure of a Ti(001) film, Phys. Rev. B, 20, 1433-1443, 1979.

Flodstrom, S. A., R. Z. Bachrach, R. S. Bauer, and S. B. M. Hagstrom, Multiple oxidation states of Al observed by photoelectron spectroscopy of substrate core level shifts, Phys. Rev. Lett., 37, 1282-1285, 1976.

Fuggle, J. C., E. Umbach, D. Menzel, K. Wandelt, and C. R. Brundle, Adsorbate line shapes and multiple lines in XPS; Comparison of theory and experiment, Solid State Commun., 27, 65-69, 1978.

Garner, C. D., I. H. Hillier, M. F. Guest, J. C. Green, and A. W. Coleman, The nature of the metal-metal interaction in tetra-μ-carboxylato-chromium (II) systems, Chem. Phys. Lett., 41, 91-94, 1976.

Hay, P. J., T. H. Dunning, Jr., and W. A. Goddard III, Configuration interaction studies of O_3 and O_3^+. Ground and excited states, J. Chem. Phys., 62, 3912-3924, 1975.

Johnson, K. H., Scattered-wave theory of the chemical bond, Advan. Quantum Chem., 7, 143-185, 1973.

Kittel, C., Quantum Theory of Solids, John Wiley and Sons, New York, 1963.

Kohn, W., and N. Rostoker, Solution of the Schroedinger equation periodic lattices with an application to metallic lithium, Phys. Rev., 94, 1111-1120, 1954.

Korringa, J., On the calculation of the energy of a Bloch wave in a metal, J. Physica Grav., 13, 392-400, 1974.

Messmer, R. P., K. Knudson, K. H. Johnson, J. B. Diamond, and C. Y. Yang, Molecular-orbital studies of transition- and noble-metal clusters by the self-consistent-field Xα scattered-wave method, Phys. Rev. B, 13, 1396-1415, 1976.

Messmer, R. P., S. H. Lamson, and D. R. Salahub, Co on Cu: Assignment of the O1s and C1s X-ray photoelectron core spectra, Solid State Commun., 36, 265-270, 1980.

Messmer, R. P., S. H. Lamson, and D. R. Salahub, Interpretation of satellite structure in the X-ray photoelectron spectra of CO adsorbed on Cu(100), Phys. Rev. B, in press, 1982.

Messmer, R. P., and D. R. Salahub, Molecular orbital study of the ground and excited states of ozone, J. Chem. Phys., 65, 779-784, 1976.

Messmer, R. P., and D. R. Salahub, Chemisorption of oxygen atoms on aluminum(100): A molecular-orbital cluster study, Phys. Rev. B, 16, 3415-3427, 1977.

Monkhorst, H. J., Hartree-Fock density of states for extended systems, Phys. Rev. B, 20, 1504-1513, 1979.

Roesch, N., and T. N. Rhodin, Bonding of ethylene to diatomic nickel according to a self-consistent-field, Xα, scattered-wave model, Phys. Rev. Lett., 32, 1189-1192, 1974.

Salahub, D. R., and R. P. Messmer, Molecular-orbital study of aluminum clusters containing up to 43 atoms, Phys. Rev. B, 16, 2526, 1977.

Salahub, D. R., and R. P. Messmer, Magnetic order in transition metal clusters: A molecular orbital study, Surface Sci., 106, 415, 1981.

Salahub, D. R., M. Roche, and R. P. Messmer, Chemisorption of oxygen atoms on aluminum(111): A molecular-orbital cluster study, Phys. Rev. B, 18, 6495-6505, 1978.

Sattler, K., J. Muehlbach, and E. Rechnagel, Generation of metal clusters containing from 2 to 500 atoms, Phys. Rev. Lett., 45, 821-824, 1980.

Shull, C. G., and H. A. Mook, Distribution of internal magnetization in iron, Phys. Rev. Lett., 16, 184, 1966.

Slater, J. C., Quantum Theory of Molecules and Solids, Vol. 2, Symmetry and Energy Bands in Crystals, McGraw-Hill, New York, 1965.

Slater, J. C., Stastical exchange-correlation in the self-consistent field, Advan. Quantum Chem., 6, 1-92, 1972.

Slater, J. C., The Self-Consistent Field for Molecules and Solids, Vol. 4, McGraw-Hill, New York, 1974.

Upton, T. H., and W. A. Goddard III, The electronic states of Ni_2 and Ni_2^+, J. Am. Chem. Soc., 100, 5659-5668, 1978.

Yang, Chiang Y., K. H. Johnson, D. R. Salahub, J. Kaspar, and R. P. Messmer, Iron clusters: Electron structure and magnetism, Phys. Rev. B, 24, 5673-5692, 1981.

Gas-Solid Interactions

THE SURFACE SCIENCE OF HETEROGENEOUS CATALYSIS: POSSIBLE APPLICATIONS IN ATMOSPHERIC SCIENCES

G. A. Somorjai

Materials and Molecular Research Division, Lawrence Berkeley Laboratory, and
Department of Chemistry, University of California, Berkeley, CA 94720

Abstract. Surface science has been increasingly utilized in recent years to explore the molecular details of heterogeneous catalytic processes. A large number of techniques have been developed which determine the atomic surface structure, the composition, and oxidation states in the surface monolayer. Correlations of studies of catalytic reaction rates and product distributions with atomic scale surface properties revealed many important ingredients of surface reactivity. The structure of the catalytic surface markedly influences the surface chemical bonds of adsorbed molecules. Additives which are often electron donors (alkali metals) or electron acceptors (halogens) change the oxidation state of surface atoms, block sites, or change the surface structure and markedly influence the surface reactivity. The catalyzed reactions of carbonaceous deposits and water vapor are discussed along with the possible importance of trace metals and high-surface-area alumina silicates in atmospheric chemistry.

Introduction

Research in heterogeneous catalysis is practiced in a passive or active mode. The passive approach is aimed at understanding the catalytic process, which is the basis of a working chemical technology. This is performed by using kinetic studies to obtain accurate data of reaction rates and their pressure and temperature dependence. Catalytic studies of this type often utilize the various techniques of modern surface science to analyze the atomic surface structure and composition of the working catalyst and relate these atomic scale parameters to the macroscopic kinetic information, activity, and selectivity (Somorjai, 1981). The active approach attempts (1) to use all available data and molecular level understanding to build new catalytic systems that carry out the desired reaction with greater rate and selectivity, and (2) to design a formulation or preparation sequence to assure greater catalyst stability against deactivation. Atmospheric chemistry research, it appears, has focused mostly on the passive mode of research with one notable exception, the development of the catalytic converter which cleans up the car exhaust by oxidizing unburned hydrocarbons and carbon monoxide to water and CO_2. This very successful system uses platinum, palladium, and rhodium noble metals as catalyst surfaces and is one of the largest volume catalyst systems in use at present in the United States (Watson and Somorjai, 1982; Wei, 1975).

Some researchers in atmospheric chemistry attempt to uncover the sources of nitrogen oxide production and hydrocarbon emission, including carbonaceous particulates that are the results of energy conversion reactions. It should be remembered, however, that another group of chemists in equally large numbers is working on the reverse reactions - the catalytic fixation of N_2 or NO to produce fertilizers and the production of synthetic hydrocarbon fuels from CO and water, or CO_2 and hydrogen. In fact, it is not unusual that both the forward and reverse chemical reactions are of unique importance in different segments of chemical technology.

Let us review some of the principles of heterogeneous catalysis. Figure 1 indicates that catalysis is a complex and cyclic process that begins with the adsorption of molecules on the surface, followed first by their diffusion, rearrangement, and chemical transformation on the surface, and finally by desorption of new reactant molecules (Somorjai, 1981; Davis and Somorjai, 1982). Figure 2 shows the turnover rates and reaction probabilities for several types of hydrocarbon conversion reactions at different temperatures of catalytic interest. The turnover rates are defined as the number of molecules produced per exposed surface site per second. For a turnover rate of 10^{-3} one molecule is produced per site in every 1000 seconds, a long time indeed. In order to qualify as a catalyst, the surface site must produce more than one molecule during its lifetime. Thus the reaction with a turnover rate of 10^{-3} must be carried out for times longer than 1000 seconds to prove that it indeed is catalytic and not a stoichiometric process that would, of

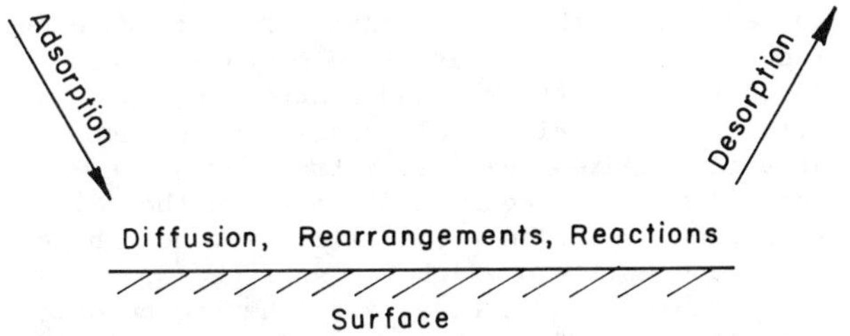

Fig. 1. Scheme of elementary processes during gas-surface interactions.

course, stop after the formation of one product molecule or some small finite number indicated by the reaction stoichiometry. Therefore, one of the important experiments is to ascertain that the reaction is catalytic.

The turnover rate of 10^{-3} molecules/site/second also indicates a long surface residence time (over 1 second). During its lifetime on the surface, the adsorbed species can visit many surface sites. Assuming a lower limit for its diffusion coefficient of 10^{-8} cm^2 sec and a residence time of about 1 second, simple random walk arguments would indicate a diffusion length of $\langle X \rangle = \sqrt{2Dt}$, about 10^{-4} cm, which is similar to or larger than the size of the catalyst particles. Thus the molecule visits many surface sites within its residence time to undergo the complex chemical processes needed to form the product molecules.

The reaction probability is obtained by dividing the turnover rate by the rate of molecular incidence J (molecule/site/second) which is obtained from knowledge of the reactant pressure $J = P/(\sqrt{2\pi MRT})$. It can be seen from Figure 2 that the reaction probability is very low, less than 10^{-6}, for the reactions shown. This is not surprising and is caused by the long molecular residence times that keep most of the catalytic surface fully occupied with adsorbates. Thus most reactant molecules incident on the catalyst surface cannot find a reactive site and desorb intact. For this reason extremely high surface area catalysts (~100 m^2/gm) can be used without fear of further chemical reactions as the molecules migrate through the catalyst bed while suffering multiple collisions with the catalyst surface. As long as the reaction probability is low for both reactant and product molecules, the catalyzed reaction may be run at high conversion (a large fraction, over 50 percent of the reactants converted to products). This is the case for many hydrocarbon conversion reactions where the reactants and products have roughly equal reaction probabilities. The circumstances for high conversion conditions are even better for hydrocarbon or carbon monoxide oxidation to carbon dioxide and water, or for the hydrogenation of olefins to alkanes. In these reactions the products have reaction probabilities which are much smaller than those of the reactants. In the opposite case, when the reaction probability of the product molecules is high, only low-conversion conditions can be tolerated unless the secondary reactions between the product molecules and the catalytic surface are acceptable. This is the case during the hydrogenation of carbon monoxide over iron catalyst surfaces, which yields ethylene and other olefins during the primary reaction step. The olefins then readsorb and react with more CO and hydrogen molecules again to produce longer chain hydrocarbon polymers.

Another important feature of the working catalyst is that it is able to atomize large-binding-energy diatomic molecules with ease (Somorjai, 1981), thereby producing a high concentration of atoms on the surface that are ready to participate in chemical reactions. H_2, O_2, and NO are readily atomized by most transition metal surfaces, while the higher-binding-energy CO and N_2 molecules are atomized by some of the transition metals (Fe, Ru, and Re).

Most catalytic reactions carried out in the chemical technology, in either reducing (in hydrogen) or oxidizing (in oxygen) media, are carried out in the temperature range of 100° to 400°C to obtain reasonable turnover rates ($>10^{-4}$). This is because the activation energies for these reactions are in the range of 15 to 45 kcal/mol. In many cases the reaction rate is desorption limited. Desorption is always an endothermic reaction step in the sequence of elementary surface processes (adsorption, surface diffusion, surface reaction, and desorption) necessary to carry out the catalytic reaction cycle. Thus it is likely that these chemical reactions (hydrocarbon conversion or partial oxidation) are not very important in atmospheric chemistry around 25°C, where their rates become negligibly slow. However, certain exothermic reactions, including the oxidation of CO and some of the hydrocarbons, may occur at an appreciable rate even at room

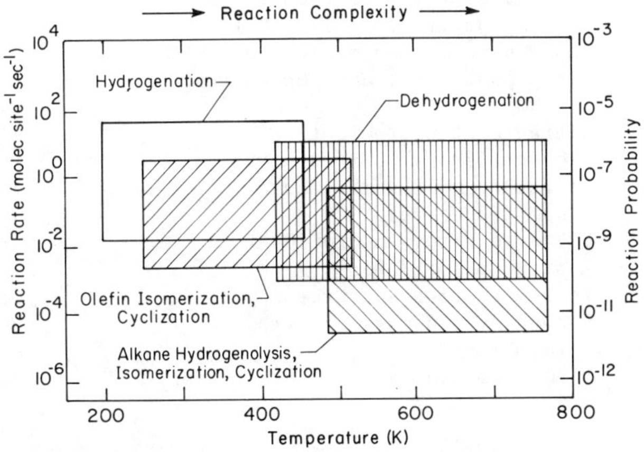

Fig. 2. Block diagram for hydrocarbon conversion over platinum catalysts showing the approximate range of turnover rates, reaction probabilities, and temperature regimes most commonly studied.

temperature. The photodissociation of water is one of the reactions that can be carried out at 25°C on an appropriate oxide surface (TiO_2 or $SrTiO_3$) with a turnover rate of about unity.

We shall first enumerate several of the techniques of surface science research that could be useful in atmospheric chemistry. Then we shall review some of the reactions of hydrocarbons and water, the two important ingredients which are likely to be participants in most atmospheric chemical reactions, along with the possible trace metals that are likely to play a role in facilitating the chemical processes in the atmosphere.

Selected Techniques for Surface Studies

The surface compositon and oxidation state of surface atoms are usually monitored by electron spectroscopies, for example Auger electron spectroscopy (Carlson, 1975) and X-ray photoelectron spectroscopy (Roberts and McKee, 1978). The vibrational spectra of adsorbate monolayers or small particles can be determined by HREELS (high-resolution electron energy loss spectroscopy) (Dubois and Somorjai, 1980). Low-energy electron diffraction (LEED) is used to determine the atomic surface structure of ordered monolayers (Somorjai and Van Hove, 1979). Thermal desorption spectroscopy (TDS) is utilized to determine the binding energy of adsorbates and their sequential fragmentation via temperature-programmed thermal desorption (Gower, 1975). ^{14}C isotope labeling of adsorbed hydrocarbons and other carbon-containing molecules permits the detection of their whereabouts, residence times, and reaction paths (Davis et al., 1981). Molecular beam surface scattering studies the energy transfer which occurs in gas-surface interactions and the sticking and reaction probabilities of adsorbates on a single collision with the surface (Somorjai, 1981). Table 1 lists these and other techniques along with some of their important applications.

The Prevalence of Carbonaceous Deposits

We live in an organic world, and as a consequence most practical surfaces are covered with a carbonaceous deposit. One reason for this is that thermodynamic organic monolayers have lower surface-free energies than metals, oxides, or water-covered surfaces; thus they tend to segregate to the surface in most circumstances. The hydrogen-to-carbon ratio in these organic deposits depends on the conditions of formation and the temperature; the higher the temperature, the less hydrogen is contained in the deposit. Above 400°C the carbonaceous layer is likely to be graphitized (Somorjai, 1981; Davis and Somorjai, 1982). If the organic deposit is formed at 25°C or below it may contain a large fraction of the intact gas-phase organic molecule.

Figure 3 shows a model of the working platinum catalyst that is used to carry out hydrocarbon conversion reactions of many types. Under catal-

TABLE 1. Some of the More Frequently Applied Techniques for Atomic Scale Studies of Surfaces

Technique	Application	Surface Area Needed
Low-energy electron diffraction (LEED)	Atomic structure Bonding	Small (≤1 cm^2)
Auger electron spectroscopy (AES)	Composition	Small
High-resolution electron energy loss spectroscopy (HREELS)	Vibrational structure	Small
Scanning electron microscopy (SEM)	Topology	Small
X-ray photoelectron spectroscopy (XPS)	Oxidation state Electronic structure	Small
Ultraviolet photoelectron spectroscopy (UPS)	Electronic structure	Small
Infrared and laser Raman spectroscopies	Vibrational structure	Small or large
Extended X-ray absorption fine structure (EXAFS)	Atomic structure	Large
Ion scattering spectroscopy (ISS)	Composition	Small
Secondary ion mass spectroscopy (SIMS)	Composition	Small
Molecular beam surface scattering	Reaction rate parameters Energy transfer	Small
Helium diffraction	Atomic structure	Small
Neutron scattering	Atomic structure Vibrational structure	Large (~10 m^2/gm)
Thermal desorption	Heat of adsorption Bonding	Small

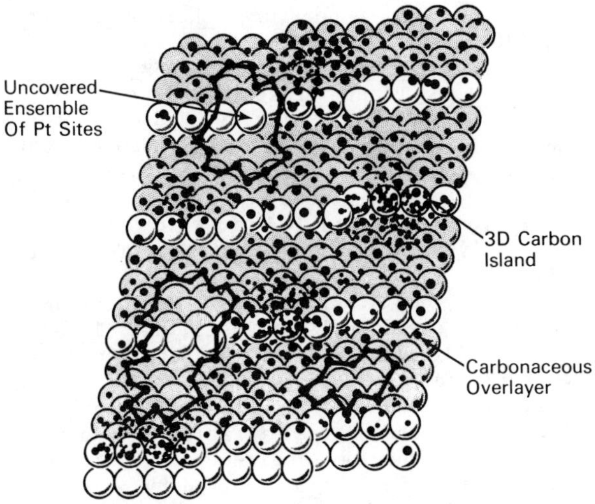

Fig. 3. Model for the working platinum catalyst during hydrocarbon reactions.

ytic reaction conditions, 90 to 95 percent of the transition metal is covered with a carbonaceous deposit. No more than 5 to 10 percent of the metal is exposed during the reaction conditions in excess hydrogen. These active metal sites participate in the surface reaction of adsorbed molecules, which then migrate over to the carbonaceous deposit to undergo further possible reactions and then desorb (Davis and Somorjai, 1982).

Trace Metals

Soot and other carbonaceous particulates may contain trace metals that are present in the atmosphere as a result of combustion processes. These are most commonly vanadium, iron, and nickel, which are produced either from metal-organic molecules present in hydrocarbon fuels or from inorganic sources. The presence of even small concentrations of these metals can lead to increased catalytic activity, as it is frequently found that the turnover rates are much greater on metal catalysts than over oxide or other catalytic compound surfaces.

The Water-Hydrocarbon Interaction

When graphite is heated in water vapor in the presence of alkali hydroxides or carbonates as catalysts in the temperature range of 200° to 300°C, methane and CO_2 evolve with a turnover rate of 10^{-3}/site/sec (Cabrera et al., 1981). At higher temperatures, CO and hydrogen are produced catalytically by the process used for coal gasification in the chemical technology. The interaction of water vapor with hydrocarbon on the surface instead of graphite is very similar and yields gaseous products in the presence of various catalysts of alkali and transition metal compounds.

The Importance of Alumina Silicates as Catalysts

Alumina silicates are in great abundance in the Earth's mantle, and as a result are likely to be major constituents of dust particles produced from fine sand deposits. Synthetic alumina silicates are important catalysts which have very high internal surface areas, called zeolites (Rabo, 1976). This high surface area comes from their unique crystal and pore structure. One of the synthetic zeolites, mordenite, is shown in Figure 4. Hydrocarbon cracking and rearrangement can take place in the internal pores as long as the molecular shape and size are compatible with the cage size. These materials are also used for gas adsorption and separation, as they can adsorb a large amount of gas due to their enormous internal surface area, about 300 to 500 m^2/g.

The Catalyzed Photodissociation of Water

When $SrTiO_3$, strontium titanate, is illuminated by light of greater than band gap energy in the presence of liquid or gaseous water, it catalyzes the dissociation of the molecules to hydrogen and oxygen, H_2 and O_2 (Wagner and Somorjai, 1980; Carr and Somorjai, 1981). The presence of alkali hydroxide and transition metals further accelerates this reaction. The process occurs at 25°C at a turnover rate of about unity under the proper experimental conditions. Titanium oxide, TiO_2, which is an important ingredient in most paints, also catalyzes this reaction. The hydrogen and oxygen products then may undergo further reactions with adsorbates or in the gas phase.

Discussion and Conclusion

The principles of operation of many heterogeneous catalysts and the ingredients needed for

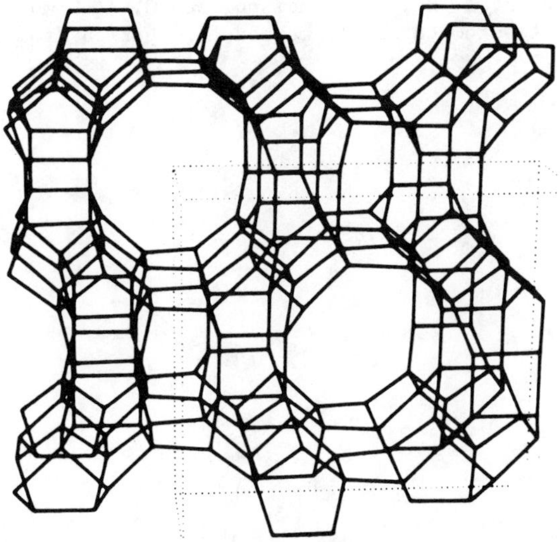

Fig. 4. One of the important ziolites, mordenite, $Na_8Al_8Si_{40}O_{96} \cdot 24H_2O$ viewed along the (001) axis.

catalytic processes to exist have been determined by modern surface science techniques. It is essential that careful studies of atmospheric reactions which may take place on particulates identify whether the process is catalytic or stoichiometric. Most catalytic reactions involving hydrocarbons and water where these compounds are thought to be important constituents of the particulate surfaces occur at appreciable rates only at temperatures higher than 25°C, and therefore are perhaps not of great importance under ambient conditions. However, other catalyzed-surfaces reactions which may be present and have not been explored might be important in atmospheric chemistry. It is apparent that using catalyst surfaces at higher temperatures should aid in the removal of pollutants, undesirable particulates, or gaseous species, as the outstanding success of the automobile catalytic converter demonstrates.

Photochemical reactions can also be catalyzed by surfaces of various types, as the water photodissociation reaction demonstrates. Photochemical surface processes should also be explored to determine the possibility of catalyzed formation or the removal of molecules by this route.

Alumina silicates may be important high-surface-area catalysts and adsorbents of atmospheric reactants of various types. Their chemistry and possible catalytic behavior should be studied to identify their possible participation in atmospheric reactions.

Acknowledgement. This work was supported by the Director, Office of Energy Research, Office of Basic Energy Sciences, Materials Sciences Division of the U. S. Department of Energy under contract W-7405-ENG-48.

References

Cabrera, A. L., H. Heinemann, and G. A. Somorjai, Potassium-catalyzed methane production from graphite at low temperature, Chem. Phys. Lett., 81, 420, 1981.

Carlson, T. A., Photoelectron and Auger Spectroscopy, Plenum Press, New York, 1975.

Carr, R. G., and G. A. Somorjai, Hydrogen production from photolysis of steam adsorbed onto platinized $SrTiO_3$, Nature, 290, 577, 1981.

Davis, S. M., B. E. Gordon, M. Press, and G. A. Somorjai, Radiotracer technique for adsorption and catalysis studies, J. Vac. Sci. Tech., 19, 231, 1981.

Davis, S. M., and G. A. Somorjai, Hydrocarbon conversion over metal catalysts, Encyclopedia of Science, edited by D. A. King, Pergamon Press, New York, in press, 1982.

Dubois, L. H., and G. A. Somorjai, The application of HREELS to the characterization of adsorbed molecules on rhodium single-crystal surfaces, ACS Symposium Series 137, edited by A. T. Bell and M. L. Hair, Amer. Chem. Soc., Washington, D.C., pp. 1-27, 1980.

Gower, R. (Ed.), Interactions on Metal Surfaces, Springer-Verlag, New York, pp. 102-142, 1975.

Rabo, J. A. (Ed.), Zeolite Chemistry and Catalysis, ACS Monograph 171, Amer. Chem. Soc., Washington, D.C., 1976.

Roberts, M. W., and C. S. McKee, Chemistry of the Metal-Gas Interface, Clarendon Press, Oxford, 1978.

Somorjai, G. A., Chemistry in Two Dimensions: Surfaces, Cornell Univ. Press, Ithaca, N.Y., 1981.

Somorjai, G. A., and M. A. Van Hove, Structure and Bonding. Adsorbed Monolayers on Solid Surfaces, Vol. 38, Springer-Verlag, Heidelberg, 1979.

Wagner, F. T., and G. A. Somorjai, Photocatalytic and photoelectro chemical hydrogen production on strontium titanate single crystals, J. Am. Chem. Soc., 102, 5494-5502, 1980.

Watson, P. R., and G. A. Somorjai, Recent advances in heterogeneous catalysis, Editrice di Chimica, Italian Chemical Society, in press, 1982.

Wei, J., Catalysis for motor vehicle emissions, Adv. Catal., 24, 57-129, 1975.

THE REMOVAL OF ATMOSPHERIC GASES BY PARTICULATE MATTER

Julian Heicklen

Department of Chemistry and Center for Air Environment Studies,
Pennsylvania State University, University Park, Pennsylvania 16802

Abstract. The conditions for removal of an atmospheric pollutant by water droplets or on the surface of an aerosol have been investigated. For pollutants which do not react in water droplets or on aerosol surfaces, the removal lifetimes of the pollutant equal the atmospheric turnover lifetimes ($\sim 10^6$ sec) of the droplets or aerosol if (1) the ratio of the Henry's law constant to the fraction of air occupied by water droplets is less than 1 for water droplets, or (2) the ratio of the vapor-phase concentration of the pollutant to the number of aerosol sites occupied at equilibrium is less than 1 for aerosols.

For pollutants that do undergo reactions in water droplets or on aerosol surfaces, the removal rate can be enhanced over the nonreactive removal rate only if the effective removal rate coefficient for the pollutant in the water droplet or on the aerosol surface exceeds the reciprocal turnover lifetime of the particulate matter (i.e., for removal rate coefficients greater than $\sim 10^{-6}$ sec^{-1}). For the chemical reaction to be rate-controlling it is also necessary that chemical reaction be sufficiently slow so that diffusion of the gaseous pollutant to the droplet or aerosol is not rate controlling. Chemical reaction is rate controlling if the chemical reaction rate coefficient lies between 10^{-2} and 10^{-6} sec^{-1}.

Introduction

Atmospheric gases may be removed by water droplets or suspended solid particulate matter in the air. Though droplets of liquids other than water may be present, generally they will be of sufficiently small concentration so that they will not be important sinks for atmospheric gases. We will confine our discussion to water droplets or aerosols (suspended solid particulate matter).

The atmospheric gases may dissolve in water or be adsorbed on the aerosol surface without undergoing further reaction. In this case the partitioning of the species will depend on the appropriate equilibrium constant, and the removal rate will be related to the turnover rate of the atmospheric particulate matter. The lifetime for removal in this manner will be derived, and it will be shown that it equals the atmospheric turnover lifetime for the particulate matter ($\sim 10^6$ sec) only for species with vapor pressures less than $\sim 10^{11}$ molecules cm^{-3} (~ 4 ppb). For higher vapor pressure species, the removal lifetimes are longer.

The chemical species may react either in the water droplet or on the aerosol surface. If so, removal can be enhanced, and species with vapor pressures greater than $\sim 10^{11}$ molecules cm^{-3} can be removed efficiently. For this enhancement to occur requires reaction rate coefficients which exceed the reciprocal atmospheric turnover lifetime for the particulate matter.

Removal by Water Droplets

In this section we examine removal of a gaseous species in the atmosphere by water droplets. First we shall derive the general rate law, then apply it for various reactivities in water droplets, and finally estimate the Henry's law constant which is needed to evaluate the rate law.

General Rate Law

For the case where removal of an atmospheric pollutant can occur by reaction in a droplet, let us consider the reactions involved:

$$A_{vapor} + \text{droplets} \rightleftarrows A_{solu} \qquad (a)$$

$$A_{solu} \rightarrow \text{products} \qquad (b)$$

where A represents the pollutant. The steady-state value of total concentration of A in the droplets (including any fraction dissociated), C_{solu}, is

$$C_{solu} = k_1 C_{vap}/(k_{-1} + k_2)v \qquad (1)$$

where k_1 and k_{-1} are the forward and reverse rate coefficients for reaction (a) and k_2 is the rate coefficient for reaction (b), v is the average droplet volume, and c_{vap} is the vapor-phase concentration of the pollutant.

The rate coefficients k_1 and k_{-1} are related through the Henry's law constant κ and the average droplet volume v

$$k_{-1} = k_1\kappa/v = k_1 n\kappa/f \qquad (2)$$

where n is the particle number density and f is the fractional volume of air occupied by the mist, i.e.,

$$f \equiv nv \qquad (3)$$

The Henry's law constant κ is just the ratio C_{vap}/C_{solu} at equilibrium, i.e., when reaction (b) is negligible.

The pollutant is removed from the system by reaction (b). It can also be removed if the droplets settle out (rain out) at the surface of the Earth and the pollutant enters the soil or the oceans. This rate of droplet removal by settling out of the atmosphere will be considered to occur with a lifetime $\tau_0\{solu\}$. (For simplicity we assume an average lifetime $\tau_0\{solu\}$ or $\tau_0\{surf\}$. This is an oversimplification because removal by settling occurs only at the Earth's surface. The exact derivation is given in the Appendix). Then the total rate of pollutant removal R_{solu} will be given

$$R_{solu} = (k_2 + \tau_0\{solu\}^{-1})C_{solu}f \qquad (4)$$

The lifetime of the pollutant for removal by the water droplets, $\tau\{solu\}$,is

$$\tau\{solu\} \equiv (C_{vap} + fC_{solu})/R_{solu} \qquad (5)$$

which becomes, when equation (2) is substituted into equation (1), equation (1) is substituted into equations (4) and (5), and equation (4) is substituted into equation (5),

$$\tau\{solu\} = (1 + \kappa/f + k_2/k_1 n)\tau_0\{solu\}/(1 + k_2\tau_0\{solu\}) \qquad (6)$$

Equation (6) is the general equation which gives the lifetime of removal of the pollutant by water droplets. If the parameters are known, the computation can be made directly. The quantities $\tau_0\{solu\}$, f, and n are parameters of the atmosphere. They can vary widely, and we will discuss their effect later. The rate coefficient k_1 depends mainly on both v = f/n and on the molecular weight of the pollutant. However, it can be computed easily if v is known. The two parameters which depend on the pollutant itself and not on the atmospheric conditions are k_2 and κ. In the rest of this section we will discuss the effect of k_2 on the rate law and the evaluation of κ.

Rate Laws for Various Values of k_2

There are three general classifications of reactions that will give different limiting forms of $\tau\{solu\}$ from equation (6). In the first of these k_2 is sufficiently large so that $k_2/k_1n > 1 + \kappa/f$ and $k_2\tau_0\{solu\} > 1$. Then equation (6) reduces to

$$\tau\{solu\} \simeq (k_1 n)^{-1} \qquad (7)$$

For practical purposes this always will be the case when $k_2 > 1\ sec^{-1}$, though it could occur for lower rate coefficients, even as low as $10^{-3}\ sec^{-1}$. Reactions with rate coefficients this large include acid-base and hydrolysis reactions. Thus we expect equation (7) to be valid for any pollutant that is readily hydrolyzed or any base that reacts with acid rain.

The other extreme limiting case is when k_2 is too small to be of any consequence. Typically this means $k_2 < 10^{-6}\ sec^{-1}$. Under these conditions $\tau\{solu\}$ is either so small as to be of no concern or equation (6) reduces to

$$\tau\{solu\} \simeq (1 + \kappa/f)\tau_0\{solu\} \qquad (8)$$

The intermediate case occurs when $k_2/k_1n < 1$ and $k_2\tau_0\{solu\} > 1$. The reverse situation is not important because under all atmospheric conditions $k_1n > \tau_0\{solu\}^{-1}$. Equation (6) reduces to

$$\tau\{solu\} \simeq (1 + \kappa/f)/k_2 \qquad (9)$$

This intermediate situation occurs typically for $10^{-2} < k_2 < 10^{-6}\ sec^{-1}$. Reactions with these magnitudes of rate coefficients are either solution oxidations with dissolved oxygen, often catalyzed by some metal ion, or slow hydrolysis reactions.

It can be seen that to evaluate equations (8) and (9) it will be necessary to know κ as well as the atmospheric parameters f and $\tau_0\{solu\}$.

Evaluation of κ

For a pollutant to have a significant rate of removal by water droplets when equation (8) or (9) is applicable, it will need to have a small Henry's law constant (i.e., $\kappa/f < 1$). For compounds with small Henry's law constants, the values are not known. Thus it is necessary for an estimate to be made.

If the pollutant is completely miscible with H_2O, the approximate Henry's law constant can be obtained by assuming an ideal solution so that Raoult's law will be obeyed:

$$C_{vap} = XC^0_{vap} \qquad (10)$$

where X is the mole fraction of the pollutant in solution and C^0_{vap} is the equilibrium vapor pressure of pure liquid pollutant. The more usual case will be when the pollutant is not completely miscible with water. In this case we estimate the Henry's law constant by assuming ideality up to the solubility limit, i.e.,

$$C_{vap} = (C^0_{vap}/C^0_{solu})C_{solu} \qquad (11)$$

where C^0_{solu} is the total concentration of pollutant (including any dissociated species) in a saturated solution. Thus

$$\kappa = C^0_{vap}/C^0_{solu} \quad (12)$$

Removal by Aerosol

In this section we repeat the analysis of the last section, except that instead of the pollutant being dissolved in the mist it is adsorbed on the surface. Thus there are some differences in the derivation, though the general conclusions will be the same.

General Rate Law

For the case where removal of the pollutant can occur by reaction on the aerosol surface, the reactions are

$$A_{vapor} + \text{surface} \rightleftarrows A_{surface} \quad (c)$$

$$A_{surface} \rightarrow \text{products} \quad (d)$$

The steady-state expression on $A_{surface}$ gives

$$k_{15}C_{vap}nN(1-\theta) = (k_{-15} + k_{16})nN\theta \quad (13)$$

where N is the total number of adsorption sites (occupied plus unoccupied) per particle, θ is the fraction of sites occupied, k_{15} and k_{-15} are the forward and reverse rate coefficients for reaction (c), and k_{16} is the rate coefficient for reaction (d). Solving equation (13) for θ gives

$$\theta = k_{15}C_{vap}/(k_{-15} + k_{16} + k_{15}C_{vap}) \quad (14)$$

The rate coefficients k_{15} and k_{-15} are related through the equilibrium expression

$$k_{15}/k_{-15} = \theta_{eq}/(1-\theta_{eq})C_{vap} \quad (15)$$

where θ_{eq} is the fraction of sites occupied at equilibrium. This fraction θ_{eq} is a function of C_{vap}, but for the time being it is convenient to leave equation (14) as is. The total rate of pollutant removal by the aerosol, R_{surf}, is

$$R\{surf\} = (k_{16} + \tau_0\{surf\}^{-1})nN\theta \quad (16)$$

where $\tau_0\{surf\}$ is the lifetime for removal of the aerosol by settling at the Earth's surface. (For simplicity we assume an average lifetime $\tau_0\{X\}$. This is an oversimplification because removal by settling occurs only at the Earth's surface. The exact derivation is given in the Appendix.) The lifetime for pollutant removal by the aerosol, $\tau\{surf\}$, is

$$\tau\{surf\} = (C_{vap} + nN\theta)/R_{surf} \quad (17)$$

When equations (14) through (17) are combined the final expression for $\tau\{surf\}$ becomes

$$\tau\{surf\} = (1 + k_{16}/k_{15}nN + C_{vap}/nN\theta_{eq})/ (k_{16} + \tau_0\{surf\}^{-1}) \quad (18)$$

Equation (18) is the general equation which gives the lifetime of removal by the aerosol. It is analogous to equation (6) for water droplets except that k_{16} replaces k_2, and $C_{vap}/nN\theta_{eq}$ replaces κ/f. The quantity $k_{15}nN$ is equivalent to k_1n. Thus the two parameters which depend on the pollutant itself are k_{16} and θ_{eq}. In the rest of this section we will discuss the effect of k_{16} on the rate law and the evaluation of θ_{eq}.

Rate Laws for Various Values of k_{16}

The three general classifications of reactions are the same as for reactions in water droplets. The first one occurs when k_{16} is sufficiently large (>1 sec^{-1}) that $k_{16}/k_{15}nN > 1 + C_{vap}/nN\theta_{eq}$ and $k_{16} > \tau_0\{surf\}^{-1}$, in which case

$$\tau\{surf\} \simeq k_{15}nN)^{-1} \quad (19)$$

when k_{16} is sufficiently small to be unimportant ($k_{16} < 10^{-6}$ sec^{-1}), so that

$$\tau\{surf\} \simeq (1 + C_{vap}/nN\theta_{eq})\tau_0\{surf\} \quad (20)$$

or when k_{16} is in the intermediate regime (10^{-2} to 10^{-6} sec^{-1}), so that

$$\tau\{surf\} = (1 + C_{vap}/nN\theta_{eq})/k_{16} \quad (21)$$

The first case will only occur if the pollutant reacts directly with the aerosol. Either the surface will become inactivated by product formation or the aerosol will be consumed. In either case the reaction will soon cease. For all practical purposes this limiting case can be ignored. The other two cases correspond to no reaction or to moderate reaction rates which would be expected of a surface decomposition or oxidation. In either of these cases it will be necessary to estimate θ_{eq} as well as the atmospheric parameters n, N, and $\tau_0\{surf\}$ in order to evaluate $\tau\{surf\}$.

Evaluation of θ_{eq}

The quantity θ_{eq} is related to the adsorption isotherm of the pollutant on the aerosol. In general this is not known, so it is necessary to estimate it. Fortunately for many adsorption isotherms, $\theta_{eq} = 0.5$ when $C_{vap}/C^0_{vap} \simeq 0.1$. For example, this is true with CCl_4 adsorbed on Teflon T-6 at 20°C (Wade, 1974), with methyl acetate adsorbed on carbon black at 10° to 60°C (Mueez, 1974), and with n-butylamine, epichlohydrin, or methanol adsorbed on Cab-O-Sil at 23°C (Clark-Monks et al., 1970). Thus for practical purposes we can solve equation (15) to give $k_{15}/k_{-15} \simeq 10/C^0_{vap}$, and thus

$$\theta_{eq} \simeq 10(C_{vap}/C^0_{vap})(1 + 10C_{vap}/C^0_{vap})^{-1} \quad (22)$$

Evaluation of Atmospheric Parameters

In order to use the general equations (6) and (18) it is necessary to evaluate the atmospheric parameters k_1, v, n, and $\tau_0\{solu\}$ for equation (6) and k_{15}, N, and $\tau_0\{surf\}$ for equation (18). These parameters vary with atmospheric conditions. Thus during precipitation or dust storms the particles are large and numerous, leading to large values of k_1, f, n, or k_{15} and N, and low values of $\tau_0\{solu\}$ or $\tau_0\{surf\}$. Even moderately soluble or adsorbing gases will be removed. However, these are not typical conditions, and our interest here will be to evaluate typical values of the parameters to estimate average global lifetimes $\tau\{solu\}$ or $\tau\{surf\}$.

The amount of particulate matter in background air is about 10 μg/m^3, corresponding to $\simeq 10^3$ particles/cm^3 near the surface of the Earth. There are about 10 times as many particles in urban air and about 100 times as many for polluted air (Heicklen, 1976a). The average mass particle radius is $\simeq 0.13$ μm, so that the particle volume is $\simeq 4 \times 10^{-15}$ cm^3 and the total surface area per particle using this radius would be $\simeq 2.1 \times 10^7$ Å^2/cm^3 for background air in the low troposphere. Actually the surface area is considerably larger than this because of the greater preponderance of small particles. However this is offset by the fact that all the particles will not be efficient adsorbents. If a molecule occupies 20 Å, then the number of sites N per particle corresponds to $\simeq 1 \times 10^6$ sites/particle. Furthermore for particles of this size in the atmosphere $k_1 \simeq 10^{-6}$ cm^3 sec^{-1} (Heicklen, 1976b).

If we assume the above conditions, i.e., an average particle radius of 0.13 μm and a density near the surface of the Earth $n_0 \simeq 3 \times 10^3$ particles/cm^3, then the atmospheric parameters can be estimated with the realization that n will drop off exponentially with altitude. The parameters are: $N \simeq 1 \times 10^6$ sites/particle; $k_1 \simeq 1 \times 10^{-6}$ cm^3 sec^{-1}; $k_{15} = k_1/N \simeq 1 \times 10^{-12}$ cm^3 sec^{-1}; $n_0 \simeq 3 \times 10^3$ particles/cm^3; and $v \simeq 4 \times 10^{-15}$ cm^3. Furthermore,

$$n = n_0 \exp\{-Z/h\} \quad (23)$$

where Z is the altitude and h is the scale height. At 7 km the concentration of Aitken nuclei (radii ≤0.1μm) is $\simeq 300$ cm^{-3} (Heicklen, 1976a). Thus $h \simeq 3 \times 10^5$ cm. Also

$$f \equiv nv \simeq 1.2 \times 10^{-11} \exp\{-Z/h\} \quad (24)$$

The turnover lifetimes of the particulate matter can be computed from the ratio of their total column concentration $\int_0^\infty n dZ$ and the flux at the surface of the Earth $-D(dn/dZ)_0$, where D is the eddy diffusion coefficient of the atmosphere which varies between 10^4 and 2.5×10^5 cm^2/sec. Since

$$-D(dn/dZ)_0 = Dn_0/h \quad (25)$$

and

$$\int_0^\infty n dZ = n_0 h \quad (26)$$

then the turnover lifetime $\tau_0\{X\}$ is

$$\tau_0\{X\} = h^2/D \quad (27)$$

With $h \simeq 3 \times 10^5$ cm and D taken to be 10^5 cm^2 sec^{-1} on the average, $\tau_0\{X\} \simeq 1 \times 10^6$ sec, and this value will be accurate to much better than a factor of 10 for a global average.

Conclusion

The general expression for the lifetime of removal of a pollutant is given by equation (6) for removal by water droplets and by equation (18) for removal by aerosols. For these processes to be significant in the atmosphere, these lifetimes should not exceed $\simeq 10^6$ sec (i.e. $\simeq 11.5$ days). Since $\tau_0\{solu\}$ or $\tau_0\{surf\}$ is approximately equal to 10^6 sec, then for this condition to be met it will be necessary either that the rate coefficients k_2 or k_{16} exceed 10^{-6} sec^{-1} or that

$$1 > \kappa/f \simeq (C^0_{vap}/C^0_{solu})/1.2 \times 10^{-11} \exp\{-Z/h\} \quad \text{mist} \quad (28)$$

$$1 > C_{vap}/nN\theta_{eq} \simeq (C^0_{vap}/3 \times 10^{10} \exp\{-Z/h\})(1 + 10C_{vap}/C^0_{vap}) \quad \text{aerosol} \quad (29)$$

Now C^0_{solu} cannot exceed 10^{22} molec/cm^3, since the molecular volume is greater than 10^{-22} cm^3. Also $\exp\{-Z/h\} \leq 1$, and $(1 + 10C_{vap}/C^0_{vap}) \geq 1$. Thus we find that unless the reactive rate coefficients k_2 and k_{16} exceed 10^{-6} sec^{-1}, heterogeneous removal by either water mist or aerosol can only be important if C^0_{vap} is less than about 10^{11} molec/cm^3 4 ppb for average atmospheric conditions.

The conditions for chemical reaction to be controlling are k_2 or k_{16} greater than 10^{-6} sec^{-1} but small enough so that k_1 or k_{15} is not rate controlling. These conditions are met if

$$k_2/k_1 n > 1 + \kappa/f \quad \text{mist}$$

$$k_{16}/k_{15} nN > 1 + C_{vap}/nN\theta_{eq} \quad \text{aerosol}$$

Since $k_1 n \simeq k_{15} nN \simeq 10^{-2}\ sec^{-1}$, for k_2 or $k_{16} > 10^{-2}\ sec^{-1}$ chemical reaction may not be rate controlling.

Appendix

In the text the expressions for the removal lifetimes $\tau\{X\}$ were derived assuming some intrinsic lifetime of removal by settling for the particle itself $\tau_0\{X\}$, which applied everywhere in space. Actually this is a simplification to facilitate the derivation. In reality, the removal by particle settling occurs only at the surface of the Earth with a flux $-D_0(dn/dZ)_0$ where the subscript 0 refers to zero altitude. With this correct form, we now derive the accurate expressions for $\tau\{X\}$.

Removal by Water Droplets

The total rate of removal of the pollutant in a column of air above a unit area of the Earth's surface, R^T_{solu}, is

$$R^T_{solu} = \int_0^\infty k_2 C_{solu} f dZ - D_0(dn/dZ)_0 (C_{solu})_0 v \tag{A1}$$

The lifetime $\tau\{solu\}$ is given by the ratio of the total amount of pollutant in the column (both in the gas phase and in solution) to the total rate of removal

$$\tau\{solu\} = \frac{\int_0^\infty C_{solu} f dZ + \int_0^\infty C_{vap} dZ}{\int_0^\infty k_2 C_{solu} f dZ - D_0(dn/dZ)_0 (C_{solu})_0 v} \tag{A2}$$

In order to perform the integration, it is necessary to know the functional form of the parameters. These can be taken to be

$$C_{solu} = (C_{solu})_0\ g\{Z\} \tag{A3}$$

$$f = nv = n_0 v e^{-Z/h} \tag{A4}$$

$$C_{vap} = (C_{vap})_0 e^{-Z/\zeta} \tag{A5}$$

$$k_2 = (k_2)_0 \phi\{Z\} \tag{A6}$$

where $g\{Z\}$ and $\phi\{Z\}$ are for the moment some unspecified functions of altitude with the end condition that they are unity at the surface of the Earth. The parameter ζ is the scale height for the vapor-phase species, which in the absence of reactive removal is 7×10^5 cm (Heicklen, 1976a).

By performing the integrations and realizing that C_{solu} is given by equation (1), we can rearrange equation (A2) to read

$$\tau\{solu\} = \frac{I_1 + (h/\zeta)[(k_2)_0/n_0(k_1)_0 + \kappa_0/f_0]}{(k_2)_0 I_2 + D_0/h^2} \tag{A7}$$

where the subscript 0 still refers to the surface of the Earth and I_1 and I_2 are defined as

$$I_1 \equiv \int_0^\infty g\{Z\} e^{-Z/h} dZ/h \tag{A8}$$

$$I_2 \equiv \int_0^\infty \phi\{Z\} g\{Z\} e^{-Z/h} dZ/h \tag{A9}$$

Except for the integrals I_1 and I_2 and the ratio of scale heights h/ζ, equation (A7) is identical to equation (6) with the parameters evaluated at the Earth's surface. All that is needed is to evaluate $g\{Z\}$ and $\phi\{Z\}$ and integrate equations (A8) and (A9). Combining equations (1) and (A3) gives

$$g\{Z\} = k_1 C_{vap}/(k_{-1} + k_2) v (C_{solu})_0 \tag{A10}$$

Although C_{vap} decreases with altitude, as given by equation (A5), k_1 and v are not very dependent on altitude. Because of the decrease in temperature with altitude, both k_{-1} and k_2 will drop with increasing altitude, the magnitude of the drop depending on the activation energies for the reaction and the temperature profile of the atmosphere. In order to evaluate equation (A10) accurately, the exact dependencies must be known. However, the trend in C_{vap} is in the same direction as the trend in $k_{-1} + k_2$, and the two will correspond if $k_{-1} + k_2$ has an effective activation energy of $\simeq$6 kcal/mol. This is a reasonable typical value and we set $g\{Z\} \simeq 1$ as a zeroth order approximation. The value of $\phi\{Z\}$ is also not known, but it is always less than 1; thus $I_2 < I_1$.

With the above approximation, equation (A7) reduces to

$$\tau\{solu\} \gtrsim \frac{1 + (h/\zeta)[(k_2)_0/n_0(k_1)_0 + \kappa_0/f_0]}{(k_2)_0 + D_0/h^2} \tag{A11}$$

Removal by Aerosol

For an aerosol, the quantity fC_{solu} is replaced by $nN\theta$, and $\tau\{surf\}$ becomes

$$\tau\{surf\} = \frac{\int_0^\infty nN\theta dZ + \int_0^\infty C_{vap} dZ}{\int_0^\infty k_{16} nN\theta dZ - D_0(dn/dZ)_0 N\theta} \tag{A12}$$

with C_{vap} still given by equation (A5). The functional forms for the other parameters are

$$N\theta = N_0\theta_0\rho\{Z\} \tag{A13}$$

$$k_{16} = (k_{16})_0 \sigma\{Z\} \tag{A14}$$

where $\rho\{Z\}$ and $\sigma\{Z\}$ give the altitude dependence of $N\theta$ and k_{16}, respectively.

By performing the integrations and realizing that θ is given by equation (14) and k_{-15} by equation (15), equation (A12) can be rearranged to give

$$\tau\{surf\} = \frac{I_3 + (h/\zeta)[(k_{16})_0/(k_{15})_0 n_0 N_0 + (C_{vap})_0/(\theta_{eq})_0 n_0 N_0]}{(k_{16})_0 I_4 + D_0/h} \quad \text{(A15)}$$

where I_3 and I_4 are defined as

$$I_3 \equiv \int_0^\infty \rho\{Z\} e^{-Z/h} dZ/h \quad \text{(A16)}$$

$$I_4 \equiv \int_0^\infty \sigma\{Z\}\rho\{Z\} e^{-Z/h} dZ/h \quad \text{(A17)}$$

Except for the integrals I_3 and I_4 and the ratio of scale heights h/ζ, equation (A15) is identical to equation (18) with the parameters evaluated at the Earth's surface. All that is needed is to evaluate $\rho\{Z\}$ and $\sigma\{Z\}$ and integrate equations (A16) and (A17). Combining equations (14) and (A13) gives

$$\rho\{Z\} = Nk_{15}C_{vap}/N_0\theta_0(k_{-15} + k_{16} + k_{15}C_{vap}) \quad \text{(A18)}$$

This function will behave similarly to $g\{Z\}$, and as a zeroth order approximation can be set approximately equal to 1. Also, $\sigma\{Z\}$ is analogous to $\phi\{Z\}$, so that $I_4 < I_3$.

With the above approximation, equation (A15) reduces to

$$\tau\{surf\} \gtrsim \frac{1 + (h/\zeta)[(k_{16})_0/(k_{15})_0 n_0 N_0 + (C_{vap})_0/(\theta_{eq})_0 n_0 N_0]}{(k_{16})_0 + D_0/h^2} \quad \text{(A19)}$$

Acknowledgement. This paper is adapted from one published in Atmospheric Environment, volume 15, 781 (1981) and appears with permission of Pergamon Press. This work was supported by the Environmental Protection Agency through Subcontract No. T-6414 (7197)-028 with Battelle Columbus Laboratories.

References

Clark-Monks, C., B. Ellis, and K. Rowan, The estimation of irreversible adsorption from sequential adsorption isotherms, J. Coll. Inter. Sci., 32, 628-632, 1970.

Heicklen, J., Atmospheric Chemistry, Academic Press, N. Y., 1976a.

Heicklen, J., Colloid Formation and Growth: A Chemical Kinetics Approach, Academic Press, N.Y., 1976b.

Mueez, M. A., The adsorption isotherms and the heat of adsorption of methyl acetate on carbon black, Pakistan J. Sci. Ind. Res., 17, 1-4, 1974.

Wade, W. H., The adsorption kinetics for CCl_4 on Teflon, J. Coll. Inter. Sci., 47, 676-681, 1974.

REACTIONS OF GASES ON PROTOTYPE AEROSOL PARTICLE SURFACES

Alan C. Baldwin

Department of Chemical Kinetics, SRI International, Menlo Park, California 94025

Abstract. The interactions between many common atmospheric gas-phase species and two prototype aerosol particle surfaces, carbon and sulfuric acid, have been studied. The results show that heterogeneous reaction may be a significant sink for some atmospheric species.

Introduction

There is a growing awareness, as evidenced by this conference, that heterogeneous reactions can be important in the chemistry of the atmosphere, and that heterogeneous processes must be considered in any realistic atmospheric model. At present, the complex physical and chemical interactions that occur in the atmosphere can only be understood in terms of models containing parametric representations of the various processes. However, incorporating heterogeneous processes into atmospheric models is a formidable task. It is well known that the rates of heterogeneous reactions depend sensitively on the chemical and physical nature of the surface, and that the surface characteristics may be drastically modified by the reaction or by the presence of other chemical species. Thus, the rate expression for a heterogeneous reaction must take into account not only the concentration of the heterogeneous surface, but also its physical and chemical structure, and allowance must be made for the rate to change with time as the surface characteristics change.

Laboratory studies of heterogeneous atmospheric reactions are difficult to perform and characterize, and extrapolating laboratory data to atmospheric conditions is hampered by the lack of data on the physical and chemical properties of atmospheric particulate matter. In this paper, we describe some basic studies of the interaction of common atmospheric gas-phase species with two prototype aerosol particle surfaces, carbon and sulfuric acid. The experiments were conducted at low pressure under controlled collisional conditions, and yield results that we believe may be conveniently extrapolated to atmospheric conditions, and are suitable for incorporation into models of atmospheric chemical transformations.

Experimental Technique

The experimental technique has been described in detail (Golden et al., 1973; Baldwin and Golden, 1979). A controlled flow of reactant gases enters a Knudsen cell reactor, shown diagramatically in Figure 1. The reactor has a volume of 360 cm^3 and the residence time of the gases in the reactor can be varied by a factor of ~9 by varying the diameter of the escape aperture between 0.1 cm and 0.3 cm. The unreacted material and the reaction products leave the reactor, are formed into a molecular beam, and are detected by a differentially pumped, phase-sensitive mass spectrometer. The pressure in the reactor is $<10^{-3}$ torr, and molecular flow conditions prevail.

The reactor contains two chambers connected by a large, sealable aperture. The lower chamber contains the reactive surface, and the upper chamber contains the reactant inlets and escape aperture. When the connecting aperture is opened, the extent of reaction or adsorption on the surface may be easily measured from the change in the mass spectral signal of the reactant. This measurement can then be simply converted to an absolute rate constant or the probability of reaction per collision with the reactive surface.

At the low pressures used, there are no complications due to bulk mass transfer, or homogeneous secondary chemical reactions, and the results are a true measure of the elementary collisional process. A disadvantage of this system is that we cannot conveniently test the effect of the high concentration atmospheric constituents such as water or oxygen. This will be addressed in a separate series of experiments.

Results for Prototype Reaction Systems

Atmospheric aerosol particles show great variability in their chemical and physical structure (Hidy, 1972; Benson, 1975). We have chosen two prototype surfaces for our initial experiments. Graphitic carbon (Novakov et al., 1974) is a common component of atmospheric aerosols,

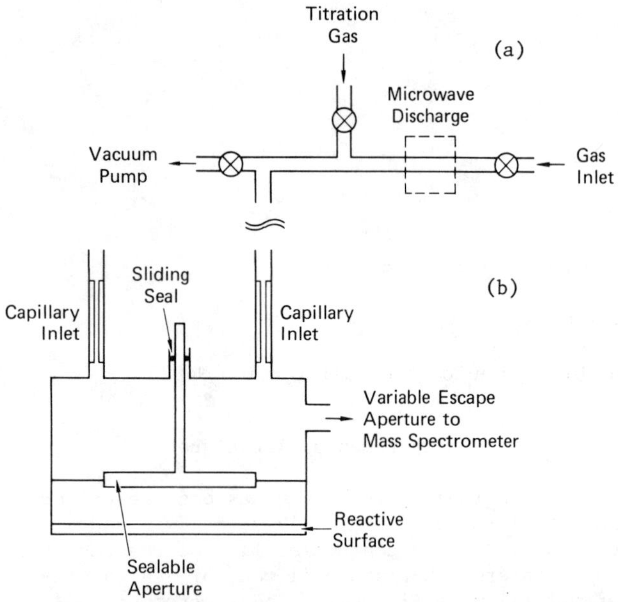

Fig. 1. (a) Inlet system used to generate atomic and free radical species. (b) The two-chamber Knudsen cell reactor.

and has been extensively studied, particularly with regard to SO_2 oxidation (Novakov et al., 1974; Judeikis et al., 1978; Liberti et al., 1978; Tartarelli et al., 1978; Cofer et al., 1980; Britton and Clarke, 1980). Sulfuric acid (presumably formed from the oxidation of SO_2) is also found as a common coating on aerosols (Hidy, 1972), and has been previously studied (Harker and Ho, 1979). These two surfaces were therefore chosen for our experiments. One problem in characterizing heterogeneous reactions is the definition of the reactive surface area, a problem which is obviated with a liquid surface, such as sulfuric acid, which was used for an extensive series of measurements.

Reactions With a Sulfuric Acid Surface

We have looked at the interaction of many stable, atomic, and free radical atmospheric species with a bulk sulfuric acid surface. A typical experimental result is shown in Figure 2. Initially we observe a steady-state signal due to the reactant. When the connecting aperture is opened, the signal drops to a new steady state; when the aperture is closed, the signal returns to its previous level. The difference between the two signals represents the amount that has undergone reaction. A number of results are shown in Table 1 as the probability of reaction γ per collision between a gas-phase molecule and the sulfuric acid surface. The measured values of γ were independent of the surface area of the sulfuric acid, the bulk volume of the acid, the concentration of the gaseous species, and the number of collisions, indicating that there is no effect due to the reactor walls. The surface was sulfuric acid that had been evacuated at 10^{-7} torr, and therefore contained less than 5 percent water.

The species in the lower half of the table showed no detectable reactivity in our system. For the species in the upper half of the table, gas-phase products were only detected for HNO_3 and N_2O_5. In both these cases, the product formed had a mass peak at m/e = 46, the same peak used to monitor the reactant species (HNO_3 and N_2O_5 have no molecular ion peak). The evolution of this product, presumably NO_2, gave experimental results as shown in Figure 3.

We also attempted to discover whether the sulfuric acid surface would act as a catalyst for the following bimolecular reactions: $O_3 + NO_2$, $O_3 + SO_2$, O_3 + alkenes, and $O_3 + NO$. No reaction was observed with $\sim 10^{-3}$ torr of reactants for the first three reactions, and no change in the homogeneous rate was observed for the last reaction.

A simple assessment of the significance of these results can be made by taking typical aerosol concentration data and computing the collision frequency between a gas-phase species and an aerosol particle. This gives values (Baldwin and Golden, 1979) of $\sim 10^{-5}$ s^{-1} for the stratosphere and ~ 1 s^{-1} for the polluted troposphere. Taking a typical γ of about 10^{-5} for the reactive species in Table 1, we calculate a first-order loss rate constant of $\sim 10^{-10}$ s^{-1} in the stratosphere, too slow to be of importance. However, the tropospheric first-order loss rate constant is $\sim 10^{-5}$ s^{-1}. This is slower than the homogeneous processes for reactive species like OH, but for stable species such as HNO_3, HO_2NO_2, H_2O_2, and N_2O_5, heterogeneous reaction is comparable to homogeneous loss mechanisms such as photolysis (maximum first-order rate constant for these species $\sim 10^{-5}$ s^{-1}). Therefore, sulfuric acid-coated aerosols may be significant sinks for the above species.

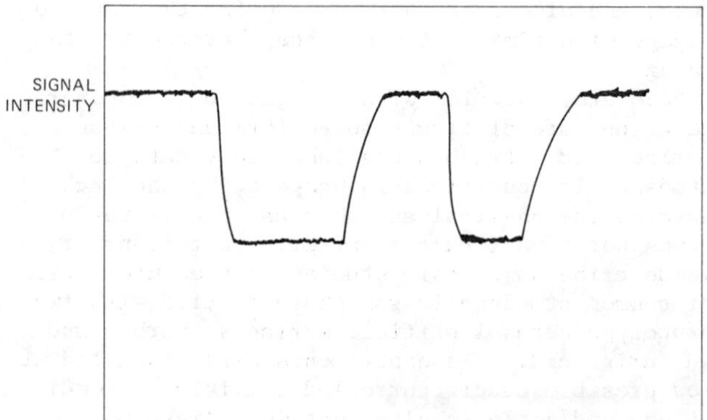

Fig. 2. A typical experimental result.

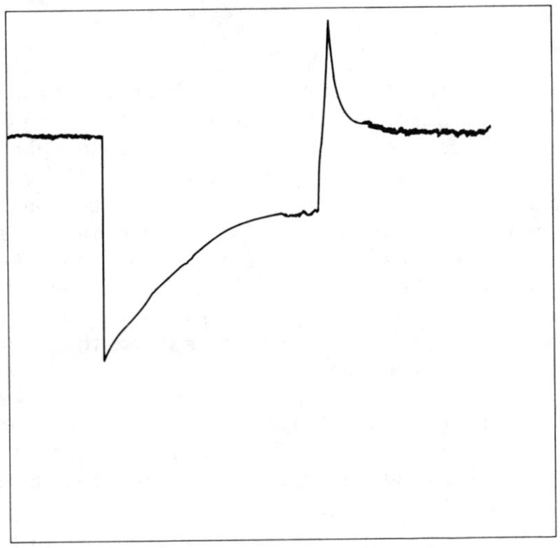

Fig. 3. Experimental results for the case when a gas-phase product is formed having a mass peak at the monitoring peak of the reactant.

Reactions With Carbonaceous Surfaces

The oxidation of SO_2 to sulfate in the atmosphere is of considerable interest. Many studies have been made of possible heterogeneous components in this process, especially the role of carbonaceous particles (Novakov et al., 1974; Judeikis et al., 1978; Liberti et al., 1978; Tartarelli et al., 1978; Cofer et al., 1980; Britton and Clarke, 1980). The results of these studies reflect the complexity of heterogeneous reaction studies; an array of experimental variables has been seen to affect the rate and mechanism of the carbon-catalyzed oxidation of SO_2,

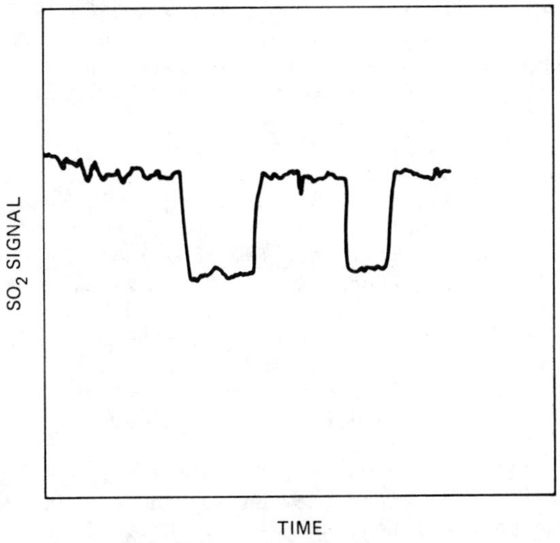

Fig. 4. Experimental result for the adsorption of SO_2 on a carbon surface.

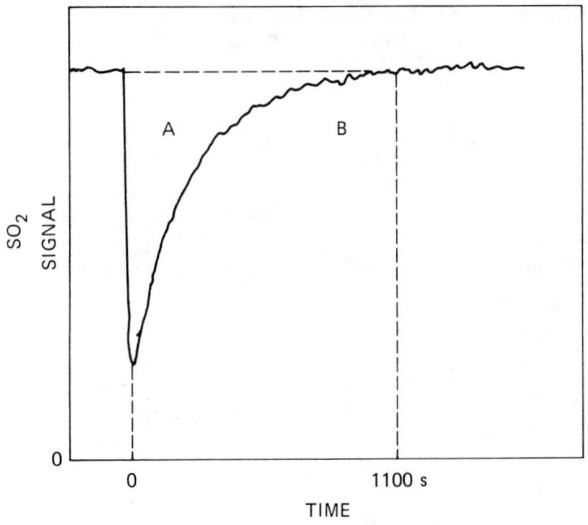

Fig. 5. Experimental result showing saturation of the carbon surface with SO_2.

including the surface composition, H_2O concentration, NO_2 concentration, and SO_2 concentration.

We attempted to supply some baseline data on the interaction of SO_2 with carbon surfaces to further our understanding of this important reaction. As a representative prototype carbonaceous surface, we selected a commercial ground charcoal (Norit A) which had a measured BET surface area of 37 m^2-g^{-1}. In all experiments, except those to measure the dependence of the rate on the surface concentration, 50-mg samples of the carbon were used. The samples are pumped at 10^{-7} torr before use. In the experiments of Judeikis et al. (1978), a charcoal with a BET surface area of 40 m^2-g^{-1} was used and behaved roughly as the median of a number of atmospheric particulates. The experiments of Liberti et al. (1978), and Tartarelli et al. (1978), also used atmospheric particulate matter, principally of BET surface area 10 to 100 m^2-g^{-1}. Our surface can therefore be considered representative. More detailed studies of the dependence of the interaction on the detailed surface structure are unjustified in view of the lack of data characterizing atmospheric particulate matter.

Figure 4 shows the result obtained with SO_2 exposed to a fresh carbon surface. The adsorption of SO_2 was found to be first order in SO_2 and carbon. The average first-order rate constant for loss of SO_2 on the carbon surface was 5 s^{-1}. To avoid difficulties in defining the active surface area of the particles, we can express this result in terms of the mass concentration of the carbon, or 3×10^{-2} s^{-1} [μg C $cm^{-3}]^{-1}$.

This result is for the initial rate of reaction on a fresh surface. The rate of adsorption of SO_2 was found to decrease monotonically with exposure of the surface to SO_2 until saturation at which point no more SO_2 was observed. Figure 5 shows this behavior, where a fresh surface was

TABLE 1. Collisional Reaction Probabilities on a H_2SO_4 Surface at 300 K

Species	Collisional Reaction Probability, γ
H_2O_2	7.8×10^{-4}
HNO_3	$\geq 2.4 \times 10^{-4}$
HO_2NO_2	2.7×10^{-5}
$ClONO_2$	1.0×10^{-5}
N_2O_5	$\geq 3.8 \times 10^{-5}$
H_2O	$\sim 2.0 \times 10^{-3}$
NH_3	$>1.0 \times 10^{-3}$
OH	4.9×10^{-4}
O_3	$<1.0 \times 10^{-6}$
NO	$<1.0 \times 10^{-6}$
NO_2	$<1.0 \times 10^{-6}$
SO_2	$<1.0 \times 10^{-6}$
Alkenes	$<1.0 \times 10^{-6}$
Alkanes	$<1.0 \times 10^{-6}$
CF_4	$<1.0 \times 10^{-6}$
CCl_2F_2	$<1.0 \times 10^{-6}$
$O(^3P)$	$<1.0 \times 10^{-6}$
N	$<1.0 \times 10^{-6}$

exposed to a constant high concentration of SO_2. The rate of adsorption declines until after 1100 sec the surface is completely saturated. The amount of SO_2 required to saturate the surface was $2 \pm 1 \times 10^{-4}$ molecules SO_2 per molecule of carbon or 1.1 ± 0.6 mg SO_2 g^{-1} C.

The initial rate of adsorption we measure, which corresponds to an atmospheric loss rate of 1.2 percent per hour with a typical atmospheric particle burden of 100 $\mu g\text{-}m^{-3}$, is comparable to homogeneous reaction of SO_2 (Altshuller, 1979). However, the rapid saturation of the surface indicates that this process cannot be a major sink for SO_2. If a surface-catalyzed reaction can occur, however, leading to regeneration of the surface and further adsorption, then this heterogeneous process may be important. We examined the effect of NO_2 on the rate of adsorption of SO_2 and the saturation behavior and found no effect. Indeed, NO_2 and SO_2 seem to be adsorbed independently on different types of site on the carbon surface. Unfortunately, we have not yet been able to investigate the effect of water vapor on this reaction system as SO_2 and H_2O undergo a fast reaction (presumably catalyzed by the reaction vessel walls because the measured rate constant is faster than collision frequency) in our reactor. Experiments are planned to investigate this effect and the rate of adsorption of water by carbon particles, as there is growing evidence that droplet-phase reactions catalyzed by carbon particles may be an important atmospheric sink.

Conclusions

Experimental studies of heterogeneous atmospheric reactions are still in their infancy. The understanding of complex reactions that may involve gas, solid, and liquid-droplet phases is a challenging task, and one that will require much effort. We hope that some of the basic experimental data here can be used in the same way as studies of elementary homogeneous reactions are used - to build models of ever-increasing complexity to explain the overall process of chemical transformation that occurs in a heterogeneous reaction.

Acknowledgement. This work was supported by the U.S. Department of Energy, Division of Biomedical and Environmental Research, under Contract EP-78-C-03-2109.

References

Altshuller, A. P., Model predictions of the rates of homogeneous oxidation of sulfur dioxide to sulfate in the troposphere, Atmos. Environ., 13, 1653-1661, 1979.

Baldwin, A. C., and D. M. Golden, Heterogeneous atmospheric reactions: Sulfuric acid aerosols as tropospheric sinks, Science, 206, 562-563, 1979.

Benson, S. W. (Ed.), Proceedings of the Symposium on Chemical Kinetics Data for the Upper and Lower Atmosphere, John Wiley, New York, 1975.

Britton, L. G., and A. G. Clarke, Heterogeneous reactions of sulphur dioxide and SO_2/NO_2 mixtures with a carbon soot aerosol, Atmos. Environ., 14, 829-839, 1980.

Cofer, W. R. III, D. R. Schryer, and R. S. Rogowski, The enhanced oxidation of SO_2 by NO_2 on carbon particulates, Atmos. Environ., 14, 571-575, 1980.

Golden, D. M., G. N. Spokes, and S. W. Benson, Very low-pressure pyrolysis (VLPP); A versatile kinetic tool, Angew. Chem., 12, 534-546, 1973.

Harker, A. B., and W. W. Ho, Heterogeneous ozone decomposition on sulfuric acid surfaces at stratospheric temperatures, Atmos. Environ., 13, 1005-1010, 1979.

Hidy, G. M. (Ed.), Aerosols and Atmospheric Chemistry, Academic Press, New York, 1972.

Judeikis, H. S., T. B. Stewart, and A. G. Wren, Laboratory studies of heterogeneous reactions of SO_2, Atmos. Environ., 12, 1633-1641, 1978.

Liberti, A., D. Brocco, and M. Possanzini, Adsorption and oxidation of sulfur dioxide on particles, Atmos. Environ., 12, 255-261, 1978.

Novakov, T., S. G. Chang, and A. B. Harker, Sulfates as pollution particulates: Catalytic formation on carbon (soot) particles, Science, 186, 259-261, 1974.

Tartarelli, R., P. Davini, F. Morelli, and P. Corsi, Interactions between SO_2 and carbonaceous particulates, Atmos. Environ., 12, 289-293, 1978.

LABORATORY MEASUREMENTS OF DRY DEPOSITION OF ACETONE OVER ADOBE CLAY SOIL

Henry S. Judeikis

The Aerospace Corporation, P. O. Box 92957, Los Angeles, California 90009

Abstract. The deposition of a selected atmospheric ketone (acetone) over a representative (adobe clay) soil from the Los Angeles area has been measured in the laboratory. Deposition from dry gas mixtures, even in the presence of simulated sunlight where evidence for photodesorption is observed, occurs at sufficiently high rates to be of environmental interest. However, the presence of water vapor, even at relative humidities as low as 12 percent, leads to no acetone uptake to within experimental error, presumably due to a competition for surface adsorption sites between the water vapor and acetone. Consequently it is unlikely that dry deposition over land surfaces will represent an important sink for atmospheric ketones.

Introduction

Dry deposition on environmental surfaces represents an important mode for removal of trace atmospheric gases. Considerable efforts have been expended in the measurement of deposition velocities (V_g, defined as the flux of a trace gas to a surface divided by its concentration above the surface) for SO_2 and a number of other gases. (For a recent review, see McMahon and Denison (1979).)

Recent attention has turned toward the measurement of deposition velocities for organic materials, notably formaldehyde (Thompson, 1980; Zafiriou et al., 1980). This has resulted from measurements of atmospheric aldehydes, particularly formaldehyde (Zafiriou et al., 1980; Tuazon et al., 1978; Warneck et al., 1978; Platt et al., 1979; Barbe et al., 1979; Yokouchi et al., 1979; Lowe et al., 1980; Kok et al., 1980), due to its presence in auto exhaust (Mansfield et al., 1977), and its importance in photochemical air pollution (Demerjian et al., 1974; Morris and Niki, 1974; Hanst and Gay, 1977). Of course, mechanisms for removal of these species from the atmosphere by photochemical and gas phase reactions (Calvert et al., 1972; Finlayson and Pitts, 1976; Niki et al., 1978; Clark et al., 1964) are of considerable interest. The literature data (Thompson, 1980; Zafiriou et al., 1980) indicate deposition velocities of 0.73 and 0.50 cm/sec, respectively, for formaldehyde at the air/sea surface, corresponding to fluxes of 5 and 5.7 $\mu g/cm^2 \cdot yr$. These fluxes correspond to ~1 to 2 percent of the total organic carbon input into the sea (Thompson, 1980; Zafiriou et al., 1980).

Some recent measurements (Pellizarri et al., 1976; Johansson, 1978) have also detected acetone in polluted air, at concentrations of up to 20 $\mu g/m^3$ (Johansson, 1978). However, little is known about the potential removal of atmospheric ketones by deposition on environmental surfaces. Here we report on the deposition of acetone, as a model for atmospheric ketones, onto a representative soil in the presence and absence of water vapor and simulated sunlight.

Experimental

The apparatus used in these experiments was a modified version of a cylindrical flow reactor described by Judeikis and Stewart (1976). In this reactor the environmental surface of interest was coated on the outside of a cylindrical pyrex tube, generally using a water slurry of the solid material. The coated cylinder was air dried, inserted concentrically into a larger, cylindrical-flow reactor, and subsequently vacuum dried (10^{-4} torr) overnight. The reactor was surrounded by a bank of daylight-type fluorescent lights which gave a representation (wavelength and intensity) of the sea-level solar spectrum (Hedgpeth et al., 1974).

Operationally, an appropriately prepared gas mixture (using research-grade materials) containing the trace gas of interest was made to flow through the reactor. Typical conditions were ~100 torr total pressure, flow rates of 10 to 25 cm^3/sec (corresponding to linear velocities of 0.5 to 1.2 cm/sec), O_2:N_2 carrier gas with O_2:N_2 ratios of 0 to 0.2, and trace gas partial pressures of 50 millitorr to 30 torr. (Experimental results were independent of the O_2 and trace gas concentrations to within experimental error.) For humidified mixtures, a portion of the O_2:N_2 carrier gas flow passed through a bubbler containing distilled water to achieve the desired relative humidity.

As the reaction mixture flowed through the annular space between the two concentric cylinders, the trace gas diffused to the coated walls where it was heterogeneously removed. Flow in the annular region was laminar (Reynolds numbers less than 50), and approximately the first 10 cm of the coated cylinder on the inlet side of the reactor were left uncoated to permit full development of laminar flow before heterogeneous wall removal occurred.

Wall reaction led to concentration gradients for the reacting species in the annular region in both the radial and axial directions. The concentration gradient in the axial direction was sampled by means of a small probe held against the inner wall of the reactor (opposite the coated cylinder), but moveable in the axial direction. The gas mixture sampled by the probe was analyzed by mass spectrometry.

Analysis of the data was carried out using a model that specifically accounted for mass transport by diffusion and flow (Judeikis, 1980). In this model the diffusive flux to the coated wall is related to the rate of removal at that wall by the expression (Judeikis and Stewart, 1976; Judeikis, 1980)

$$D\left(\frac{\partial c}{\partial r}\right) = -\gamma k_r C = -V_g C \qquad (1)$$

where k_r is the velocity of the reacting species in the radial direction ($k_r = (RT/2\pi M)^{1/2}$, R being the gas constant, T the absolute temperature, and M the molecular weight), γ is the wall reaction coefficient (the fraction of gas-wall collisions that lead to removal of the reacting species), and V_g is the deposition velocity. Values of V_g determined in this fashion can be used to determine dry deposition rates over related environmental surfaces, given atmospheric concentrations of the reacting species. Since mass transport by diffusion and flow has been specifically accounted for in the data analysis, the values of V_g reported here would apply to the environment under turbulent atmospheric conditions and represent the maximum rates of dry deposition.

Results

A representative plot for acetone removal by adobe clay soil (representative sample from the Los Angeles area) is shown in Figure 1. Shown in the figure is the acetone concentration as a function of distance from the leading edge of the inner cylinder (partially) coated with adobe clay soil. This experiment was run in the dark at 100 torr total pressure with an O_2:N_2 carrier gas (0.16:1) containing 50 millitorr acetone and no water vapor. The flow rate was 22 cm^3/sec (linear velocity of 1.1 cm/sec), and the soil coating began 10 cm from the leading edge of the inner cylinder. The circles in the figure indicate the measured concentrations at various distances, while the solid line represents the asymptotic exponential decay with distance. Analysis of this data using the previously described models yields

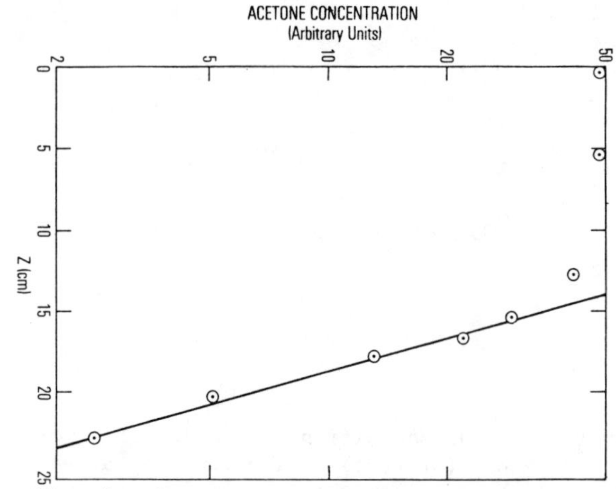

Fig. 1. Acetone concentration versus distance from leading edge of inner cylinder (partially) coated with adobe clay soil.

$\gamma = 3.1 \times 10^{-4}$ and $V_g = 2.6$ cm/sec. Repeat experiments with some six different samples yielded average values of $\gamma = 3.7 \times 10^{-4}$ and $V_g = 3.1$ cm/sec (±30 percent), compared to blank runs (uncoated inner cylinder with no clay soil) of $\gamma \leq 5.4 \times 10^{-7}$ and $V_g \leq 0.004$ cm/sec.

In the presence of simulated sunlight, somewhat lower values were obtained for dry reaction mixtures; $\gamma = 1.9 \times 10^{-4}$ and $V_g = 1.6$ cm/sec (±30 percent). These lower values suggested the possibility of photodesorption of adsorbed acetone from the clay soil surface. Separate experiments demonstrated that this was indeed occurring. In these experiments the sampling probe was set at a fixed position in the reactor, at a point where the acetone was reduced to approximately one-third of its inlet value due to adsorption by the soil in the dark. Subsequently, the fluorescent lamps were turned on and the acetone concentration rapidly increased to about two-thirds of the inlet concentration. Extinction of the lamps led to a rapid decrease to its original dark value of about one-third of the inlet concentration.

The adsorbed acetone appeared to be moderately strongly adsorbed on the surface, since removal of acetone from the gas mixture led to a decrease in the measured acetone concentration with a time-constant characteristic of the flow parameters. For a weak adsorption, we would have expected a much longer time constant due to desorption of acetone from the surface into the acetone-poor flowing carrier gas.

All of the experiments described thus far were carried out in the absence of water vapor and suggest a significant environmental sink for acetone (ketones). However, in the presence of water vapor, even at relative humidities as low as 12 percent, no acetone uptake by the clay soil was observed, to within experimental error. Thus, $\gamma \leq 5.4 \times 10^{-7}$ and $V_g \leq 0.004$ cm/sec. Similar results were obtained at higher relative humidities, up to 90 percent. These results suggest a competition between water and acetone for available adsorption sites on the clay soil, with water vapor winning out. Support for this conclusion comes from a separate experiment in which a clay soil was exposed to a dry reaction mixture containing acetone until substantial uptake had occurred. The reaction mixture was then humidified (52 percent relative humidity), with the result that acetone desorption was observed; i.e., the acetone concentration rapidly rose to a value 2½ times that of the inlet concentration and then gradually fell back to the inlet concentration. Integration of the latter data indicated that, to within experimental error, all of the acetone adsorbed during exposure to the dry reaction mixture was desorbed when the reaction mixture was humidified. Similar results were obtained from a related experiment carried out at 12 percent relative humidity.

Discussion

The high uptake of acetone by adobe clay soil from dry reaction mixtures and the negligible uptake in the presence of water vapor are in marked contrast to results previously obtained for reactive gases such as SO_2, NO_2, and NO (Judeikis and Stewart, 1976; Judeikis and Wren, 1978). In the latter cases we found that those species were permanently removed from the gas stream, although with prolonged exposure we observed $NO_2 \rightarrow NO$ conversion in the case of NO_2. For SO_2, NO_2, and NO, we found that deposition velocities over freshly prepared samples were essentially unaffected by the presence of water vapor; however, the total uptake of these gases over prolonged exposures was significantly enhanced in humidified systems.

The acetone results obtained here are qualitatively similar to those previously obtained for dimenthyl sulfide (DMS) deposition over similar materials (Judeikis and Wren, 1977). Thus, the deposition velocities in (dark) dry systems (3.1 and 0.28 cm/sec for acetone and DMS) were significantly reduced in the presence of water vapor ($\leq$0.004 and <0.003 cm/sec, respectively). As in the case of DMS, we ascribe the acetone results here to a reversible physical adsorption process that cannot compete with adsorption by water vapor.

Although the high deposition velocities observed for acetone over adobe clay soil in dry systems (even in the presence of simulated sunlight where we see evidence for photodesorption) look environmentally interesting, it appears that dry deposition over land surfaces is not likely to be an important process for acetone (ketones). The presence of even small amounts of water vapor, which are well below levels generally found in the atmosphere, produces an effective competition with and even displacement of acetone by water on the soil surface.

Quantitatively, removal of acetone (ketones) by dry deposition can be compared to other processes such as photolysis or reaction with OH. The time constant for photolysis of acetone (solar zenith angle of 40°) has been estimated at 20 hours (Demerjian et al., 1974). In addition, an estimate of $\sim 8.6 \times 10^3$ ppm^{-1} min^{-1} is given in the same reference for abstraction of a proton from the CH_2 group in methyl ethyl ketone. Abstraction of a proton from the CH_3 groups in acetone would be more difficult. From the relative rates of CH_2 and CH_3 proton abstraction (Demerjian et al., 1974; Greiner, 1970), the OH reaction with acetone can be estimated to be about six times slower than the reaction with methyl ethyl ketone, or $\sim 1.4 \times 10^3$ ppm^{-1} min^{-1}. For an OH concentration of $\sim 3 \times 10^6$ cm^{-3}, this would give a time constant for the OH + acetone reaction of about 95 hours. If we consider a 1000-m by 1-cm^2 column of air containing 20 $\mu g/m^3$ of acetone, photolysis would lead to removal of 100 ng/hr, while reaction with OH would amount to an acetone loss of 21 ng/hr. On the other hand, removal by dry deposition ($V_g \leq 0.004$ cm/sec) would yield a loss of only 0.3 ng/hr. Although these figures suggest a minimal removal rate by dry deposition over land surfaces, this conclusion may not prevail for dry deposition over bodies of water. The high solubility of acetone in water and the substantial deposition velocities already noted for formaldehyde over the oceans (Thompson, 1980; Zafiriou et al., 1980) suggest that dry deposition of acetone (ketones) over the oceans may represent an important global sink.

Acknowledgements. The author gratefully acknowledges the efforts of James E. Foster in carrying out the laboratory experiments and the National Science Foundation for its support under grant nos. ATM 77-23435 and ATM 79-17862.

References

Barbe, A., P. Marché, C. Secroun, and P. Jouve, Measurements of tropospheric and stratospheric H_2CO by an infrared high resolution technique, Geophys. Res. Lett., 6, 463-465, 1979.

Calvert, J. G., J. A. Kerr, K. L. Demerjian, and R. D. McQuigg, Photolysis of formaldehyde as a hydrogen atom source in the lower atmosphere, Science, 175, 751-752, 1972.

Clark, J. H., C. B. Moore, and N. S. Nogar, The photochemistry of formaldehyde: absolute quantum yields, radical reactions, and NO reactions, J. Chem. Phys., 68, 1264-1271, 1978.

Demerjian, K. L., J. A. Kerr, and J. G. Calvert, The mechanism of photochemical smog formation, in Advances in Environmental Science and Technology, Vol. 4, edited by J. N. Pitts, Jr. and R. T. Metcalf, John Wiley and Sons, New York, pp. 1-262, 1974.

Finlayson, B. J., and J. N. Pitts, Jr., Photochemistry of the polluted troposphere, Science, 192, 111-119, 1976.

Greiner, N. R., Hydroxyl radical kinetics by kinetic spectroscopy. IV. Reactions with alkanes in the range 300-500 K, J. Chem Phys., 53, 1070-1076, 1970.

Hanst, P. L., and B. W. Gay, Jr., Photochemical reactions among formaldehyde, chlorine, and nitrogen dioxide in air, Environ. Sci. Tech., 11, 1105-1109, 1977.

Hedgpeth, H., S. Siegel, T. Stewart, and H. S. Judeikis, Cylindrical flow reactor for the study of heterogeneous reactions of possible importance in polluted atmospheres, Rev. Sci. Instrm., 45, 344-347, 1974.

Johansson, I., Determination of organic compounds in indoor air with potential reference to air quality, Atmos. Environ., 12, 1371-1377, 1978.

Judeikis, H. S., Mass transport and chemical reaction in cylindrical and annular flow tubes, J. Phys. Chem., 84, 2481-2484, 1980.

Judeikis, H. S., and T. B. Stewart, Laboratory measurement of SO_2 deposition velocities on selected building materials and soils, Atmos. Environ., 10, 769-776, 1976.

Judeikis, H. S., and A. G. Wren, Deposition of H_2S and dimethyl sulfide on selected soil materials, Atmos. Environ., 11, 1221-1224, 1977.

Judeikis, H. S., and A. G. Wren, Laboratory measurements of NO and NO_2 depositions onto soil and cement surfaces, Atmos. Environ., 12, 2315-2319, 1978.

Kok, G. L., J. G. Nuttall, and B. J. Pierce, Formaldehyde levels in the Los Angeles air basin, (abstract) EOS Trans. AGU, 61, 967,1980.

Lowe, D. C., U. Schmidt, and D. H. Enhalt, A new technique for measuring tropospheric formaldehyde $(CH_2O)^2$, Geophys. Res. Lett., 7, 825-828, 1980.

Mansfield, C. T., B. T. Hodge, R. B. Hege, Jr., and W. C. Hamlin, Analysis of formaldehyde in tobacco smoke by high-performance liquid chromomatography, J. Chromatogr. Sci., 15, 301-302, 1977.

McMahon, T. A., and P. J. Denison, Empirical atmospheric deposition parameters - a survey, Amos. Environ., 13, 571-585, 1979.

Morris, E. D., Jr., and H. Niki, Reaction of the nitrate radical with acetaldehyde and propylene, J. Phys. Chem., 78, 1337-1338, 1974.

Niki, H., P. D. Maker, C. M. Savage, and L. P. Breitenbach, Relative rate constants for the reaction of hydroxyl radical with aldehydes, J. Phys. Chem., 82, 132-134, 1978.

Pellizarri, E. D., J. E. Bunch, R. E. Berkley, and J. McRae, Determination of trace hazardous organic vapor pollutants in ambient atmospheres by gas chromatography/mass spectrometry computer, Anal. Chem., 48, 803-807, 1976.

Platt, U., D. Perner, and H. W. Patz, Simultaneous measurement of atmospheric CH_2O, O_3, and NO_2 by differential optical absorption, J. Geophys. Res., 84, 6329-6335, 1979.

Thompson, A. M., Wet and dry removal of tropospheric formaldehyde at a coastal site, Tellus, 32, 376-383, 1980.

Tuazon, E. C., R. A. Graham, A. M. Winer, R. R. Easton, J. N. Pitts, Jr., and P. L. Hanst, A kilometer pathlength Fourier-transform infrared system for the study of trace pollutants in ambient and synthetic atmospheres, Atmos. Environ., 12, 865-875, 1978.

Warneck, P., W. Klippel, and G. K. Moortgat, Formaldehyd in troposphärischer Reinluft, Ber. Bunsenges. Phys. Chem., 82, 1136-1142, 1978.

Yokouchi, Y., T. Fujii, Y. Ambe, and K. Fuwa, Gas chromatography - mass spectrometric analysis of formaldehyde in ambient air using a sampling tube, J. Chromatog., 180, 133-138, 1979.

Zafiriou, O. C., J. Alford, M. Herrara, E. T. Peltzer, R. B. Gagosian, and S. C. Liu, Formaldehyde in remote marine air and rain: Flux measurements and estimates, Geophys. Res. Lett., 7, 341-344, 1980.

KINETICS OF REACTIONS BETWEEN FREE RADICALS AND SURFACES (AEROSOLS) APPLICABLE TO ATMOSPHERIC CHEMISTRY

Daryl D. Jech, Patrick G. Easley, and Barbara B. Krieger

Department of Chemical Engineering, University of Washington, Seattle, Washington 98195

Abstract. The reactions between free radicals of atmospheric interest and solid surfaces were experimentally investigated using a fast-flow tubular reactor. The pure surfaces were in the form of dried aerosols and coatings. The OH radical was detected by UV resonance fluorescence. At room temperature, the scavenging rate ranged from 0.01 to 0.40 of the kinetic collision frequency. Oxygen and hydrogen atoms were reacted with actual aerosols and were detected using chemiluminescence. Rates for these atoms and HO_2 were generally much slower than for OH unless the atoms reacted with the surface to form volatile products. When this occurred (e.g., O + NH_4Cl), the fraction of collisions resulting in loss of the atom was nearly 1.0.

The chosen surfaces were: (1) "natural" aerosols (NH_4NO_3, $(NH_4)_2SO_4$, NH_4Cl, Na_2SO_4, $NaNO_3$), (2) "man-made" surfaces typical of fly ash and auto emissions composition ($FeSO_4$-FeO_x, $Zn(NO_3)_2$, $Pb(NO_3)_2$, K_2CO_3), and (3) organic surfaces (malonic acid, glycine, sodium propionate). For OH, the second group of surfaces was somewhat more reactive than the other two; however, exceptions occurred (e.g., malonic acid).

The data were compared to results of other researchers engaged in modeling atmospheric chemistry. The present results support the concept that heterogeneous reactions of certain free radicals have rates that are composition-dependent and should be included in models of the local urban troposphere (near a specific source). For models over a larger area, it is felt that unreactive aerosols will "dilute" reactive aerosols so that heterogeneous losses of free radicals will be less important.

Objectives and Introduction

It is well established that free radicals play an important part in the chemistry of the atmosphere. While we are currently able to characterize size and composition of atmospheric particulate matter, the role that these surfaces play in altering photochemical smog cycles is largely unknown. Modeling work has been done in an attempt to understand the interactions between free radicals and particles in the urban atmosphere (Graedel et al., 1975, 1976). Some studies indicate that the free radical deactivation process has to be reasonably fast in order to assert that aerosol surfaces do indeed affect the photochemical smog cycles (Levy, 1971). A loss coefficient of 1 (γ = 1) is generally assumed. (This loss coefficient describes the fraction of collisions with a surface that are effective in destroying the free radical.) An assumption of γ = 1 is at variance with the heterogeneous rate coefficients that have been obtained in flow tube reactor experiments, which are generally found to be very low (compilations appear in Tables 1 to 3). Therefore it is of interest to determine the reaction rate of radicals with actual aerosols, particularly as a function of composition.

Atmospheric particle composition falls into three general classes. Condensation (secondary) aerosols, such as sulfates and nitrates, are formed by gas-to-particle conversion processes where the gas is often a pollutant (McMurray and Friedlander, 1978). Other compositions of particles reflect the nearby activities of man (primary aerosols), such as lead from auto emissions and soil and rubber components. The third class includes organic substances which ultimately condense (Grosjean, 1977).

The scope of this investigation was to study a diverse set of pure surfaces in all of the above classes to determine if the chemical composition was a major contributor to the destruction rate of free radicals of atmospheric interest. Several radicals were studied to determine if their composition affected the scavenging rate. Particle compositions were chosen to represent abundant components in urban aerosols and to assess realistically their scavenging ability without excessive experimental effort or cost (see Table 4). The cations chosen represent only a few of those found from elemental analyses of urban aerosols, yet they reflect a variety of potential aerosol sources. The organic substances were chosen to have functional groups that might be present in photochemical smog (Grosjean, 1977). Sulfates and nitrates are likely to be present on aerosol surfaces. Both oxygen atoms and hydrogen

TABLE 1. Surface Loss Coefficients for O and N Atoms (room temperature)

Surface	Surface Loss Coefficient, γ O Atoms	N Atoms
LiCl	0.0019[a]	
KBr	0.0013[a]	
NaCl	0.00094[a]	
KF	0.00092[a]	
KCl	0.00078[a]	
KI	0.00074-0.0035[a]	
$BaCl_2$	0.00057-0.0019[a]	
RbCl	0.00045-0.0024[a]	
Soda glass	0.00034[a]	
Glass	0.00002[b]	0.000017[c]
Quartz (fused)	0.000024-0.00007[d]	
"	0.00004[e]	
"	0.0001[f]	
Vitreosil quartz	0.00004-0.00008[g]	
"	0.00016[h]	
Silica	0.00071[h]	
"	0.000017-0.00071[d]	
Vycor glass	0.00007[i]	
Pyrex	0.00005[j]	0.00003[k]
"	0.00002-0.00012[d]	
"	0.00002-0.00050[l]	
"	0.00011[m]	
"	0.000077[n]	
"	0.00021[o]	
"	0.00013[p]	
"(with O_2)	0.00039[q]	
" " O_2, He)	0.00058[q]	
" " O_2, He, Ar)	0.00051[q]	
Carbon	0.000005[r]	
Surface/Coating (if mentioned)		
H_2SO_4	0.000023[g]	
H_2SO_4/Pyrex	0.00002[j]	
$(NH_4)_2Cl$/Pyrex	0.000019	
NH_4Cl	~1[j]	
H_3PO_4/Pyrex	0.000001-0.000050[d]	
" "	0.00004[g]	
" "	"less than quartz"[s]	
HF/quartz	0.000078-0.0001[t]	
Teflon	0.000041[g]	0.000029[c]
"	"less than quartz"[s]	
Adsorbed water	"less than quartz"[s]	0.00001[k]
Glass, poisoned with metaphosphoric acid		0.00001[k]
Metaphosphoric acid		0.0000014[c]
Molybdenum/glass		0.04-0.22[u]

[a]Greaves & Linnett (1958)
[b]Kaufman and Kelso (1967)
[c]Young (1961)
[d]Venugopalan (1968)
[e]Hacker et al. (1961)
[f]Smith et al. (1980)
[g]Williams & Mulcahy (1966)
[h]Greaves & Linnett (1959a, b, c)
[i]Kaufman (1961)
[j]Akers & Wightman (1976)
[k]Herron et al. (1959)
[l]Wise and Wood (1967)
[m]Herron & Schiff (1958)
[n]Elias et al. (1959)
[o]Wise & Rosser (1963)
[p]Linnett & Marsden (1956a, b)
[q]Yolles and Wise (1968)
[r]Otterbein and Bonnetain (1969)
[s]Mearns and Morris (1970)
[t]Azatyan (1972)
[u]Lund and Oskam (1968)

atoms were studied to interpret the laboratory results from OH and HO$_2$. Only preliminary results of HO$_2$ are given since measurement of rates for this radical proved to be difficult, as is widely recognized.

Although information on the chemical mechanism of radical loss at particle surfaces would be desirable, these experiments are difficult and require surface species characterization, single-particle methods (J. Davis, Inst. of Paper Chemistry, private communication, 1981; S. Arnold, private communication, 1981; Arnold and Lewittes, this volume) or use of single-crystal surfaces (R. Madix, Stanford Univ., private communication, 1980), all of which utilize high-vacuum techniques. Since atmospheric aerosols are covered with an equilibrium layer of water (Graedel et al., 1976; Graedel, 1979) surfaces generated from water solution are deemed to be realistic models for our purposes, yet are incompatible with the above methods. Our aim, then, is to determine relative reaction rates among those surfaces and free radicals of atmospheric interest in order to determine which of these warrant further study.

Previous Work

Compiled rate coefficients pertaining to the reactions of free radicals with various surface compositions appear in Tables 1 to 3. The reaction rate of atoms with reactor materials is in general quite low. Based on these data for atomic species, it is generally accepted that atom reactions with particles cannot compete with homogeneous losses in the atmospheric chemistry smog cycles. Many of the surfaces show similar reactivity to both OH and HO$_2$. An apparent trend (at room and combustion temperatures) is that alkaline surfaces are more reactive than acid surfaces; hence the common practice of acid-washing flow tube reactors.

Available mechanistic data reveal that the general concentration dependence is first order in the gas-phase radical species, but a two-stage mechanism is involved:

$$A + S \longrightarrow S\text{-}A \qquad (a)$$

$$A + S\text{-}A \longrightarrow S + A_2 \qquad (b)$$

where A is an atom and S is a surface site. It is generally assumed that surface coverage is fast, so reaction (b) controls the overall rate. Nonetheless, if either reaction controls the overall rate, the process is first order, as required by observations. Other mechanisms, such as the Hinshelwood mechanism which postulates a migration of atoms on the surface, do not agree with order-of-reaction observations as well as the above Rideal mechanism (Hardy and Linnett, 1966). Wood and Wise (1962) explain how the Rideal mechanism might be replaced by second-order mechanisms at low or high temperature.

Several authors have suggested that the adsorbed

TABLE 2. Surface Loss Coefficients for H, OH, and HO$_2$ (room temperature)

Surface	Surface Loss Coefficient, γ		
	H	OH	HO$_2$
Pt	1.0[a]		
Ti	0.38[a]		
Al$_2$O$_3$	0.33[a]		
Al	0.27[b]		
Cu	0.11[b]		
KOH	>0.10[a]		
Au	0.08[b]		
Ni	0.08[b]		
Na$_3$PO$_4$	$\geq$0.07[a]		
Al$_2$O$_3$ moist	$\geq$0.07[a]		
K$_2$SiO$_3$	0.07[a]		
Pd	0.07[b]		
W	0.06[b]		
K$_2$CO$_3$	$\geq$0.05[a]		
Pt	0.02[b]		
C	0.7[c]		
C	0.009[b]	0.005[d]	
K$_2$SiO$_3$ moist	$\leq$0.0017[a]		
K$_2$CO$_3$ moist	<0.001[a]		
Na$_3$PO$_4$ moist	<0.001[a]		
KOH moist	<0.001[a]		
KCl	0.00002[a]		
H$_3$PO$_4$ moist	0.00002[a]		
H$_3$PO$_4$ baked		0.012[d]	0.0003[e]
" "		0.002[f]	<0.0003[g]
" "		<0.001[e]	
" "		<0.0002[h]	
B$_2$O$_3$		0.003[i]	
"		<0.0001[j]	

[a]Smith (1943)
[b]Wise and Rosser (1963)
[c]Thrush (1965)
[d]Mulcahy and Young (1975)
[e]Chang and Kaufman (1978)
[f]Anderson and Kaufman (1972)
[g]Howard and Evenson (1977)
[h]Leu (1979)
[i]Breen and Glass (1970)
[j]Westenberg and de Haas (1973)

atom should not be too tightly bound in order to favor catalytic recombination (Hacker et al., 1961; Melin and Madix, 1971; Manella and Harteck, 1961). Cheaney et al. (1959) proposed that the surface reactivity depends on the ability of the surfaces to donate an electron to the radical (i.e., its Lewis basicity). Presumably this mechanism would provide a temporary bond strong enough to capture the radical until it permanently gains another electron by recombining with a second gaseous radical. This mechanism is supported by the observed trends for acidic versus basic surfaces mentioned earlier. It might also explain why the relative reactivities of surfaces appear to be the same regardless of the radical composition, since the proposed mechanism ad-

TABLE 3. Surface Loss Rates for H, OH, and HO_2 (relative values)

Surface	H Radical[a] (inferred)	O, OH Radicals[a] (inferred)	HO_2 Radical[a] (inferred)	HO_2[b]
	Relative Loss Rate			
MnO_2	high	high		
MgO	high	high		
$CuCl_2$/CuO	high	high		
Ag	high	high		
NaOH	1050	high	high	
PbO	1100	60		
KOH	1020	55		
$MnCl_2$/MnO	1000	70		
Graphite	1000	70		
Al_2O_3	910	56		
K_2HPO_4	680	40	971	
$BaBr_2$				2.5
KBr				2.1
KCl	100	49	1205	1.9
$Na_2B_4O_7$	70	28		
CsCl	60	49		
NaCl	55	30		
$BaCl_2$	15	13		
KH_2PO_4	3	very low		
NaH_2PO_4	low	low	862	
H_3PO_4	very low	very low	low	
B_2O_3	very low	very low	low	1.0
$K_2B_4O_7$				1.02

[a]From Warren (1957). These data were obtained from study of chain carriers in the $H_2 + O_2$ reaction. Compositions were inferred from the data. The values in one column show qualitative differences (non-linear, however), but bear no relationship to values in other columns.

[b]From Vardanyan et al. (1971).

dresses itself to a common aspect of free radicals (their unshared electron) rather than to their specific chemical form.

Although interest in atmospheric heterogeneous mechanisms is growing (Judeikis and Siegel, 1973; Baldwin and Golden, 1979; Liberti, 1970), most studies are concerned with reactions of stable gases, such as surface-catalyzed oxidation of SO_2 (Judeikis et al., 1978). To our knowledge there is only one published experimental study dealing with the kinetics of actual aerosol-free radical reactions (Akers and Wightman, 1976).

Experimental

General Description

The apparatus consisted of the aerosol generation, drying, and measurement system and the discharge-flow tube reactor and detection system (Figures 1 and 2). The generation of a dense aerosol proved to be difficult; details can be found in Jech (1981) and Easley (1980).

The cylindrical reactor was optical-quality quartz; several diameters were used and the effective volume (length) varied owing to location of the injection ports for reactant gases. The reactor was treated with phosphoric acid to make the walls unreactive. The reactor surface was conditioned by a steady flow of radicals for more than an hour prior to data collection. A critical flow orifice was used to allow aerosol generation at high pressure and operation of the flow tube at low pressure. For studies with low reaction rates, a rod coated with the surface of interest was inserted into the reactor.

About 1 to 5 percent of the oxygen and hydrogen was dissociated, as verified by titration with NO_2. A mixture of 0.5-percent NO_2 in helium was used to minimize corrosion and handling problems.

TABLE 4. Surfaces Chosen for Study Grouped According to Aerosol Source

Natural	Fly Ash, Auto	Organic
$(NH_4)_2SO_4$	$FeSO_4$	Malonic acid
NH_4NO_3	FeO_x	Glycine
NH_4Cl	$Zn(NO_3)_2$	Sodium propionate
Na_2SO_4	$Pb(NO_3)_2$	
$NaNO_3$		

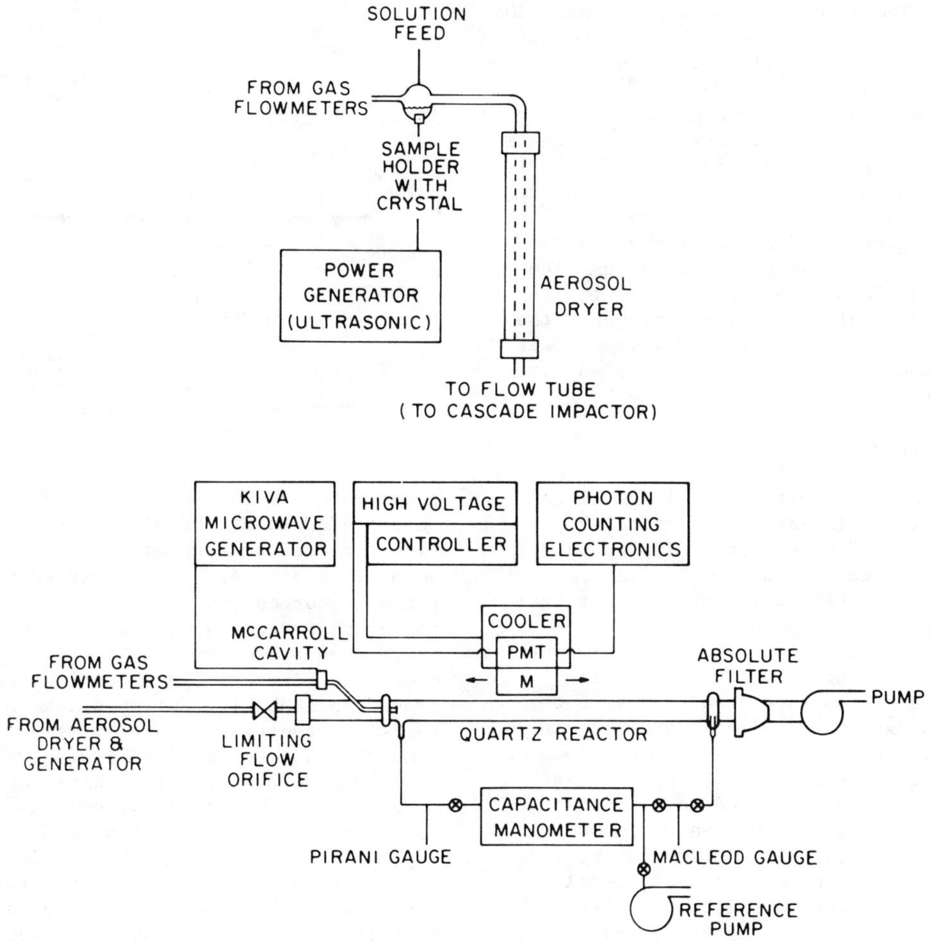

Fig. 1. Schematic diagram of aerosol generation and flowtube reactor system for studying aerosol/free-radical reaction rates. M = monochromator; PMT = photomultiplier.

For oxygen atom reactions with aerosols, NO was added with the aerosol. It is believed no reaction occurred between the particles and NO owing to short contact times and experimental findings which are described further in this paper. Reactants were injected at an angle into the larger particle-containing flow (50 to 80 percent) via a perforated Teflon injection tube, which increased mixing.

Hydroxyl radicals were continuously generated upstream of the reactor via the fast reaction, $H + NO_2 \rightarrow OH + NO$, with a slight excess of NO_2 (Figure 2; Clyne and Down, 1974; Anderson and Kaufman, 1972). After sufficient residence time (~1 msec) to convert H to OH, the flow was pumped along a cylindrical Pyrex reactor tube (19 mm I.D.). Typical OH concentrations entering the reactor upstream were $<1.6 \times 10^{13}$ cm^{-3}, with H concentrations $\simeq 0.1$ [OH]. The absolute H concentration was determined by chemical titration, while OH concentration was estimated from a kinetic model. Reactant (OH) entered the reactor through four equally spaced orifices in the reactor wall (see Figure 2). Mixing of the reactant with the bulk stream was determined to be very rapid using the criteria of Brown (1970) and from photos of chemiluminescence (Jech, 1981); in fact, some back diffusion occurred.

Gas concentrations and velocities were determined from absolute pressure and the pressure drop across the reactor before and after the addition of each gas (1) using calibrated rotameters on the reactant streams, and (2) from differential pressure measurements assuming Poiseuille flow over the length of the reactor tube. Residence time in the reactor was typically 10 to 40 msec. Photon-counting detection of chemiluminescence decay as a function of position was used to measure the surface deactivation rate (Krieger and Kummler, 1977; Easley, 1980). Relative concentration of the OH radical was measured by resonance fluorescence (3090 Å) at a fixed point downstream as described by Stuhl and Niki (1972) with modifications described by Jech (1981). Clyne and Down (1974) demonstrated that the intensity should be proportional to the OH concentration for OH <

5×10^{14} cm^{-3}. The signal-to-noise ratio was >10.

Low reactor pressure was used to eliminate diffusion rate limitations. Criteria of Walker (1961), Smith et al. (1980), Ogren (1975), and Kaufman (1961) were used to ensure that the plug flow assumption could be applied to the reactants even though the bulk velocity profile was clearly in the laminar region (Reynolds number was about 1 to 10). The most severe case occurred when the rod was in the reactor. These criteria generally estimated the error in the plug flow assumption to be about 20 to 30 percent for a 4-mm-diameter rod with $\gamma = 1.0$. This condition represents the worst case used in the present study; for slower surface reactions (smaller γ) the estimated error was correspondingly less.

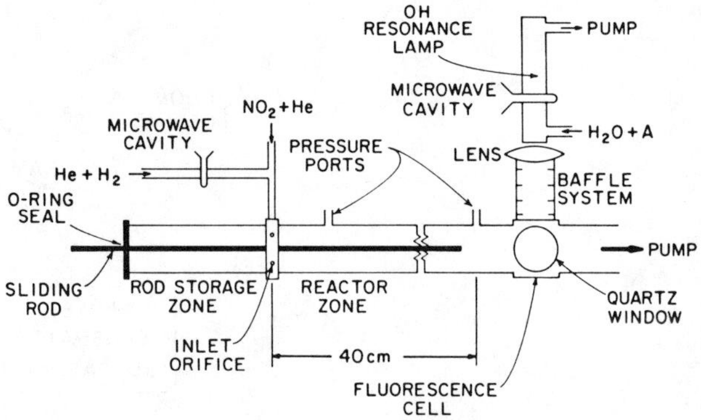

Fig. 2. Schematic of flow system, OH resonance lamp, and fluorescence cell.

Aerosol Generation

In order to generate particle surfaces of well-defined composition, intensive effort developed three "dense aerosol" generators: (1) an ultrasonic nebulizer based on a design by Ames Laboratory; (2) a Plasma-Therm Ultrasonic nebulizer chamber with an in-house power supply; and (3) a hospital-type ultrasonic nebulizer. In all methods, the aerosol was dried using a diffusion dryer. Although the surface area from a monodisperse aerosol generated with a Bergland-Liu vibrating orifice was maximized, the surface area was 1 to 2 orders of magnitude lower than that generated by other means. Based on previous detailed work by Easley (1980), an ultrasonic nebulizer was modified and optimized (Jech, 1981).

The aerosol system consisted of a salt-solution feed pump, the nebulizer, a silica gel diffusion dryer, a five-stage cascade impactor, and a 50-W, 1.4-MHz power source (Olsen et al., 1977). The heart of the nebulizer was a piezoelectric crystal transducer, which converted electrical power to high-frequency mechanical vibration. A thin layer of salt solution applied to the vibrating crystal was thus transformed into tiny droplets. A flow of argon swept the aerosols out of the nebulizer unit, through the dryer, and then either through the impactor or into the reactor via a flow-limiting orifice.

Optimum nebulizer performance is described in Jech (1981), and nebulization efficiency depended on the specific salt used and on its concentration. Attainment of the optimum required extremely delicate adjustments. Nonetheless, fairly stable performance could be attained, and a 25-fold increase in aerosol surface area was achieved compared to that obtained by Easley (1980).

Aerosol Characterization

The characterization of aerosols is potentially important since atmospheric aerosols are known to be highly porous with uneven surfaces, while flat surfaces such as coated reactor walls are likely to be smooth. Thus the effective surface area of an aerosol particle is probably much larger than the area based on its apparent spherical surface. Generated aerosols have an extremely short retention time in the reactor compared to atmospheric aerosol lifetimes. Thus measured rate coefficients correspond to fresh, almost dry aerosols rather than aged, atmospheric aerosols.

Electron photomicrographs of the aerosols were made (see Figure 3), although the preparation procedure probably destroyed some of the original aerosol properties. Results were similar to Pinnick and Auverman (1979) with respect to shape and clumping of crystals; there was little liquid associated with the particles. As shown in Figure 3, there are no sharp edges or distinct crystals typical of dry aerosols (Matijević, 1981). Calculations in Easley (1980) show droplet evaporation time to be 18 percent of the actual residence time in the dryer, and bulk diffusion is rapid enough to remove the vapor in one-half the dryer length. Use of an impaction plate to remove large droplets was found to be helpful in preventing overload of the dryer. In any case, the particles mimic atmospheric particles if not dry, or if dry, probably show upper-limit destruction rates for radicals; either case is suited to our purposes.

A Brink five-stage cascade impactor was used to measure the aerosol surface area. Although the particle size distribution was not especially stable from any of the tested methods, the total surface area for all fractions was very reproducible for given nebulizer conditions, as demonstrated in Jech (1981). A substantial improvement in the experimental technique could be achieved if the aerosol could be generated and dried under vacuum conditions. Other improvements are suggested in Jech et al. (1981).

Low Deactivation Coefficient Measurement

For certain radical species it was impossible to quantitatively measure a rate of reaction with aerosol surfaces. Rather, the solid surfaces were introduced into the reactor as a coating on a glass rod. The use of resonance fluorescence to detect OH precluded the presence of aerosols owing to scattered light interference. The axial position of the rod was adjustable, which varied reaction time and surface-to-volume ratio. The coatings were applied by dipping and rapidly removing the rod and allowing the warm, aqueous saturated solution to crystallize by cooling.

Experimental Procedure

In a typical experiment using actual aerosols, a solution of appropriate composition was introduced to the nebulizer and the surface area of the resulting dried particles was measured using the cascade impactor. A sans-aerosol experiment was conducted before and after each aerosol experiment to observe any signal changes due to pressure drift or conditioning of the reactor walls.

For experiments with a rod, the portion not exposed to the reaction zone was stored upstream of the reactant inlet port. The wall of the storage cell was treated with NaOH so it would rapidly scavenge free radicals and thus prevent them from diffusing far into the storage zone.

In a typical experiment, H_2 and NO_2 flow rates were set and the relative OH concentration was measured (after steady operation) at a fixed point downstream for decreasing lengths of the rod. Data were not taken for lengths of the rod inserted less than 5 cm into the reaction zone

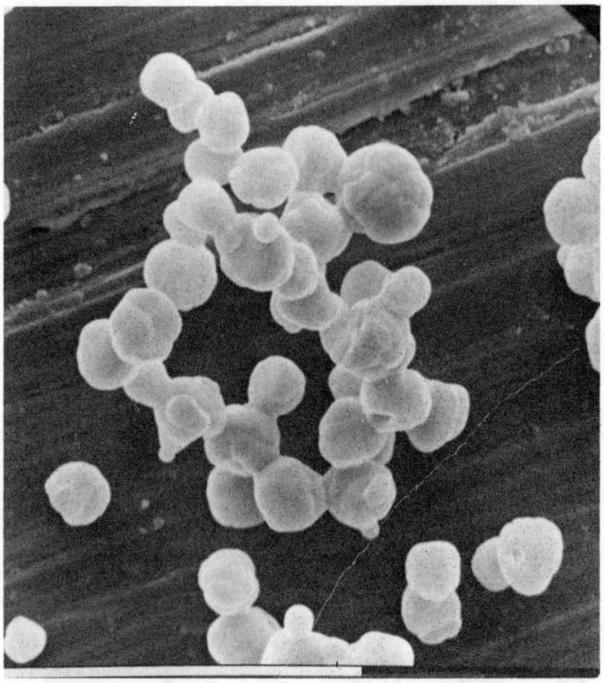

Fig. 3. Scanning electron photomicrograph of $(NH_4)_2SO_4$ aerosol particles impacted on plate 5 of a Brink impactor. The white bar at bottom is 10 μm long. Note the uneven impaction of the aerosols.

TABLE 5. Deactivation Coefficients Measured on Pure Aerosols

Surface	Loss Coefficient, γ
Oxygen Atoms	
NH_4Cl	1
NH_4NO_3	<0.03
$NaNO_3$	<0.09
$Fe(NO_3)_3 \cdot 9H_2O$	<0.03
$Fe_2(SO_4)_3 \cdot nH_2O$	<0.05
Hydrogen Atoms	
$NaNO_3$	<0.04
K_2CO_3	<0.006
Na_2SO_4	<0.004

because of the mixing length uncertainties. It was assumed that these uncertainties would nonetheless be a constant factor for longer rod lengths, and thus would not affect the rate constant analysis.

Results and Discussion

Atom Deactivation on Aerosols

In Table 5 results are given for the deactivation rates of oxygen atoms and hydrogen atoms on actual pure aerosol surfaces. As reported by Easley (1980), only NH_4Cl appeared to have a rapid O-atom destruction rate. The upper limit for loss coefficients on other aerosol compositions was determined by assuming that an increased decay of oxygen atom chemiluminescence (due to aerosol reactions) of 6 percent (experimental error + 1 percent) was detectable. Using the measured aerosol surface area, a maximum residence time in the flowtube, and the kinetic theory velocity of oxygen atoms, an upper limit to the deactivation rate is reported. The variability in the upper limits is entirely caused by the variation in surface area for the particular aerosol composition. For residence times longer than about 13 msec, homogeneous reactions at the 7-torr pressure of the study consume the oxygen atoms.

Similar experiments were performed on hydrogen atoms after optimizing the nebulizer performance. To minimize diffusion of H atoms to the wall, a higher reactor pressure was used (10 to 16 torr) and the surface area per volume for the aerosols was increased (Jech, 1981). In spite of this, the aerosol surface area was still too small to provide a measurable scavenging rate for H atoms, as can be seen in Table 5.

Radical Destruction on Coated Rod Surfaces

Owing to the low values of the loss coefficients for H atoms and to the upper-bound nature of the experiments, the use of a coated rod as the active surface was compared to destruction on actual

TABLE 6. Rod Surface Loss Coefficients (γ) Measured for H Atoms

Surface	Rod Number	Loss Coefficient Measurements	Mean Loss Coefficient	Estimated[a] Error (%)
$NaNO_3$[b]	1	0.05 - 0.20	0.013	50
K_2CO_3	1	0.054	0.039	50
	2	0.023		
NH_4Cl	1	8.6×10^{-4}	7.7×10^{-4}	20
	2	7.0×10^{-4}, 7.5×10^{-4}		
Na_2SO_4	1	5×10^{-4}	5×10^{-4}	50
$(NH_4)_2SO_4$	1	3.4×10^{-4}	2.5×10^{-4}	20
	2	1.5×10^{-4}		
NH_4NO_3	1	1.2×10^{-4}	1.7×10^{-4}	20
	2	1.7×10^{-4}		
	3	2.1×10^{-4}		

[a]Estimate of error is based on reproducibility of the data.
[b]See section entitled Comments Concerning Surface Reaction Mechanisms for pertinent discussion concerning this surface.

aerosol surfaces. Typical conditions were about 10^{15} atoms/cc at about 1 torr total pressure, and homogeneous H atom recombination was verified to be negligible by simulation calculations (Jech, 1981). Relative H concentration was determined at a fixed point downstream using the method of Clyne (1973) with a signal-to-noise ratio greater than 10. Loss coefficients are summarized in Table 6.

The surface deactivation rate of OH was determined using a coated rod and low enough concentration to ensure first-order OH loss rates. The kinetic analysis applied to this system appears in Jech (1981). Typical data appear in Figure 4. The two rods were identically treated; both were coated with NH_4NO_3. The analysis necessary to convert the apparent intensity decay (slopes) to deactivation coefficients appears in Jech (1981) and the results are shown in Table 7.

Discussion of Atom Destruction on Aerosol Surfaces

Simulation of well-established oxygen atom kinetics without aerosol reactions (and with them using $\gamma = 1$) showed that an apparent one-step surface deactivation mechanism was inadequate to describe the signal variation with time when aerosols were present. A two-step mechanism (Rideal) was postulated in which an estimate to the cross-sectional area of an O atom was used to convert aerosol area available per gas volume to surface sites. With this estimate and reaction at every collision, the model showed that the surface reached steady state in a far shorter residence time than could be observed in the experiment. Therefore, the failure to approach steady state could not account for the experimental trends. The experimental signal with aerosols present is reduced beyond that predicted by $\gamma = 1$ in the simulation (see Figure 5). Since the sans-aerosol intensity is adequately predicted, it is reasonable to assume that the deactivation coefficient for the aerosol is also 1, and that the initial aerosol surface area might be larger than was measured owing to effects mentioned in the section entitled Aerosol Characterization. Akers and Wightman (1976) also found a deactivation coefficient of 1 for O atoms on NH_4Cl.

Salt surfaces are poor catalysts for oxygen atom destruction, as can be seen in Table 1. The large reactivity associated with the NH_4Cl surface is consistent with the data if it behaves as

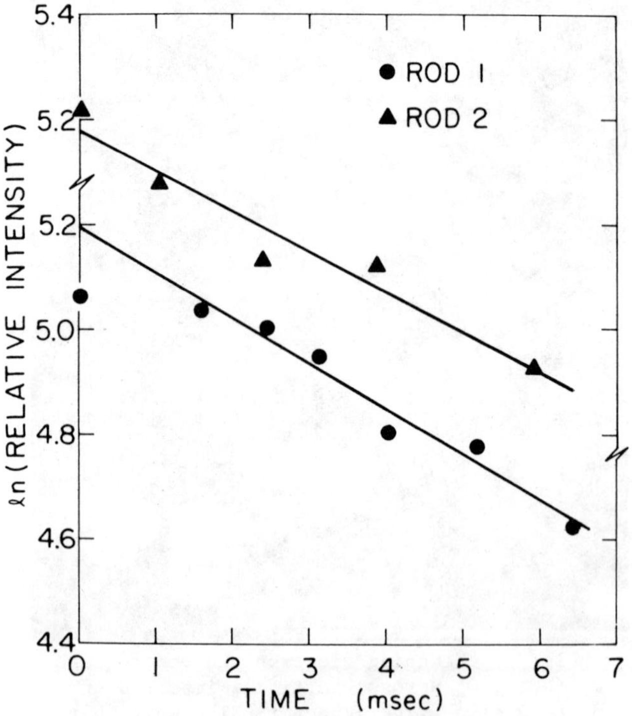

Fig. 4. Hydroxyl radical decay for ammonium nitrate surfaces.

TABLE 7. Surface Loss Coefficients (γ) Measured for OH Radical

Surface	Rod Number	Loss Coefficient Measurements	Mean Loss Coefficient	Estimated[a] Error (%)
$Pb(NO_3)_2$	1	0.25, 0.23	0.24	50
NH_4Cl	1	0.125, 0.135, 0.114	0.13	20
	2	0.126, 0.144		
$Fe(NO_3)_3$-FeO_x	1	0.16, 0.95	0.4	200
	2	0.058		
$FeSO_4 \cdot nH_2O$	1	0.095	0.12	20
	2	0.135		
$Zn(NO_3)_3$	1	0.026	0.045	50
	2	0.063		
$NaNO_3$	1	0.018	0.036	50
	2	0.053		
Na_2SO_4	1	0.013, 0.023	0.018	50
NH_4NO_3	1	0.0081	0.0099	20
	2	0.0109		
$(NH_4)_2SO_4$	1	0.0089, 0.0082	0.0085	20
$CH_2(COOH)_2$	1	0.17	0.17	20
NH_2CH_2COOH	1	0.030	0.020	50
	2	0.010		
CH_3CH_2COONa	1	0.013	0.013	20
Uncoated rod	1	0.0028	0.0027	20
	2	0.0026		

[a]Estimated error is based on reproducibility of the data, as explained in the text.

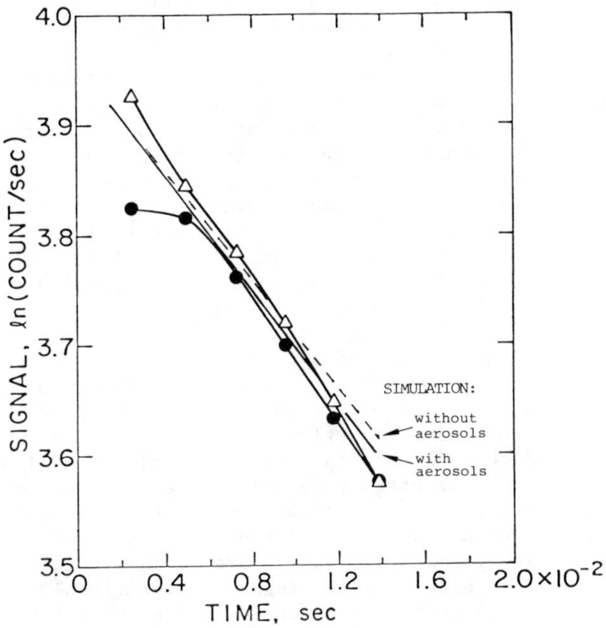

Fig. 5. Comparison of experimental results and simulation for O atom recombination on NH_4Cl aerosols. Dashed line indicates simulation without aerosols; solid line indicates simulation with aerosols; Δ = experimental results without aerosols; ● = experimental results with aerosols.

a reactant. Akers and Wightman (1976) found that NH_3, NO, H_2O, and HCl were the products of oxygen atom reaction with an NH_4Cl coating at long reaction times; they also found that the reactor surface became warm. These products imply the reactor coating was consumed to form volatile products. It has been shown (Krieger, 1975) that in the presence of excess oxygen atoms NH_3 produces chemiluminescence typical of the OH* Meinel bands, NO_2*, and HNO*. Thus it is reasonable to conclude that NO and H_2O measured by Akers and Wightman may be the result of further gas-phase oxidation of the ammonia, and HCl and NH_3 are probable reaction products directly from the surface reaction, causing consumption of the aerosol.

Similar behavior was observed in the rapid scavenging of H atoms by the $NaNO_3$ coating. Visual inspection revealed an initial color change from white to grayish-silver, followed by rapid loss of H atoms. (NaH is silver-gray, $Na_2O_2 \cdot H_2O$ is a white solid, and Na_2O is gray and deliquescent.) During the surface color change, emission at 760 nm (probably HNO*) was observed, indicating that NO_x is probably involved. Reexposure to the atmosphere returned the surface to white. A plausible multistep, thermodynamically favorable (exothermic) reaction mechanism for O atoms is proposed by Jech et al. (1981) which employs volatile reaction products such as NH_3 and HCl, consistent with the products found by Akers and Wightman (1976). A similar scheme is presented for H atoms with the last two steps constituting the Rideal mechanism for atoms.

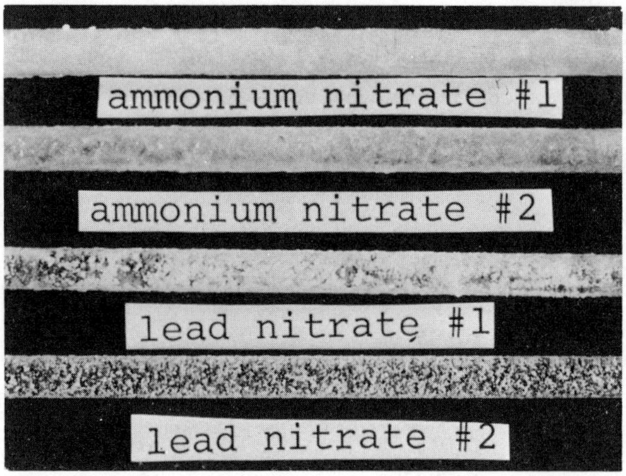

Fig. 6. Reproducibility of rod surface coatings. NH_4NO_3 rods represent surfaces easy to apply; $Pb(NO_3)_2$ rods represent surfaces difficult to apply.

In Figure 5 the experimental curves at long reaction times with and without aerosols converge. This behavior is attributed to the superposition of chemiluminescence derived from the original indicator reactions and "delayed" chemiluminescent reactions of the volatile products formed from the aerosol consumption. This type of behavior is supported by the H-atom/$NaNO_3$ results cited previously. The interference cannot be quantified by measurements at other wavelengths, since it is due to NO_x and the stoichiometry of the aerosol ablation reactions is unknown.

The issue of the wall condition versus axial distance remains to be discussed in light of Figure 5. Although the reactor was treated to reduce surface activity (Badachhape et al., 1976), it still represented a large removal rate for O atoms judging from the simulation and experiments without aerosols present. For the short contact time in the flow tube, arguments presented by Friedlander (1977) suggest that aerosol accumulation on the wall should be small, and visual inspection verified this. It is possible that water, although not reacting in the gas phase with species of interest, could react with the wall and decrease its activity as a function of distance. It is expected that if water were to affect the experiment in this fashion, it would also do so in the other unreactive salt experiments. Since the data show no change in signal decay, it is reasonable to conclude that it did not.

Discussion of Deactivation on Rod Coatings

The estimate of error in Tables 6 and 7 was based on the reproducibility of the data. The major error appeared to be in reproducing and characterizing the surface area of the rod coating. Reproducibility for a given rod was reasonably good (10 percent) based on results for NH_4Cl, $Pb(NO_3)_2$, and $(NH_4)_2SO_4$. Reproducibility of different rods coated with the same surface was not as good, and depended on the ease with which the coating was applied. These differences in surface reproducibility are illustrated in Figure 6. Shown are two NH_4NO_3 rods, typical of surfaces easily applied, for which reproducibility was 20 percent (based on data for NH_4NO_3 and $FeSO_4 \cdot nH_2O$). The two $Pb(NO_3)_2$ rods represent surfaces difficult to apply, which were reproducible to 50 percent (based on additional data from glycine and zinc nitrate).

The ferric nitrate surface was especially hard to apply and appeared to be gradually converted to the oxide by exposure to OH in the reactor. (The coating turned reddish brown, then progressively darker, and finally began to flake off the rod with increased exposure time.) Because of these factors, confidence in the loss coefficient value for this surface is poor.

A variety of surface coating characteristics were achieved with this method. Few coatings were smooth and many rods were incompletely covered. Although coverage was incomplete on about half the rods, the degree of coverage was reasonably uniform along the length of the rods, which was the critical parameter in the experimental method. For partially covered rods, the degree of coverage was estimated by visual observation under a microscope and a correction applied to the loss coefficient based on the coefficient for an uncoated rod. The OH loss coefficients reported reflect these corrections.

While the data are not very precise, they are nonetheless adequate for the purpose of this study, which was to screen various types of surfaces for their ability to scavenge OH radicals. Emphasis was placed on obtaining estimates of the loss coefficient for many surfaces rather than improvement of the accuracy for a few surfaces. Also, the values reported here are about as accurate as similar heterogeneous data typically found in the literature (e.g., Tables 1 to 3).

The accuracy of the present method could be significantly improved by developing better techniques for coating the rod with the desired surface and for characterizing the actual surface area. Some of the partially covered rods exhibited considerable three-dimensional crystalline character. In estimating the degree of coverage the coating was considered to be completely flat, so the surface area associated with the three-dimensional character was ignored. This effect, as well as uncertainties regarding additional surface due to porosity, could be accounted for by measuring the surface area by gas adsorption techniques, such as BET method. However, presently used models for aerosol scavenging of free radicals consider only the apparent (spherical) surface of the aerosol, ignoring surface roughness and porosity (Graedel et al., 1975; Warneck, 1974). Therefore, the grosser estimate of fractional coverage used in this study may be more appropriate.

Significance of the Results to Atmospheric Chemistry

As suggested in Zoller et al. (1977), rapid scavenging of radicals by aerosols could influence local levels of ozone in the stratosphere during major volcanic eruptions. The results of the present study suggest that such mechanisms are rapid and perhaps should be included in certain modeling studies (e.g., Kasting and Donahue, 1980). The results of the present study - especially those for OH - also relate to the chemistry of the troposphere as modeled by Graedel et al. (1976).

The pertinent experimental OH data obtained in this study are summarized in Table 8. Because of the uncertainty of the data, the surfaces were merely categorized according to whether scavenging of OH radicals was found to be relatively fast ($\gamma > 0.1$), medium, or slow ($\gamma < 0.01$). These results may be directly used in atmospheric models where the surroundings of the particle are assumed to be locally well mixed, since the experiments (at low pressure) are free of slow mass-transfer rate limitations.

Some trends were observed in these data. As expected, the loss coefficient depended strongly on the surface composition. The fly-ash-type surfaces (Pb, Fe, and Zn salts) were generally more active than the natural aerosol-type surfaces (NH_4 and Na salts). NH_4Cl, considered to be an aerosol derived from natural sources, was a notable exception. Among the organic surfaces, the activity decreased according to the functional group in the order acid, amine, acid salt. This trend will be discussed further in this paper. To the authors' knowledge, no data exist in the literature for direct comparison to these data.

Graedel et al. (1976) showed aerosol scavenging to be "important" for $\gamma > 0.033$ for HO_2 radicals. They obtained significant improvement in matching model-calculated pollutant concentrations to field data for $0.033 < \gamma < 1.0$; the calculated concentrations of O_3, NO, and NO_2 were only 10 percent different for the $\gamma = 0.033$ versus $\gamma = 1.0$ cases. On this basis, aerosol-HO_2 sinks were deemed to be important constituents of chemical models of the urban troposphere.

Graedel also concluded that aerosol scavenging of OH would be unimportant even for $\gamma = 1$ since aerosol scavenging of OH must compete with other OH sinks in order to be significant (Graedel et al., 1976). One competing sink mechanism is the scavenging of HO_2 by aerosols, since OH and HO_2 form a catalytic cycle, and thus a sink for one radical is actually a sink for the other.

However, an important modification should be applied to Graedel's model (Graedel et al., 1976), based on a revised rate constant for the key reaction:

TABLE 8. Loss Coefficients (γ) for OH Radical on Various Surfaces

$\gamma > 0.1$ (fast OH scavenging surfaces)	$0.1 > \gamma > 0.01$ (medium OH scavenging surfaces)	$\gamma < 0.01$ (slow OH scavenging surfaces)
$Pb(NO_3)_2$	$Zn(NO_3)_2$	Na_2SO_4
NH_4Cl	$NaNO_3$	NH_4NO_3
$FeSO_4 \cdot nH_2O$	NH_2CH_2COOH (glycine)	$(NH_4)_2SO_4$
$Fe(NO_3)_3 - FeO_x$		CH_3CH_2COONa (sodium propionate)
$CH_2(COOH)_2$ (malonic acid)		

$$HO_2 + \longrightarrow OH + NO_2 \qquad (c)$$

The newly accepted value is 40 times faster than the rate constant used by Graedel. Since this reaction is the major channel for converting HO_2 to OH, the faster rate constant would substantially increase the OH concentration predicted by this model. Although computer calculations are needed to accurately assess this effect (Luther and Peters, this volume), a rough calculation (Jech, 1981) showed that the relative importance of OH aerosol scavenging increased such that the OH loss coefficient need only be ~0.3 to 0.5 in order to effect the model improvement noted by Graedel.

Several of the surfaces shown in Table 8 were found to have an OH loss coefficient near this adjusted "critical" value. Also, the few preliminary data obtained for HO_2 (see Table 9) are within the critical value for HO_2. The results of this study and details in Jech et al. (1981) therefore support the concept that aerosol sink mechanisms for free radicals might be rapid in the polluted troposphere. However, urban aerosols do not have a single composition. It is felt that over a large area, unreactive aerosols will "dilute" reactive aerosols. Heterogeneous reactions with free radicals will be important only locally, i.e., near an aerosol source. This conclusion hinges on the previous argument that Graedel's model (Graedel et al., 1976) substantially underestimated urban OH concentrations.

Uncertainty presently exists concerning measured ambient concentrations of free radicals (D. D. Davis, private communication, 1981; Jech, 1981). Isaksen and Crutzen (1977) have shown that model-calculated OH concentrations vary by a factor of 100 because of uncertainties in CH_4 oxidation, HO_2 scavenging by aerosols, and the rate of the homogeneous sink $OH + HO_2 \rightarrow H_2O + O_2$. Larger uncertainties were shown to exist for HO_2 concentrations. Further, recent work by Selzer and Wang (1979) indicates that the direct measurement value using resonance fluorescence should be ~10 times larger than previously thought. This correction

TABLE 9. Loss Coefficient Estimates Measured for HO_2

Surface	Loss Coefficient, γ	Confidence in Value
NH_4Cl	0.023	poor[a]
$(NH_4)_2SO_4$	0.013	fair

[a]Atom generation from homogeneous reactions of HO_2 complicates kinetics.

arises because of new measurements of quenching of $OH(A^2\Sigma^+)$ by ambient molecules (N_2 and O_2), as well as other factors relating to the OH fluorescence efficiency. If valid, such a correction would further strengthen the relative importance of aerosol sinks for OH radicals.

The assessment of the importance of free-radical scavenging by aerosols also depends on the assumed aerosol concentration and size distribution. Graedel et al. (1976) used the size distribution of Whitby et al. (1972), which results in a total aerosol surface of 1×10^{-5} cm^2/cm^3. Size distributions given by other authors result in comparable but slightly higher surface areas: 2×10^{-5} cm^2/cm^3 (Husar et al., 1972) and 3.8×10^{-5} cm^2/cm^3 (Ensor and Charlson, 1972). If urban aerosol concentrations do exceed the value used by Graedel, or are allowed to increase in the future, the critical OH loss coefficient would be further reduced from the adjusted 0.5 value given previously. Aerosol scavenging of free radicals might then be of importance in local tropospheric models.

Comments Concerning Surface Reaction Mechanisms

The main purpose of this investigation was not to elucidate the detailed mechanisms of the heterogeneous reactions; rather, priority was given to the determination of the relative scavenging ability for a reasonably large set of surfaces for use with atmospheric models. Accordingly, products were not identified nor surface changes quantified. Although detailed mechanisms were not determined, some comments can be made.

Three general types of surfaces were noted: (1) reactive (stoichiometric), (2) "catalytic", and (3) reactive, then "catalytic". (The term catalytic is used in its loosest sense, without regard for turnover number considerations.) The first type of surface is represented by the system $O + NH_4Cl$, in which the surface rapidly reacts with the radical to form volatile products which escape, exposing fresh surface to react further. This type of surface is gradually consumed. The second type of surface is the most commonly observed. Here, the surface appears to remain unchanged while the radical is "destroyed," probably via the Rideal mechanism.

The third surface type was characterized by a relatively slow initial reaction, followed by (usually) faster "catalytic" removal of the free radical. An example is provided by the system $OH + Fe(NO_3)_3$, in which the surface was gradually converted to a nearly black substance which gradually flaked off (probably an oxide of iron). The new surface, labeled FeO_x in Table 8, appeared to be an extremely rapid scavenger of OH radicals. More work is needed in order to remove these remarks from the realm of speculation.

It is interesting to compare the OH scavenging data obtained for organic surfaces with the mechanisms proposed by Cheaney et al. (1959). These surfaces were chosen to have functional groups that have increasing acidic character in the sequence sodium propionate, glycine, malonic acid (i.e., acid salt, amine, acid). The activity of these surfaces was found to increase as the acidity increased, which is opposite to the postulation of Cheaney and coworkers. More data, perhaps including surface sequences as used by Cheaney (e.g., KF, KCl, KBr, KI), are needed in order to determine whether OH behaves differently than other radicals or whether these particular organic surfaces are fundamentally different from the surface sequences investigated by previous researchers.

Conclusions

Several of the chemical components known to be abundant in urban aerosols have been shown to be active surfaces for the removal of hydroxyl (OH) radicals from the gas phase. In particular, the results of this study suggest that aerosols derived from fly ash, photochemical smog, and certain natural sources are likely to be relatively active sinks for OH radicals. Most of the abundant natural aerosols, such as $(NH_4)_2SO_4$, NH_4NO_3, and $NaNO_3$, appear to be relatively inactive. Scavenging of hydroperoxyl radicals (HO_2) by aerosols also appears to be fairly rapid, based on a few preliminary data. Since HO_2 is also abundant, it is probably the most important free radical scavenged by aerosols, and additional data for HO_2 are very desirable.

The most active surfaces studied were found to have slower scavenging rates for OH than those used in atmospheric models to show these mechanisms to be important. In addition, the wide variety of compositions in actual atmospheric aerosols suggests that unreactive surfaces would "dilute" the reactive ones.

At present, urban aerosol concentrations are low enough that the effect of aerosol scavenging on the chemical behavior of major atmospheric pollutants is subtle. Therefore, other considerations will probably dominate in the development of air pollution control strategies, such as carcinogenicity, toxicity, or ozone layer destruction. However, with the expected increased use of coal-fired power plants and increased political pressure to relax emission standards for such plants, aerosol concentrations may rise in the future. If this occurs, heterogeneous mechanisms would become more important, especially locally where a single reactive aerosol composition might dominate, and the overall chemical behavior of the local atmosphere might be more profoundly affected.

Acknowledgements. The authors would like to acknowledge the financial support of the National Science Foundation (Engineering - ENG 78-11698) and the University of Washington Graduate School Research Fund. In addition, the assistance of Drs. Robert Charlson, Alan Waggoner, David Covert, John Rosen, and Tim Larson is gratefully acknowledged. The comments of the reviewers aided in clarification of the manuscript and are appreciated.

References

Akers, F. I., and J. P. Wightman, Kinetic study of the interaction between atomic oxygen and aerosols, J. Phys. Chem., 80, 835, 1976.

Anderson, J. G., and F. Kaufman, Kinetics of the reaction $OH + NO_2 + M \rightarrow HNO_3 + M^*$, Chem. Phys. Lett., 16, 375, 1972.

Arnold, S., and M. Lewittes, Photophoretic spectroscopy: A search for the composition of a single aerosol particle, this volume.

Azatyan, V. V., F. A. Grigoryan, and S. B. Filippov, Study of heterogeneous recombination of hydrogen and oxygen atoms by means of an ESR method, Kinet. Catal., 13, 1241, 1972.

Badachhape, R. B., P. Kamarchik, A. P. Conroy, G. P. Glass, and J. L. Margrave, Preparation of nonreactive surfaces for reaction-rate studies, Int. J. Chem. Kin., 8, 23, 1976.

Baldwin, A. C., and D. M. Golden, Heterogeneous atmospheric reactions: Sulfuric acid aerosols as tropospheric sinks, Science, 206, 562, 1979.

Breen, J. E., and G. P. Glass, Rate of some hydroxyl radical reactions, J. Chem. Phys., 52, 1082, 1970.

Brown, R. L., Diffusion of a trace gas into a flowing carrier, Int. J. Chem. Kin., 2, 475, 1970.

Chang, J. S., and F. Kaufman, Upper boundaries and probable value of the rate constant of the reaction $OH + HO_2 \rightarrow H_2O + O_2$, J. Phys. Chem., 82, 1683, 1978.

Cheaney, D. E., D. Davies, A. Davis, D. E. Hoare, J. Protheroe, and A. D. Walsh, Effects of surfaces on combustion of methane and mode of action of anti-knocks containing metals, Seventh Symposium (Int.) on Combustion, The Combustion Institute, Pittsburgh, 1959.

Clyne, M. A. A., Reactions of atoms and free radicals studied in discharge-flow systems, in Physical Chemistry of Fast Reactions, Plenum Press, New York, 1973.

Clyne, M. A. A., and S. Down, Kinetics behavior of hydroxyl radical $X^2\pi$ and $A^2\Sigma^+$ using molecular resonance fluorescence spectrometry, Trans. Faraday Soc., 70, 253-266, 1974.

Easley, P. G., Kinetics of free-radical scavenging by aerosols, M. S. Thesis, University of Washington, Seattle, 1980.

Elias, L., E. A. Ogryzlo, and H. I. Schiff, The study of electrically discharged O_2 by means of an isothermal calorimetric detector, Can. J. Chem., 37, 1680, 1959.

Ensor, D. S., and R. J. Charlson, Multiwavelength nephelometer measurements in Los Angeles smog aerosol, J. Coll. Inter. Sci., 39, 242, 1972.

Friedlander, S. K., Smoke, Dust, and Haze: Fundamentals of Aerosol Behavior, John Wiley & Sons, New York, 1977.

Graedel, T. E., The kinetic photochemistry of the marine atmosphere, J. Geophys. Res., 84, 223, 1979.

Graedel, T. E., L. A. Farrow, and T. A. Weber, The influence of aerosols on the chemistry of the troposphere, Proc. Sym. Chem. Kin. Upper and Lower Atm., Int. J. Chem. Kin. Symp., 1, 581, 1975.

Graedel, T. E., L. A. Farrow, and T. A. Weber, Kinetic studies of the photochemistry of the urban troposphere, Atmos. Environ., 10, 1095, 1976.

Greaves, J. C., and J. W. Linnett, The recombination of oxygen atoms at surfaces, Trans. Faraday Soc., 54, 1323, 1958.

Greaves, J. C., and J. W. Linnett, Recombination of atoms at surfaces. Parts 4-6, Trans. Faraday Soc., 55, 1338-1361, 1959.

Grosjean, D., Aerosols, in Ozone and Other Photochemical Oxidants, Chapter 2, Nat. Acad. Sci., Washington, D.C., 1977.

Hacker, D. S., S. A. Marshall, and M. Steinberg, Recombination of atomic oxygen on surfaces, J. Chem. Phys., 35, 1788, 1961.

Hardy, W. A., and J. W. Linnett, Mechanisms of atom recombination on surfaces, Eleventh Symp. (Int.) on Combustion, p. 167, Academic Press, New York, 1966.

Herron, J. T., J. L. Franklin, P. Bradt, and V. H. Dibeler, Kinetics of nitrogen atom recombination, J. Chem. Phys., 30, 879, 1959.

Herron, J. T., and H. I. Schiff, A mass spectrometric study of normal oxygen and oxygen subjected to electrical discharge, Can. J. Chem., 36, 1159, 1958.

Howard, C. J., and K. M. Evenson, Kinetics of the reactions of HO_2 with NO, Geophys. Res. Lett., 4, 437, 1977.

Husar, R. B., K. T. Whitby, and B. Y. H. Liu, Physical mechanisms governing the dynamics of Los Angeles smog aerosol, J. Coll. Inter. Sci., 39, 211, 1972.

Isaksen, I. S. A., and P. J. Crutzen, Uncertainties in calculated hydroxyl radical densities in the troposphere and stratosphere, Geophys. Nor., 31, 1, 1977.

Jech, D. D., Kinetics of reactions between free radicals and surfaces applicable to atmospheric chemistry, M.S. Thesis, University of Washington, Seattle, 1981.

Jech, D. D., P. G. Easley, and B. B. Krieger, Kinetics of Free Radical Scavenging by Aero-

sols, National Science Foundation, Washington, D.C., 1981.
Judeikis, H. S., and S. Siegel, Particle-catalyzed oxidation of atmospheric pollutants, Atmos. Environ., 7, 619, 1973.
Judeikis, H. S., T. B. Stewart, and A. G. Wren, Laboratory studies of heterogeneous reactions of SO_2, Atmos. Environ., 12, 1633, 1978.
Kasting, J. F., and T. M. Donahue, The evolution of atmospheric ozone, J. Geophys. Res., 85, 3255, 1980.
Kaufman, F., Reactions of oxygen atoms, Progr. Reac. Kin., 1, 1-39, 1961.
Kaufman, F., and John R. Kelso, M effect in the gas-phase recombination of O with O_2, J. Chem. Phys., 46, 4541, 1967.
Krieger, B. B., Oxygen atom and hydrocarbon chemiluminescence: The mechanism and rate constant for the production and loss of OH (v = 9) in the reaction of oxygen atoms and ethylene, Ph. D. Thesis, Wayne State University, Detroit, Michigan, 1975.
Krieger, Barbara B., and Ralph H. Kummler, Use of mathematical model and experiments to determine the mechanism and rate constant of OH chemiluminescence from oxygen atom attack on ethylene, J. Phys. Chem., 81, 2493, 1977.
Leu, M. T., Rate constant for the reaction $HO_2 + NO \rightarrow OH + NO_2$, J. Chem. Phys., 70, 1662, 1979.
Levy, H., Normal atmosphere large radical and formaldehyde concentrations predicted, Science, 173, 141, 1971.
Liberti, Arnaldo, The nature of particulate matter, Pure Appl. Chem., 24, 631, 1970.
Linnett, J. W., and D. G. H. Marsden, The kinetics of the recombination of oxygen atoms at a glass surface, Proc. Roy. Soc. A, 234, 489, 1956a.
Linnett, J. W., and D. G. H. Marsden, The recombination of oxygen atoms at salt and oxide surfaces, Proc. Roy. Soc. A, 234, 504, 1956b.
Lund, R. E., and H. J. Oskam, Studies of N_2 afterglow. I. Surface catalytic efficiency and diffusion-controlled decay of atomic nitrogen, J. Chem. Phys., 48, 109, 1968.
Luther, C. J., and L. K. Peters, The possible role of heterogeneous aerosol processes in the chemistry of CH_4 and CO in the troposphere, this volume.
Manella, G., and P. Harteck, Surface-catalyzed excitations in the oxygen system, J. Chem. Phys., 34, 2177, 1961.
Matijević, E., Monodispersed metal (hydrous) oxides - A fascinating field of colloid science, Acct. Chem. Res., 14, 22-29, 1981.
McMurry, P. H., and S. K. Friedlander, Aerosol formation in reacting gases: Relation of surface area to ratio of gas-to-particle conversion, J. Coll. Inter. Sci., 64, 248-257, 1978.
Mearns, A. M., and A. J. Morris, Production of oxygen atoms in quartz and polytetrafluorethylene discharge tubes, Nature, 225, 59, 1970.
Melin, G. A., and R. J. Madix, Energy Accommodation during oxygen atom recombination on metal surfaces, Trans. Faraday Soc., 67, 205, 1971.
Mulcahy, M. F. R., and B. C. Young, Heterogeneous reactions of OH radicals, Int. J. Chem. Kinet., 7, (Symp. 1), 595-609, 1974.
Ogren, P. J., Analytical results for first-order kinetics in flow-tube reactors with wall reactions, J. Phys. Chem., 79, 1749, 1975.
Olsen, K. W., W. J. Haas, and V. A. Fassel, Multielement detection limits and sample nebulization efficiencies of an improved ultrasonic nebulizer and a conventional pneumatic nebulizer in inductively coupled plasma-atomic emission spectrometry, Anal. Chem., 49, 632, 1977.
Otterbein, M., and L. Bonnetain, Reaction du carbone avec l'oxygene atomique, Carbon, 7, 539, 1969.
Pinnick, R. G., and H. J. Auverman, Response characteristics of Knollenberg light-scattering aerosol counters, J. Aerosol Sci., 10, 55, 1979.
Selzer, P. M., and C. C. Wang, Quenching rates and fluorescence efficiency in the $A^2\Sigma^+$ state of hydroxyl radical, J. Chem. Phys., 71, 3786, 1979.
Smith, G. K., B. B. Krieger, and P. M. Herzog, Experimental and analytical study of wall reaction and transport effects in fast reaction systems, AIChE J., 26, 567, 1980.
Smith, W. V., The surface recombination of H atoms and OH radicals, J. Chem. Phys., 11, 110, 1943.
Stuhl, F., and H. Niki, Pulsed vacuum-photochemical study of reactions of OH with H_2, O_2 and CO using a resonance-fluorescent detection method, J. Chem. Phys., 57, 3671, 1972.
Thrush, B. A., Reactions of hydrogen atoms in the gas phase, Prog. React. Kin., 3, 73, 1965.
Vardanyan, I. A., G. A. Sachyan, and A. B. Nalbandyan, Kinetics and mechanisms of formaldehyde oxidation, Comb. Flame, 17, 315, 1971.
Venugopalan, M., Chemistry of Dissociation of Water Vapor and Related Systems, Interscience Publishers, New York, 1968.
Walker, R. E., Chemical reactions and diffusion in a catalytic tubular reactor, Phys. Fluid, 4, 1211, 1961.
Warneck, P., On the role of OH and HO_2 radicals in the troposphere, Tellus, 26, 40, 1974.
Warren, D. R., Surface effects in combustion reactions, Part 1 and 2, Trans. Faraday Soc., 53, 199, 1957.
Westenberg, A. A., and N. de Haas, Rate of the reaction $OH + OH \rightarrow H_2O + O$, J. Chem. Phys., 58, 4066, 1973.
Whitby, K. T., R. B. Husar, and B. Y. H. Liu, The aerosol size distribution of Los Angeles smog, J. Coll. Inter. Sci., 39, 177, 1972.
Williams, D. J., and M. F. R. Mulcahy, Effects of various coatings on the recombination coefficients of oxygen atoms at glass surfaces, Aust. J. Chem., 19, 2163, 1966.
Wise, H., and W. A. Rosser, Homogeneous and heterogeneous reactions of flame intermediates,

Ninth Symposium on Combustion, pp. 733-746, Academic Press, New York, 1963.
Wise, H., and B. J. Wood, Gas and surface atom reactive collisions, Adv. Atom. Molec. Phys., 3, 291-348, 1967.
Wood, B. J., and H. Wise, The kinetics of hydrogen atom recombination on Pyrex glass and fused quartz, J. Phys. Chem., 66, 1049, 1962.
Yolles, R. S., and H. Wise, Diffusion and heterogeneous reaction. X. Diffusion coefficient measurements of atomic oxygen through various gases, J. Chem. Phys., 48, 5109, 1968.
Young, R. A., Pressure dependence of the absolute catalytic efficiency of surfaces for removal of atomic nitrogen, J. Chem. Phys., 34, 1292, 1961.
Zoller, W. H., E. A. Lepel, E. J. Mroz, and K. M. Stefansson, Trace elements from volcanoes: Augustine, 1976, in Air Pollution Measurement Techniques, WMO Publ. No. 460, World Meterological Organization, 1977.

PHOTOASSISTED REACTIONS ON DOPED METAL OXIDE PARTICLES

J. M. White

Department of Chemistry, University of Texas, Austin, Texas 78712

Abstract. Photoassisted reactions involving doped metal oxide semiconductor powders are reviewed. The mechanisms, insofar as they are now understood, are discussed for several reactions involving gas-phase water near its equilibrium vapor pressure in contact with the oxide. Experimental results are presented along with some discussion of how these processes may be involved in heterogeneous processes in the atmosphere.

Introduction

Solid-state photochemistry and photophysics have been studied for a very long time. One outstanding example is the photographic process. The study of gas-solid and liquid-solid photoassisted processes is being pursued with renewed vigor today and must be considered as one of the very exciting topics of surface science and catalysis. Interest has developed along several fronts; one of the significant historical markers is the report in 1972 by Fujishima and Honda (1972) that water can be decomposed to H_2 and O_2 in a photoelectrochemical cell using a Pt cathode and an irradiated TiO_2 anode. Prior to that, a significant amount of work on heterogeneous photocatalysis at the gas-solid interface had been reported. Much of this work, and some closely related work, has been recently reviewed (Formenti and Teichner, 1978; Bickley, 1978; Morrison, 1977). Reviews of photoelectrochemistry (Bard, 1979; Nozik, 1978; Butler and Ginley, 1980) and the surface science of irradiated surfaces (Somorjai, 1981) have also appeared and a recent symposium (Wrighton, 1980) dealt with interfacial photoprocesses. These and several other reports from across the international scientific community indicate a surge in activity, not unlike the chemical activity we hope to see at an irradiated interface. My own interest dates to the mid-sixties when, in the course of a photochemical hot-hydrogen-atom gas-phase reaction study (Kuppermann and White, 1966), I noticed some interesting iodine atom storage properties of Teflon. This, along with reports that CO could be photodesorbed from nickel surfaces, led to our first work on photoeffects at the gas-solid interface; McAllister showed that the cross section for direct photodesorption of CO from polycrystalline nickel was less than 10^{-23} cm^2 (McAllister and White, 1973). In other words, thermal effects are responsible for most of the observed desorption. To follow up this work, I proposed to study photoassisted reactions on small oxide particles as models for atmospheric reactions. This proposal was delayed until about 3 years ago, when Dr. Shinri Sato from the Catalysis Institute at Hokkaido joined our research group and set out to search for gas-phase analogs of what had been observed in photoelectrochemistry. Much of the work discussed here is due to his research.

The purpose of this paper is to review briefly the present status of our research on photoassisted reactions at the gas-solid interface. There are a number of very interesting papers in the literature on this topic; only a few of them are mentioned here.

General Considerations

With the possible exception of the conversion of methanol to formaldehyde over silver (Fenstik et al., 1972), photoassisted heterogeneous catalytic reactions involve compound semiconductors and band-gap irradiation. The essential features for an n-type semiconducting oxide such as TiO_2 are shown in Figure 1. Absorption of a photon excites an electron from the filled valence band into the conduction band. As shown in the figure, the bands will typically bend upward at the interface between the n-type semiconductor and either a metal (Schottky diode junction) or an electrolyte. The presence of this electrical field causes spatial separation of the electron in the conduction band and the hole in the valence band. As a result, two excited and charged species exist with which charge transfer (redox) chemistry can be carried out. Although band bending is well established for many systems, the situation is not so clear for small metal oxide particles with or without even smaller metal particles present on their surfaces. In fact, and particularly under illumination, the bands may be nearly flat. Under such conditions electron-hole recombination will compete more favorably with charge transfer reac-

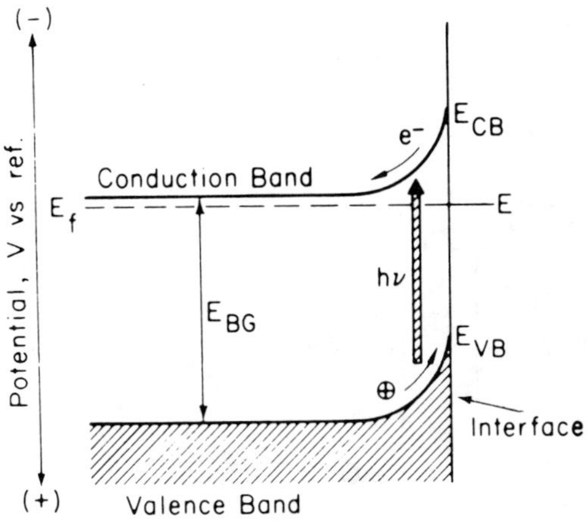

Fig. 1. Schematic energy level diagram for an n-type semiconductor surface.

tion channels, and there is more heat produced than chemistry. This does not mean that charge transfer reactions will not occur; it does mean that they must proceed rapidly to be competitive.

The charge transfer reactions will involve oxidation and reduction components and, energetically, the levels of the oxidant and reductant must be appropriately placed with respect to the bands of the solid. The proper placement is shown in Figure 2; the oxidation step involves a level placed above the top of the valence band while the reduction step involves a level lying below the bottom of the conduction band.

The actual situation for several compound semiconductors is shown in Figure 3 as it relates to charge transfer reactions leading to the decomposition of water into hydrogen and oxygen. The more stable materials, ZnO, TiO_2, WO_3 and SnO_2, all have band gaps which are much larger than the 1.23 eV needed to meet the thermodynamic requirements of the water decomposition reaction. The problem is that the potential of the H^+/H reduction reaction lies above the bottom of the conduction band, so that electron transfer from this band through the interface to reduce a proton is not very probable and the overall rate will drop to an insignificantly low value. Other materials which have narrower band gaps (which would allow better utilization of the solar spectrum) and proper placement of the bands on the redox energy scale (for example, CdS and CdSe) typically corrode. That is, the lattice is itself reacted away through redox processes involving the cations and anions of the lattice. In photoelectrochemical cells, the problem with the placement of the bands can be overcome by using an external power supply to bias the semiconductor with respect to the redox levels of interest. In the case of TiO_2, such a bias allows the photoassisted decomposition of water (Fujishima and Honda, 1972).

In examining Figure 3 it is important to keep in mind that some of these compounds exist in more than one crystalline form and these may have different band gaps. For example, TiO_2 exists in two forms, anatase and rutile. The gap for rutile is shown in Figure 3, and there is some evidence that anatase has a slightly lower gap, with the bottom of the valence band lying above the H^+/H potential (Tomkiewicz, 1979) which makes the water decomposition reaction thermodynamically feasible.

Of course, the number of compound semiconductors is large and their properties can not be reviewed in detail here. Those that have been studied in photoelectrochemistry are listed in Table 1 (Kung et al., 1977). Iron oxide, with a band gap of 2.0 ±0.1 eV, may be a compound of interest in atmospheric processes. Its conduction band is located at a voltage too low for the water decomposition reaction. However, many other reactions, particularly those involving oxidation of carbon sources, might occur on irradiated iron oxide. Even if the iron oxide is destroyed in the process, so that reactions become heterogeneous stoichiometric processes as opposed to heterogeneous catalytic processes, the source of the iron oxide will keep its concentration at a steady-state value. The same considerations will apply to other particulates in the atmosphere.

This discussion has focused on compound semiconductors. Charge transfer reactions on metals are unlikely because the lifetimes of metal excited states are too short to allow chemical reactions to become competitive. Photoassisted reactions could occur by direct excitation of an adsorbed or gas-phase species which then reacts or decays through surface interactions. This topic lies outside the scope of this report.

Whether photoassisted heterogeneous reactions are catalytic or stoichiometric is a very important question, particularly if the goal is catalytic activity requires that the measured turnover

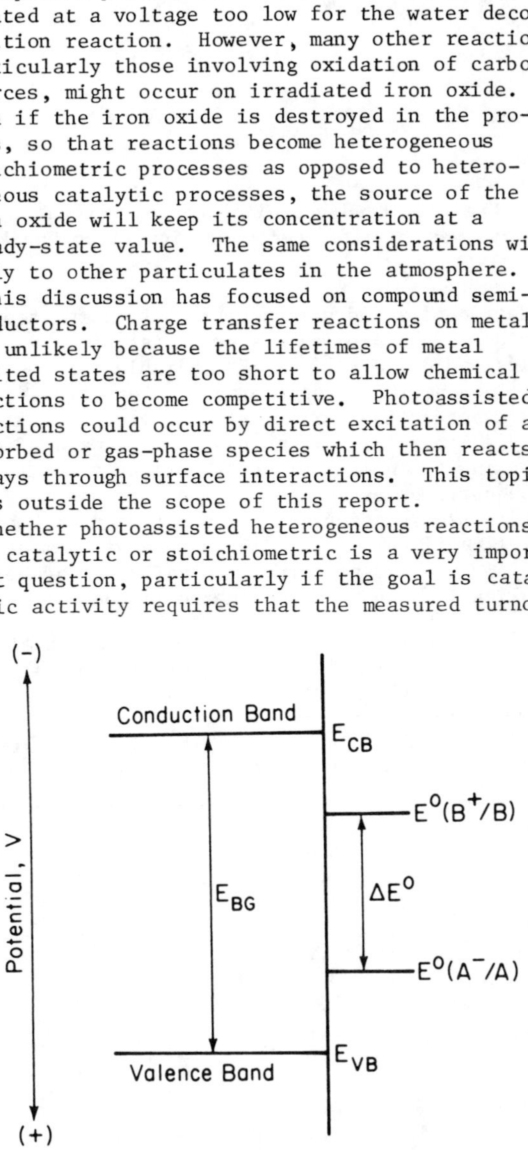

Fig. 2. Energy level diagram showing desired placement of redox levels with respect to band gap in a semiconducting material.

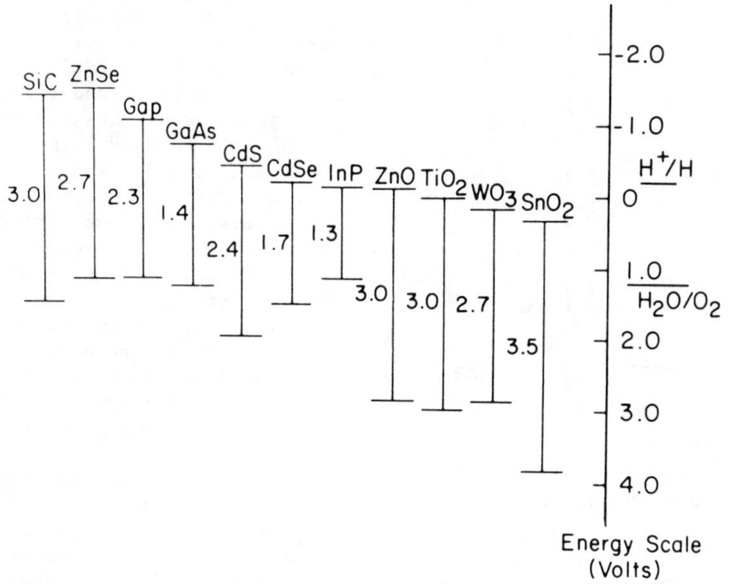

Fig. 3. Band gap placement of several compound semiconductors with respect to the redox levels involved in water decomposition (from Nozik, 1978).

oxidation of solid carbon sources to form carbon dioxide. This problem has been discussed by Childs and Ollis (1980). Establishing photocatalytic production of some product, for example, the number per active surface site be greater than unity. Since there is no unambiguous method for measuring the number of active sites at a metal oxide surface, estimates must be made and these must be used cautiously. As pointed out by Childs and Ollis (1980), a calculated turnover number of 100 is certainly preferable to unity. A number of complications must be considered in demonstrating

TABLE 1. Band Gaps and Flat Band Potentials of Oxide Anodes (from Kung et al., 1977)

Material	Band Gap Energy (eV, ±0.1 eV)	Flat-Band Potential V_{fb} (V, ±0.1 V)
SnO_2	3.5	0.5
ZrO_2	5.0	-1.0
Ta_2O_5	4.0	-0.4
Wb_2O_5	3.4	0.0
$KTaO_3$	3.5	-0.2
$SrTiO_3$	3.2	-0.2
$BaTiO_3$	3.1	0.15
TiO_2	2.9	0.2
WO_3	2.4	0.5
$Pb_4Ti_3WO_{13}$	2.4	0.4
CdO	2.1	0.8
$CdFe_2O_4$	2.3	0.8
Fe_2O_3	2.0	0.7
$PbFe_{12}O_{19}$	2.3	1.0
$Hg_2Nb_2O_7$	1.8	1.1
Hg_2TaO_7	1.8	1.2

that more than a stoichiometric reaction is being studied. As one example, consider studies involving TiO_2 substrates. Typically, these substrates are partially reduced prior to their use. The reduction leads to oxygen vacancies and Ti^{+3} species at the surface and in the subsurface region. There is considerable evidence to suggest that Ti^{+3} is intimately involved in photodriven processes (Somorjai, 1981). How should we determine the concentration of these species to be used in turnover number calculations? If those Ti^{+3} species are located in the first few layers of the solid participate, then the number of moles of product produced must exceed the total Ti^{+3} concentration, not just the surface Ti^{+3}, before a nonstoichiometric reaction can be unambiguously demonstrated.

Experimental

All of the experimental work reported here was carried out in a glass high-vacuum system which contained a sample loop as shown in Figure 4. The sample loop had a volume of 180 cc and was designed so that gases were circulated through the loop by a circulating pump. The catalyst, consisting of a fine powder, was placed in the bottom of a quartz reaction vessel. Light from a 200-W high-pressure mercury arc lamp, which is rich in ultraviolet and visible radiation, was passed through a filter containing nickel sulfate to take out the infrared and was reflected off the mirror into the reaction cell. The reaction cell itself was dipped in a constant-temperature water bath. This constant-temperature bath as well as thermal conduction by means of the gas phase assured that the input energy did not overly heat the solid.

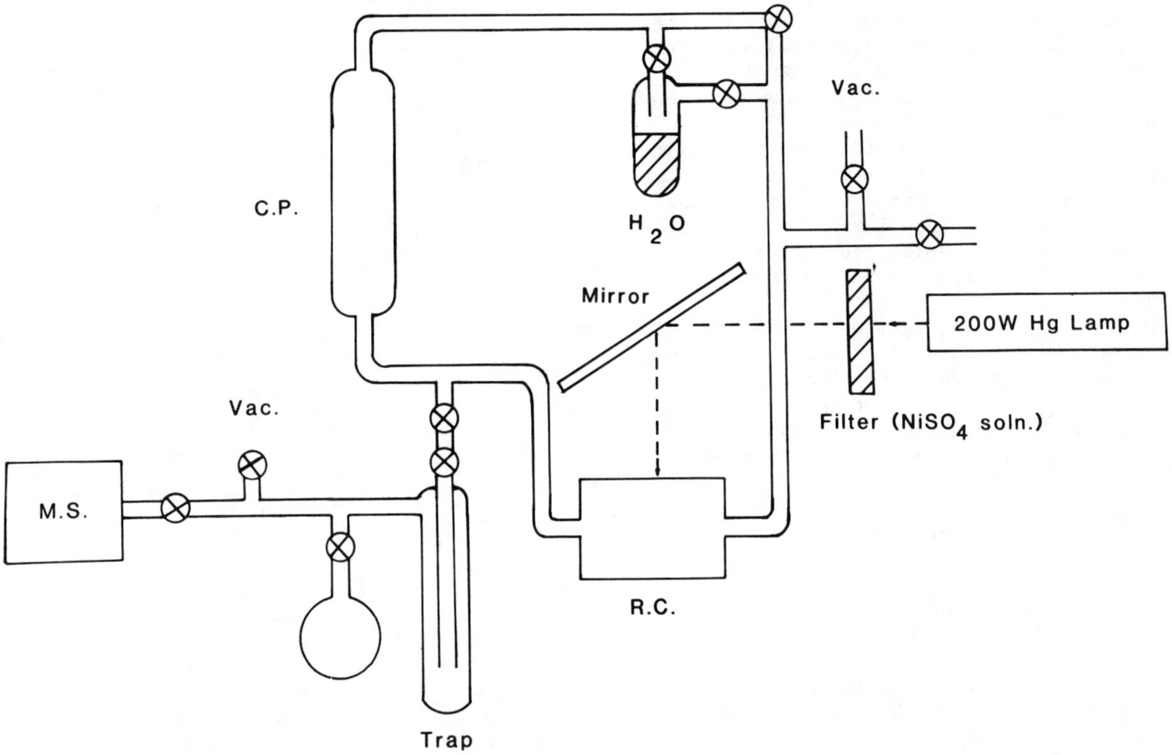

Fig. 4. Apparatus for studying photoassisted reactions involving H_2O and titania-based powders. C.P. = circulating pump; R.C. = reaction chamber; M.S. = mass spectrometer.

Water vapor was stored in a degasable tube and connected into the circulation system through valves, as shown in the upper right part of the figure. To work with gas-phase water this valve assembly was simply opened momentarily and the vapor pressure of water was allowed to expand into the circulation system. When we worked with liquid-phase water the reaction cell temperature was lowered to 0°C with an ice bath, the water vessel was opened, and vapor was condensed into the reaction cell. Once the reaction had started, samples were periodically taken into the small space between the two valves shown at lower left. Expansion from this small volume into the trap, where water vapor was removed by a suitable liquid nitrogen bath or other cold-temperature sources, was used to prepare the sample for analysis by the mass spectrometer shown schematically at the far left. A typical experiment took several hours, so the samples were taken periodically, say at the rate of no more than one every 10 minutes.

The powder was typically about 0.25 g of finely divided and platinized titanium dioxide prepared by following a photochemical method developed at the University of Texas by Professor Bard and co-workers (Kraeutler and Bard, 1978b). It consists first of doping anatase by heating in flowing hydrogen at 700°C for 6 hours. Once this doped TiO_2 has been prepared it can be stored rather indefinitely and then used to prepare the platinized titanium dioxide. This is achieved by the photodecomposition of chloroplatinic acid in an acetate buffer solution. Temperature of the reaction is 55°C and nitrogen is bubbled through the solution. After a certain irradiation time the material is filtered and carefully washed and the platinized titanium dioxide is then dried. The total BET surface area was typically 11 m^2/gm and the Pt composition was typically 2 wt %. Other preparation procedures were sometimes used; these will be described, as necessary, in the following text.

Photoassisted Hydrogen Production from Titania and Water

Schrauzer and Guth (1977) concluded that water adsorbed on TiO_2 or Fe_2O_3-doped TiO_2 was catalytically photolyzed in their system, whereas Van Damme and Hall (1979), on the basis of finding only a trace of H_2, concluded that H_2 formation arose from the noncatalytic photodecomposition of hydroxyl groups originally present on TiO_2. Kawai and Sakata (1980), on the other hand, found that D_2 was formed in the dark when gaseous D_2O was contacted with TiO_2 reduced by CO under UV irradiation. The evolution of D_2 was accelerated by illumination and continued even after evacuating D_2O, but no O_2 was observed. The acceleration was ascribed to the photodecomposition of D_2O over TiO_2 on the assumption that oxygen formed was held at the TiO_2 surface. This assumption is based on the fact that O_2 as well as H_2 was formed by the addition of RuO_2, a good electrode material for O_2 evolution, to TiO_2. Rao et al. (1980) have recently reported that H_2 and H_2O_2 are produced when reduced TiO_2 (in flowing H_2 for 6 hr at 700° to 800°C) is suspended in liquid water and illuminated in bubbling N_2. Unreduced TiO_2 produces neither H_2 nor H_2O_2. We have investigated the activity of reduced anatase for water decomposition (Sato and White, 1981b) and find evidence for a stoichiometric reaction involving oxygen defects.

For these experiments, reduced TiO_2 (0.25 g) was spread on the flat bottom of a quartz reaction cell and outgassed at 200°C for 3 hr. After introducing water vapor at room temperature, the sample was illuminated by a 200-W high-pressure Hg lamp and the products were analyzed by a mass spectrometer.

In every case studied, only H_2 was observed in the gas phase and its formation rate dropped to almost zero after a few hours of illumination. The maximum amount of H_2 formed increased with the reduction temperature and time, and it was larger for H_2-reduced TiO_2 than for CO-reduced samples prepared under the same conditions. The results described below were obtained for H_2-reduced TiO_2. For substrates reduced at temperatures above 700°C, H_2 was formed even in the dark, in agreement with Kawai and Sakata (1980), but its formation stopped within 30 min. When D_2O instead of H_2O was used, the products were dominated by D_2. Since the amount of HD formed did not exceed the value expected from the isotopic purity of D_2O, the hydrogen evolved is believed to come from water added and not from preexisting surface hydroxyl groups. Support for this also comes from the observations that no increase in HD was observed when D_2 (0.12 torr) was added to the H_2O-reduced TiO_2 system under illumination and no products were formed when TiO_2 samples were illuminated in vacuo.

Light of energy less than the band gap of TiO_2 produced no H_2, suggesting that photogenerated electrons and/or holes play an important role. The addition of O_2 (3.2×10^{-3} torr) completely inhibited H_2 formation and its pressure dropped by a factor of 2 after 1 hr of illumination. The addition of ^{13}CO (0.25 torr), on the other hand, had no effect and no $^{13}CO_2$ was observed. This is significant since CO is oxidized over TiO_2 in the presence of band-gap light and oxygen. We find that H_2 is also formed when reduced TiO_2 samples are heated in gaseous water at temperatures higher than 200°C.

When reduced TiO_2 was immersed in liquid water and illuminated, the amount of H_2 formed was larger than observed in the gas-phase process. The liquid-water/TiO_2 system was prepared by cooling the bottom of the reaction cell to 0°C in order to cryogenically pump water from the reservoir to the cell. After the sample was covered with 0.2 to 0.3 ml of water, the cell was warmed to ≃ 23°C and then illuminated. The results are

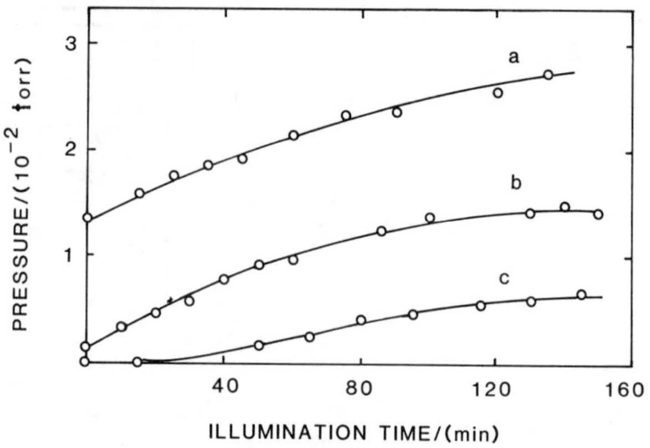

Fig. 5. Evolution of H_2 from illuminated, reduced TiO_2 immersed in liquid water: (a) TiO_2 reduced by H_2 at 750°C for 4 hr, (b) at 700°C for 6 hr, and (c) at 600°C for 3 hr (0.1 torr = 1 μmol).

shown in Figure 5 for variously reduced TiO_2 samples. Just as in the gas-phase process, the H_2 evolution rate dropped to zero after a few hours and no O_2 was detected. The TiO_2 sample reduced at 750°C for 4 hours produced 1.3×10^{-2} torr (0.13 μmol) of H_2 in the dark (the pressure at time zero of curve (a) in Figure 5 is due to this), and H_2 formation was accelerated by illumination.

Although the formation of H_2O_2 was not checked in our experiments, its concentration is limited by photodecomposition to O_2 and H_2O over TiO_2. Rao et al. (1980) observed that the addition of H_2O_2 (5 μmol) to their reaction mixture (700 ml) followed by 1 hr of illumination brought about a twofold decrease in the H_2O_2 concentration. This implies that the maximum achievable concentration of H_2O_2 over illuminated TiO_2 is very low, less than 4 μmol/l. Applying this to our system, at most 1×10^{-3} μmol of H_2O_2 could exist in the water. This is much less than the amount of H_2 formed (0.1 μmol).

All of these results are consistent with a mechanism in which a reaction between H_2O and oxygen vacancies of reduced TiO_2 is photoassisted by the production of electron-hole pairs in the solid. This reaction is thermodynamically downhill and not catalytic. In the process proposed here, water reacts slowly with surface oxygen vacancies to evolve H_2 and remove the vacancies by filling them with oxygen or hydroxyl species. This noncatalytic process is significantly accelerated by band-gap irradiation. Bulk oxygen vacancies are retained during the photoprocess. Experimental support for this proposal comes from isotope tracing, the effects of reduction temperature and time, and the effects of added O_2 and CO. In the photoprocess, photogenerated holes probably oxidize water to produce some oxygen-containing species which react with the oxygen vacancies at the surface.

The fact that TiO_2 alone is inactive for water

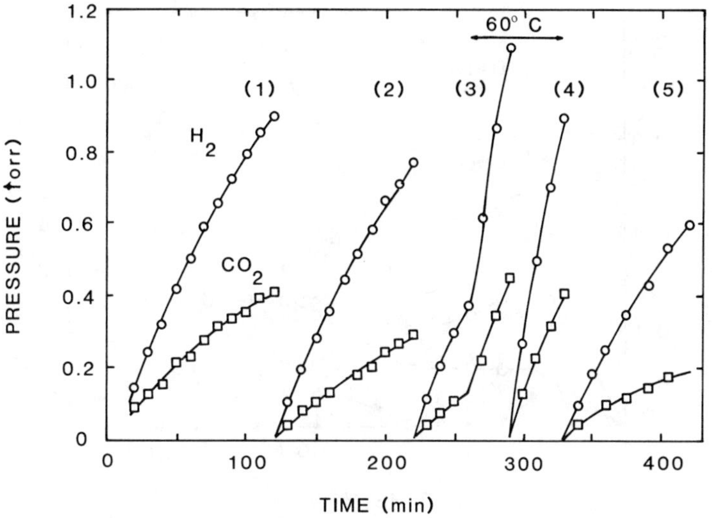

Fig. 6. Evolution of H_2 and CO_2 when a mixture of active carbon and Pt/TiO_2 catalyst is illuminated at room temperature in the presence of gas-phase H_2O. The reaction was repeated after evacuating the system briefly.

photolysis can be described, as discussed earlier, in terms of the energy band diagram of illuminated TiO_2 (Rajeshwar et al., 1978). According to a recent study (Tomkiewicz, 1979) in this area, the flat-band potential (electron Fermi level) of TiO_2 (rutile) is about 100 mV more negative than the H^+/H_2 redox potential. This implies that the water photolysis in PEC (photoelectrochemical) cells with a TiO_2 photoanode is energetically possible under open circuit conditions. However, there are some potential drops, for example, across the Helmholtz layer, so that the overvoltage available for H_2 evolution becomes lower. Even if anatase has a somewhat more negative flat-band potential than rutile, the overvoltage would be too low for efficient evolution of H_2 at the TiO_2 surface (Kraeutler and Bard, 1978c). Similar overvoltage requirements for the reduction of protons are found with $SrTiO_3$ (Wrighton et al., 1977). Consequently, these semiconductor catalysts show increased photocatalytic activity for water decomposition when a material such as Pt is added which readily evolves H_2 at a lower overvoltage.

The Photocatalytic Reaction of Water with Carbon over Platinized Titania

Platinized titania has been used in a large number of photoassisted reactions (Rajeshwar et al., 1978; Kraeutler and Bard, 1978a, c; Wrighton et al., 1977; Sato and White, 1981a; Bulatov and Khidekel, 1976; Lehn et al., 1980; Kawai and Sakata, 1979b; Kalyanasundaram and Graetzel, 1979). One of our more detailed studies involved active charcoal and Texas lignite (Sato and White, 1980b, 1981a). The general conclusion is that solid-phase, liquid-phase, and gas-phase carbon-hydrogen sources with water as a coreactant can be converted into CO_2 and H_2 when platinized titania is irradiated. We have not achieved partial oxidation, for example, to CO, but there is one such report using solid carbon with a RuO_2/TiO_2 semiconductor powder (Kawai and Sakata, 1979a).

In these experiments, active charcoal (0.05 g) and Pt/TiO_2 (0.2 g) were physically mixed and the mixture spread uniformly on the flat bottom of a quartz reaction cell. The cell was then connected to the circulation system and the sample outgassed at 200°C for 2 hours before introducing water vapor and starting the reaction.

The evolution of H_2 and CO_2 is shown in Figure 6 when the sample at room temperature was illuminated by UV light. Small amounts of O_2 and CH_4 (less than 10^{-2} torr) were the only other products detected. Five repetitions are shown and in each the reaction was repeated after evacuating the reaction system in the dark for 15 minutes at the reaction temperature to check reproducibility. Both H_2 and CO_2 formation became slower with time in every run; for example, in the first run the H_2 formation rate dropped from 0.6 torr (1.8×10^{-4} mol/hr) initially to 0.4 torr/hr after 2 hours illumination. The initial rate of H_2 formation, however, was reproduced in the second and third runs. During the third run the reaction temperature was raised to 60°C to measure the temperature dependence. The H_2 formation rate increased by a factor of 2.5, from which the activation energy is estimated to be 5 kcal/mol. This rate was reproduced in the fourth run. Returning the cell to room temperature for the fifth run, the initial rate of H_2 formation was significantly lower and was not restored by outgassing the sample at 200°C. The sample was taken out after the H_2 formation rate had dropped to 0.2 torr/hr and was physically remixed and returned to the reaction cell. The reaction rate was fully restored by

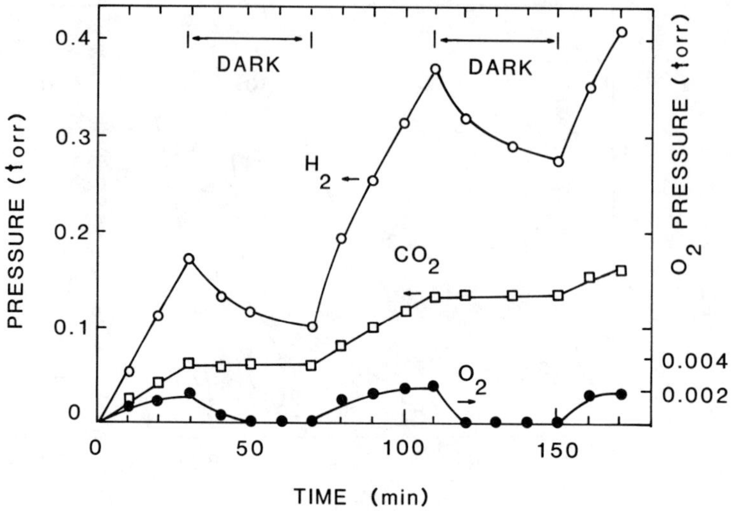

Fig. 7. Change of H_2/CO_2 ratio during the reaction of active carbon with gas-phase H_2O.

this procedure. Therefore, the observed decrease in the rate was not due to intrinsic activity loss, but probably rather to a decrease in the contact area between the carbon particles and the oxidation sites of the catalyst.

By stoichiometry ($C + 2H_2O \rightarrow 2H_2 + CO_2$) the H_2/CO_2 ratio should be 2, but it was always larger for the five runs shown in Figure 6. The ratio in the initial stage of the first run is less than 2, due to CO_2 desorption from the solid. The formation of O_2 cannot account for an H_2/CO_2 ratio greater than 2 because it is formed in such small amounts (less than 10^{-2} torr).

The relative excess of H_2 probably arises because some oxygenated reaction intermediates accumulate on the sample surface. Support for this comes from the fact that an appreciable amount of CO_2 desorbed when the sample was heated at 200°C after the reaction. During heating, CO_2 desorption maximized between room temperature and 80°C and increased again at temperatures above 120°C. The CO_2 desorption above 120°C may be the result of the decomposition of reaction intermediates, such as carboxyl, or the reaction of adsorbed oxygen with carbon. The desorption of H_2 was also observed, but in amounts much less than CO_2.

As shown in Figure 7, we observed a decrease in the H_2 pressure when the UV illumination was stopped during the reaction at room temperature. The sample was outgassed at 200°C prior to this experiment to remove adsorbed species accumulated in the preceding runs. The loss of H_2 is thought to arise mainly from its reaction with adsorbed oxygen species located on TiO_2 and carbon. Adsorption of H_2 may also be involved, but only to a small extent since very little H_2 was noted in subsequent thermal desorption. The H_2/CO_2 ratio fell below 2 in the first dark period (Figure 7), probably because of some H_2 adsorption. However, in the second dark period where the reaction proceeded further, the ratio approached 2 but would have fallen slightly below it had the dark period been extended. These results suggest that, in the dark, the H_2/CO_2 ratio would attain the stoichiometric value under conditions where H_2 adsorption can be neglected.

The formation of O_2 during the reaction is tiny but thought to be important because it indicates that the photodecomposition of H_2O may play an important role in the present system. The amount of O_2 formed tends to increase with repetition of the reaction but never exceeds 1×10^{-2} torr under the conditions in Figure 6. When the light is turned off, O_2 disappears promptly from the gas phase, as shown in Figure 7.

In this connection the reactivity of gas-phase O_2 was examined in the absence of H_2O and under illumination. The pressure decrease of O_2 was very slow in the dark and increased somewhat upon illumination, but no appreciable amount of CO_2 was formed (see Figure 8). After introducing H_2O into the system the O_2 pressure fell rather sharply to below 10^{-2} torr and CO_2 appeared in the gas phase. This result shows that gas-phase O_2 alone is much less active for the oxidation of carbon than oxygen species produced by the photodecomposition of H_2O. The rapid decrease in the O_2 pressure after introducing H_2O is probably due to its reaction with H_2 formed in the reaction of H_2O with carbon.

The formation of CH_4 was small. It accumulated to a pressure of 6×10^{-4} torr during the first run with a fresh catalyst, but after a few hours of illumination its formation rate dropped to zero.

Although no appreciable CO was observed in the reaction products, adsorbed CO is one of the most probable intermediates of the present reaction. In this connection the effect of the presence of CO on the reaction rate was examined using ^{13}CO to discriminate the products. The water-gas shift

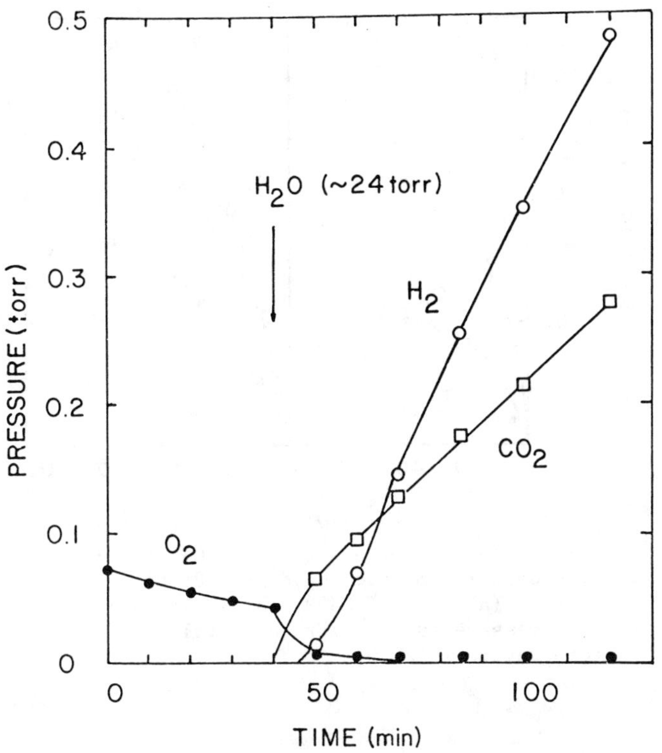

Fig. 8. Change of O_2 pressure in the absence and the presence of gas-phase H_2O over the illuminated active carbon Pt/TiO_2 mixture. H_2O (≃24 torr) was introduced at 40 min.

reaction, $^{13}CO + H_2O \rightarrow {}^{13}CO_2 + H_2$, took place simultaneously with the oxidation of carbon, but H_2 formation was considerably suppressed as compared to that observed in the absence of CO. After the CO pressure fell to about 1×10^{-2} torr H_2 formation accelerated, indicating the inhibitory effect of CO. At the same time, the O_2 pressure abruptly increased and then slowly decreased to expected values. We have observed similar phenomena in the photoassisted water-gas shift reaction over Pt/TiO_2 (Sato and White, 1980c).

It is still not clear from this result whether adsorbed CO is an intermediate in the carbon/water reaction. However, it is worth noting that adsorbed CO formed in oxidizing carbon would be bound to a carbon particle and would not migrate to a Pt surface where CO inhibits the H_2 evolution (Sato and White, 1980c).

The H_2 formation rate was only slightly dependent on H_2O pressure. Lowering the pressure from 25 to 5 torr caused a rate increase of no more than 5 percent.

The wavelength dependence of the reaction rate was qualitatively measured using three cutoff filters (Figure 9). A commercial UV cutoff filter (415-nm cutoff) and a Plexiglass filter (≃380-nm cutoff) completely eliminated H_2 formation, while a Pyrex glass filter (≃275-nm cutoff) reduced its rate to about two-thirds of that found with no filter. This measurement includes some experimental error because the reaction rate declines slowly even in the same run.

In another set of experiments, liquid H_2O was used instead of gas-phase H_2O. The results are shown in Figure 10. The formation of H_2 was much slower than in the reaction with gas-phase H_2O, and O_2 exceeded CO_2, indicating that the photodecomposition of H_2O dominates the reaction. The relatively rapid CO_2 formation at the beginning is probably just the desorption of CO_2 accumulated in the previous runs, since the H_2/O_2 ratio is nearly 2. The rate of product formation became faster after 1 hr illumination, for reasons which are not clear. We speculate that local warming by the UV light source may have caused some of the water to evaporate and allowed the rate to increase because the H_2 and O_2 could escape more readily to the gas phase. No CO was detected in this reaction.

The reason the oxidation of carbon is inhibited in liquid H_2O is not well understood. The liquid H_2O layer may retard the migration of adsorbed oxygen or hydroxyl radicals formed on TiO_2 to the carbon surface, and as a result these species would become O_2 before oxidizing carbon.

Although the mechanism of the oxidation of active carbon is not clear, we assume it involves intermediates similar to those involved in the electrochemical oxidation of graphite electrodes, where hydroxyl, carbonyl, and carboxyl groups are typical surface compounds proposed (Panzer and Elving, 1975; Coughlin, 1969). On the basis of the above discussion one mechanism describing the present system can be written as:

$$H_2O + h^+ \xrightarrow{TiO_2} \cdot OH + H^+ \qquad (1)$$

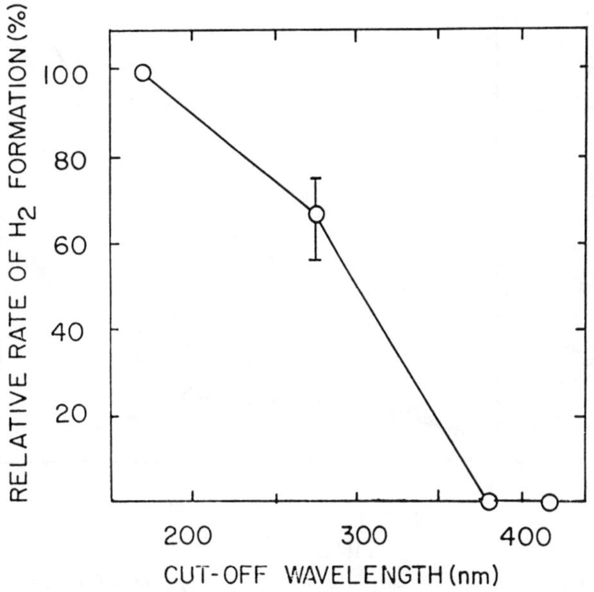

Fig. 9. Effects of cutoff filters on the rate of H_2 formation.

$$2\ \cdot OH \xrightarrow{TiO_2} H_2O + O(a) \quad (2)$$

$$2\ O(a) \xrightarrow{TiO_2} O_2 \quad (3)$$

$$H^+ + e^- \xrightarrow{Pt} O_2 \quad (4)$$

$$2\ H(a) \underset{\leftarrow}{\overset{Pt}{\rightarrow}} H(a) \quad (5)$$

$$2\ H(a) + O(a) \xrightarrow{Pt,TiO_2} H_2O \quad (6)$$

$$H(a) + \cdot OH \xrightarrow{Pt,TiO_2} H_2O \quad (7)$$

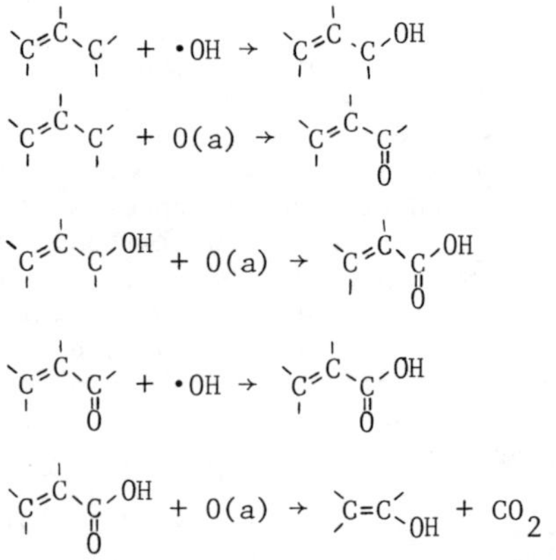

(8)

(9)

(10)

(11)

(12)

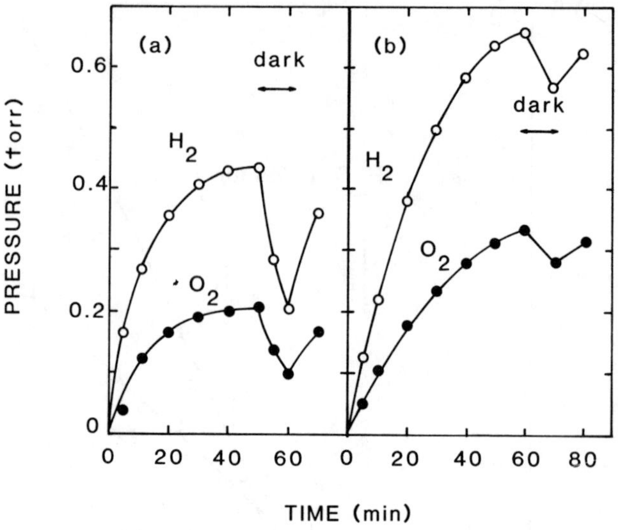

Fig. 11. Time courses of the GWD reaction over illuminated NaOH/Pt/TiO_2 catalysts: (a) 7 wt. % NaOH; (b) 14 wt. % NaOH. H_2O pressure is ≃24 torr and catalyst temperature is 25°C.

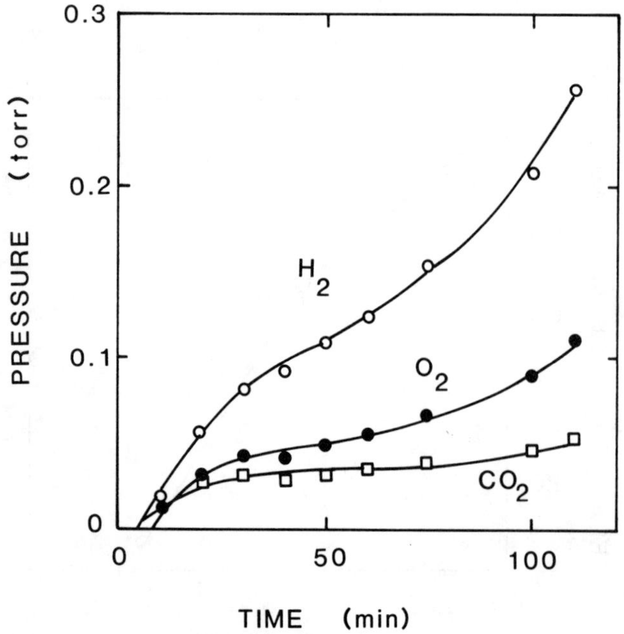

Fig. 10. Evolution of H_2, O_2, and CO_2 when carbon, liquid H_2O, and Pt/TiO_2 are illuminated.

where (a) indicates adsorbed species.

According to this mechanism, the surface of carbon would be covered with various oxides of carbon and adsorbed oxygen during the reaction at room temperature. Consequently, the H_2/CO_2 ratio in gas phase would exceed the stoichiometric ratio. At 60°C the coverage of these surface species would be lower than at room temperature since they would be decomposed or oxidized effectively as the temperature is raised. Products other than H_2 and CO_2, such as HCOOH or HCHO, might be formed on the surface but would not appear at detectable levels in the gas phase since they would be rapidly oxidized.

The kinetics of the present reaction are similar to those of the photoassisted water-gas shift reaction over Pt/TiO_2 (Sato and White, 1980c). The almost zero-order dependence of the rate on H_2O pressure is the same as in the latter, and the activation energy (≃5 kcal/mol) is close to the 7.5 kcal/mol of the latter. As for the wavelength dependence, the present reaction shows a slightly shorter onset than the shift reaction, but the reason is not clear.

The decline of the H_2 formation rate in a given run is probably due to the accumulation of H_2, which competes with carbon for oxygen species. The long-term decline arises from the loss of a good contact between the catalyst and carbon, since the initial reaction rate can be reproduced by remixing the sample.

The formation of O_2 increases with time, but in any run its maximum amount is less than observed in the reaction with lignite (Sato and White, 1980b). Although the reaction of H_2 with O_2 occurs rapidly on a clean Pt/TiO_2 even in the presence of gas-phase H_2O (Sato and White, 1980a), CO inhibits this reaction to some extent as observed in the water-gas shift reaction (Sato and White, 1980c). Since CO was not detected in the gas phase of the present reaction, its inhibitory effect is not established. Our results show that a decrease in the amount of adsorbed H_2O on the sample results in a decrease in the O_2 pressure. Therefore, it is reasonable to assume that since active carbon adsorbs a large amount of H_2O, the H_2O layer on the Pt that is in contact with carbon is thicker than in the absence of carbon. When the H_2O layer is thick, the surface reaction between H_2 and O_2 is inhibited. As the contact area between the catalyst and carbon decreases with the consumption of carbon, oxygen species would have to migrate longer distances to react with carbon; consequently, they would tend to desorb as O_2.

Assuming, as an upper limit, a flux of 10^{17} photons/sec (White, 1966) with energy greater than the band-gap energy of TiO_2 ($\simeq$3.0 eV), the quantum yield of the H_2 production is about 2 percent at the beginning of the reaction at room temperature and increases with increasing temperature.

The Role of Surface Hydroxyl Groups

In photoassisted reactions involving water, surface hydroxyl groups are thought to play a central role. Their presence, state, reactivity, and replaceability become important in any effort to characterize them. To date results are largely empirical. We have examined the water decomposition and water-gas shift reactions over NaOH-coated platinized TiO_2 (Sato and White, 1981c). Very interesting related work on $SrTiO_3$ has been reported by Wagner and Somorjai (1980).

The time evolution of the water decomposition reaction over 2 wt % Pt on TiO_2 coated with 7 and 14 wt % NaOH is shown in Figure 11. Although gas-phase water was introduced, the deliquescent nature of NaOH guarantees that the surface of the powders became wet in these experiments. In Figure 11 it can be seen that the initial rate is lower and the photostationary state higher for the larger NaOH. Moreover, the dark back-reaction is slowed down for the heavier loading. The reactions were both run at room temperature.

The variation of the initial rate of H_2 formation with NaOH loading is shown in Figure 12 for three reactions, the water-gas shift (WGS), gas-phase water decomposition (GWD), and liquid water decomposition (LWD) reactions. The difference in the GWD and LWD reactions is that in the latter, 0.25 ml of water was condensed to cover the catalyst. A 1-wt-% NaOH loading yields a 0.25 M solution. Interestingly, both the LWD and GWD rates are inhibited at high loadings of NaOH.

The mechanism of the water photolysis over Pt/TiO_2 can be described in analogy with PEC cells. In an acidic solution, a suitable mechanism is

$$H_2O + h^+ \xrightarrow{TiO_2} \cdot OH + H^+ \qquad (13)$$

$$2\ \cdot OH \xrightarrow{TiO_2} H_2O + O(a) \qquad (14)$$

$$2\ O(a) \overset{TiO_2}{\rightleftarrows} O_2 \qquad (15)$$

$$H^+ + e^- \xrightarrow{Pt} H(a) \qquad (16)$$

$$2\ H(a) \overset{Pt}{\rightleftarrows} H_2 \qquad (17)$$

In neutral and alkaline solutions instead of steps (13) and (16) use

$$OH^- + h^+ \xrightarrow{TiO_2} \cdot OH \qquad (18)$$

$$H_2O + e^- \xrightarrow{Pt} H(a) + OH^- \qquad (19)$$

where h^+ denotes a hole. In the photoassisted water-gas shift reaction, CO reacts with O(a) formed in step (14) to give CO_2.

When water electrolysis (or photolysis) is carried out in neutral water, i.e., with no electrolyte, the diffusion of OH^- from cathode to anode would limit the reaction rate. In our other studies (Sato and White, 1980a, c, 1981a, b) the rates of most photoreactions over Pt/TiO_2 in which pure water was used were temperature dependent (activation energy of 5 to 8 kcal/mol), while the rates of the GWD and the WGS reactions over illuminated $NaOH/Pt/TiO_2$ are independent of temperature between 18° and 50°C. This difference in the temperature dependence leads us to suppose that the reactions involving pure water are rate limited by the diffusion of OH^- from Pt to TiO_2 and the activation energy observed comes mainly from this diffusion step.

The water photolysis reaction over Pt/TiO_2 catalysts competes with the thermal back-reaction over Pt. Therefore the observed rate, V^{H_2}, of the formation is given by the equation

$$V^{H_2} = 2V^{O_2} = V_p - V_t$$

where V_p is the rate of the photocatalytic production of H_2 and V_t is the rate of the thermal back reaction. The rate V_p would be proportional to the intensity of light with energy greater than band-gap energy and independent of the partial pressures of the products under the present experimental conditions. In alkaline solution V_p would be independent of temperature but would depend on the pH of the solution. In the GWD reaction over $NaOH/Pt/TiO_2$ the NaOH coating somewhat reduces V_p, while in the LWD reaction the water (solution)

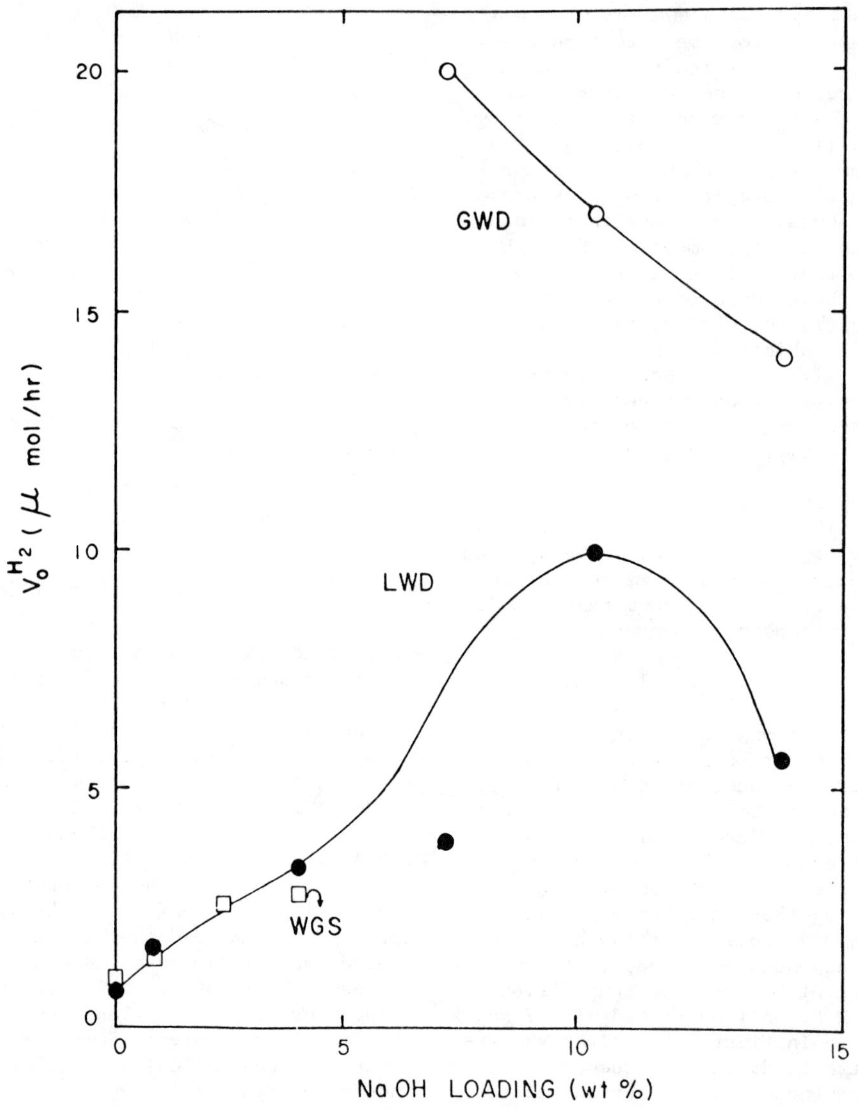

Fig. 12. Initial H_2 formation rates, $V_o^{H_2}$, in the GWD (○), WGS (□), and LDW (●) reactions over illuminated Pt/TiO_2 as a function of NaOH loading. (See text for the definition of the abscissa for the LWD reaction.)

layer over the catalyst also limits the escape of the products to the gas phase. On the other hand, V_t is proportional to the Pt surface area and the product pressures, and is temperature dependent. The NaOH coating and liquid water layers considerably suppress V_t.

In the GWD reaction over pure Pt/TiO_2, neither H_2 nor O_2 is observed because the back reaction occurs even before the products desorb from the catalyst. The LWD reaction takes place because V_t is much suppressed by the water layer. However, V_t increases with product pressures so that V^{H_2} decays with time and a photostationary state is eventually reached. The present experiments show that the GWD reaction becomes detectable even in the presence of Pt when Pt/TiO_2 is coated with more than 7 wt % of NaOH. This apparently arises because V_p is enhanced by the addition of alkaline electrolyte while V_t is suppressed by the NaOH coating. Increasing NaOH above 7 wt % gives rise to decreases in both V_p and V_t, but the former is less pronounced than the latter, as seen from Figure 11. As a result, the maximum pressures of H_2 and O_2 increase. One may suppose here that the liquid electrolyte layer, formed by coating the catalyst with a deliquescent material and exposing it to gas-phase water, is essential to gaseous water photolysis. Although this deliquescent film is important, we find that gaseous water photolysis does occur over Pt/TiO_2 without such a coating. When the photodecomposition of NO over Pt/TiO_2 is carried out in the presence of gaseous water, H_2, O_2, N_2O, and a small amount of N_2 are formed (S. Sato and J. M. White, unpublished data). Since no H_2 is produced in a control experiment involving no water, the product H_2 is attributed to water decomposition. The formation of O_2 during the water-gas shift reaction also gives further evidence for gaseous water photolysis.

In the WGS reaction V_t is very small since O_2 formed is removed by its reaction with CO to form CO_2, so that V^{H2} remains constant until the CO pressure falls to about 0.05 torr (see Figure 3). The reaction of O_2 with CO, however, is suppressed to some extent when the NaOH coating exceeds 7 wt %.

Our results may be compared to those obtained by Wagner and Somorjai (1980) on single-crystal (111) samples of platinized and metal-free $SrTiO_3$. Like our Pt/TiO_2 results, they found that on $Pt/SrTiO_3$ a NaOH film gave rise to the activity for gas-phase water photolysis. As they properly point out, the deliquescent character of NaOH makes the surfaces wet in these experiments. Our results involving an aqueous solution of NaOH differ from theirs in that we find a local maximum in the variation of rate with NaOH concentration (around 2.5 M), whereas they find that the rate increases throughout the range 0 to 20 M. Most significantly, they find catalytic activity on NaOH-coated metal-free $SrTiO_3$, while we find no activity for the analogous system based on powdered TiO_2. Yoneyama et al. (1979) have also reported the activity of $SrTiO_3$ powder when its aqueous suspension is illuminated by a 1-kW Xe lamp. These results show that metal-free $SrTiO_3$, unlike TiO_2, is active in the photoassisted decomposition of H_2O. This property of $SrTiO_3$ may come from its higher flat-band potential (i.e., 200 mV more negative than the H^+/H_2 potential (Maruska and Ghosh, 1978).) Although the mechanism involved is not well understood, a part of the $SrTiO_3$ surface could become the cathode and another part the photoanode. The photoactivity of powdered $SrTiO_3$, however, seems to depend on its preparation method. In our preliminary experiments, powdered $SrTiO_3$ coated with NaOH showed no activity, while platinized forms (with and without NaOH coating) were active.

Summary

These results may be summarized by the following points.

We conclude that there is no catalytic reaction of water with any form of titanium dioxide, doped or undoped, unless that titanium dioxide is mixed with some transition metal or other chemical component. This holds for both the gas-phase and liquid-phase water reactions.

In the case of platinized titanium dioxide, we find some photoassisted catalytic activity for water decomposition in the liquid phase. This activity occurs for both reduced and unreduced titanium dioxide, but the rates are much higher when a reduced form of titanium dioxide is used. This enhancement may have to do with the increased ability of reduced titanium dioxide to transfer charge from one region to another because of its increased conductivity.

The water decomposition reaction cannot be detected over platinized titanium dioxide when gas-phase water is used. We believe this is because the back reaction of hydrogen and oxygen is so facile over platinum. It appears that the forward reaction to decompose water actually does take place.

The water-gas shift reaction, which in essence prevents the back reaction of hydrogen with oxygen, occurs readily and catalytically. Interesting effects of CO pressure at low values are observed which indicate that poisoning the Pt surface by small amounts of CO allows the forward reaction of water decomposition to occur to a slight extent.

Gas-phase water also reacts with carbon sources, both gas-phase and solid, to form carbon dioxide and hydrogen.

In both the water decomposition reaction and the water-gas shift reaction there is a definite wavelength dependence and a definite activation energy. The wavelength dependence indicates that this reaction depends on electron-hole pair formation in titanium dioxide. The temperature dependence indicates that there is some kind of thermal barrier occuring with one of the photogenerated intermediates.

Adding sodium hydroxide as a coating or in solution enhances the rate of the water decomposition reaction. This can be understood in terms of the ease with which the charge transfer reactions occur in such media.

A simple mechanism has been proposed which is analogous to the mechanisms that have been used in photoelectrochemical cells.

Some possible applications of these concepts to atmospheric processes have been mentioned in the section entitled General Considerations. In such processes stoichiometric reactions may contribute as much as catalytic processes, provided a source of particulates is available. Moreover, corrosive processes which plague many catalytic processes and typically involve narrow-band-gap semiconductors should be considered as likely candidates for photoassisted heterogeneous reactions. The particle size of the materials used in our studies has been between 100 and 250 μm in diameter. However, much smaller particles are known to be active and use light much more efficiently than our results would indicate. For example, Gratzel and coworkers (Kalyanasundaram et al., 1981; Borgarello et al., 1981) have shown that colloidal particles of bulk semiconductor powders are very active in water decomposition.

Acknowledgement. This work was supported in part by the Office of Naval Research.

References

Bard, A. J., Photoelectrochemistry and heterogeneous photocatalysis at semiconductors, *J.*

Photochem., 10, 59-75, 1979.

Bickley, R. I., Photo-induced reactivity at oxide surfaces, Chemical Physics of Solids and Their Surfaces, Vol. 7, Specialists Periodical Reports, The Chemical Society, Burlington House, London, 118-156, 1978.

Borgarello, E., J. Kiwi, E. Pelizzetti, M. Visca, and M. Graetzel, Photochemical cleavage of water by photocatalysis, Nature, 289, 158-160, 1981.

Bulatov, A. V., and M. L. Khidekel, Decomposition of water under the effect of UV irradiation in the presence of platinized titanium dioxide, Izv. Akad. Nauk. SSSR, Ser. Khim., 1902, 1976.

Butler, M. A., and S. D. Ginley, Principles of photoelectrochemical solar energy conversion, J. Mater. Sci., 15, 1-19, 1980.

Childs, L. P., and D. F. Ollis, Is photocatalysis catalytic, J. Catal., 66, 383-390, 1980.

Coughlin, R. W., Carbon as adsorbent and catalyst, Ind. Eng. Chem. Prod. Res. Dev., 8, 12-23, 1969.

Fenstik, V. P., L. V. Fenstik, and P. M. Stachnik, Oxidation of methanol on silver with ultraviolet rays, Dopovidi Akad. Nauk. Ukrain. S.S.R., 34B, 738-740, 1972.

Formenti, M., and S. J. Teichner, Heterogeneous photocatalysis, Specialist Periodical Reports, Catalysis, 2, 87-106, 1978.

Fujishima, A., and K. Honda, Electrochemical photolysis of water at a semiconductor electrode, Nature, 238, 37-38, 1972.

Kalyanasundaram, K., E. Borgarello, and M. Graetzel, Visible light-induced water cleavage in cadmium sulfide dispersions loaded with platinum and ruthenium oxide (RuO_2), hole scavenging by RuO_2, Helv. Chim. Acta., 64, 362, 1981.

Kalyanasundaram, K., and M. Graetzel, Cyclic water splitting into hydrogen and oxygen by visible light with coupled Redox catalysts, Angew. Chem. Intern. Ed., 18, 701, 1979.

Kawai, T., and T. Sakata, Production of hydrogen and carbon monoxide from liquid water and carbon using solar energy, J.C.S. Chem. Comm., (23), 1047-1048, 1979a.

Kawai, T., and T. Sakata, Hydrogen evolution from water using solid carbon and light energy, Nature, 282, 283-284, 1979b.

Kawai, T., and T. Sakata, Photocatalytic decomposition of gaseous water over titanium dioxide and titanium dioxide-ruthenium dioxide surfaces, Chem. Phys Lett., 72, 87, 1980.

Kraeutler, B., and A. J. Bard, Heterogeneous photocatalytic synthesis of methane from acetic acid - new Kolbe reaction pathway, J. Am. Chem. Soc., 100, 2239-2240, 1978a.

Kraeutler, B, and A. J. Bard, Heterogeneous photocatalytic preparation of supported catalysts. Photodeposition of platinum on TiO_2 powder and other substrates, J. Am. Chem. Soc., 100, 4317-4318, 1978b.

Kraeutler, B., and A. J. Bard, Heterogeneous photocatalytic decomposition of saturated carboxylic acids on TiO_2 powder. Decarboxylative route to alkanes, J. Am. Chem. Soc., 100, 5985-5992, 1978c.

Kung, H. H., H. S. Jarrett, A. W. Sleight, and A. Ferretti, Semiconducting oxide anodes in photoassisted electrolysis of water, J. Appl. Phys., 48, 2463-2469, 1977.

Kupperman, A., and J. M. White, Energy threshold for $D + H_2 \rightarrow DH + H$ reaction, J. Chem. Phys., 44, 4352-4354, 1966.

Lehn, J. M., J. P. Sauvage, and R. Ziessel, Zeolite supported metal oxide catalysis for the photoinduced oxygen generation from water, Nov. J. Chim., 4, 355-358, 1980.

Maruska, H. P., and A. K. Ghosh, Photocatalytic decomposition of water at semiconductor electrodes, Solar Energy, 20, 443-458, 1978.

McAllister, J. W., and J. M. White, Photodesorption of carbon monoxide from polycrystalline nickel, J. Chem. Phys., 58, 1496-1504, 1973.

Morrison, S. R., The Chemical Physics of Surfaces, p. 358, Plenum Press, N.Y., 1977.

Nozik, A. J., Photoelectrochemistry: Applications to solar energy conversion, Ann. Rev. Phys. Chem., 29, 189-222, 1978.

Panzer, R. E., and P. J. Elving, Nature of the surface compounds and reactions observed on graphite electrodes, Electrochim. Acta, 20, 635-647, 1975.

Rajeshwar, K., P. Sinch, and J. DuBow, Energy conversion in photochemical systems - A review, Electrochim. Acta, 23, 1117-1143, 1978.

Rao, M. V., K. Rajeshwar, V. R. Pal Verneker, and J. DuBow, Photosynthetic production of H_2 and H_2O_2 on semiconducting oxide grains in aqueous solutions, J. Phys. Chem., 84, 1987-1991, 1980.

Sato, S., and J. M. White, Photodecomposition of waste over platinum/titanium dioxide catalysts, Chem. Phys. Lett., 72, 83, 1980a.

Sato, S., and J. M. White, Photocatalytic production of hydrogen from water and Texas lignite by use of a platinized titania catalyst, Ind. Eng. Chem. Prod. Res. Dev., 19, 542, 1980b.

Sato, S., and J. M. White, Photoassisted water-gas shift reaction over platinized TiO_2 catalysts, J. Am. Chem. Soc., 102, 7206-7210, 1980c.

Sato, S., and J. M. White, Photoassisted hydrogen production from titania and water, J. Phys. Chem., 81, 592-594, 1981a.

Sato, S., and J. M. White, Photocatalytic reaction of water with carbon over platinized titania, J. Phys. Chem., 85, 336-341, 1981b.

Sato, S., and J. M. White, Photocatalytic water decomposition and water-gas shift reaction over NaOH-coated platinized TiO_2, J. Catal., 69, 128, 1981c.

Schrauzer, G. N., and T. D. Guth, Photolysis of water and photoreduction of nitrogen on titanium dioxide, J. Am. Chem. Soc., 99, 7189-7193, 1977.

Somorjai, G. A., Chemistry in Two Dimensions: Surfaces, Chapter 11, Cornell Univ. Press, Ithaca, 1981.

Tomkiewicz, M., The potential distribution at the titanium dioxide aqueous electrolyte interface,

J. Electrochem. Soc., 126, 1505-1510, 1979.

Van Damme, H., and W. K. Hall, On the photoassisted decomposition of water at the gas-solid interface on TiO_2, J. Am. Chem. Soc., 101, 4373-4374, 1979.

Wagner, F. T., and G. A. Somorjai, Photocatalytic and photoelectrochemical hydrogen production on strontium titanate single crystals, J. Am. Chem. Soc., 102, 5494, 1980.

White, J. M., Reaction of mono-energetic deuterium atoms with hydrogen molecules, Ph. D. Thesis, Univ. of Illinois, Urbana-Champaign, 1966.

Wrighton, M. S. (Ed.), Interfacial photoprocesses: Energy conversion and synthesis, ACS Adv. in Chemistry Series, 184, American Chemical Society, 1980.

Wrighton, M. S., P. T. Wolczanski, and A. B. Ellis, Photoelectrolysis of water by irradiation of platinized n-type semiconducting metal oxides, J. Solid State Chem., 22, 17-29, 1977.

Yoneyama, H., M. Koizumi, and H. Tamura, photolysis of water on illuminated strontium titanium trioxide, Bull. Chem. Soc. Japan, 52, 3449-3450, 1981.

PHOTOASSISTED HETEROGENEOUS CATALYSIS: DEFINITION AND HYDROCARBON AND CHLOROCARBON OXIDATIONS

David F. Ollis and Ann Lorette Pruden*

Department of Chemical Engineering, University of California, Davis, California 95616

Abstract. The existence of heterogeneous photoassisted catalysis is defined by a simple criterion: the demonstrated formation of more than 100 monolayer equivalents of product. The kinetics and possible mechanism for the TiO_2-photocatalyzed selective partial oxidation of alkanes to oxygenates (aldehydes and ketones) are reviewed, and data establishing the photocatalyzed mineralization of the chloromethane $CHCl_3$ (chloroform) and the chloroethylene $Cl_2C{=}CClH$ (trichloroethylene) are presented. A brief final discussion concerns the possible formation of photocatalytically active solid phases in fly ash, and the relevance of these hydrocarbon and halocarbon conversions and other photocatalyzed conversions to atmospheric chemistry.

Introduction: Definition and Reactions

Heterogeneous photochemistry may include a number of distinctly different phenomena, namely (1) catalyzed conversion of an adsorbed molecule A to B by the surface of a repeatedly photoexcited solid, (2) noncatalytic reaction of a photoactivated solid surface with an adsorbed species, (3) direct photoactivation of the adsorbed species, and (4) surface reaction with photoproduced species originating in the bulk (gas) or liquid phases. By heterogeneous photoassisted catalysis, or more compactly, heterogeneous photocatalysis, we shall mean only the first process mentioned, although the latter processes may occur, and may even be of greater importance to atmospheric chemistry. In various atmospheric conditions we may encounter gas-solid, liquid-solid, and solid (ice)-solid interactions. Thus, we shall be interested in all reports of heterogeneous photocatalysis concerning species of atmospheric occurrence, irrespective of whether a gas, liquid, or solid phase adjoins the solid photocatalyst particle.

Heterogeneous photocatalysis is only now becoming a nascent field of inquiry. Consequently, we must first of all be concerned with the question of whether each newly reported heterogeneous photoreaction is photocatalytic (example (1)) or photochemical (examples (2), (3), and (4)). A simple criterion allowing clear discrimination between examples (1) and (2) is the demonstrated formation of much more than several monolayer equivalents of product. We have elsewhere argued arbitrarily that a factor of 100 serves this purpose. By this criterion, we have established (Childs and Ollis, 1980) that titanium dioxide (TiO_2)-mediated oxidations of CN^- to CNO^- (Frank and Bard, 1977b), of isobutane (C_4H_{10}) to acetone (CH_3COCH_3) and CO_2 (Formenti et al., 1974; Djeghri et al., 1974), and the decarboxylation of acetic acid (CH_3COOH) to methane (CH_4) and CO_2 (Kraeutler and Bard, 1978) are all photocatalytic (Childs and Ollis, 1980), as is the reaction of water photolysis ($H_2O \rightarrow H_2 + 1/2\ O_2$) over NaOH-coated platinum (Pt)/strontium titanate ($SrTiO_3$) surfaces (Wagner et al., 1979; Wagner and Somorjai, 1980) (Figure 1) and Pt/TiO_2 surfaces (Sato and White, 1981; White, this volume).

Other reactions of interest in atmospheric chemistry have been observed in the presence of illuminated semiconductor oxides, e.g.,

$$CO + \tfrac{1}{2}O_2 \xrightarrow[TiO_2 \text{ or } ZnO]{h\nu} CO_2 \qquad (a)$$

(Frank and Bard, 1977a, b)

$$SO_3 + \tfrac{1}{2}O_2 \xrightarrow[TiO_2 \text{ or } Fe_2O_3]{h\nu} SO_4^{=} \qquad (b)$$

(Frank and Bard, 1977b)

$$N_2 + 3H_2O \xrightarrow[TiO_2]{h\nu} 2NH + 3/2\ O_2 \qquad (c)$$

(Schrauzer and Guth, 1977)

$$N_2 + 2H_2O \xrightarrow[TiO_2]{h\nu} N_2H_4 + O_2 \qquad (d)$$

(Schrauzer and Guth, 1977)

$$\text{dichlorobiphenyl} \xrightarrow[TiO_2]{h\nu} Cl^- + \text{other products} \qquad (e)$$

(Carey et al., 1976)

---

*Presently at Mobil Research and Development Corp., P.O. Box 1025, Princeton, New Jersey 08540.

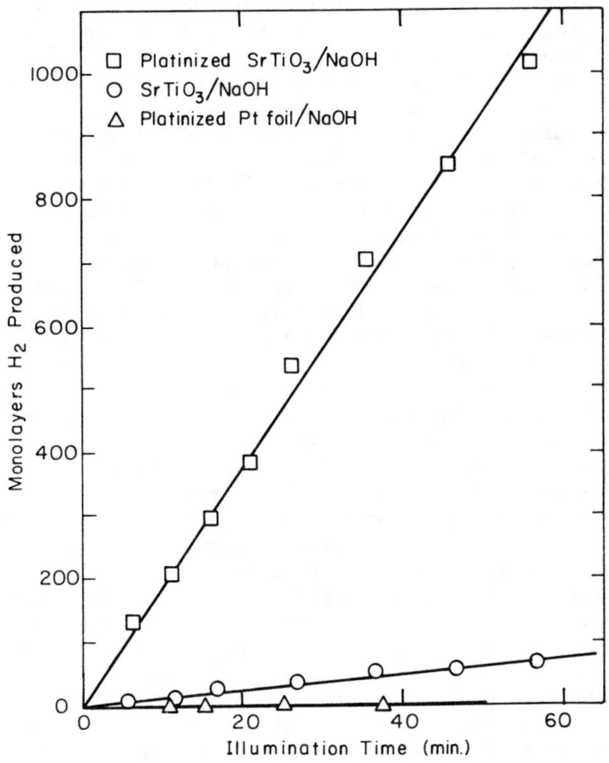

Fig. 1. Hydrogen production from platinized and metal-free n-type $SrTiO_3(111)$ single-crystal surfaces. Both $SrTiO_3$ crystals were coated with >30 μm NaOH and illuminated in a saturation pressure (~20 mm Hg_2) of water vapor (1 monolayer = 10^{15} molec H_2/cm^2 of illuminated surface) (from Wagner et al., 1979).

$$CO_2 + 2H_2O \xrightarrow[Pt/SrTiO_3]{h\nu} CH_4 + 2O_2 \qquad \text{(f)}$$

(Hemminger et al., 1978)

Among these reactions, the fixation of nitrogen to ammonia or hydrazine has not been found to be catalytic (Van Damme and Hall, 1979), the dehalogenation reaction converted only about 2×10^{-3} monolayers of material, and the remainder provided only modest demonstrated conversions (0.1 to 10 monolayer equivalents) (Childs and Ollis, 1980) and thus remain unproven as photocatalytic processes. (A site density of $5 \times 10^{14}/cm^2$ was assumed in converting product obtained to monolayer equivalents.)

Kinetics of Photocatalyzed Processes

The kinetics of heterogeneous photocatalyzed processes is distinguished by a near absence in the literature. Two clear examples are provided by the oxidation of isobutane to acetone and CO_2 (Formenti et al., 1976; Childs and Ollis, 1981a) and by the dehalogenation of chloroform ($CHCl_3$) (Childs, 1981; Childs and Ollis, 1981b) and of trichloroethylene (CCl_2CHCl) (Childs, 1981; Childs and Ollis, 1981c). As these partial oxidations and dehalogenations are of atmospheric interest in connection with smog formation and the fate of chlorofluorocarbons, we summarize the kinetics and mechanistic implications of these photocatalyzed reactions in the remainder of this paper.

Photocatalyzed Alkane Partial Oxidation

Isobutane oxidation (Formenti et al., 1976a) provides a clear case illustrative of the photocatalyzed oxidations of branched and linear alkanes reported by Teichner's group in France (Formenti et al., 1972, 1976a, b; Gravelle et al., 1971; Djeghri et al., 1974; Juillet et al., 1973; Walker et al., 1977; Hermann et al., 1979; Formenti and Teichner, 1978) and Stone's in England (Bickley et al., 1973; Gravelle et al., 1971). The French group has demonstrated the activity of high-surface-area submicron particles (50 Å to 1000 Å) of many oxides, including TiO_2, ZnO, and oxides of iron. A proposed mechanism from Formenti et al. (1976a) is shown in Figure 2. The characteristic features of partial oxidation of three and higher-carbon-number alkanes (Formenti and Teichner, 1978) are evident as catalyst photoactivation (light is a reagent), oxygen activation and dissociation, and alcohol formation. At this point, oxygen addition and dehydration yield an oxygenate of the same carbon number (butaraldehyde) as the original reactant hydrocarbon (isobutane), or more likely an alcohol dehydration to give an olefin, followed by oxygen addition to cleave the double bond and give oxygenate plus CO_2 (for terminal olefins) or two oxygenates (for internal olefins).

A rate equation consistent with dissociated oxygen involvement in the partial oxidation has been derived (Childs and Ollis, 1981a) for formation of acetone, the predominant product in Figure 2:

$$r_{acetone} = \frac{k' P_{O_2}^{1/2} P_{HC}}{\left[1 + \left(K_{O_2} P_{O_2}\right)^{1/2} + K_{HC} P_{HC} + K_A P_{O_2}^{1/2} P_{HC}\right]^2} \qquad (1)$$

where P_{O_2} and P_{HC} are oxygen and alkane pressures, respectively, and K_{O_2}, K_{HC}, and K_A are binding constants for oxygen, hydrocarbon, and alcohol, respectively.

The comparison of model (equation (1)) (Childs and Ollis, 1981a) and data (Formenti et al., 1976a) is shown in Figure 3. The agreement is best at the higher hydrocarbon pressures. At low alkane pressures the oxygenate product is not protected from further oxidation to CO_2, water, and lower-carbon-number oxygenates.

Photoexcitation (to produce hole (h^+) and electron (e^-)):

$TiO_2 + h\nu \rightarrow TiO_2 + h^+ + e^-$

Oxygen activation (at surface):

$e^- + O_2(g) \rightarrow O_2^-(\text{adsorbed})$

Dissociation of O_2^-:

$O_2^- + h^+ \rightarrow 2O(\text{adsorbed})$

Reaction with adsorbed hydrocarbon:

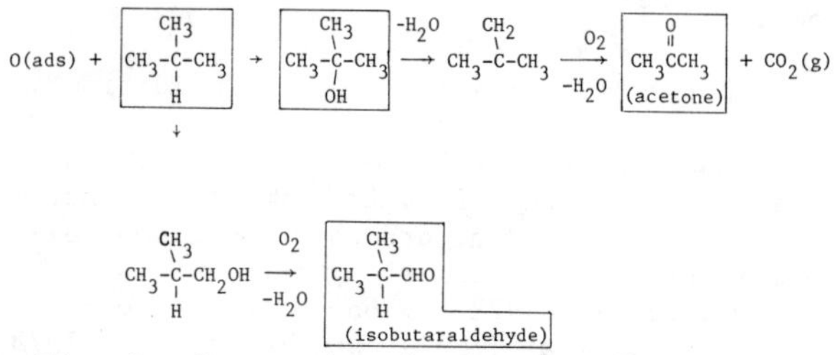

Fig. 2. Proposed mechanism for photocatalyzed oxidation of isobutane to aldehydes over TiO_2 (from Formenti et al., 1976a). (Boxed species observed in gas phase; O_2 = lattice oxygen.)

Photocatalyzed Dechlorinations

Chlorinated hydrocarbons are found frequently in natural and drinking waters of Europe and North America (Symons et al., 1975; Foley and Missingham, 1976; Rook, 1976). Two halocarbons characteristic of these contaminants are chloroform ($CHCl_3$) and trichloroethylene ($CHCl\text{-}CCl_2$), the former being the predominant halomethane arising from water chlorination practices (Rook, 1974) and the latter a common industrial degreasing solvent.

Our reaction experimental studies with these halocarbons were carried out in aqueous phase containing a deaerated 0.1 wt % slurry of TiO_2 (anatase) particles (surface area = 7 m^2/g TiO_2). During an experiment the slurry and the trace halocarbon solution were continuously recirculated in a closed loop system containing a quartz annular photoreactor with seven 15-W black-light illuminators (one in the reactor center and six outside) arranged parallel to the reactor axis. Black lights emit predominantly 320- to 400-nm wavelengths, with virtually no contributions shorter than 300 nm or longer than 500 nm. The short wavelength limit avoids energies known to allow the presence of homogeneous dechlorinations (Calvert and Pitts, 1966); recirculation through a water bath (25°C) and the absence of longer wavelengths in the light sources allowed operation at 38° to 40°C in the photoreactor.

Chloroform conversion versus time and reaction conditions is shown in Figure 4. The simultaneous presence of illumination and TiO_2 slurry (region III) produces dehalogenation of chloroform, yielding chloride ions as detected by a specific chloride ion electrode located in the recirculation loop. Neither illumination alone (region I) nor TiO_2 slurry alone (region II) produced an appreciable reaction. Complete dechlorination of the chloroform is eventually achieved: at 180 minutes (65 minutes of illumination plus catalyst) 99 ppm chloride ion product is detected, compared with the 100 ppm of chlorine present in the initial chloroform loading. The chloride ion electrode determination was within 2 percent of values obtained by direct silver nitrate titration (Mohr's method) (Kolthoff and Sandell, 1952; Am. Public Health Assoc., 1971). Following catalyzed photodechlorination, a nitrogen stream stripped the reacted solution of soluble product gases. The immediate formation of barium carbonate precipitate when the stripping gas was subsequently passed through a barium hydroxide solution verified the complete mineralization of chloroform to CO_2 and (H)Cl. (No organic products

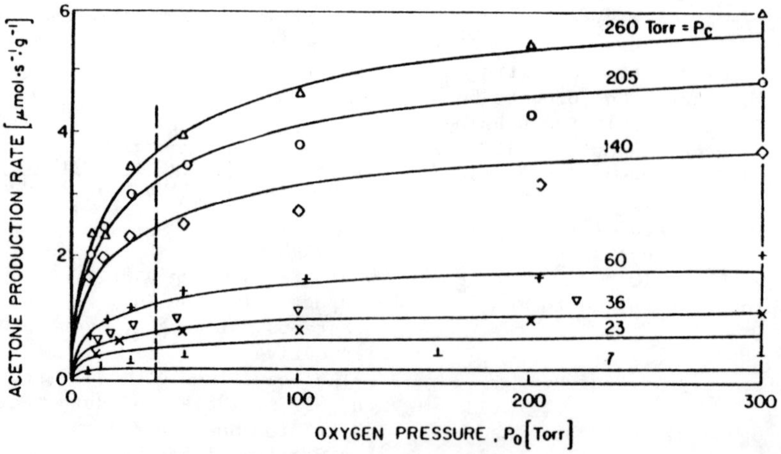

Fig. 3. Partial oxidation of isobutane to acetone. Comparison of model rate equation (solid line) with data of Formenti et al. (1976a) (from Childs and Ollis, 1981). P_c = isobutane pressure (torr): 260 (Δ), 205 (O), 140 (◇), 60 (+), 36 (∇), 23 (×), 7 (⊥).

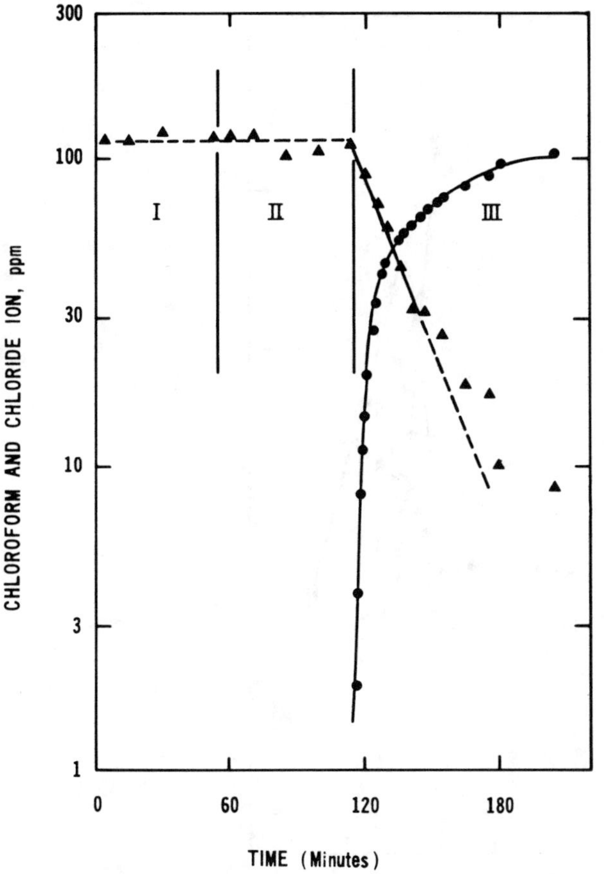

Fig. 4. Chloroform conversion versus time: chloroform (Δ) and chloride (•). Conditions: I - illumination only; II - TiO_2 photocatalyst only; III - illumination and photocatalyst simultaneously present (from Childs, 1981; Childs and Ollis, 1981a).

or intermediates were noted in gas-chromatograph/mass-spectrometer (GC-MS) analysis of sample liquids or associated sample headspace vapors.)

Photocatalyzed dechlorination of trichloroethylene (TCE) (Figure 5) also eventually produced complete mineralization of the reactant to CO_2 and (H)Cl. At 10 ppm TCE no other product species were noted, but at subsequent TCE injections which provided 22 ppm and 50 ppm starting reactant levels (Figure 5), one transient organic intermediate was observed. GC-MS analysis indicated that this organic intermediate was dichloroacetaldehyde (CCl_2CHO) (Childs, 1981). Chlorine mass balances (final chloride ion versus initial reactant chlorine content) indicated 97 to 98 percent chlorine recovery, and again N_2 stripping of the reacted liquid phase produced barium carbonate precipitate. Thus, TCE was eventually completely mineralized to CO_2 and (H)Cl.

The linear slopes of log (reactant) versus time plots (Figures 4 and 5) provide first-order apparent rate constants, specific activities (based on total catalyst surface areas), and turnover rates (columns 3, 4, and 5 of Table 1). The characteristic turnover number for conversions of chloroform, trichloroethylene, and dichloroacetaldehyde is $\sim 10^{-2}$ molec/site-sec.

The pathway of TCE conversion to dichloroacetaldehyde is unproven, but we may postulate a mechanism by considering a known homogeneous gas-phase route for tetrachloroethylene. In the presence of hydroxyl radical, tetrachloroethylene was converted to dichloroacetylchloride (D. G. Crosby, Univ. of California, private communication, 1980):

$$Cl_2C{=}CCl_2 + \cdot OH \rightarrow Cl_2C{=}CCl_2 + \cdot Cl \qquad (g)$$

As TiO_2 may well provide $\cdot OH$ by the oxidation of OH^- with a photoproduced hole (h^+)

$$OH^- + (h^+) \xrightarrow{TiO_2} \cdot OH \qquad (h)$$

the analogous replacement of chlorine by hydroxyl for the photocatalyzed TCE conversion would be

$$Cl_2C{=}CClH + \cdot OH \rightarrow Cl_2C{=}C(OH)H \rightarrow H\text{-}CCl_2\text{-}C({=}O)\text{-}H \qquad (i)$$

(trichloroethylene) (dichloroacetaldehyde)

In any event, it is clearly indicated in Figures 4 and 5 that a saturated halocarbon ($CHCl_3$) and an unsaturated halocarbon ($Cl_2C = CHCl$) are both completely mineralized by TiO_2 in the presence of near-UV illumination. Total demonstration conversions were in excess of 100 for both reactants based on chloride ions formed, thus demonstrating the truly catalytic nature of these photoassisted reactions.

Heterogeneous Photocatalysis in Atmospheric Chemistry?

The most commonly photoactivated solids appearing in the photocatalysis literature are TiO_2 (anatase and rutile forms), ZnO, and Fe_2O_3, with other oxides such as Ga_2O_3 and SnO_2 occasionally reported. A few papers have found some activity with Al_2O_3, SiO_2, and MgO, but as these are large-band-gap materials, we are most likely dealing either with photoexcitation from impurity states or with noncatalytic processes such as examples (2), (3), and (4) mentioned in the introduction. Thus, if heterogeneous photocatalysis is of atmospheric importance, we think first of TiO_2, ZnO, and Fe_2O_3.

Let us first consider anthropogenic sources, as characterized by power plant emissions. According to A. Sarofim (private communication,(1981), coal combustion produces fly ash with major constituents SiO_2, MgO, CaO, FeO, Al_2O_3, Na_2O, and P_2O_5 and minor constituents containing the metals Mn, V, Cr, Zn, Co, Ag, La, and Sc. Thus iron

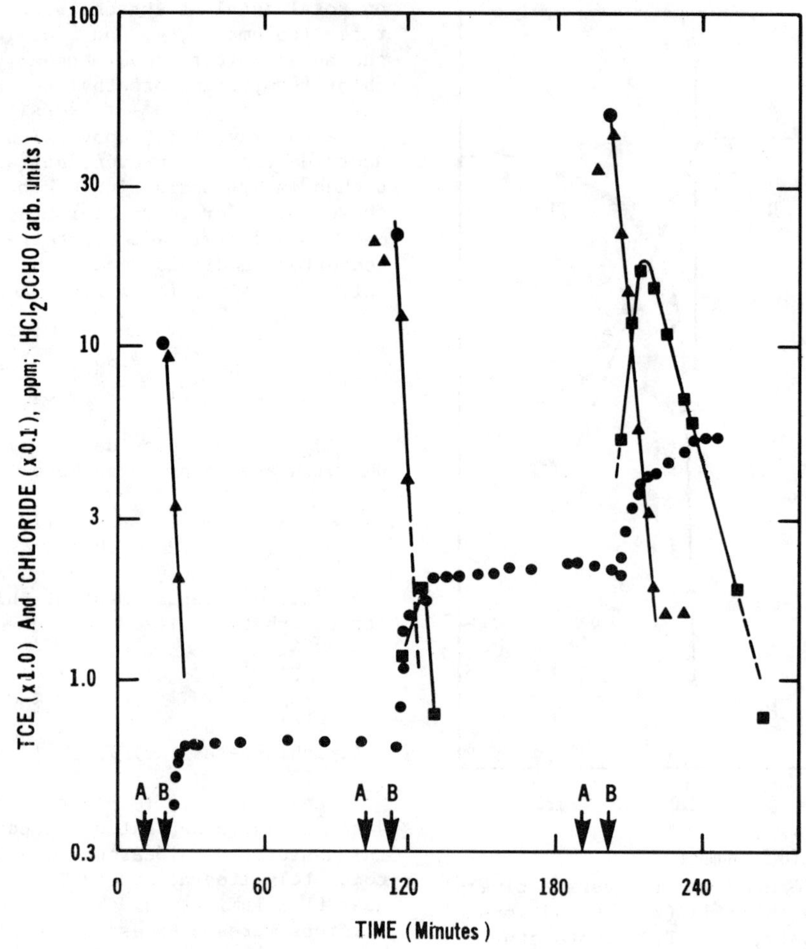

Fig. 5. Trichloroethylene conversion versus time: trichloroethylene (Δ), dichloroacetaldehyde (■), chloride (•). Conditions: A - catalyst only, TCE added; B - catalyst and illumination (from Childs and Ollis, 1981a, b).

oxides may exist, although the presence of other components of lower surface free energies may cover any such iron phases. Similarly, solid solutions of iron oxides in other ash components may eliminate any modest bandgap (2.0 to 3.0 eV) semiconductor phases which might otherwise form in fly ash.

The minor components, according to Sarofim, may

TABLE 1. Initial Rates of Photocatalytic Conversion of TCE and $CHCl_3$ (from Childs, 1981)

	Initial Concentration C_o (ppm)	First-Order Rate Constant k (min^{-1}) Disappearance	Specific Activity (μmol/g cat •sec)	Turnover No. (molec/ site•sec)
$CHCl_3$	122	0.044	0.73	1.3×10^{-2}
TCE	10.3	0.365	0.48	0.9×10^{-2}
	23.6	0.352	1.1	1.9×10^{-2}
	45.4	0.355	2.1	3.7×19^{-2}
Cl_2HCCHO	$(25)^b$	(0.065)	(0.20)	(0.4×10^{-2})

[a]Calculated from initial slopes in Figures 4 and 5.
[b]Estimated.

dominate the surface compositions of fly ash, as some of these are the more volatile components and condense upon the major constituents after the latter form the fly ash "core." Thus, ZnO might conceivably be formed on fly ash surfaces, and, given near-UV illumination, could give rise to photocatalyzed reactions such as hydrocarbon oxidation. Indeed, Buseck and Bradley (this volume) found ZnO particles on the external surface of fly ash from copper smelter operation. Even were ZnO to become established as a general and active photocatalyst, the presence of large concentrations of SO_3 in power plant atmospheres may dominate the solid chemistry. Thus particulate sulfate emissions in the form of $(NH_4)_2SO_4$, H_2SO_4, and refractory metal sulfates (noted in cold weather) from power plants were mentioned (Buseck and Bradley, this volume; A. Sarofim, private communication, 1981). Ultimately, whether a number of the more reactive metals exist in fly ash as sulfates or oxides may depend, as it did so strikingly in auto exhaust catalysts using active base metal components, on the fly ash temperature while in the presence of SO_3 or H_2SO_4. In the automotive catalyst instance, base metal sulfates stable below 425° to 540°C were decomposed to SO_3 (vapor) and metal oxides at temperatures above 540° to 650°C.

We have discussed herein the photocatalyzed formation of aldehydes and ketones from simple hydrocarbons and the dehalogenation of a saturated but not completely halogenated chloroform molecule, $CHCl_3$. The relevance, if any, of this TiO_2-photocatalyzed reaction to the eventual fate of the halomethanes Freon 11 ($CFCl_3$) and Freon 12 (CF_2Cl_2) in the atmosphere is unknown. While these fully saturated molecules are unreactive toward ·OH radicals (Seinfeld, 1980), we note that attack of hydroxyl radical on unsaturated haloethylenes appears to be at the halogen, not the hydrogen (thus allowing dichloroacetaldehyde as previously shown). Thus photocatalyst tests of such known halocarbons as CH_3CCl_3, CH_3Cl, and CCl_2CCl_2 with pure ZnO, TiO, and iron oxide aerosols are suggested, as are similar tests with coal and smelter fly ash containing known semiconductor oxides (Buseck and Bradley, this volume).

As SO_2 may be oxidized to SO_3 by hydroxyl radical participation (Seinfeld, 1980), photocatalytic surfaces which can produce OH radicals (ZnO, TiO, and probably other semiconductor oxides) deserve further study with SO_2 present.

The partial oxidation of C_4 and higher hydrocarbons to aldehydes and ketones discussed earlier may deserve further study in appropriate air chamber simulators, such as the aerosol reactor configurations treated by Crump and Seinfeld (1980) and S. Friedlander (private communication, June 1981). Indeed, while such oxygenates may be formed only near ground level, where appreciable particulates with photocatalytically active phases might exist, such near-ground locations are the prime regions for smog formation in cities.

Finally, ZnO and probably TiO_2 may photocatalytically produce H_2O_2 in liquid-phase systems. Whether such hydroxyl radical production surfaces could be, under any conditions, an atmospheric source of H_2O_2 or hydroxyl radical is uncertain.

Altogether, the ingredients needed for heterogeneous photocatalysis in the atmosphere are present in certain low-altitude situations, but further quantification is needed to determine whether such photoassisted conversions are present, and, if so, whether such conversions are significant.

Acknowledgements. The authors acknowledge the support of the National Science Foundation and the Engineering Foundation.

References

American Public Health Association, Standard Methods for the Examination of Water and Wastewater, 13th ed., p. 96, APHA, Washington, D.C., 1971.

Bickley, R. I., G. Munuera, and F. S. Stone, Photoadsorption and photocatalysis at rutile surfaces. II. Photocatalytic oxidation of isopropanol, J. Catal., 31, 398-407, 1973.

Bickley, R. I., and F. S. Stone, Photoadsorption and photocatalysis at rutile surfaces. I. Photoadsorption of oxygen, J. Catal., 31, 389-397, 1973.

Buseck, P. R., and J. P. Bradley, Electron beam studies of individual natural and anthropogenic microparticles: Composition, structures, and surface reactions, this volume.

Calvert, J. C., and N. Pitts, Photochemistry, John Wiley and Sons, New York, 522-528, 1966.

Carey, J. H., J. Lawrence, and H. M. Tosine, Photodechlorination of PCB's in the presence of titanium dioxide in aqueous suspensions, Bull. Environ. Contam. Toxicol., 16, 697-701, 1976.

Childs, L. P., Studies in photoassisted heterogeneous catalysis: Rate equations for 2-methyl-2-butyl alcohol, Ph.D. Thesis, Princeton Univ., Princeton, N.J., 1981.

Childs, L. P., and D. F. Ollis, Is photocatalysis catalytic? J. Catal., 66, 383-390, 1980.

Childs, L. P., and D. F. Ollis, Photoassisted heterogeneous catalysis: Rate equations for the oxidation of 2-methyl-2-butyl-alcohol and isobutyl, J. Catal., 67, 35-48, 1981a.

Childs, L. P., and D. F. Ollis, Photoassisted heterogeneous catalysis: Mineralization of trichloromethane, submitted to J. Water Poll. Contr. Fed., 1981b.

Childs, L. P., and D. F. Ollis, Photoassisted heterogeneous catalysis: The degradation of trichloroethylene in water, submitted to J. Catal., 1981c.

Crump, J. G., and J. H. Seinfeld, Aerosol behavior in the continuous stirred tank reactor, AIChE J., 26, 610, 1980.

Djeghri, N., M. Formenti, F. Juillet, and S. J. Teichner, Photointeraction on the surface of titanium dioxide between oxygen and alkanes, Disc. Faraday Soc., 58, 185-193, 1974.

Foley, P. D., and G. A. Missingham, Monitoring of community water supplies, J. Am. Water Works Assoc., 68, 105-111, 1976.

Formenti, M., F. Juillet, P. Meriandeau, and S. J. Teichner, Heterogeneous photocatalysis for partial oxidation of paraffins, Chem Technol., 1, 680-686, 1971.

Formenti, M., F. Juillet, P. Meriandeau, and S. J. Teichner, Partial oxidation of paraffins and olefins by a heterogeneous photocatalysis process, Bull. Soc. Chim. France, (1), 69-76, 1972.

Formenti, M., F. Juillet, and S. J. Teichner, Photocatalytic oxidation mechanism of alkanes over titanium dioxide, Bull. Soc. Chim. France, 7-8, 1031-1036, 1976a.

Formenti, M., F. Juillet, and S. J. Teichner, Mechanism of the photocatalytic oxidation of isobutane over titanium(IV) oxide, Bull. Soc. Chim. France, 9-10, 1315-1320, 1976b.

Formenti, M., and S. J. Teichner, Heterogeneous photocatalysis, The Chemical Society: Catalysis, 2, 87-106, 1978.

Frank, S. N., and A. J. Bard, Heterogeneous photocatalytic oxidation of cyanide ion in aqueous solutions of TiO_2 powder, J. Am. Chem. Soc., 99, 303-304, 1977a.

Frank, S. N., and A. J. Bard, Heterogeneous photocatalytic oxidation of cyanide and sulfite in aqueous solutions at semiconductor powders, J. Phys. Chem., 81, 1484-1488, 1977b.

Gravelle, P., F. Juillet, P. Meriandeau, and S. J. Teichner, Surface reactivity of reduced titanium dioxide, Disc. Faraday Soc., 52, 140-148, 1971.

Hemminger, J. C., R. Carr, and G. A. Somorjai, The photoassisted reaction of gaseous water and carbon dioxide adsorbed on the $SrTiO_3$(III) crystal face to form methane, Chem. Phys. Lett., 57, 100-104, 1978.

Herrmann, J. M., J. Disdier, M. Mozzanega, and P. Pichat, Heterogeneous photocatalysis: In situ photoconductivity study of titanium dioxide during oxidation of isobutane into acetone, J. Catal., 60, 369-377, 1979.

Juillet, F., F. LeComte, H. Mozzanega, S. J. Teichner, A. Thevenet, and P. Vergnon, Inorganic oxide aerosols of controlled submicronic dimensions, Faraday Symp. Chem. Soc., 7, 57-62, 1973.

Kolthoff, J. M., and F. B. Sandell, Textbook of Quantitative Analysis, 3rd edition, p. 542, MacMillan, New York, 1952.

Kraeutler, B., and A. J. Bard, Heterogeneous Photocatalytic decomposition of saturated carboxylic acids on TiO_2 powder. Decarboxylative route to alkanes, J. Am. Chem. Soc., 100, 5985-5992, 1978.

Rook, J. J., Formation of haloforms during chlorination of natural waters, Water Treat. Exam., 23, 234-243, 1974.

Rook, J. J., Haloforms in drinking water, J. Am. Water Works Assoc., 68, 168-172, 1976.

Sato, S., and J. M. White, Photocatalytic water decomposition and water-gas shift reaction over NaOH-coated, platinized TiO_2, J. Catal., 69, 128-139, 1981.

Schrauzer, G. N., and T. D. Guth, Photolysis of water and photoreduction of nitrogen on titanium dioxide, J. Am. Chem. Soc., 99, 7189-7193, 1977.

Seinfeld, J., Lectures in atmospheric chemistry, AIChE Monograph Series, Vol. 76, 1980.

Symons, J. M., T. A. Bellar, J. K. Carswell, J. DeMarco, K. L. Kropp, G. G. Robeck, D. R. Seeger, C. J. Slocum, B. L. Smith, and A. A. Stevens, National organics reconnaissance survey for halogenated organics, J. Am. Water Works Assoc., 67, 634-647, 1975.

Van Damme, H., and W. K. Hall, On the photoassisted decomposition of water at the gas-solid interface on TiO_2, J. Am. Chem. Soc., 101, 4373-4374, 1979.

Wagner, F. T., S. Ferrar, and G. A. Somorjai, Photocatalytic Hydrogen Production from Water over $SrTiO_3$ Crystal Surfaces, LBL preprint 9942, Lawrence Berkeley Laboratory, Univ. of California, Berkeley, 1979.

Wagner, F. T., and G. A. Somorjai, Photocatalyzed Production of Hydrogen from Water on Pt-free $SrTiO_3$ Single Crystals in the Presence of Alkali Hydroxides, LBL preprint 10434, Lawrence Berkeley Laboratory, Univ. of California, Berkeley, 1980.

Walker, A., M. Formenti, P. Meriandeau, and S. J. Teichner, Heterogeneous photocatalysis: Photooxidation of methylbutanols, J. Catal., 50, 237-243, 1977.

White, T. M., Photoassisted reactions on doped metal oxide particles, this volume.

HETEROGENEOUS CATALYZED PHOTOLYSIS VIA PHOTOACOUSTIC SPECTROSCOPY

L. Robbin Martin and Marilyn Wun-Fogle

The Aerospace Corporation, Laboratory Operations, P. O. Box 92957, Los Angeles, California 90009

Abstract. We have measured the ultraviolet absorption spectra of $CFCl_3$ and N_2O physically adsorbed on lithium fluoride (LiF). The technique employed was photoacoustic spectroscopy, which is sensitive to small amounts of material adsorbed on powdered solids. Preliminary results indicate that the optical absorption band of $CFCl_3$ is broadened by contact with the solid LiF. Broadening of this kind may be responsible for reported long-wavelength photolysis of these molecules on transparent solids. Broadening of the N_2O absorption band was not seen on LiF.

Introduction

While the existence of heterogeneously catalyzed photolysis and photochemistry has been known for some time, the potential for environmental significance of the process has only recently been appreciated. Our interest stems from the work in 1977 by Ausloos and Rebbert (Ausloos et al., 1978) of the National Bureau of Standards on the photolysis of chlorofluoromethanes adsorbed on solids. They reported that $CFCl_3$, $CFCl_2$, and CCl_4 will photolyze at much longer wavelengths when adsorbed on sand or quartz particles than when they are in the gas phase. This result suggests that such photochemically stable molecules, previously thought to undergo photolysis only in the stratosphere, may in fact undergo heterogeneous photocatalytic degradation in the troposphere.

Rebbert and Ausloos (1978) further reported that N_2O decomposed at long wavelengths when adsorbed on sand. Also in 1978, Gäb and Korte (1978) in Germany reported evidence of the photolysis of CF_2Cl_2 and $CFCl_3$ adsorbed on silica gel in the presence of ground level sunlight. Discussions of the possible environmental effects of this kind of process have been given by Pierotti et al. (1978) and Alyea et al. (1978).

The purpose of the present study was to search for weak optical absorptions that might account for long wavelength photolysis. For experimental reasons, the study was confined to "transparent" solids such as SiO_2 and LiF. There have been numerous studies of photocatalytic effects in the semiconducting materials TiO_2, ZnO, and ZrO_2 (Krasnovskii and Brin, 1966; Cunningham et al., 1971; Tanaka and Blyholder, 1971). The mechanism of photocatalysis in these materials has been ascribed to the semiconducting properties (Cunningham and Penny, 1974), and it is unlikely that a similar mechanism is operative in the case of silica, which is highly transparent in the wavelength region of interest (200 to 400 nm).

Our approach was to take photoacoustic spectra of the gases and solids together and separately. Photoacoustic spectroscopy is uniquely suited to reveal weak optical absorptions of physically adsorbed molecules. It is a highly sensitive technique and does not respond to scattered light.

Experimental

Apparatus Description

A block diagram of the apparatus is shown in Figure 1. The photoacoustic spectrometer consists of an Oriel 1000-W xenon arc lamp filtered through distilled water to remove infrared radiation and dispersed through a J-Y Model DH-10 UV double monochromator with holographic gratings (f/3.5). The beam passes through a variable-speed chopper and enters the cells in an airtight sound isolation chamber. All optical elements are made of Suprasil quartz to permit operation down to 200 nm. The cells are equipped with high-sensitivity Brüel and Kjaer condenser microphones, Model 4165. These microphones have built-in preamplifiers. The sample cell preamplifier drives a Princeton Applied Research (P.A.R.) Model 5204 lock-in analyzer, and the reference preamplifier drives a P.A.R. Model 186 lock-in amplifier. The microprocessor, which is of our own design, controls the wavelength drive on the monochromator, stores the spectra, subtracts backgrounds, and averages and ratios the signals.

All gases are chromatographically pure and the lithium fluoride is Suprapur (MCB Manufacturing Chemist, Inc.). Krypton is used as the reference gas because its acoustic properties are similar to those of the sample gases. The photoacoustic cells are Helmholtz resonators of our own design, and the chopping frequency is at or near the krypton acoustic resonance at 250 Hz. The reference cell contains carbon black, and both cells are run at 1 atm absolute pressure.

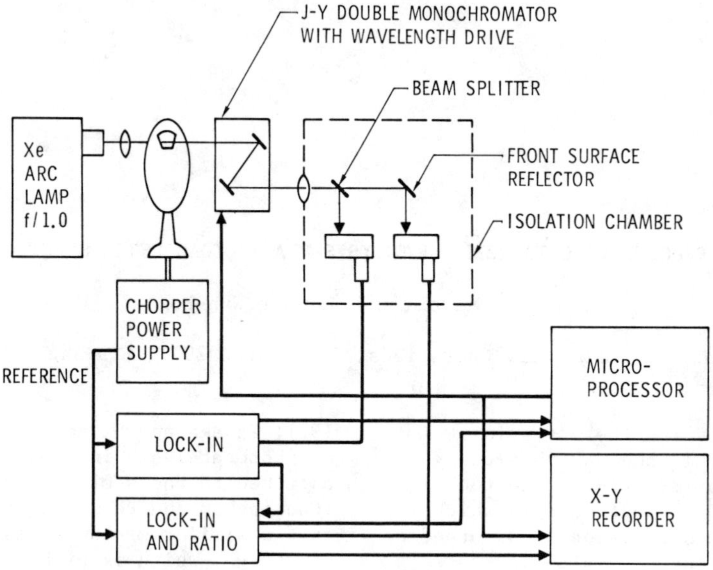

Fig. 1. Photoacoustic spectrometer.

Spectra are typically taken by stepping through the optical region of interest (200 to 400 nm) in 2-nm steps and storing the signal digitally on the computer. Five spectra of the sample and reference are taken with the beam on and five with the beam blocked (acoustic background). The microprocessor then averages the runs and subtracts the background. This procedure is necessary because the microphone occasionally picks up "siren" noise from the light chopper that is in phase with the lock-in amplifier. Next, the microprocessor takes the ratio of the sample signal to the reference signal at each point. It is these ratio spectra that are shown in figures 2 to 9.

Results

Ratio spectra are shown in Figure 2 for a cell containing powdered lithium fluoride (LiF) and two different gases. The circles represent krypton and the triangles Freon 11 ($CFCl_3$). Since the absolute magnitude of photoacoustic spectra is arbitrary, i.e., it is a function of many extra-

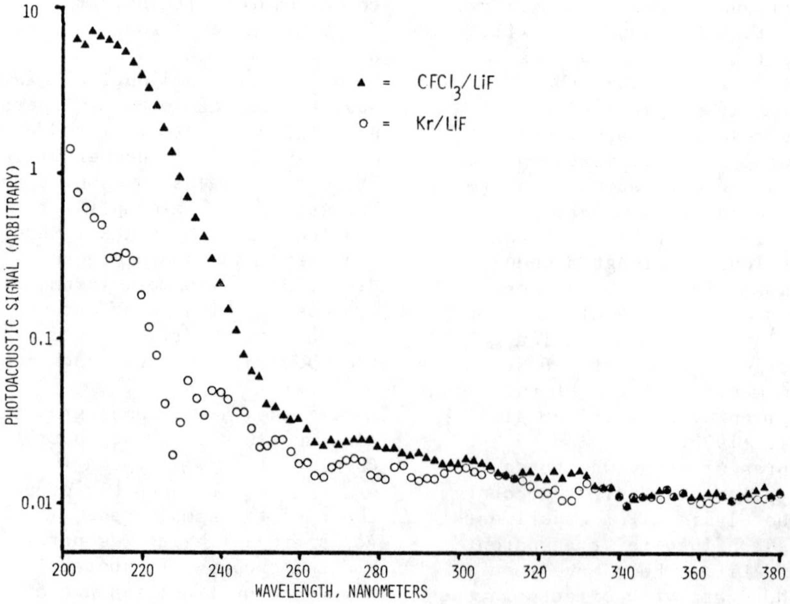

Fig. 2. Ratio spectra for a cell containing powdered LiF for $CFCl_3$ and Kr.

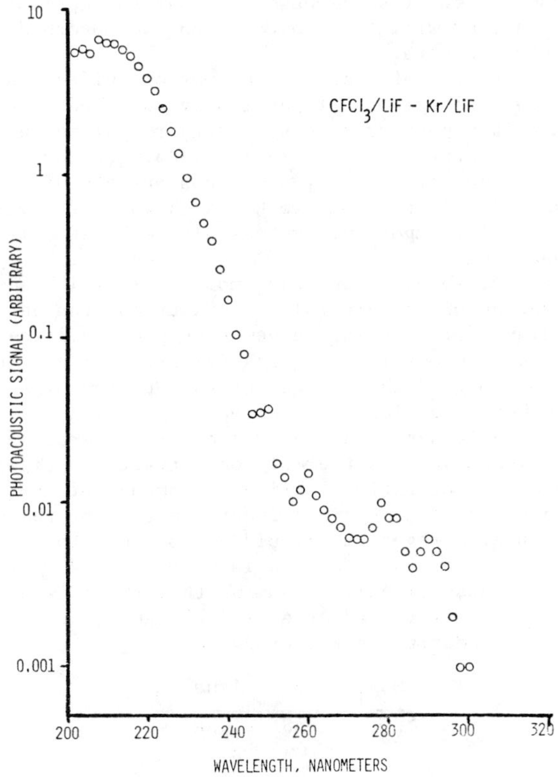

Fig. 3. Difference between adjusted spectra for $CFCl_3$/LiF and Kr/LiF.

neous variables such as grain size and the acoustic properties of the gas, it is necessary to adjust spectra by a constant factor in order to make comparisons. In the spectra shown in Figure 2, the krypton spectrum has been adjusted to bring it into agreement with the $CFCl_3$ spectrum at long wavelengths where no absorptions are expected other than the LiF background. The difference between these two adjusted spectra is shown in Figure 3. This difference spectrum represents the gas-phase spectrum of $CFCl_3$ plus a contribution from the adsorbed molecules.

In Figure 4 the ratio spectra is shown for an "empty" cell containing only a front surface mirror on the bottom instead of a powdered sample. **The circles represent krypton and the triangles Freon 11 ($CFCl_3$). Again, the two spectra have** been brought into agreement at long wavelengths. The difference between these two spectra is shown in Figure 5. Since there is a minimum of surface area available in this cell, the spectrum should represent mainly the pure gas-phase spectrum of $CFCl_3$, with possibly a small contribution from adsorbed material on the windows and walls of the cell.

Allowing for some saturation of the signal (and a high noise level) at wavelengths below 210 nm, the agreement in the shape of the spectrum and the cross section for gaseous $CFCl_3$ is reasonably good (Hampson, 1980). On the other hand, the difference spectrum in Figure 3 is noticeably broadened, which we believe shows a broadened spectrum for physically adsorbed molecules. It was necessary to use LiF, which is highly transparent at

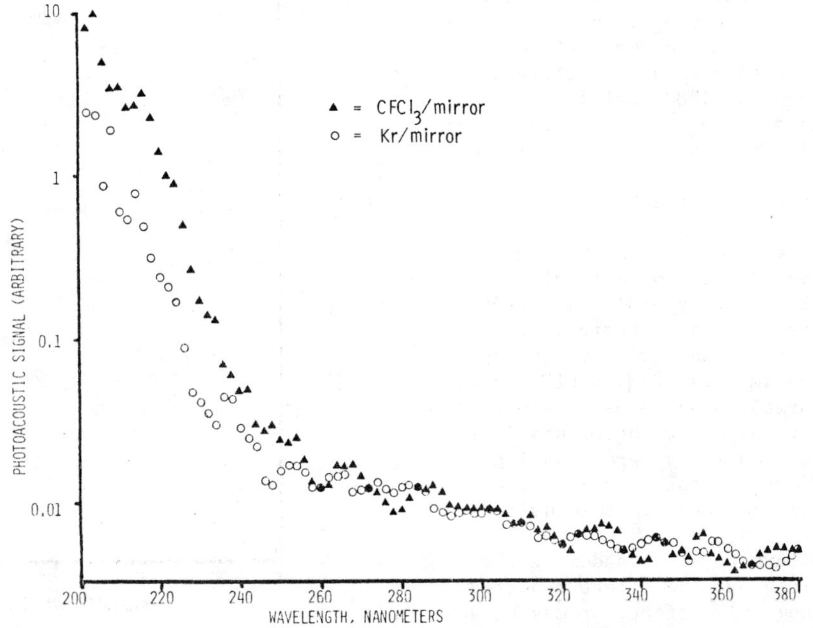

Fig. 4. Ratio spectra for $CFCl_3$ and Kr in a cell with front surface mirror in place of powdered sample.

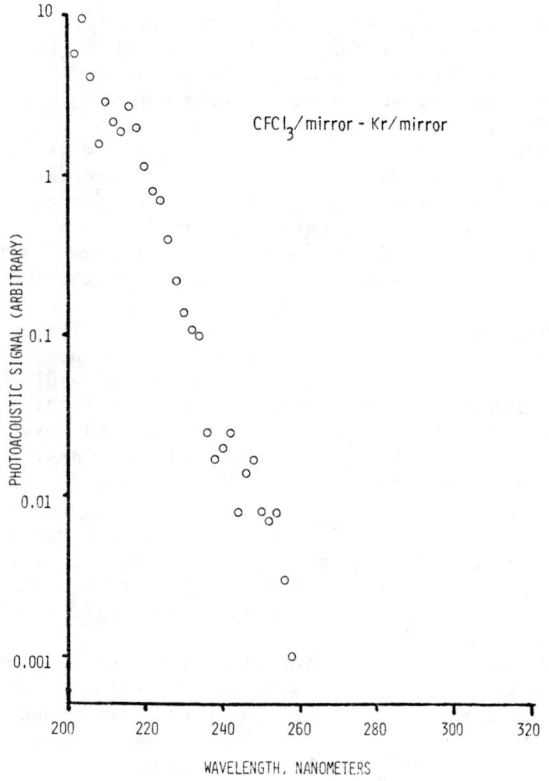

Fig. 5. Difference between adjusted spectra for $CFCl_3$ and Kr with front surface mirror.

these short wavelengths, in order to observe this small broadening effect.

An analogous series of data is shown in Figures 6 through 9 for nitrous oxide, N_2O, in a cell with LiF and an empty cell. Here the two difference spectra are slightly broader than the literature gas-phase spectra (Hampson, 1980) but do not differ appreciably from one another. We take this to be a null result for N_2O on LiF.

Interpretation of the Data

Observation of optical absorptions at longer wavelengths is a necessary but not sufficient condition for long wavelength photolysis. Proof of photolysis must rest with the chemical experiments. Nevertheless, the observation of absorptions out to about 300 nm for $CFCl_3$ on LiF, which is as far as the apparatus signal-to-noise permits, suggests that tropospheric photolysis is a possibility. Here, of course, we are assuming that LiF shows an effect typical of crystalline ionic solids that might be found in the environment.

Conversely, the failure to see broadening for N_2O on LiF does not rule out the effect reported on sand. Sand may show this effect, or may cause photolysis by a different mechanism. In any case, sand will be difficult to study by the photoacoustic technique because its inherent optical absorptions will tend to cover up the adsorbed species spectra.

Observation of a broadening, or of shifts, in the optical spectra of physically adsorbed materials is reported for several other systems in the literature (Gerischer, 1974; Bach and Breuer, 1974), and may be attributed to a variety of physical effects. Our work has shown that environmentally important systems also exhibit this broadening.

Lastly, we point out that more work will be needed before meaningful cross sections can be obtained for calculating new tropospheric lifetimes. The observed signals are proportional to the product of the optical cross sections of the adsorbed molecules and their concentration, and the signals depend as well on other properties of the system such as thermal conductivity, optical depth, and acoustics. Since the amount of adsorbed material is a nonlinear function of the gas pressure, however, it should be possible to vary the pressure and to extrapolate to a "pure" adsorbed phase signal. Assuming that there is no saturation of the adsorbed phase signal, we may assume linearity in the cross section

$$\frac{\sigma_{Ma\lambda_1}}{\sigma_{Ma\lambda_2}} = \frac{\text{PAS signal } \lambda_1}{\text{PAS signal } \lambda_2} \tag{1}$$

where $\sigma_{Ma\lambda}$ is the cross section of adsorbed mol-

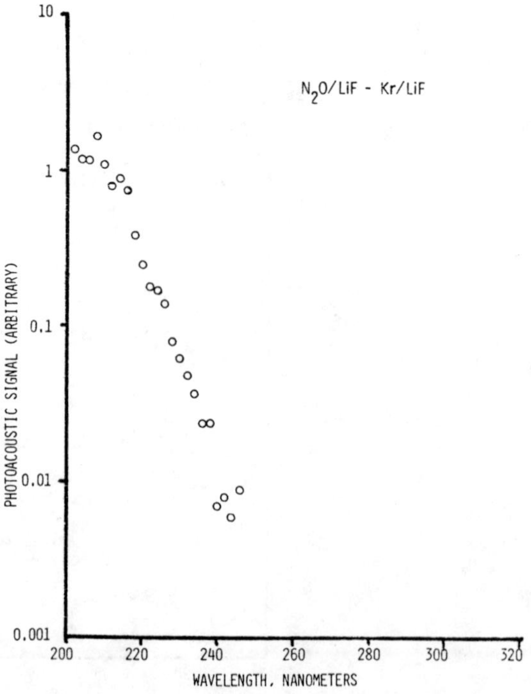

Fig. 6. Ratio spectra for a cell containing powdered LiF for N_2O and Kr.

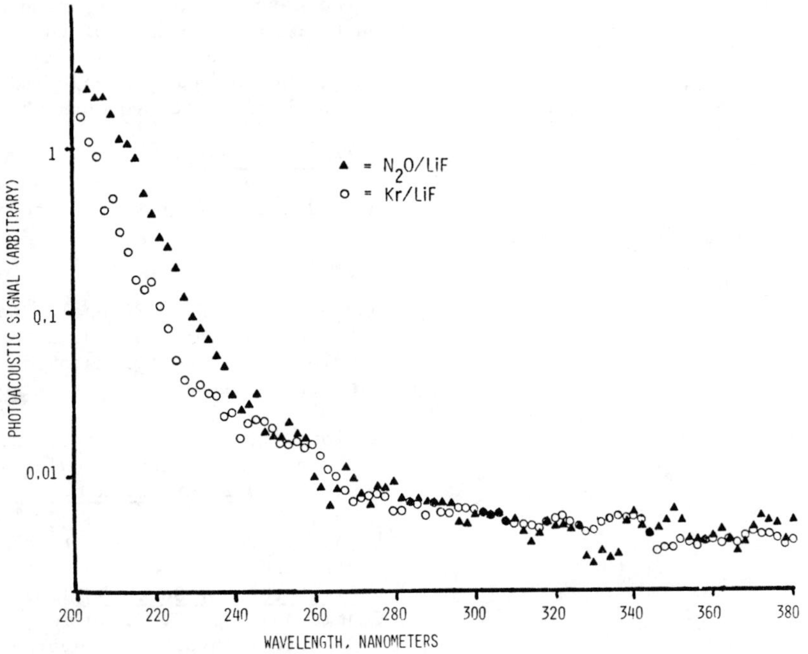

Fig. 7. Difference between adjusted spectra for N_2O and Kr.

ecules per wavelength λ, cm^2, and PAS denotes photoacoustic spectroscopy. This relationship would allow estimation of an adsorbed species cross section from the gas-phase cross section at a wavelength at which the gas-phase cross section is known. Recent work has shown, however, that adsorption may alter a molecule's band strength as well as band shape, at least in the case of electrically conductive substrates (Nitzan and Brus, 1981). In such a case, information about the surface concentration of adsorbed material will also be required.

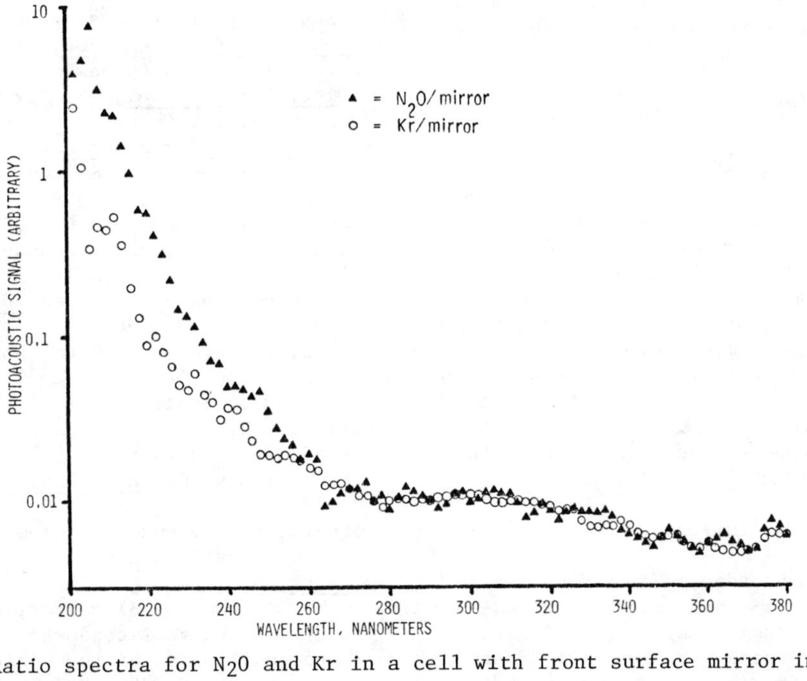

Fig. 8. Ratio spectra for N_2O and Kr in a cell with front surface mirror in place of powdered sample.

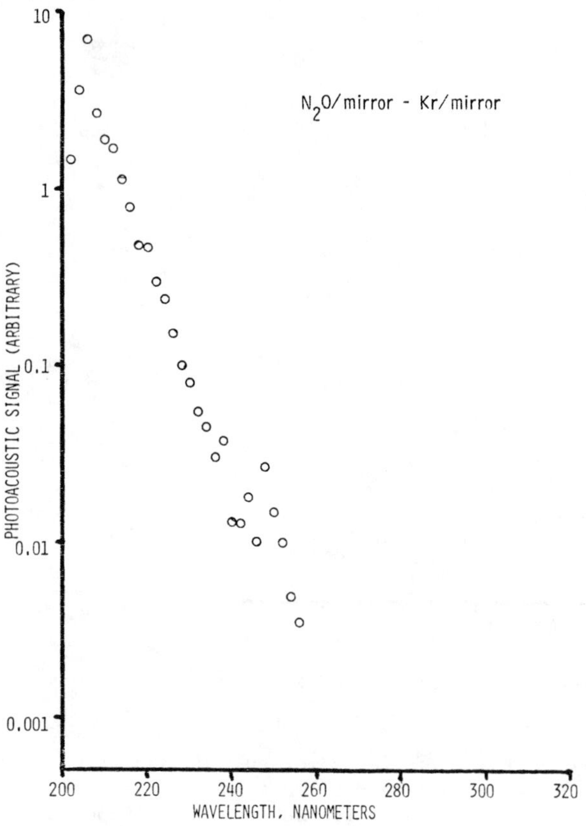

Fig. 9. Difference between adjusted spectra for N_2O and Kr with front surface mirror.

Calculation of the environmental decomposition rate per unit volume of atmosphere would be done from a relationship such as

$$d[M_a]/dt = -[M_a]\sigma_{Ma\lambda}I_\lambda$$

$$[M_a] = [S]kP_M \tag{2}$$

where $[M_a]$ is the concentration of adsorbed molecules cm^{-3}; [S] is the concentration of substrate, g cm^{-3}; I_λ is the photon flux in the bandwidth near λ, cm^{-2} sec^{-1}; P_M is the gas-phase pressure of molecule M, atm; and k is a low pressure limit of the isotherm slope, g^{-1} atm^{-1}. Note that the adsorption in the photoacoustic experiments we have done is in a high pressure limit (1 atm), so values of $[M_a]$ as a function of pressure will be needed over a wide range in order to apply the relationship correctly.

Conclusions

We have observed a broadening of the optical absorption band of $CFCl_3$ when adsorbed on powdered LiF. This broadening may account for the reports of long wavelength photolysis of adsorbed chlorofluorocarbons in similar systems. We did not observe significant broadening of the N_2O band on LiF. Measurements of this kind, coupled with chemical product analysis and adsorption isotherm data, should make possible calculations of tropospheric heterogeneous photolytic lifetimes.

Acknowledgement. Support from the National Science Foundation under grants ATM 77-23435 and ATM 79-17862 is gratefully acknowledged.

References

Alyea, F. N., D. M. Cunnold, and R. G. Prinn, Meteorological constraints on tropospheric halocarbon and nitrous oxide destructions by siliceous land surfaces, Atmos. Environ., 12, 1009-1011, 1978.

Ausloos, P., R. E. Rebbert, and L. Glasgow, Photodecomposition of chloromethanes adsorbed on silica surfaces, J. Res. Natl. Bur. Stds., 82, 1-8, 1977.

Bach, W., and H. D. Breuer, Absorption spectroscopy and photo-reactions of adsorbed molecules, Disc. Faraday Soc., 58, 237-243, 1974.

Cunningham, J., J. J. Kelley, and A. L. Penny, Reactions involving electron transfer at semiconductor surfaces. II. Photoassisted dissociation of nitrous oxide over illuminated ferric oxide and zinc oxides, J. Phys. Chem., 75, 617-625, 1971.

Cunningham, J., and A. L. Penny, Reactions involving electron transfer at semiconducting surfaces. V. Reactivity and electron paramagnetic resonance of electron transfer sites on rutile, J. Phys. Chem., 78, 870-875, 1974.

Gäb, S., and F. Korte, Photomineralisierung von fluorchlorkohlen-wasserstoffen unter simulierten troposphärischen bedingungen, Ber. Bunsen. Phys. Chem., 82, 1151-1154, 1978.

Gerischer, H., Photochemistry of adsorbed species, Disc. Faraday Soc., 58, 219-236, 1974.

Hampson, R. F., Chemical Kinetic and Photochemical Data Sheets for Atmospheric Reactions, Report No. FAA-EE-80-17, U.S. Dept. of Transportation, Washington, D.C., 1980.

Krasnovskii, A. A., and G. P. Brin, Photochemical liberation of oxygen in aqueous solutions of ferric compounds; sensitization by tungsten, titanium, and zinc oxides, Dok. Adad. Nauk SSSR, 168, 1100-1103, 1966.

Nitzan, A., and L. E. Brus, Can photochemistry be enhanced on rough surfaces?, J. Chem. Phys., 74, 5321-5322, 1981.

Pierotti, D., L. E. Rasmussen, and R. A. Rasmussen, The Sahara as a possible sink for trace gases, Geophys. Res. Lett., 5, 1001-1004, 1978.

Rebbert, R. E., and P. Ausloos, Decomposition of N_2O over particulate matter, Geophys. Res. Lett., 5, 761-764, 1978.

Tanaka, K., and G. Blyholder, Photocatalytic reactions on semi-conductor surfaces. I. Decomposition of nitrous oxide on zinc oxide, J. Phys. Chem., 75, 1037-1043, 1971.

OBSERVATION OF SULFATE COMPOUNDS ON FILTER SUBSTRATES BY MEANS OF X-RAY DIFFRACTION

Briant L. Davis and L. Ronald Johnson

Institute of Atmospheric Sciences, South Dakota School of Mines and Technology, Rapid City, South Dakota 57701

Robert K. Stevens

Environmental Sciences Research Laboratory, U.S. Environmental Protection Agency, Research Triangle Park, North Carolina 27111

Donald F. Gatz and Gary J. Stensland

State Water Survey Division, Illinois Institute of Natural Resources, Urbana, Illinois 61801

Abstract. Nearly 200 quantitative compound analyses of airborne particulates have been completed over the past several years using the reference intensity X-ray diffraction technique. This technique combined with direct beam transmission measurements provides quantitative evaluation of both crystalline and amorphous components. Some of the important features of these analyses include the dominance of the minerals calcite and quartz in particulates of the more arid regions of the United States, especially in the coarse (>2.5 μm diameter) particulate fraction, whereas clay minerals and several sulfate compounds dominate the fine fraction (<2.5 μm diameter), especially in midwestern and eastern urban areas. A number of simple and complex sulfate compounds have been observed by this means in the ambient particulate matter of major urban areas. These include mascagnite $(NH_4)_2SO_4$, gypsum $CaSO_4 \cdot 2H_2$, thenardite Na_2SO_4, and the "double salt" compounds $(NH_4)_2Ca(SO_4)_2 \cdot H_2O$, $(NH_4)_2Fe(SO_4)_2 \cdot 6H_2O$, and $(NH_4)_2Pb(SO_4)_2$. Although evidence suggests that some sulfate compounds actually form as atmospheric reaction products, it is also possible that some sulfate compounds, especially $CaSO_4 \cdot 2H_2O$, occur primarily as a reaction artifact on the filter substrate.

Introduction and Background

Much interest in the composition and source identification of atmospheric aerosols has developed over recent years. In most studies prior to the last decade, much of the compositional characteristics of particulates was inferred from wet chemical analyses, bulk-sample X-ray fluorescence analyses, and optical scattering characteristics. Some work with optical polarizing microscopy yielded results for the major components but only for those particulates at the coarse end of the size distribution.

During the past decade much more effort has been expended toward development of techniques yielding direct compound identification. An example of a quantitative study of this type is found in the report of Bradway and Record (1976), wherein optical polarizing analysis of mineral particulates trapped on glass fiber (hi-vol) filters is given for 14 urban areas. In the work of Stevens et al. (1978), particulate compositions were deduced from ion equivalents required by stoichiometry of the compounds with ion chromatography, thorin spectrophotometry, and X-ray fluorescence being employed as the analytical methods. Although not a direct observational technique, infrared absorption analysis does provide evidence of the presence of certain compounds according to the observed bond frequencies. Work of Adler and Kerr (1965), and Moharram and Sowelim (1980) provides infrared information on the occurrence of sulfate ions in various sulfate minerals. Some particulate compounds may be identified by morphological characteristics and the composition deduced by single-particle microbeam scanning electron techniques (usually energy-dispersive X-ray fluorescence). Perhaps the most powerful tool for direct compound identification, both from a qualitative and quantitative standpoint, is X-ray diffraction. Difficulty in application of this technique to aerosols comes primarily from the low mass coverage of the particulates on the collecting filter media. Nevertheless, good qualitative studies reporting identification of atmospheric aerosols from filters have been completed by Biggins and Harrison (1979a, b), Moharram and Sowelim (1980), O'Connor and Jaklevic (1981), and Brosset et al. (1975). Several of these studies included a separation of the particulate masses into size fractions to identify the dominant compounds in each.

During the past five years, the Institute of Atmospheric Sciences' (IAS) Cloud Physics Laboratory has worked toward the development of a quantitative technique for analysis of particulate matter collected on various filter substrates. The results of these studies have been reported in Davis and Cho (1977), Davis (1978, 1981a, b), and Davis and Johnson (1981, 1982). The basis for the technique is the reference intensity method combined with direct beam X-ray transmission measurements, which allows the simultaneous determination of crystalline components and the major classes of amorphous components present in thin aerosol layers collected on filter substrates. Analyses may be completed using Teflon, glass fiber, cellulose acetate, quartz, and polycarbonate filter media.

These studies have included a formal "variance error" analysis based on uncertainties known to exist or anticipated in the various measurements and physical parameters. In general, weight quantities are known to within 10 percent for those major components with a stated analyzed amount of 50 percent or greater. For minor components (i.e., less than 10 percent from the stated analyzed amount), errors are quite variable but may be as high as 100 percent or greater in some cases. The detection limit, or minimum amount detectable, is less than 1 percent for nearly all compounds. In the analyses described below, the individual errors are not given since nearly all of the data are averages of several separate analyses. The standard deviations represented in the several sets of analyses averaged are presented, however. The data are reported to one decimal place because of the detection limit being, in general, less than 1 percent, although actual absolute uncertainties associated with each component may be much greater. In the tables to follow, the dashes (-) signify that the given component was present in amounts less than the detectable limit for the method. The lowest detectable limit for any of the components discussed in this paper (for a sample load of 100 $\mu g\text{-}cm^{-2}$) is for the lead ammonium sulphate (PS) equal to approximately 0.05 weight percent, and the highest is for muscovite at a value of approximately 1 weight percent.

From the nearly 200 quantitative analyses now completed by the IAS group we have observed certain characteristics of particulate composition over the continental United States, both in urban and in rural areas, and find many interesting relationships between the observed components and the environmental characteristics of the site. In this paper we present a series of tabulated X-ray analyses of filter particulates designed to illustrate these differences, and specifically address the presence of various sulfate compounds and their mode of formation.

General Features of the Analyses

An obvious and anticipated feature of the analyses is the dominance of the mineralogic compounds in arid regions of the western Great Plains and inter-mountain area, and dominance of clay-type minerals and sulfate compounds in large urban areas, particularly those of the eastern United States. Table 1 presents the average mineralogic analysis of hi-vol filter particulates taken from six western Great Plains and intermountain cities. Although we do not have data in the far western part of the country at this time, we suggest, that on the basis of known geologic exposures in the west, these types of minerals would also be present. The most obvious features of Table 1 are the dominance of quartz and calcite and secondary importance of the phyllosilicates-biotite, muscovite (or illite), chlorite, kaolinite, and montmorillonite, many of which are secondary alteration products of primary igneous or metamorphic bedrock. The only significant sulfate for this part of the analysis is gypsum which in these cases can be explained by the exposure and wind erosion of natural gypsiferous formations. Although gypsum formed as an artifact reaction on these filters is possible, it is not likely in view of the semiarid conditions prevalent during sampling. (In our usage of the phrase "artifact reaction" we include reactions occurring on the

TABLE 1. Average Mineralogical Analysis of Airborne Particulate Matter From Six Great Plains and Intermountain Cities[a]

	Weight Percent	Standard Deviation
Quartz	23	14
Calcite	22	15
Dolomite	2	3
Biotite	0.7	0.8
Muscovite (illite)	8	6
Chlorite	3	3
Kaolinite	0.4	0.6
Montmorillonite	4	7
Plagioclase	4	6
Microcline	3	8
Gypsum	4	3
Halite	3	5
Analcite	0.8	2
Stilbite	0.7	2
Hematite	0.5	1
Carbonaceous matter	16	16
Silicious fly ash	6	14

[a]Hi-vol filter particulates from Bismarck, ND; Rapid City, SD; Provo, UT; Devils Tower, WY; Ekalaka, MT; Denver, CO (X-ray diffraction reference intensity method).

filter and reactions of trapped particulates with filter fibers themselves (Stevens et al., 1978).)

On the other hand, in major urban and rural areas in the eastern United States we find the dominance of clays and sulfates, many of which form from precursory components which are anthropogenic in origin. Table 2 presents a list of sulfate compounds that have been observed by direct methods such as X-ray diffraction or infrared absorption. The occurrence and distribution of these sulfates and comments on their possible mode of origin will now be discussed.

Specific Case Studies

Inter-Mountain Area (Utah/Montana)

Table 3 presents averages of analyses from sites in Utah and Montana. The Orem-Utah analyses (low-volume sampler-glass fiber filters) came from two closely adjacent sites, one approximately 0.8 km east of the Geneva steel mill complex and the other on the campus of the Brigham Young University (BYU) located 8 km to the southeast of the steel complex. Important features of these sites include the Utah Valley basin topography lending to air entrapment during the winter months, as well as sources for the carbonates, carbon, and sulfur-bearing material in the local industries. The Missoula, Montana, hi-vol site is located in downtown Missoula in the northern Rocky Mountains where several wood processing industries exist (including a Kraft paper mill) and where wood burning is very popular as a means of space heating during the wintertime. At all three of these sites, a high carbonaceous content is observed in the particulates of the filter. Industrial sources of sulfur dioxide or sulfate are also present at all sites.

TABLE 2. Sulfate Compounds Observed by X-ray Diffraction and Infrared Absorption Methods

Name or Symbol	Composition	Observers[a]
Anglesite	$PbSO_4$	OJ
Anhydrite	$CaSO_4$	DJ
Gypsum (GY)	$CaSO_4 \cdot 2H_2O$	DJ, MS, BH
Thenardite (TH)	Na_2SO_4	DJ, BH
NHS	NH_4HSO_4	B
Letovicite	$(NH_4)_3H(SO_4)_2$	OJ, B
Mascagnite (MS)	$(NH_4)_2SO_4$	DJ, BH, OJ, B
FS	$(NH_4)_2Fe(SO_4)_2 \cdot 6H_2O$	DJ
ZS	$(NH_4)_2Zn(SO_4)_2 \cdot 6H_2O$	OJ
CS	$(NH_4)_2Ca(SO_4)_3$	BH
Koktaite (KO)	$(NH_4)_2Ca(SO_4)_2 \cdot H_2O$	DJ
PS	$(NH_4)_2Pb(SO_4)_2$	DJ, BH, OJ
Natrojarosite	$NaFe_3(OH)_6(SO_4)_2$	DJ

[a]The designations in this column are defined as follows:

OJ O'Connor and Jaklevic (1981)
DJ This paper and Davis (1978)
MS Moharram and Sowelim (1980)
BH Biggins and Harrison (1979a, b)
B Brosset et al. (1975)

TABLE 3. Quantitative Compound Analyses of Aerosols from Utah and Montana

	Weight Percent (Standard Deviation)	
Compound	Orem, UT[a]	Missoula, MT[b]
Quartz	15 (4)	15 (10)
Calcite	12 (4)	0.1 (0.3)
Dolomite	2 (2)	------
Biotite	0.1 (0.3)	------
Muscovite-illite	4 (4)	15 (6)
Chlorite	------	4 (2)
Kaolinite	2 (3)	------
Gypsum	2 (3)	2 (3)
Microcline	------	6 (5)
Plagioclase	1 (3)	6 (4)
Hematite	4 (2)	------
Magnetite	1 (2)	------
Silicious fly ash	32 (18)	------
Carbonaceous fly ash	22 (11)	53[c] (22)
Slag glass	3 (6)	------

[a]Average of eight analyses (fall and winter, 1979-80), Davis (1981b).
[b]Average of 18 analyses (fall and winter, 1979-80), Davis et al. (1980).
[c]Carbonaceous matter from space heating and wood processing industries.

The Orem site presents an interesting problem because of the possibility of either atmospheric reaction or artifactual reaction of sulfuric acid mists with the abundant calcium carbonate of the region. Natural gypsum deposits are located over 80 km to the south and the meteorological conditions attendant to collection of these particulates suggest that this is not the source of primary $CaSO_4 \cdot 2H_2O$. At the BYU site the only time during which calcium sulfate was observed on the filters was during the winter months when humidities were high and stagnated air masses existed. It should be noted that the $CaCO_3$ aerosol, ubiquitous in the Utah Valley area, is unusual because of the nature of the weathering process in that area which yields the so-called "beach rock," an almost clay-size particulate $CaCO_3$. Studies during the winter months by one of us (BLD) at BYU points to the formation of $CaSO_4 \cdot 2H_2O$ as an artifact on the filter. The major evidence was the lower than expected diffraction angles for the peaks of this compound which appear to result from a recrystallized liquid or partial liquid suspension pulled to the backside of the filter. (This conclusion is based on the well-known diffraction phenomenon of shift in peak position with departure of the sample surface from the focusing circle of the instrument. Using quartz as an

TABLE 4. Soil, Road Dust, and Filter Analyses; Champaign/Urbana Area, Illinois

	Weight Percent (Standard Deviation)			
	Soil[a]	Roads[b]	Filters	
	Spring 1980		Spring 1980[c]	Fall 1980[d]
Quartz	39 (4)	15 (12)	5 (3)	17 (10)
Calcite	------	44 (31)	------	5 (6)
Dolomite	------	22 (4)	------	4 (6)
Illite	5 (1)	3 (2)	3 (3)	13 (8)
Kaolinite	3 (1)	1 (1)	0.2 (0.2)	0.1 (0.2)
Montmorillonite	25 (5)	3 (2)	------	------
Orthoclase	14 (2)	6 (6)	------	2 (3)
Plagioclase	14 (3)	6 (4)	------	4 (6)
Hornblende	0.8 (0.5)	------	------	------
Mascagnite	------	------	1 (0.2)	------
Koktaite	------	------	2 (3)	------
Gypsum	------	------	2 (0.4)	24 (22)
Carbonaceous	------	------	87 (10)	31 (22)

[a]Average of five samples from a 1 km^2 area.
[b]Average of material from two localities separated by 5 km (crushed limestone and glacial outwash are major roadbed materials).
[c]Average of two samples from the same locality, each collected over a three-day period.
[d]Average of five samples from a single location, each collected over a three-day period.

internal standard, the gypsum peaks were corrected for other factors but were still observed to fall below the theoretical Bragg position.) At Missoula the source of calcium carbonate is much less plentiful and the gypsum content of the aerosol also more minor. The roll of carbonaceous matter as a catalyst for production of calcium sulfate appears to be a possibility at these sites.

Champaign-Urbana Area

The analyses of Table 4 are part of a particle characterization program which includes the development of elemental and compound mass balance studies in the east central Illinois vicinity. Presented in the table are averages of five soil source samples taken from a 1 km^2 area in the rural Champaign-Urbana area, as well as an average analysis of two road surface samples. Analysis of particulate material was completed for two samples collected in the spring of 1980 and five samples collected in the fall of 1980. The average analyses for these filters are presented in the table. Also it should be noted that large carbonaceous quantities occur in the filter samples, and optical examination shows this material to consist of pollen, starch grains, trigomes, and other organic debris presumably derived from agricultural activities of the area. It is noteworthy that montmorillonite, though very common in the soil of the surrounding area, is apparently not readily suspended into the air masses of this area. This is an observation that is consistent with those of the western Great Plains and results from the structure of the montmorillonite and its water absorption characteristics.

These analyses reveal the presence of the ammonium sulfate mascagnite and the unusual double-salt koktaite (see Table 2) in the spring aerosols analyzed but not in the aerosols collected in the fall. When considering the crystalline components only of these analyses, the gypsum is still a major component of the spring aerosol as well as the fall aerosol. Because of the lack of a natural gypsum source in the area it is believed that the gypsum is either an atmospheric reactant or a filter artifact reactant. We also suggest that the formation of mascagnite and koktaite results from a relatively high ammonia concentration during the spring months where there is widespread use of agricultural fertilizers. For some of these samples, the carbonaceous matter appears to be a consequence of long range transport. The possibility of catalytic action of carbonaceous matter in the sulfate conversion must also be considered for these samples.

Analyses From Urban Areas

We have completed 58 analyses of both fine and coarse fractions from dichotomous filter sites in various urban areas throughout the United States, as far west as Denver, Colorado. Samples were obtained from the various EPA field networks following standard X-ray fluorescence analyses by EPA. Table 5 presents a summary of these analyses

TABLE 5. Sulphate and Carbon Contents of Dichotomous Filter Loads From Six U.S. Cities[a]

Site	Filter	Weight Percent (Standard Deviation)				
		PS	FS	MS	GY	Carbon
Denver	Coarse	------	------	0.3 (0.8)	1 (0.7)	29 (26)
(6)[b]	Fine	7 (5)	------	11 (21)	3 (3)	49 (44)
Houston	Coarse	0.1 (0.2)	------	0.8 (1)	0.5 (1)	40 (15)
(4)	Fine	1 (1)	------	17 (8)	------	33 (27)
St. Louis[c]	Coarse	------	------	10 (7)	1 (2)	26 (24)
(4)	Fine	1 (2)	------	60 (17)	------	34 (23)
Steubenville[d]	Coarse	------	0.7 (1)	11 (9)	6 (3)	46 (9)
(3)	Fine	------	5 (9)	54 (19)	0.6 (1)	30 (22)
Boston	Coarse	1 (0.2)	------	3 (4)	------	16 (23)
(2)	Fine	6 (3)	------	56 (38)	2 (2)	31 (44)
Philadelphia (10)	Coarse	1-3	------	12 (6)	2 (4)	35 (19)

[a]PS = $(NH_4)_2Pb(SO_4)_2$
MS = $(NH_4)_2SO_4$
FS = $(NH_4)_2Fe(SO_4)_2 \cdot 6H_2O$
GY = $CaSO_4 \cdot 2H_2O$
[b]Numbers in parentheses refer to the number of samples in the set.
[c]Includes 1.88 (3.8) KO.
[d]Includes 0.6 (1.0) TH.

from six urban localities (with the number of filters analyzed given in parentheses). It is in these analyses that we find the influence from anthropogenic sources most significant. In all but the Steubenville site we observe a small amount of the lead ammonium sulfate with nearly all such observations occurring in the fine fraction. Mascagnite was observed in all filters, both coarse and fine, with the greatest amounts occurring in the fines of St. Louis, Steubenville, and Boston. Steubenville represents a unique situation because of the presence of the iron-ammonium double-salt and thenardite (Na_2SO_4). Steubenville is located in between bluffs of the Ohio River. The area contains heavy industry, including iron and steel, coke ovens, and power facilities. The lack of the lead-ammonia double-salt at Steubenville is not unexpected in view of the smaller concentration of vehicular traffic here in comparison to the other large metropolitan areas.

Distribution of gypsum occurs primarily in the coarse mode, although exceptions to this are noted, such as Denver. All sites contain significant carbonaceous matter. The total sulfate concentration in some of the more populated areas can be very significant as illustrated by the data for Philadelphia, summarized in Table 6. These data are taken from filters of two sites, and in both cases show a good correspondence between the independently determined sulfate concentration as reduced from the mineral components observed by X-ray diffraction (XRD) and the direct sulfate measurement by X-ray fluorescence (XRF) spectroscopy (EPA analyses). Since only the coarse material was analyzed by XRD, the total sulfate listed is a sum of the coarse and fine sulfate observed by the EPA X-ray fluorescence analyses. The sulfates in the Philadelphia analyses are primarily mascagnite and gypsum, although some sulfate could also occur as adsorbed films on carbonaceous matter and thus detectable only by X-ray fluorescence. In these data all values were corrected for presence of 10 percent of the fine fraction (particles less than 2.5 μm) which becomes entrained into the "coarse" side of the dichotomous sampler ductwork. The 10-percent correction was applied to make a more rigorous comparison with the corresponding XRF data of EPA.

Discussion

Many of the double-salt compounds observed in these analyses have been reported by others as indicated in Table 2. The zinc analog to the iron-ammonium double-salt, however, was not observed in any of our studies. It is quite apparent that the presence of the lead-ammonium double-salt is related to the presence of high-density vehicular traffic in major urban areas. The double-salts involving iron and calcium, however, seem to occur only under peculiar atmospheric conditions apparently related to the balance of H_2SO_4, NH_3, H_2O, and calcium-bearing compounds, the total process of which is undoubtedly complicated whether or not it occurs as an atmospheric reactant or as an artifact reaction on the filter itself.

It would be of great interest, if not practical importance from a health standpoint, to know if these compounds actually occur as atmospheric reactants or not. It appears very likely that

TABLE 6. Total Filter and Ambient Sulphate Concentration - Philadelphia

Site/Filter/Date	Filter, Weight Percent[a]			Total Ambient $SO_4^{=}$(XRF)
	Coarse-XRD	Coarse-XRF	Fines-XRF	$\mu g\text{-}m^{-3}$
Broad Street				
121-024 (8-5-80)	13	13	42	33
121-027 (8-8-80)	9	7	48	26
121-028 (8-8-80)	10	6	49	19
121-030 (8-10-80)	7	7	48	17
St. Johns Street				
121-006 (7-31-80)	4	6	41	21
121-007 (7-31-80)	8	6	45	29
121-010 (8-5-80)	9	11	42	27
121-011 (8-5-80)	4	5	41	17
121-012 (8-6-80)	4	9	31	13
121-014 (8-8-80)	8	7	46	26

[a]All values corrected for 10-percent fines in coarse fraction.

mascagnite occurs in large part as an atmospheric reaction product, mainly because the process is a heterogeneous gas-phase reaction. Additional evidence that $(NH_4)_2SO_4$ is a primary atmospheric reaction comes from aerosol scattering studies such as those reported by Charlson et al. (1974). It would appear reasonable to expect the ammonium-lead and possibly the ammonium-iron compounds to occur as a natural atmospheric reactant. In the former case the primary (unaged) lead aerosol particle size is very small (<0.05 μm diameter) and would participate in a reaction with the other components by means of diffusion mechanisms as well as from stochastic collision processes. Thus the PbBrCl involved would most likely exist in the "Transient Nuclei Mode" (Willeke and Whitby, 1975) during peak conversion rates. The origin of the iron compound leading to the iron-ammonium double-salt observed at Steubenville is not known, although the presence of hematite (Fe_2O_3) and magnatite (Fe_3O_4) in the analyses for this site would suggest these to be potential reactants. Hematite particles from open hearth operations can be of very small particle size (even to the extent - less than 0.1 μm - that X-ray broadening may be observed) such that the diffusional properties of this material could also be significant in the reaction process.

The very common occurrence of $CaSO_4 \cdot 2H_2O$ aerosols presents a more difficult challenge as to the nature of its origin. Two likely chemical precursors to this compound are $CaCO_3$ or $CaMg(CO_3)_2$, the minerals calcite and dolomite, respectively, which are very commonly used as SO_2 scrubbing materials in power plants and for flux additives in iron ore smelting. Some $CaSO_4 \cdot 2H_2O$ particulate by-product of SO_2 scrubbing is undoubtedly present in the atmosphere around heavily industrialized regions. In these cases the particle size of the aerosol emitted should be relatively fine, with most mass occurring below the 2.5 μm cutoff point in the dichotomous sampler.

If it is considered that a primary $CaCO_3$ aerosol reacts with H_2SO_4 particulate or only partially neutralized sulfate compounds, one might suspect that their reaction rate would be dependent primarily on particle collision processes. Under the concentrations expected in natural atmospheres, very low rates of reaction are predicted on the basis of collision as the controlling kinetic factor when employing Brownian and shear coagulation calculations. (See Friedlander (1977) for a treatment of these techniques.)

If, in fact, $CaSO_4 \cdot 2H_2O$ does occur as an atmospheric reactant, and because of the high sensitivity of X-ray diffraction as an analytical tool for aerosols in excess of 200 $\mu g\text{-}cm^{-2}$ coverage, it would appear to be possible to experimentally determine the nature of the origin of this material by first neutralizing all of the acid sulfate or H_2SO_4 particulate in the sampler inlet system prior to collection of the aerosol on the filter. Preliminary studies of this nature are being conducted by the IAS group.

A notable aspect of most of these analyses is the significant quantity of carbonaceous matter, some of which is undoubtedly of a sooty nature. Characteristic high surface to volume ratio of such particles is well known as are the physical adsorption characteristics. It was suggested by Del Monte et al. (1981) that carbon particles act as catalysts in converting $CaCO_3$ to $CaSO_4 \cdot 2H_2O$ in the presence of acid aerosol on marble art treasures of the Mediterranean area. They suggest that these particles carry sufficient sulfur to promote sulfate conversion at the surface of the marble. The good correspondence between XRD and XRF total sulfate in the Philadelphia analyses suggests that only minor amounts of sulfur in an amorphous state are retained within the carbonaceous material at the time of analysis, but we have no information on the initial sulfur or sulfate content of carbon matter prior to collection. Recent studies (Cofer et al., 1981; Chang et al., 1981) suggest that carbonaceous matter may act as a catalyst in sulfate conversion under proper moisture and oxidant conditions and therefore must be considered in the conversion processes leading to any of the compounds observed in filter collections of ambient particulate matter.

Conclusions

From an examination of the results of quantitative compound analyses of particulates collected at several urban and rural sites in the United States we may make the following general conclusions. (1) Natural mineral silicates and carbonates dominate the ambient aerosols of the arid regions of the United States. (2) Sulfate compounds and some clay minerals dominate in the fine particulate fraction of dichotomous filter samples, whereas, mineral particulates dominate in the coarse fraction. (3) As much as 6.8 weight percent complex ammonium sulfate compounds has been observed in major urban areas which are related to peculiar local atmospheric chemical conditions; in some cases (e.g., St. Louis), the entire mass of the fine fraction consisted of various sulfate compounds and carbonaceous matter. (4) Many sulfates appear to occur as natural atmospheric reactants although not all such occurrences have been proven, whereas $CaSO_4 \cdot 2H_2O$ may have an origin from scrubbing activities, natural geologic deposits, atmospheric reactions, or as a filter reaction artifact. Stochastic collision appears to be the rate-limiting factor for reactions involving relatively large particulates, such as the $CaCO_3$-H_2SO_4-H_2O reaction leading to the formation of $CaSO_4 \cdot 2H_2O$. (5) Carbonaceous material is present in nearly all analyses even to a small extent in small urban or rural areas of the west. The association of the carbonaceous matter with the sulfate compounds suggests that carbon may act as a catalyst in formation of these sulfates.

Acknowledgements. We are grateful to David Maughan, Montana State Department of Health and Environmental Services, for providing a number of samples for analysis used in this study. This research was supported by the State of South Dakota, by the U.S. Environmental Protection Agency under Cooperative Agreement No. CR806769020, and by the Illinois State Water Survey.

References

Adler, H. H., and P. F. Kerr, Variations in infrared spectra, molecular symmetry, and site symmetry of sulfate minerals, Am. Min., 50, 132-147, 1965.

Biggins, Peter D. E., and Roy M. Harrison, Atmospheric chemistry of automotive lead, Env. Sci. Tech., 13, 558-565, 1979a.

Biggins, Peter D. E., and Roy M. Harrison, The identification of specific chemical compounds in size-fractionated atmospheric particulates collected at roadside sites, Atmos. Environ., 13, 1213-1216, 1979b.

Bradway, Robert M., and Frank A. Record, National assessment of the urban particulate problem; particle characterization, vol. II, EPA Rep. 450/3-76-025, U.S. Environmental Protection Agency, Office of Air Quality Planning and Standards, Research Triangle Park, N.C., 1976.

Brosset, Cyrill, K. Andreasson, and M. Ferm, The nature and possible origin of acid particles observed at the Swedish west coast, Atmos. Environ., 9, 631-642, 1975.

Chang, S. G., R. Toossi, and T. Novakov, The importance of soot particles and nitrous acid in oxidizing SO_2 in atmospheric aqueous droplets, Atmos. Environ., 15, 1281-1286, 1981.

Charlson, R. J., A. H. Vanderpol, D. S. Covert, A. P. Waggoner, and N. C. Ahlquist, Sulfuric acid-ammonium sulfate aerosol: Optical detection in the St. Louis region, Science, 184, 155-158, 1974.

Cofer, W. R., III, D. R. Schryer, and R. S. Rogowski, The oxidation of SO_2 on carbon particles in the presence of O_3, NO_2, and N_2O, Atmos. Environ., 15, 1281-1286, 1981.

Davis, B. L., Additional suggestions for X-ray quantitative analysis of high-volume filters, Atmos. Environ., 12, 2403-2406, 1978.

Davis, B. L., An error study of X-ray diffraction quantitative analysis procedures for aerosols collected on filter media, Atmos. Environ., 15, 291-296, 1981a.

Davis, B. L., Quantitative analysis of crystalline and amorphous airborne particulates in the Provo-Orem vicinity, Utah, Atmos. Environ., 15, 613-618, 1981b.

Davis, B. L., and N. K. Cho, Theory and application of X-ray diffraction compound analysis to high-volume filter samples, Atmos. Environ., 11, 73-85, 1977.

Davis, B. L., and L. R. Johnson, The use of X-ray diffraction quantitative analysis in air quality source studies, in Electron Microscopy and X-ray Applications to Environmental and Occupational Health Analysis, vol. 2, edited by Philip A. Russell, pp. 131-152, Ann Arbor Science Press, Ann Arbor, Mich., 1981.

Davis, B. L., and L. R. Johnson, On the use of various filter substrates for quantitative particulate analysis by X-ray diffraction, Atmos. Environ., 16, 273-282, 1982.

Davis, B. L., D. Maughan, and J. H. Carlson, X-ray studies of airborne particulate matter observed during wintertime at Missoula, Montana, paper presented at the 4th Symposium on Electron Microscopy and X-ray Application to Environmental and Occupational Health Analysis, University Park, Penn., Oct. 15-17, 1980.

Del Monte, M., C. Sabbioni, and O. Vittori, Airborne carbon particles and marble deterioration, Atmos. Environ., 15, 645-652, 1981.

Friedlander, S. K., Smoke, Dust, and Haze, John Wiley, New York, 1977.

Moharram, M. A., and M. A. Sowelim, Infrared study of minerals and compounds in atmospheric dust fall in Cairo, Atmos. Environ., 14, 853-856, 1980.

O'Connor, B. H., and J. M. Jaklevic, Characterization of ambient aerosol particulate samples from the St. Louis area by X-ray powder diffractometry, Atmos. Environ., 15, 1681-1690, 1981.

Stevens, R. K., T. G. Dzubay, G. Russworm, and D. Rickel, Sampling and analysis of atmospheric sulfates and related species, Atmos. Environ., 12, 55-68, 1978.

Willeke, K., and K. T. Whitby, Atmospheric aerosols: Size distribution interpretation, J. Air Pollut. Control Assoc., 25, 529-534, 1975.

ATMOSPHERIC GASES ON COLD SURFACES - CONDENSATION, THERMAL DESORPTION, AND CHEMICAL REACTIONS

R. J. Fezza and J. M. Calo

Division of Engineering, Brown University, Providence, Rhode Island 02912

Abstract. Selected results of potential significance to heterogeneous processes in the natural atmosphere are presented from a study of the behavior of atmospheric gases and vapors on cold surfaces. In particular, the occurrence of a strong "first" condensation at relatively high temperatures (200-250 K) was observed for N_2O, CO_2, NO, Freon 11, and Freon 12, ranging from approximately 10 percent of the incident beam (e.g., for CO_2) to 90 percent (e.g., for N_2O and NO). All the data are consistent with efficient trapping of these species under highly unsaturated conditions by a continually renewing water sublayer, probably with a solid clathrate hydrate-type structure.

Temperature-programmed desorption with mass spectrometric detection and flash desorption with subsequent chemiluminescence analysis were applied to studies of the reactions of nitric oxide and ozone upon thermal desorption. The most salient results are (1) disproportionation of nitric oxide to nitrogen dioxide and nitrogen during desorption of pure nitric oxide condensate at levels exceeding 50 percent, (2) significant recombination of ozone upon flash desorption, and (3) oxidation of nitric oxide to nitrogen dioxide by ozone upon desorption of nitric-oxide/ozone condensates. Some implications of these results to heterogeneous processes in the natural atmosphere are considered in this paper.

Introduction

A study of the behavior of atmospheric species on cold surfaces was undertaken primarily to further the understanding of the phenomena that occur upon sampling stratospheric air using cryogenic samplers, as practiced, for example, by the Air Force Geophysics Laboratory (Gallagher and Pieri, 1976). In this latter program fixed metal cylinders (three, in recent configurations) with specially treated inner surfaces are immersed in a liquid helium bath surrounded by a guard volume of liquid nitrogen. This sampler is the heart of a balloon flight package which can collect one whole air sample per cylinder at various stratospheric altitudes, typically in the 12 to 30 km range. Once sampling is completed on the balloon descent leg of the flight, the sampler is parachuted to the ground where it is recovered and maintained at cryogenic temperatures until regeneration and gas-phase analysis can be performed in the laboratory to determine relative ambient stratospheric concentrations.

Even though the program is relatively simple in concept, many phenomena associated with the collection and regeneration of atmospheric constituents on and from cold surfaces of various materials are not well understood. Furthermore, the same mechanisms of multilayered deposition of these species on polycrystalline substrates and their subsequent behavior upon warming and thermal desorption, including association and possibly self-catalytic reactions, may also be relevant to heterogeneous processes occurring in various regions of the natural atmosphere as well as on the walls of the cryogenic sampler.

Some selected results from our studies of the physicochemical phenomena associated with cryogenic whole air sampling are presented (for a more complete exposition see Fezza (1981)) with an emphasis on those of potential significance to the natural atmosphere.

Experimental Procedure

The apparatus developed for these studies is illustrated schematically in Figure 1. The primary functions of the experimental system are to allow controlled deposition and subsequent desorption of the various gases and vapors of interest onto and from a cryogenic surface while monitoring gas-phase compositions with appropriate analytical techniques.

The apparatus consists of a closed-cycle cryogenic refrigerator upon which can be mounted a disk of the cryosurface material under consideration (e.g., electropolished, gold-flashed, or hexamethydisilazane (HMDS)-coated 304 stainless steel). The cryosurface temperature was monitored with a silicon diode sensor and a digital thermometer-controller capable of temperature measurements in the 1 to 400 K range with an accuracy of ±1 K. The refrigerator is capable of maintaining the cryosurface at 10 K under conditions of zero heat load in vacuo. In practice,

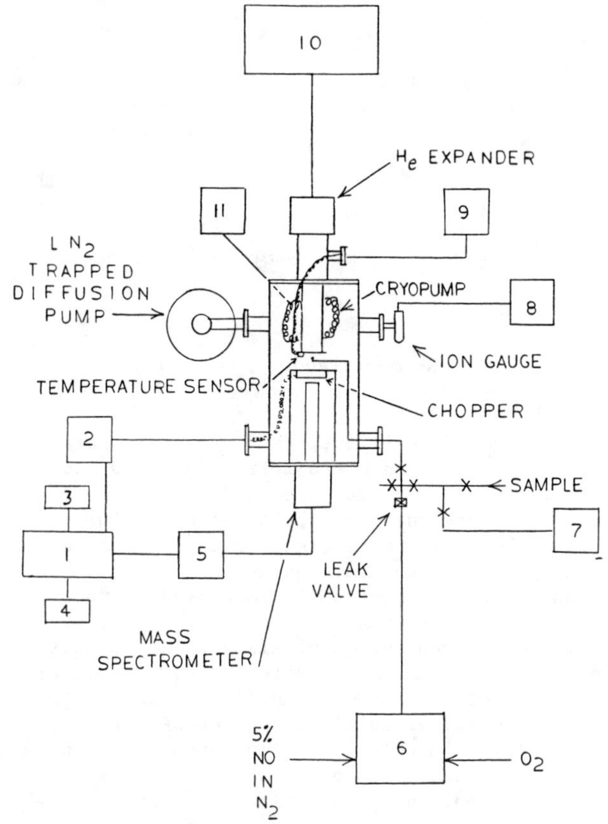

Fig. 1. Schematic of experimental apparatus: (1) mass spectrometer controller; (2) lock-in amplifier; (3) oscilloscope; (4) recorder; (5) RF/DC generator; (6) NO_x chemiluminescence analyzer; (7) ozone generator; (8) ionization gauge controller; (9) digital thermometer/controller; (10) helium compressor; (11) constant-head LN_2 reservoir.

however, a lower-temperature limit of 14 K was routinely achieved, primarily due to the absence of a radiation shield.

The cryosurface disk was juxtaposed with a quadrupole mass spectrometer positioned with its axis perpendicular to the cryosurface, centered, and located such that the extreme end of the electron impact ionizer head was 24 mm away. Both the refrigerator and the mass spectrometer were maintained in a conventional liquid-nitrogen-trapped diffusion-pumped vacuum chamber. The chamber was also equipped with ribbed stainless steel coils which could be cooled with liquid nitrogen in order to provide additional cryopumping for species condensable at these conditions, if so desired.

Gases were introduced via a variable-leak valve. The feed tubing terminated in a 1-mm I.D. tube positioned 5 mm from the cryosurface so that the centerline of the resultant molecular beam was normal to the cryosurface and aligned with its center. The flow from this tube was determined to be effusive. Most of the sample gases were introduced directly from gas cylinders, with the exception of ozone, which was generated in pure oxygen in a high-voltage DC discharge laboratory ozonator to an upper limit of 2.25 percent ozone in oxygen.

The species reflected from the cryosurface were monitored with the mass spectrometer in either an unmodulated or a modulated mode. The latter was accomplished with a 200-Hz tuning fork chopper and a lock-in amplifier. Thermal desorption spectra of cryofrost species were also determined with the mass spectrometer.

In order to circumvent signal interpretation problems caused by electron impact fragmentation in the mass spectrometer for NO_x-O_x systems, flash desorption of such cryofrosts with room-temperature nitrogen was followed by analysis using NO_x chemiluminescence. The NO_x chemiluminescence analyzer was modified to measure either NO_x using internally generated ozone or ozone using a standard 5 percent NO in nitrogen mixture.

The preceding brief description of the apparatus is intended to provide the basic orientation required to follow the various types of experiments and results discussed herein. More complete descriptions of the apparatus, experimental details, calibration procedures, etc. are given by Fezza (1981).

Condensation Phenomena

The experimental procedure for studying condensation phenomena generally involved the following steps: (1) evacuation of the chamber, (2) cooling the cryopumping coils with liquid nitrogen, if desired for the experiment, (3) admission of the species of interest via the beam inlet directed at the cryosurface, (4) startup of the refrigerator, and (5) monitoring with the mass spectrometer the flux of species reflected from the cryosurface as the surface cools. The decrease in signal as a function of temperature is taken to be indicative of condensation.

The a priori expectation for the behavior of a particular species is that condensation will occur (i.e., the reflected signal will decrease precipitously) at the point where the pressure exerted by the impinging molecules equals or exceeds the sublimation vapor pressure of the condensed phase, in accordance with thermodynamics (i.e., at the saturation point). Typical experimental results for N_2O in our apparatus are presented in Figure 2 for three beam intensities. As shown, the low-intensity curve exhibits a sharp break at a surface temperature of 250 K, followed by more than an order of magnitude decrease over the succeeding 15 K. The medium-intensity curve, on the other hand, initially exhibits a gradual decline, followed by a slight peak (probably indicative of a transition in the surface capture mechanism) prior to a more rapid decrease beginning at 248 K. The high-intensity case exhibits the same satura-

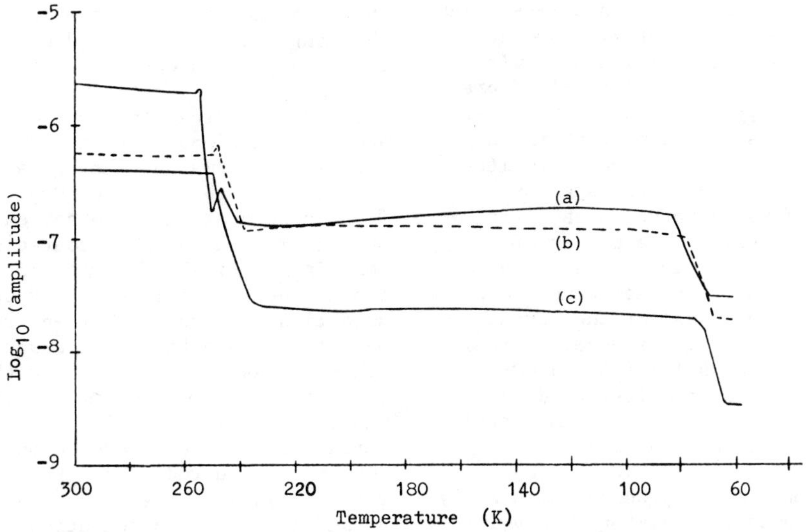

Fig. 2. N_2O condensation curves for three beam intensities on gold-flashed 304 stainless steel: (A) 8.16×10^{20} m^{-2} s^{-1}; (B) 1.36×10^{20} m^{-2} s^{-1}; and (C) 2.10×10^{19} m^{-2} s^{-1}.

tion peak followed by a second peak (possibly due to a phase change or cryofrost matrix rearrangement) towards the end of the rapid decrease beginning at 252 K. In all three cases, however, a significant second condensation occurs in the 75 to 85 K temperature range, which is the range to be expected under the experimental conditions. The high-temperature condensation which occurs where the pure species would not normally be expected to condense will be referred to as the "first" condensation. The second condensation, which occurs over the temperature range expected based on vapor pressure, will be referred to as the "true" condensation. The curves presented in Figure 2 for N_2O are, for the most part, generally characteristic of the condensation behavior of all the species investigated in the current work, i.e., "first" condensations in the 200 to 250 K range were observed for N_2O, CO_2, NO, Freon 11, and Freon 12, ranging from approximately 10 percent of the incident beam (e.g., for CO_2) to 90 percent (e.g., for N_2O and NO), depending on experimental conditions. The detailed quantitative behavior of each species and the temperature of the first condensation were found to vary significantly as functions of beam intensity, surface type (e.g., gold-flashed versus HMDS-coated stainless steel), and pumping speed (varied with auxiliary liquid-nitrogen cryopumping). The unexpectedly strong nature of the first condensation and the fact that it provided the substrate upon which the true condensation (originally of primary interest) occurred, naturally led us to an investigation of these condensation phenomena and their interrelationships.

In retrospect, what at first appeared to be unexpected high-temperature condensation is quite reminiscent of some observations in the early cryopumping literature. For example, Wang et al. (1962) found that some species which are not normally condensable at high surface temperatures are effectively trapped during the condensation of another gas species. More specifically, nitrogen was found to be quite effectively trapped by condensing carbon dioxide and water vapor. Of particular relevance to our own work, Wang and coworkers attributed the efficient trapping of nitrogen by condensing water vapor to the formation of stable water clathrate structures on the surface. More recent work along related lines includes that of Becker et al. (1972a, b) who investigated the pumping of hydrogen onto frozen substrates of Ar, C_2H_6, NH_3, and CO_2 in the 10 to 22 K range, and Bewilogua and Jackel (1974) who studied hydrogen adsorption onto CO_2 at 20 K. In the latter study it was found that once the pumping capacity of the CO_2 substrate for hydrogen was exhausted it could be regenerated to near original capacity by depositing a fresh layer of CO_2 onto the previously saturated substrate. This result implies the formation of a stable configuration of H_2 and CO_2 on the surface, for if it were not so, H_2 would diffuse along the concentration gradient into the fresh CO_2 layer, thereby significantly reducing the sorption capacity of the new layer. Cazcarra et al. (1973) found that CO_2 condensation was initiated on a prepared ice surface at temperatures between 74.46 and 76.4 K. In even more recent work, Abe and Schultz (1979) studied H_2 adsorption onto frozen noble gas (Xe, Kr, and Ar) matrices and found that the amount of H_2 condensed was a strong function of matrix porosity, and furthermore that the deposition rate of the noble gas affected this porosity significantly.

Thus the adsorption and trapping of species at temperatures significantly greater than the true condensation temperature should not have been totally unexpected in our own work, were it not for the fact that the required condensable species was not being introduced in the beam. An examination of background mass spectra taken both with the vacuum chamber at its ultimate zero-load pressure of 2.67 μPa (2×10^{-8} torr) and during actual experiments revealed no significant contaminant peaks (e.g., hydrocarbons), other than ubiquitous water vapor, as candiate condensable species. Therefore, the role of water vapor in the first condensation was investigated in a number of experiments. The most definitive of these involved thermal desorption of surface condensate just prior to and immediately after the first condensation. The water-vapor/beam-species ratio was found to be 6.61 and 6.05 for N_2O and CO_2, respectively, just prior to the first condensation, and 0.6 and 2.74, respectively, immediately after the first condensation. The fact that the first two values are similar suggests the occurrence of a stoichiometric process immediately before the first condensation, most probably involving the complexation of the beam species with water in ordered clathrate structures (e.g., see Davidson (1973), Miller (1973), and Siksna (1973a, b). Water clathrates are cage structures composed of water molecules incorporating a foreign "guest" molecule in the cage, which tends to stabilize the complex. The size and properties of the guest species essentially determine which of two types of unit cells is formed. Type I, favored by smaller guest species, has a theoretical water-to-guest ratio of 5.75. In fact, however, the actual molar ratio for CO_2 is 6 (Miller and Smythe, 1970), due to the fact that all the cages in the structure are not filled. This ratio is obviously in good agreement with our value determined immediately prior to the first condensation. The much smaller ratios observed immediately after the first condensation reflect the strong adsorption of the beam species onto the previously formed sublayer.

The instantaneous impingement rate of background water vapor onto the surface is a complex transient rate process dependent upon a number of factors, such as interaction of the pressure distribution between the beam source and the surface (i.e., scattering effects), beam dynamics (i.e., a directed beam tends to pump background species away from its centerline; e.g., see Anderson (1974)), and even the vacuum chamber operating history. Therefore, the precise rate at any time could not be determined. However, the effect of water vapor was directly explored by using auxiliary liquid nitrogen cryopumping to reduce the water vapor available to the surface. The effects of water vapor reduction were quite interesting. For example, the first condensation for Freon 11 and Freon 12 on the gold-flashed stainless-steel surface was all but completely eliminated. For N_2O and CO_2 not only was the first condensation diminished at the higher beam intensities (about 10^{20} m^{-2} s^{-1}), but it occurred at lower temperatures. For example, for N_2O impinging on the gold-flashed stainless steel with no cryopumping, the first condensation temperature, T_{fc}, was in the range 248 to 257 K and exhibited a general functional dependence, decreasing with beam intensity. With cryopumping, however, T_{fc} decreased significantly to the 224 to 240 K range, and the functional dependence of T_{fc} on beam intensity became unclear. Qualitatively identical behavior was observed for CO_2. In addition, the first condensation was observed to saturate for CO_2 and N_2O on the HMDS-coated surface at high beam intensities with cryopumping; i.e., the reflected beam signal decreased upon the first condensation as usual, but then gradually increased back to its pre-first-condensation value. This behavior was never observed without cryopumping. In fact, in one experiment without cryopumping CO_2 was directed at the gold-flashed surface and the surface temperature lowered until the first condensation occurred. When the surface temperature reached 210 K it was held constant with the beam on and the first condensation mechanism continued to pump beam species at the same rate (i.e., the reflected beam signal remained constant) for 1.5 hours without any sign of saturation.

All the preceding observations are consistent with efficient trapping and adsorption of normally noncondensable species at relatively high temperatures by condensing water vapor. Reduction of water vapor affects the first condensation directly, either diminishing it or limiting its capacity (i.e., saturation).

Based on the preceding discussion as well as additional experimental verification (Fezza, 1981), together with the relevant literature, a mechanism for the first condensation is hypothesized as follows. First, the availability of water vapor to the surface allows the adsorption of an initial water layer when energentically favorable. The distribution of this water film is probably heterogeneous and dependent on surface characteristics. During this time, guest species steadily impinge on the surface, where some are adsorbed and trapped in much the same manner as that described by Wang et al. (1962); i.e., they are covered by condensing water molecules. Once surrounded by water, enclathration may occur in a manner similar to Siksna's (1973a, b) cage development mechanism. When the surface coverage of clathrate structures becomes high enough and the surface temperature low enough, the clathrate surface, due to its polar properties, induces gross adsorption of guest species onto itself. As in the work of Abe and Schultz (1979), a thin (possibly monolayer) film of clathrate surface, hydrogen-bonded in a specific orientation and thus prevented from rearranging (i.e., annealing), becomes a powerful adsorbent. As deposition of guest species continues, the adsorbent surface can be continually renewed by constant deposition of water, as described by Bewilogua and Jackel (1974). Clathrate structures, renewed in this fashion, continue to pump guest-species molecules as successive sublayers become exhausted by saturation in a continuous manner, until the surface temperature becomes low enough for a change in condensation mechanism to occur, e.g., bulk thermodynamic condensation.

Chemical Reactions

NO Disproportionation

Deposition of NO on the cryogenic surface, followed by subsequent temperature-programmed desorption (TPD) or flash desorption (FD) with pure nitrogen, revealed a surprisingly strong and persistent disproportionation reaction. For example, in one such experiment 3.5×10^{-7} mol/sec of NO were steadily deposited on a 15-K gold-flashed stainless-steel cryosurface (i.e., condensing the entire beam) for 5 minutes, yielding a total deposit of 1.1×10^{-4} mol of NO. After approximately 3 minutes the NO cryodeposit was flash-desorbed with pure nitrogen; i.e., the entire deposit was rapidly desorbed into the gas phase with room-temperature nitrogen admitted to the vacuum chamber to a final pressure of 136 kPa such that no condensate remained on the surface. The resultant diluted concentration of NO in nitrogen should have been approximately 215 ppm. The actual results by chemiluminescence analysis were quite different, yielding 72 ppm NO and 133 ppm NO_x, or $NO/NO_x = 0.54$. Thus approximately two-thirds of the NO originally deposited apparently was converted to another NO_x species in the absence of any oxygen (O_x) species.

Based upon these results, an extensive set of experiments was conducted to confirm that these observations were not due to contamination of the gases used or to reactions occurring elsewhere in the system other than in the condensed state on the cryosurface. A summary of the most salient results from these experiments follows.

The nitrogen used to flash-desorb the cryofrost (high purity, 99.99 percent min.) was found to contain no NO or NO_x to within the sensitivity of the analyzer (i.e., to 100 ppbv). The NO used (certified $\leq$0.05 percent NO_2) was diluted to about 100 ppm with the same nitrogen analyzed previously. No deviation at all in the chemiluminescence analyzer reading could be detected in switching between the NO and NO_x modes. Since the 0 to 100-ppm scale has a detectable accuracy of $\pm$1 percent of full scale or $\pm$1 ppm, this established an upper limit of NO_2 contamination (if any at all) of less than 1 percent. (Recall that the difference between the NO_x and NO modes for the mixture generated from the cryosurface was a quite substantial 61 ppm. In order to account for this result in terms of NO_2 contamination of the pure NO, the sample gas had to be 46 percent NO_2 considerably in excess of the 1-percent detection limit.)

In order to check that NO_2 was not being generated on the walls of the feed system or on the vacuum chamber walls, the 0.5-percent NO in N_2 calibration standard (certified) was fed to the sealed-off vacuum system with the cryosurface at room temperature to a total pressure of 0.5 atm, and then brought up to atmospheric pressure with pure nitrogen. Subsequent chemiluminescence analysis of the resultant mixture yielded no difference between the NO and NO_x mode readings (i.e., 2500 ppm in both cases). Consequently, no detectable NO_2 was formed in the system under conditions which were similar to the reactive experiments except that no NO was allowed to condense on the cryosurface. It was therefore concluded that the reaction is indeed genuine and that it is a direct result of the condensation and desorption processes.

To our knowledge, these results represent the first observation of NO self-disproportionation from the solid state in the absence of a porous adsorbent. The only other observations we are aware of which are even somewhat similar are those of Addison and Barrer (1955). They observed NO disproportionation to near completion after adsorption on zeolites at low temperatures:

$$4NO \longrightarrow N_2O + N_2O_3 \qquad (a)$$

$$3NO \longrightarrow N_2O + NO_2 \qquad (b)$$

When charcoal was used as the adsorbent for NO condensed at low temperatures, considerable evolution of NO_2 was detected upon thermal desorption and the reactions

$$4NO \longrightarrow N_2 + 2NO_2 \qquad (c)$$

$$6NO \longrightarrow N_2 + N_2O_3 \qquad (d)$$

were suggested (Addison and Barrer, 1955). Yet a blank experiment (i.e., condensation of NO for 100 hours at -183°C with no adsorbent) yielded no detectable levels of disproportionation products. In this work disproportionation was attributed to catalytic activity of the porous adsorbent.

Due to the significant levels of NO_2 detected in our own experiments, the stoichiometry of reaction (c) was investigated further using a different experimental technique, thermal desorption with mass spectrometric detection (note: not flash desorption with pure nitrogen). Thermal desorption of a pure NO cryofrost while monitoring mass peak m/e = 28 yielded the desorption spectrum presented in Figure 3. (Note: no nitrogen was used in these experiments.) Assuming, for the moment, that the two peaks observed are due to nitrogen as a disproportionation product, integration of the spectrum yields approximately 9.8×10^{-7} mol N_2 generated. Knowing that 8.8×10^{-6} mol of NO were originally deposited in this experiment, and assuming that the stoichiometry of reaction (c) holds, implies that 1.96×10^{-6} mol of NO_2 should also have been formed such that the final $NO/NO_x \simeq 0.71$. This latter value compares quite favorably with the NO/NO_x ratio of 0.54 for the previous flash desorption experi-

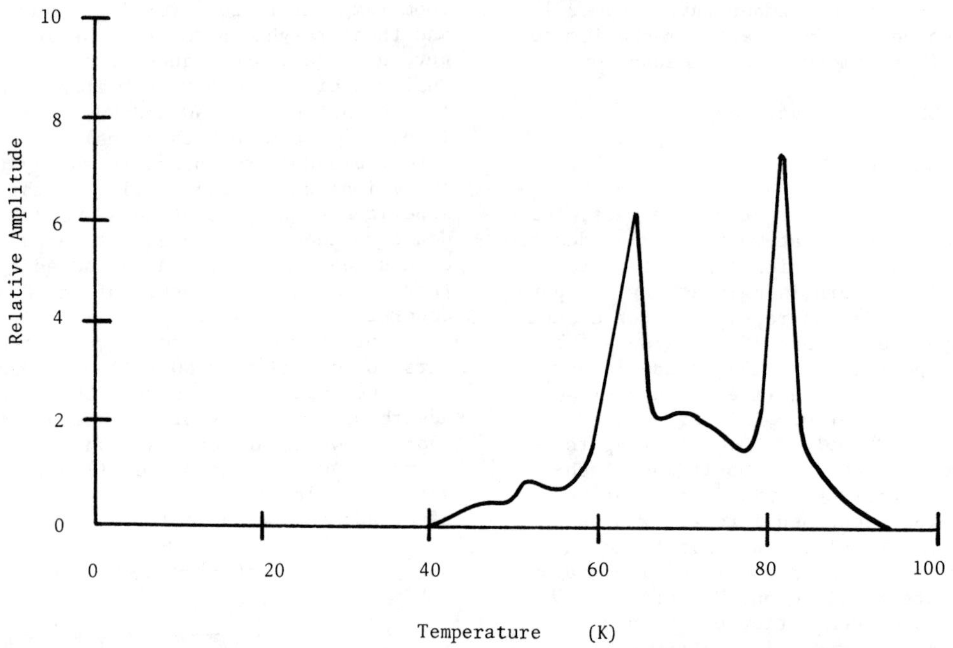

Fig. 3. Thermal desorption spectrum of peak m/e = 28 subsequent to deposition of pure NO on gold-flashed 304 stainless steel.

ments. Differences in the experimental techniques (flash desorption versus thermal desorption), analytical techniques (chemiluminescence versus mass spectrometry), and absolute quantities deposited could account for the differences in conversions actually observed. Therefore, the stoichiometry of reaction (c) accounts for at least a major fraction of the NO disproportionation. The slightly higher conversions detected directly in the flash desorption experiments with chemiluminescence may imply that reaction (c) is not the only reaction that occurs. However, the fact that no trace of the parent peaks of the nitrogen oxides N_2O and N_2O_3 could be detected with the mass spectrometer upon thermal desorption leads us to conclude that the primary products of disproportionation in our system are indeed nitrogen and nitrogen dioxide (i.e., reaction (c)). Thus, the reactions (a), (b), and (d), if they occur at all, must be significantly less important in our system.

Although not discussed here, the thermal desorption spectrum of pure condensed nitrogen is quite different from that indicated in Figure 3; i.e., pure nitrogen desorbs from the gold-flashed stainless-steel cryosurface in three closely grouped peaks appearing in the 28 to 40 K range. This strongly suggests that the nitrogen peaks at 64 K and 82 K in Figure 3 are due to spontaneous nitrogen generation at these temperatures during the desorption process. This is supported by the mass peak m/e = 30 desorption spectrum for pure NO presented in Figure 4. As shown, the 64-K and 82-K desorption peaks of nitrogen in Figure 3 correspond quite closely to the two major mass-30 desorption peaks in Figure 4. (Since mass 30 is also a fragment of NO_2 in the mass spectrometer, the fraction of either or both of the two peaks in Figure 4 attributable to NO_2 versus NO could not be determined.) This mechanism is consistent with the observations of Addison and Barrer (1955) who suspected that NO disproportionation on zeolites also occurred during the thermal desorption process.

In the present work, it is quite possible that NO disproportionation is self-catalyzed by its own frost. Abe and Schultz (1979) have shown that noble gases deposited at intensities of approximately 5×10^{20} m^{-2} s^{-1} onto low-temperature (4 to 22 K) surfaces are quite porous. These conditions correspond quite closely to those in the present work (e.g., 10^{18} to 10^{21} m^{-2} s^{-1} and 15 K). Although the pore structure probably decomposes somewhat during heating to desorption, sufficient structure could remain to provide a stabilizing environment for reaction intermediates. In addition, the frost provides a source of highly concentrated NO in the proximity required for the four-center reaction suggested by the stoichiometry of reaction (c). An estimate of the heat of reaction of reaction (c) in the solid phase at 50 K from thermodynamic data (Reid et al., 1977; Dean, 1973) yields -64 kcal/mol. The exothermicity of this reaction is sufficient to provide enough energy to allow the reaction to proceed via a chain reaction mechanism. The absence of heat transfer limitations (i.e., the energy liberated is transferred directly to other potential reactants) would tend to make this an extremely rapid process.

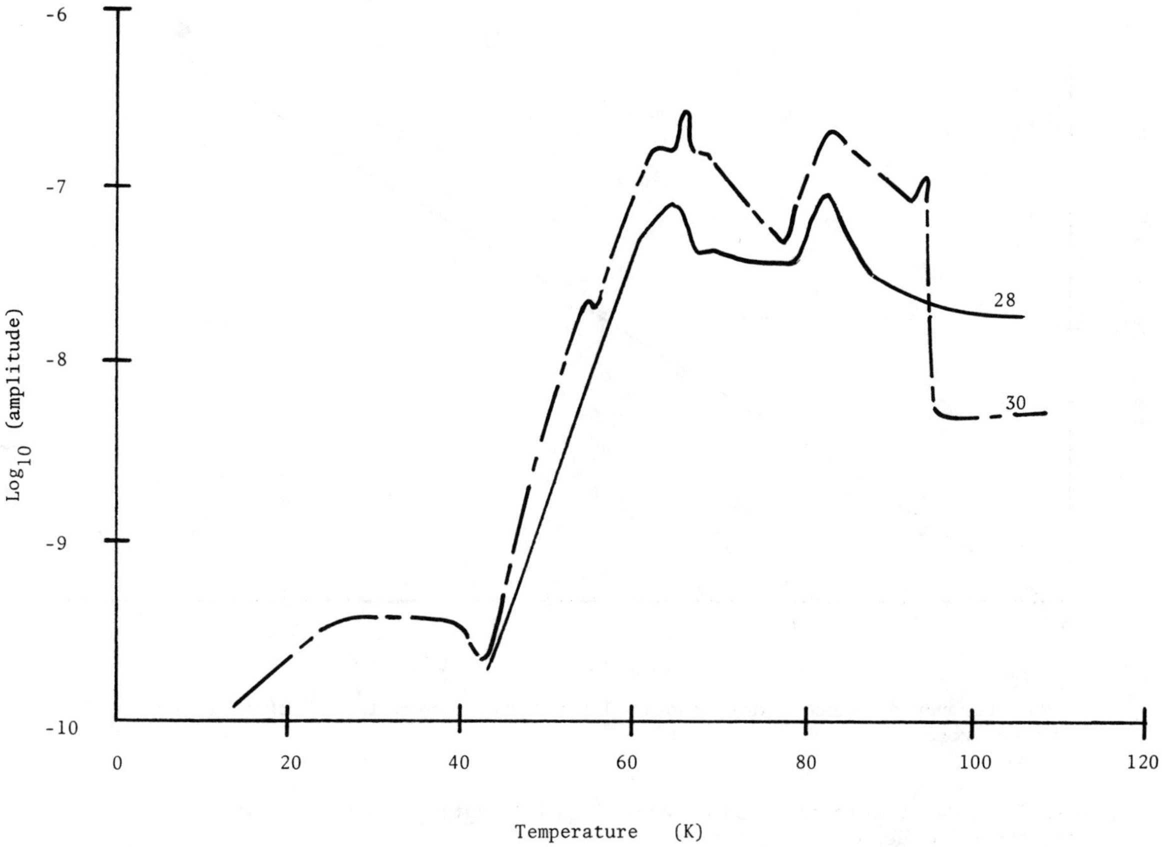

Fig. 4. Thermal desorption spectra of peaks m/e = 30 and m/e = 28 subsequent to deposition of pure NO on gold-flashed 304 stainless steel.

Ozone Recombination

As part of our work concerned with the quantitative recovery of specific species from cryofrosts, ozone regeneration was examined using flash desorption (with pure nitrogen) followed by chemiluminescence analysis. The ozone used in these studies was generated from a pure oxygen stream with a DC discharge laboratory ozonator. As indicated previously, for these studies the chemiluminescence analyzer was modified to measure ozone rather than NO/NO_x. Essentially, the internal ozone generator of the analyzer was bypassed, and the ozone-containing stream to be analyzed was reacted with a standard 5-percent NO in N_2 mixture to produce the detected chemiluminescence. (At these levels, NO was always in excess with respect to ozone in the analyzer reactor section for all the experiments.) The analyzer, modified in this manner, was calibrated with a standard wet chemical technique (i.e., reduction of ozone in KI solution and subsequent titration with sodium thiosulfate solution using starch indicator). The data from this titration, together with the total volume of the corresponding sample gas measured with a wet test meter, allowed the determination of the ozone concentration in the sample gas stream. Welsbach (1977) states that this calibration technique is accurate to within 1 percent of the ozone in the sample stream down to concentrations of approximately 28 ppm, or to 280 ppb ozone. In addition to these careful calibration procedures, care was taken with the materials used for the gas delivery system (limited to glass, stainless steel, Teflon, and Tygon) to minimize ozone loss, although no loss in the inlet system was ever detected, and exactly the same tubing system was used for the wet calibration tests, the chemiluminescence analyzer calibrations, and the actual experiments. Also, periodic recalibration verified reproducibility of the ozone concentrations from the ozonator to within ±0.1 percent ozone in oxygen, and virtually no drift in the chemiluminescence analyzer calibration.

Some typical results from our ozone experiments are presented in Figure 5. In this particular series of experiments 2.25 percent ozone in oxygen was deposited in varying total amounts on the gold-flashed stainless-steel surface at 15 K, and was subsequently flash-desorbed with pure nitrogen. There are several features of note in this figure. The most obvious is that all the ozone deposited is not recovered. The nonzero intercept

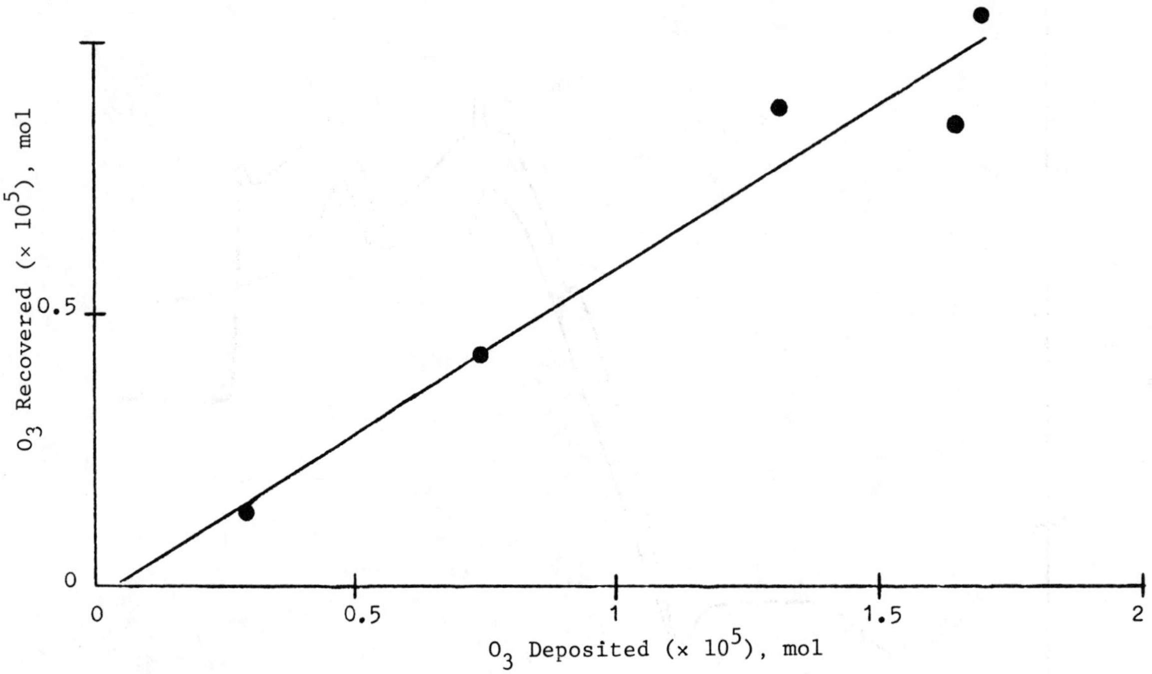

Fig. 5. Ozone recovered versus ozone deposited for flash desorption of 2.25 percent O_3 in O_2 mixtures with pure nitrogen.

on the abscissa indicates a threshold ozone level below which all the ozone deposited is lost to reaction. We have shown in other work (Fezza, 1981) that in cases where the matrix material is more volatile than the solute, phase separation will occur; i.e., the matrix material will sublimate at the appropriate temperature, leaving concentrated solute on the surface. This situation occurs in the case of ozone in oxygen (the normal boiling points are 90.2 K for oxygen and 161.9 K for ozone (Dean, 1973)), and thus small amounts of ozone, concentrated by the phase separation which occurs upon thermal desorption of oxygen, seem to recombine completely. This behavior is also supported by TPD spectra of 2.25 percent ozone in oxygen cryofrosts, which exhibit exactly the same desorption behavior as for pure oxygen. The absence of higher temperature peaks (>45 K) indicative of ozone suggests that the ozone recombines rapidly after phase-separating from oxygen as the cryofrost warms, and desorbs as molecular oxygen at the tail end of the oxygen desorption spectrum.

A linear fit of the data in Figure 5 yields a slope less than unity, indicating that although ozone is recovered at higher total deposit amounts, a fraction of the ozone deposited over and above that required to overcome the recovery appearance threshold also recombines. Since ozone recombination is exothermic (ΔH_r = -69.5 kcal/mol at 50 K in the solid phase, estimated from data given by Reid et al. (1977) and Dean (1973)), a thermal chain reaction is again possible.

NO Oxidation

Flash desorption (with pure nitrogen) of a 1/1/44.4 mixture of $NO/O_3/O_2$ (i.e., 4.6 × 10^{-5} mol NO, 4.6 × 10^{-5} mol O_3, and 2.0 × 10^{-3} mol O_2) from the gold-flashed stainless-steel cryosurface (maintained at 15 K during deposition of the mixture), followed by chemiluminescence analysis, yielded NO/NO_x = 0.224. This value is significantly less than those found for self-disporportionation of pure NO, thereby indicating additional conversion of NO to NO_2. Due to the relatively large amount of oxygen present, however, the fraction of this additional conversion attributable to oxidation by ozone as opposed to oxygen could not be immediately discerned. In order to examine this question further, blank experiments (i.e., with no ozone) were conducted. In one such experiment a 1/44.4 mixture of NO/O_2 was deposited on the 15-K gold-flashed cryosurface and flash-desorbed with pure nitrogen. Chemiluminescence analysis of the resultant mixture yielded NO/NO_x = 0.48, which is well within the range of the pure NO self-disporportionation results reported previously. Thus, molecular oxygen clearly has a minimal effect on NO conversion, if indeed any at all, and the enhanced oxidation observed for ozone-containing mixtures is attributable to ozone. The reaction

$$NO + O_3 \longrightarrow NO_2 + O_2 \quad (e)$$

is, of course, also exothermic (ΔH_r = 148 kcal/mol at 83 K in the solid phase, estimated from data given by Reid et al. (1977) and Dean (1973)). Therefore an autothermal chain reaction mechanism is also possible, just as for NO self-disproportionation and ozone recombination.

The data of Honig and Hook (1960) show that under our experimental conditions the vapor pressure of NO is less than that of O_2, thereby allowing the formation of a concentrated NO-O_3 mixture upon initial desorption of O_2. This fact could also account for the minimal effect of O_2 on NO conversion; i.e., the bulk of the O_2 may desorb prior to the NO becoming reactive enough for significant conversion to occur.

More quantitative conclusions than those included herein are currently not possible due to the complex nature of the detailed reaction mechanism of NO-O_3 mixtures. It is noted that in addition to direct oxidation of NO by ozone (reaction (e)), the competitive processes of NO self-disproportionation (reaction (c)) and ozone recombination occur as well.

The only work we are aware of which is even somewhat relevant to the preceding results is that of Lucas (1977) and Lucas and Pimentel (1979). In this work reaction (e) was studied in the solid phase (not upon thermal desorption) by IR spectrophotometry. Typically, cryofrosts of compositions such as $NO/O_3/N_2 = 1/30/250$ were deposited on a 12-K CsI window and the growth of the 1617-cm^{-1} band with time was attributed to the production of NO_2. Even though this product was clearly evident, the high dilution ratios employed resulted in barely discernible changes in the reactant concentrations. The NO-O_3 complex was hypothesized as the reaction intermediate, and the reaction rate data were found to be first order with respect to the concentration of the complex, with a preexponential factor of $1.45 \times 10^{-5}\ s^{-1}$ and an activation energy of 106 cal/mol in the solid phase at 12 K. This activation energy is of the order required for orientation rearrangement in noble-gas matrices, and is more than an order of magnitude less than the 2.3 kcal/mol reported for the gas-phase version of reaction (e). It was therefore hypothesized that matrix rearrangement could be the rate-limiting step.

Clearly, the experimental conditions and conversion levels of Lucas (1977) are quite different from our own, but the mechanism involving the NO-O_3 complex may be similar to that which occurs upon warming of the frost to desorption in our experiments. For a similar activation energy on the order of that required for matrix rearrangement, raising the temperature would accelerate reorientation as well as increase the number of potential NO-O_3 reaction pairs by increasing diffusivities. The simultaneous effect of both mechanisms could increase the overall reaction rate considerably over that observed by Lucas.

Concluding Remarks

The implications of the preceding results for cryogenic whole-air sampling are at least somewhat self-evident. However, with regard to the significance of these phenomena to natural heterogeneous processes in the atmosphere, our objective is simply to indicate their potential candidacy for further investigation. It is noted that, just as in our laboratory studies, all the necessary ingredients for the occurrence of high-temperature condensation (i.e., adsorption and trapping of noncondensable gases by condensable species such as water vapor) are also present in the natural atmosphere; i.e., it is certainly a "dirty" system with surfaces, water vapor, many other species, and temperatures in the 200 to 250 K range available as a function of altitude, latitude, and season. However, whether all the "ingredients" are present at the same time under the requisite conditions is not presently known, although the continual condensation and evaporation of water vapor suggests that this is certainly a possibility. In a similar fashion, the reactions of NO and O_3 upon thermal desorption from the solid state, although quite significant under our laboratory conditions, must be studied under conditions and concentrations more representative of the natural atmosphere. In the present work, the solid frosts for the chemical reaction studies were prepared in highly concentrated form at very low temperatures. Thus it still remains to be investigated whether such species can also react when adsorbed and trapped in the solid state at higher temperatures in a more dilute form. On the other hand, the natural atmosphere, with its considerable variety of neutral species, radicals, and ions, presents many more possibilities for reaction upon thermal desorption of an adsorbed mixture than are examined here. In summary, then, the potential exists for the participation of condensation/desorption reaction mechanisms (which are in a sense catalytic, but not in the most conventional fashion) in natural chemical cycles in the atmosphere. It is proposed that this be a subject of future investigations.

Acknowledgement. The authors gratefully acknowledge support of this work by the Air Force Geophysics Laboratory under AFGL Contract Nos. F19628-77-C-0071 and F19628-80-C-0066.

References

Abe, H., and W. Schultz, The sorption capacity of solid rare gas layers, Chem. Phys., 41, 257, 1979.

Addison, W. E., and R. M. Barrer, Sorption and reactivity of nitrous oxide with nitric oxide in crystalline and amorphous siliceous sorbents, J. Chem. Soc., 757, 1955.

Anderson, J. B., Molecular beams from nozzle sources, in Molecular Beams and Low Density Gasdynamics, edited by P. D. Wegener, Marcel Dekker, N.Y., 1974.

Becker, K., G. Klipping. W. D. Schoenherr, W. Schultz, and V. Toelle, Cryopumping of hydrogen by adsorption in condensed gases, in Proc. Int.

Cryog. Eng. Conf., 4th, pp. 319-322, IPC Sci. Technol. Press Ltd., Guildford, England, 1972a.

Becker, K., G. Klipping, W. D. Schoenherr, W. Schultz, and V. Toelle, Adsorption characteristics of condensed Ar, C_2H_6, NH_3, and CO_2 layers with respect to cryopumping, in Proc. Int. Cryog. Eng. Conf., 4th, pp. 323-326, IPC Sci. Technol. Press Ltd, Guildford, England, 1972b.

Bewilogua, L., and M. Jackel, Adsorption of hydrogen by condensed carbon dioxide at 20 K, Cryogenics, 14, 556, 1974.

Cazcarra, V., C. E. Bryson III, and L. L. Levenson, Sticking coefficient of CO_2 on solid H_2O films, J. Vac. Sci. Tech., 10, 148, 1973.

Davidson, D. W., Clathrate hydrades, in Water - A Comprehensive Treatise, Vol. 2, edited by F. Franks, pp. 115-234, Plenum Press, N.Y., 1973.

Dean, J. A. (Ed.), Lange's Handbook of Chemistry, McGraw-Hill, N.Y., 1973.

Fezza, R. J., Cryogenic deposition, desorption, and reaction studies of stratospherically relevant species, Ph.D. Dissertation, Princeton University, Princeton, N.J., 1981.

Gallagher, C. C., and R. V. Pieri, Cryogenic whole air sampler and program for stratospheric composition studies, AFGL-TR-76-0162, Air Force Geophysics Laboratory, Hanscom AFB, MA, 1976.

Honig, R. E., and H. O. Hook, Vapor pressure data for some common gases, RCA Review, 21, 360, 1960.

Lucas, D., Fast reactions, free radicals, and molecular complexes studied by the matrix isolation technique, Ph.D. Dissertation, University of California, Berkeley, 1977.

Lucas, D., and G. C. Pimentel, Reaction between nitric oxide and ozone in solid nitrogen, J. Phys. Chem., 83, 2311, 1979.

Miller, S. L., Water clathrates, in The Physics and Chemistry of Ice, edited by E. Whalley, pp. 42-50, Roy. Soc. Canada, 1973.

Miller, S. L., and W. D. Smythe, Carbon dioxide clathrates in the Martian ice cap, Science, 170, 531, 1970.

Reid, R. C., J. M. Prausnitz, and T. K. Sherwood, The Properties of Gases and Liquids, McGraw-Hill, N.Y., 1977.

Siksna, R., Water clathrates I, Uppsala University Report No. UURIE: 48-73, Stockholm, Sweden, 1973a.

Siksna, R., Water clathrates II, Uppsala University Report No. UURIE: 53-73, Stockholm, Sweden, 1973b.

Wang, E. S. J., J. A. Collins, and J. D. Haygood, General cryopumping study, in Advances in Cryogenic Engineering, Vol. 7, p. 44, Plenum Press, N.Y., 1962.

Welsbach, Basic Manual of Applications and Laboratory Ozonation Techniques, Welsbach Ozone Systems Corp., Philadelphia, PA, 1977.

INCOMPLETE ENERGY ACCOMMODATION IN SURFACE-CATALYZED REACTIONS

Bret Halpern and Daniel E. Rosner

Department of Chemical Engineering, Yale University, New Haven, Connecticut 06520

Abstract. In surface-catalyzed chemical reactions the accommodation of reaction energy is often incomplete. Evidence for incomplete energy accommodation is reviewed and its possible relevance to heterogeneous atmospheric chemistry is considered.

Introduction

In every chemical reaction, the energy of reaction is partitioned in a characteristic way among the vibrational, rotational, translational, and electronic modes of the products. This partitioning can in principle reveal the microscopic dynamics of the reactive encounter (Polyani, 1972). Gas-phase reactions have been studied for some time from this viewpoint, and the same attention is being turned to surface-catalyzed reactions. It is of both fundamental and practical interest to know how the energy of reaction is distributed among the various modes of the products and the catalyst. Do product molecules desorb, as might first be assumed, in thermal equilibrium with the catalyst?

There is accumulating evidence that in many cases they do not, and that the energy of reaction is incompletely accommodated. In this paper we review briefly some of the experimental evidence for the surface generation of excited molecules. We then examine ways in which heterogeneous reaction energy transfer may produce effects unique to atmospheric chemistry.

Evidence for Excited Molecule Generation at Surfaces

Heterogeneous energy transfer studies have been of two types. In the first, overall energy deposition in the catalyst is measured calorimetrically. In the second type, the energy content of a particular mode of the product molecule is examined.

Calorimetric experiments have been largely limited to the recombinations of H, O, and N atoms on metal or oxide-covered metal surfaces (Melin and Madix, 1971; Wise and Wood, 1963; Halpern and Rosner, 1978). These studies are relevant to atmospheric chemistry because atoms are present in the atmosphere in significant concentrations. In addition, atom recombination is highly exothermic and can represent an important energy flux to atmospheric particles. Recombination of like atoms is also the simplest system for describing surface energy transfer dynamics.

Halpern and Rosner (1978) have applied the calorimetric technique to a study of N atom recombination on metals. A metal filament is operated as an isothermal calorimeter in a fast flow of nitrogen gas at pressures near 1 torr. N atoms are easily generated by an upstream microwave discharge in concentrations sufficient to produce a measureable power input to the filament during recombination. The difference in electrical power necessary to hold the filament temperature constant in the presence and absence of atoms is equal to the energy released per second by successfully recombining atoms. Per unit area of catalyst this power Q may be written as

$$Q = \beta\gamma f\frac{D}{2}$$

The energy accommodation coefficient β is that fraction of the maximum energy release (the dissociation energy D = 225 kcal/mol) that is actually delivered to the catalyst per recombination event. The term f is the flux of atoms and γ is the probability that an atom successfully recombines, so that $\gamma\frac{f}{2}$ is the outgoing flux of surface-catalyzed recombinations. This flux may be measured, in the case of N atoms, by chemiluminescent titrations with known flows of NO in the presence and absence of the catalytic filament. Thus β can be determined by two power measurements and two titrations.

The recombination probability can also be obtained from these titrations. In the absence of diffusion limitations, the final atom concentration (or flow rate) passing through an inactive glass reactor containing S cm^2 of metal catalyst is given by

$$n = n_o e^{-\frac{\gamma}{4}\frac{cS}{vA}}$$

where n_o is the initial atom concentration (flow

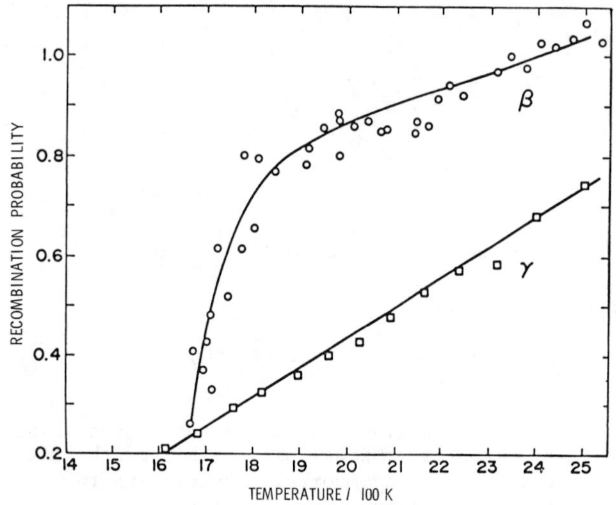

Fig. 1. Energy accommodation coefficient β and recombination probability γ for N atoms on W.

rate) in the absence of the metal catalyst, c is the thermal speed, v is the flow velocity, and A is the reactor cross section.

The behavior of N on tungsten and iridium is shown in Figures 1 and 2 (Halpern and Rosner, 1978). The case of N on W, Figure 1, seems most amenable to a simple description. In steady state, atoms are supplied to the surface either directly by atom chemisorption or by dissociation of N_2 molecules. They are removed by two recombination mechanisms. In the Rideal process, a gas-phase N atom strikes a chemisorbed N target and forms a molecule. In the Langmuir-Hinshelwood (LH) process, both partners are initially adsorbed; they may collide after surface diffusion, and product molecule desorption is then the reverse of dissociative chemisorption.

Both atom and molecule fluxes deliver energy to the tungsten catalyst. A directly chemisorbing N atom will liberate the heat of atom chemisorption $h \approx 152$ kcal/mol, while a dissociatively chemisorbed N_2 molecule liberates only the molecular heat of chemisorption $H \approx 80$ kcal/mol.

Molecules produced by recombination carry energy away from the catalyst. Those resulting from the LH process require a desorption activation energy equal to the molecular heat of adsorption, $H \approx 80$ kcal/mol. Since this is just the reverse of dissociative chemisorption, such molecules cannot be highly excited after desorption.

Molecules formed by a Rideal process, however, are initially formed with energy $D - h \approx 73$ kcal/mol. If desorption is rapid, then every Rideal recombination will generate excited molecules. This means that β can be less than unity. In fact, trajectory calculations (Tully, 1980) indicate that desorption occurs within picoseconds after the encounter. The newly formed molecule retains its excitation and does not dissipate it in collisions with the surface.

Thus at low temperatures, where the coverage of adsorbed atoms is high and surface mobility is low, the Rideal process will dominate and β will be low. This can be seen in Figure 1.

At high temperatures, however, the LH mechanism reduces the target N coverage to negligible values so that no Rideal processes can occur. All N atoms now strike empty sites and dissipate, per pair, an energy D plus the molecular heat of adsorption H. But the LH recombination and desorption of a pair of atoms removes H, so the net input is D. Thus β tends to unity, as seen in Figure 1.

This simple model does not appear to hold for N atoms recombining on iridium, for which β decreases at high temperatures as seen in Figure 2. It is conceivable that on iridium N atoms dissipate their chemisorption energy very slowly at high temperatures. They may then behave as a two-dimensional gas in which collisions result in recombination and desorption prior to energy accommodation. Since the atom chemisorption energy is not delivered to the iridium catalyst, β may be low even at high temperatures.

This very simple model shows, therefore, that a low value of β may be attributed to excited molecules that are produced by a Rideal step and desorbed without quenching.

In general, calorimetric atom recombination experiments show that the chemical energy accommodation coefficient can take values from 0.1 to unity. Two points are of special relevance to atmospheric surface chemistry: first, the low-temperature Rideal process can generate highly excited species, and second, direct chemisorption of atoms may constitute a very large, spatially localized energy input.

Calorimetric experiments can give no information about the excited product molecule. However, there exist a number of examples of surface-

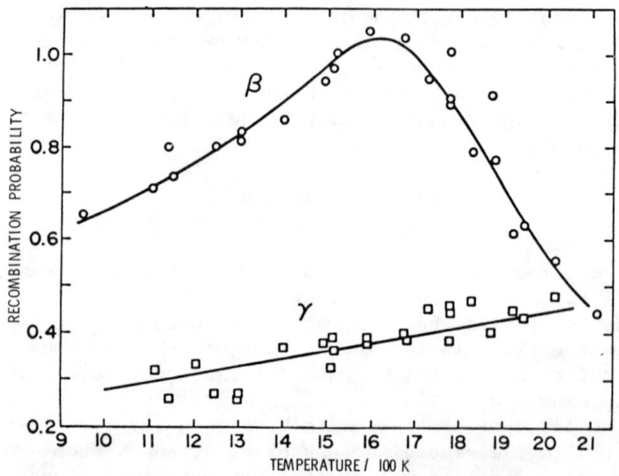

Fig. 2. Energy accommodation coefficient β and recombination probability γ for N atoms on Ir.

catalyzed excitation in which the particular mode of product excitation is known.

In one of the earliest such studies, Reeves and coworkers (Reeves et al., 1960) showed that electronically excited molecules could arise from atom recombination on metals. A stream of gas containing both N and O atoms produced electronically excited NO and N_2 when passed over foils of Ni, Co, Cu, and Ag. Nitrogen molecules were produced in the $A^3\Sigma_u^+$ state which lies ~6 eV above the ground state. Oxygen atoms alone, recombining on Ni, gave the $A^3\Sigma_u^+$ and $b^1\Sigma_g^+$ excited states.

In more recent experiments (Guillory and Shiblom, 1978), it has been shown that excited singlet molecular oxygen was thermally generated and desorbed when molecular oxygen flowed through a heated bed of lithium-tin-phosphorous catalyst.

A number of examples of translationally hot product molecules are known. H atoms recombining on Ni, Fe, Pt, and Cu desorb (Stickney, 1973) preferentially toward the surface normal, with a distribution given by $\cos^n\theta$, where n can be as high as 9. Comsa et al. (1979) showed that the translational energy of D atoms desorbing from a Ni(111) surface at 1140 K corresponded to a temperature of 1850 K. This is a small excitation when compared to that of a Rideal generated molecule; it results from a small adsorption activation barrier of several kcal/mol.

Becker et al. (1977) studied the oxidation of CO on Pt under molecular beam conditions and found, from time of flight measurements, that the translational temperature was 3650 K even when the surface temperature was 880 K.

Only a few studies have succeeded in probing the vibrational and rotational energy content of a surface reaction product. Thorman et al. (1980) allowed N atoms to permeate an Fe foil and recombine at the surface. The desorbing molecules were excited in vacuo by a high-energy (~2100 V) electron beam to the $N_2^+X^2\Sigma_g^+$ excited state from which they fluoresce. The fluorescence spectrum, at sufficient resolution, revealed the vibrational and rotational energy content of the desorbing N_2 molecules. These appeared to have a vibrational temperature of 2600 K when produced on a surface at 1150 K that had been contaminated by sulfur. It was concluded that the electronegative sulfur shifted the activation barrier for adsorption. By contrast, the rotational temperature was 400 K independent of surface temperature, indicating a steric hindrance to rotation, caused by surface topography, for the recombination-activated complex.

Bernasek and Leone (1981) have also investigated the vibrational energy of CO_2 formed in the oxidation of CO on Pt. This was done by observing the total infrared emission from the CO_2 asymmetric stretch mode. They concluded that the product CO_2 was in vibrational disequilibrium with the surface although their flow system conditions allowed some vibrational relaxation.

The limitation of collisional relaxation was not present, however, in the infrared emission study of Mantell et al. (1981), which is among the most finely detailed energy-transfer studies to date. Uncollimated jets of CO and O_2 were directed at a polycrystalline Pt surface heated to 775 K in a high-pumping-speed vacuum chamber at a pressure of 10^{-5} torr. Vibrationally excited CO_2 desorbed and radiated before undergoing collisions with background molecules. The radiation was collected and analyzed by a Nicolet 7199 Fourier Transform Infrared Spectrometer equipped with an InSb detector whose maximum sensitivity falls near the CO_2 asymmetric stretch. The resulting spectrum is displayed in Figure 3 along with a synthesized spectrum. It was possible to estimate a rotational temperature of 1150 K for the rotational lines in the (001) to (000) transition. The vibrational temperatures were less certain but were put at 1750 K for the bending mode and substantially less than 1500 K for the asymmetric stretch.

It is clear from the above examples that excited molecules, with energy contents from several kcal/mol to several eV, are characteristically generated in surface-catalyzed reactions.

Energy Partitioning in Surface Reactions: Relevance to Atmospheric Chemistry

One may ask what qualitative effects might arise in surface catalysis as a consequence of excited molecule generation or highly localized energy deposition. Which of these effects could be unique to atmospheric surface chemistry?

One feature of atmospheric catalysis that contrasts it with earthbound catalysis is the availability of gas-phase atoms such as O and N, as well as long-lived molecular excited states. A second feature is that temperatures in the atmosphere are low, so that even catalyzed reactions would proceed at a low rate. A flux of atoms or other energetic species can therefore serve as a transient energy source to drive surface reactions. This can arise in several ways.

A chemisorbing atom may liberate a large energy; that is, substrate atoms are locally disrupted and set into vibration. There is no experimental method of determining the time or distance scale over which large energies are released, and only simple one-dimensional models have been theoretically treated. At one extreme the energy deposition time may be very short, perhaps several lattice-atom vibrational periods. This energy deposition may then be regarded as a local heat spike which may thermally activate a neighboring complex of potentially reactive molecules hitherto frozen at low temperatures.

Some evidence exists for this effect. Molinari et al. (1966) noted that the rate of decomposition of NH_3 on tungsten at 1000 K was accelerated if a stream of H atoms was allowed to recombine on the filament at the same time and at the same temperature. Le Chatelier's principle would indicate that the rate should instead be diminished due to the additional hydrogen. It appears that the energy accommodated during recombination, which is

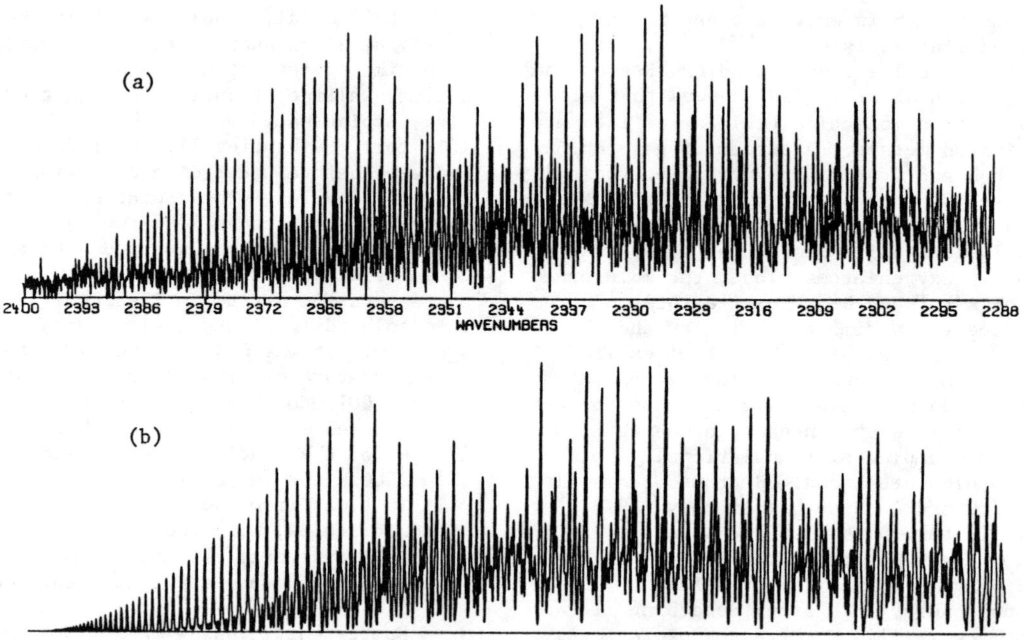

Fig. 3. Infrared emission spectrum for CO_2 in the surface-catalyzed reaction $CO_{ads} + O_2 \underset{Pt}{\rightarrow} \dot{C}O_2{}^*$; (a) experimental spectrum, (b) synthesized spectrum.

largely the atom chemisorption energy at these temperatures, is transmitted along the surface and used to activate ammonia complexes or fragments near the point of chemisorption.

At the other extreme, the atom may dissipate its chemisorption energy over a relatively long time. Recombination via two-dimensional gas collisions could then give highly excited (vibrationally or electronically) molecules which may remain weakly bound to the surface. These molecules could then serve as mobile activation energy reservoirs which activate other molecules or indeed react more readily themselves. At the high temperatures in Molinari's experiment (Molinari et al., 1966) physisorbed species would have a short surface lifetime, so this latter course would be unlikely.

Significant concentrations of fine, potentially catalytic particles exist throughout the atmosphere. An atom chemisorbing on a sufficiently small particle will raise the average temperature, even though the initial heat spike is local and of short duration. The magnitude of the average temperature rise depends strongly on the particle mass; its duration is determined by radiation cooling or gas-molecule impacts. If the temperature jump were sufficient in both magnitude and duration, then a surface reaction might become thermally activated. It is interesting to estimate roughly the particle size and pressure conditions that could permit such a reaction to occur following the chemisorption and rapid energy accommodation of a single atom.

The maximum temperature rise ΔT will occur if energy is delivered adiabatically. For an atom delivering energy h to a sphere of radius a and volume heat capacity equal to ~3 $n_o k$ (number density times Boltzmann's constant) the temperature rise would be

$$\Delta T \approx h/4\pi a^3 n_o k$$

or

$$\Delta T \approx 10^{-7} \frac{h(\text{eV})}{a^3(\mu\text{m})}$$

For h ~ 1 eV, a significant temperature rise of ~100 K would be obtained on a particle of 10 Å radius. For a 5-Å radius, the rise would be 800 K. Although energy is deposited on one side of the sphere, heat diffuses rapidly through the particle, thus leveling the temperature in a time likely to be short compared to the cooling time. If cooling is accomplished by molecular impacts in an ambient gas at pressure P whose molecules accommodate perfectly to the particle temperature, then the temperature increment decays exponentially with a time constant $\tau_{cooling}$ given approximately by

$$\tau_{\text{cooling(sec)}} \approx \frac{1}{3} \frac{n_o a(\mu\text{m})}{3 \times 10^{20} \; P(\text{torr})} \approx \frac{0.003 \; a(\mu\text{m})}{P(\text{torr})}$$

For a 10-Å particle in 1 torr (in the stratosphere) this cooling time would be in the range of 10^{-6} seconds. The maximum time available for reaction would then be of this order, assuming no other atoms adsorbed. For an unspecified first-order process the reaction time may be written as

$$\tau_{reaction} \approx 10^{-13} e^{e/kT}$$

Making the drastic simplification that the temperature remains at its maximum value during the cooling time, the condition $\tau_{reaction} \approx \tau_{cooling}$ will obtain for an activation energy $\varepsilon < 13$ kcal/mol. It is therefore marginally possible that this effect operates for very small clusters or particles at sufficiently high altitudes, or even on agglomerations of small particles which are in very poor thermal contact with their neighbors. The related problem of heat dissipation in a supported metal catalyst crystallite has been considered in great detail (Steinbruechel and Schmidt, 1973). (In that case the support constitutes the most effective heat sink.)

The possibility that highly excited molecules may have very long surface lifetimes is especially interesting in the case of oxygen atoms, which are a significant atmospheric constituent, because of the possible surface formation of electronically excited $^1\Delta$ and $^1\Sigma$ states, lying 22.5 kcal and 37.5 kcal above the ground state (Turro, 1978). The pure radiative lifetimes of both states are extremely long; at 1 atm the $^1\Delta$ state has a lifetime of 0.08 sec. Even in liquid solvents at room temperature the lifetime can vary from 2 μsec (H_2O) to several msec (CCl_4) (Turro, 1978). This suggests that singlet oxygen molecules would be inefficiently deactivated by collisions with or physisorption by a catalyst surface. Homogeneous and heterogeneous quenching of excited molecules, in particular O_2^* created by surface-catalyzed recombination, has been treated in detail by Rosner and Feng (1974).

Suppose that recombination of two O atoms to the singlet state occurs on a 1-μm particle composed of smaller 0.01-μm particles. The internal surface area of this particle will be large. Suppose that oxidizable hydrocarbons are distributed sparsely throughout this internal surface area. The reaction of singlet oxygen and hydrocarbon will be efficient if the long lifetime of the singlet state permits it to explore the internal surface for reaction partners. If the particle is regarded as containing pores of radius r (0.01 μm in this case) then the diffusion coefficient is given by

$$D \sim 2/3\ rc$$

where c is the mean thermal speed (Satterfield and Sherwood, 1963). For the case given previously $D \sim 0.1\ cm^2/sec$. If the inner surface were covered with H_2O multilayers the singlet-state lifetime τ would presumably be greater than 2 μsec, and in this time the excited oxygen molecule could diffuse $x \sim \sqrt{D\tau} \sim 10^{-4}$ cm. Thus much of the interior is accessible to a long-lived excited oxygen molecule. Even if the O_2^* diffused only on the particle's internal surfaces, the penetration depth could be significant. There is little information on the deactivation of singlet oxygen by wall collisions or physisorption. Winer and Bayes (1966) have shown that in a glass flow system at a few torr, deactivation required 10^4 wall collisions. This is not inconsistent with the long lifetimes in dense liquids if each collision involved temporary trapping for ~100 oscillations before desorption.

Apart from recombination of like atoms on metals, the direct Rideal-type reaction of a gas atom with an adsorbed target molecule has been infrequently studied. It is still likely that highly excited molecules can result. Tully (1980) has done trajectory calculations on the model system

$$O_{gas} + C_{ads} \rightarrow CO^* \qquad (a)$$

Only a few percent of the available 6-eV exothermicity appears in the catalyst, and the excited CO desorbs within picoseconds after formation.

Halpern et al. (1981) attempted to generate NO by a similar Rideal process. They combined a stream of N atoms from a microwave discharge with molecular oxygen flowing over a Pt foil. They looked for IR emission from NO, anticipating a Rideal attack of N on a high-coverage oxygen layer. The result instead was a high yield of N_2O, whose radiation was stimulated by resonant energy exchange with vibrationally excited N_2 molecules also emerging from the discharge. The production rate followed the NO desorption behavior on Pt so that adsorbed NO was the likely intermediate. The N_2O could then result from

$$N_{gas} + NO_{ads} \rightarrow N_2O \qquad (b)$$

This is an unusual reaction, since it is spin-forbidden in the gas phase, where it results instead in N_2 and O.

Evidently the surface removes the spin restrictions and acts as a third body to stabilize the N_2O formed in reaction (b). In addition, sufficient reaction energy must appear in relative translation of product and catalyst to desorb the product before it can rearrange to a favored N_2 and chemisorbed O. Thus energy partitioning can have a very powerful influence on catalytic selectivity, especially for highly exoergic reactions involving gas atoms and adsorbed target molecules.

References

Becker, C. A., J. P. Cowin, L. Wharton, and D. Auerbach, CO_2 product velocity distributions for CO oxidation on platinum, J. Chem. Phys., 67, 3394-3395, 1977.

Bernasek, S. L., and S. R. Leone, Direct detection of vibrational excitation in the CO_2 product of the oxidation of CO on a platinum surface, Chem. Phys. Lett., 84, 401, 1981.

Bradley, T. L., and R. E. Stickney, Spatial distributions of the H_2 desorbed from Fe, Pt, Cu, Nb, and stainless steel surfaces, Surface Sci., 38, 313-326, 1973.

Comsa, G., R. David, and B. J. Schumacher, The angular dependence of flux, mean energy, and speed ratio for D_2 molecules desorbing from a Ni(III) surface, Surface Sci., 85, 45-68, 1979.

Halpern, B., and D. E. Rosner, Chemical energy accommodation at catalyst surfaces. Flow reactor studies of the association of nitrogen atoms on metals at high temperatures, Trans. Faraday Soc., 74, 1883-1912, 1978.

Guillory, J. P., and C. M. Shiblom, The generation of singlet oxygen by a lithium-tin-phosphorous catalyst, J. Catal., 54, 24-30, 1978.

Halpern, B., E. J. Murphy, and J. B. Fenn, Surface-coated production of nitrous oxide from the reaction of N atoms and O_2 on platinum, J. Catal., 71, 434, 1981.

Mantell, D. A., S. B. Ryali, B. L. Halpern, G. L. Haller, and J. B. Fenn, The exciting oxidation of CO on Pt, Chem. Phys. Lett., 81, 185-187, 1981.

Melin, G. A., and R. J. Madix, Energy accommodation during oxygen atom recombination on metal surfaces, Trans. Faraday Soc., 67, 198-211, 1971.

Molinari, E., F. Cramarrossa, M. Capitelli, and A. Mercanti, Catalytic decomposition of ammonia on tungsten in the presence of hydrogen atoms, Ric. Sci., 36, 109-113, 1966.

Polyani, J. C., Concepts in reaction dynamics, Acct. Chem. Res., 5, 161-168, 1972.

Reeves, R. R., G. Manella, and P. Harteck, Formation of excited NO and N_2 by wall catalysis, J. Chem. Phys., 32, 946-947, 1960.

Rosner, D. E., and H. H. Feng, Energy transfer effects of excited molecule production by surface-catalyzed atom recombination, Trans. Faraday Soc., 70, 884-907, 1974.

Satterfield, C. N., and T. K. Sherwood, The Role of Diffusion in Catalysis, p. 16, Addison Wesley Publ. Co., Inc., Reading, Mass., 1963.

Steinbruechel, C., and L. D. Schmidt, Heat dissipation in catalytic reactions on supported crystallites, Surface Sci., 40, 693-707, 1973.

Thorman, R. P., D. Anderson, and S. L. Bernasek, Internal energy of heterogeneous reaction products: Nitrogen-atom recombination on iron, Phys. Rev. Lett., 44, 743-746, 1980.

Tully, J. C., Dynamics of gas-surface interactions: Reaction of atomic oxygen with adsorbed carbon on platinum, J. Chem. Phys., 73, 6333-6342, 1980.

Turro, N. J., Modern Molecular Photochemistry, p. 583, Benjamin/Cummings Publ. Co., Inc., Menlo Park, Ca., 1978.

Winer, A. M., and K. D. Bayes, The decay of O_2 ($a^1\Delta$) in flow experiments, J. Phys. Chem., 70, 302-304, 1966.

Wood, B. J., J. S. Mills, and H. J. Wise, Energy accommodation in exothermic heterogeneous catalysis reactions, J. Phys. Chem., 67, 1462-1465, 1963.

Aqueous Studies

OXIDATION OF SO_2 BY NO_2 AND AIR IN AN AQUEOUS SUSPENSION OF CARBON

Robert S. Rogowski, David R. Schryer, Wesley R. Cofer III, and Robert A. Edahl, Jr.

NASA Langley Research Center, Hampton, Virginia 23665

Shekhar Munavalli

Livingstone College, Salisbury, North Carolina 28144

Abstract. A series of experiments has been performed using carbon black as a surrogate for soot particles. Carbon black was suspended in water and gas mixtures were bubbled into the suspensions to observe the effect of carbon particles on the oxidation of SO_2 by air and NO_2. Identical gas mixtures were bubbled into a blank containing only pure water. After exposure each solution was analyzed for pH and sulfate. It was found that NO_2 greatly enhances the oxidation of SO_2 to sulfate in the presence of carbon black. The amount of sulfate in the blanks was significantly less. Under the conditions of our experiments no saturation of the reaction was observed and SO_2 was converted to sulfate even in a highly acid medium (pH $\geq$ 1.5).

Introduction

Many mechanisms have been proposed for the oxidation of SO_2 in the atmosphere. These include homogeneous reactions in the gas phase (Calvert et al., 1978) and heterogeneous reactions in cloud droplets or liquid-phase aerosols (Hegg and Hobbs, 1981; Bielke and Gravenhorst, 1978; Penkett et al., 1979), and on carbon particles (Novakov et al., 1974; Chang and Novakov, 1978). In order to assess the relative contribution of the various mechanisms and propose enlightened control strategies, the competing processes must be evaluated in detail.

The work described here addresses one aspect of this problem, namely heterogeneous oxidation of SO_2 by air and NO_2 on carbon (soot) particles suspended in water. Similar reactions are possible in the atmosphere in liquid droplets or aerosols containing insoluble carbon particles. Novakov (1974) has established that soot is effective in oxidizing SO_2 to sulfate in the presence of O_2. Carbon particles are found even in remote areas such as the Arctic (Rosen et al., 1981) and are virtually ubiquitous. Therefore they are widely available to provide reactive sites.

The experiments involving carbon suspended in liquid water were prompted by our original experiments performed with dry carbon particles (Cofer et al., 1980). When such particles were exposed to humidified mixtures of air, SO_2, and NO_2 they became wetted after a certain time, and it was not clear at that point whether solution chemistry was dominant or if the soot still played an important role.

Experimental Procedure

Carbon powder (commercial furnace black) was suspended in distilled water (100 mg/10 ml) and commercially prepared mixtures of gases (supplier-certified mixtures in ultra high purity carrier) were bubbled into the aqueous suspension. The same gas mixtures were also bubbled through 10 ml of pure water to serve as a blank. The gases (100 ppm SO_2 in air or N_2, and 100 ppm NO_2 in N_2) were mixed prior to splitting the flow between the two reaction vessels. The reactors were held at a constant temperature of 23°C in a water bath. The apparatus is shown schematically in Figure 1. This configuration allowed the blank and the aqueous carbon suspension to be exposed to identical conditions, so that the difference in amount of sulfate produced in the two reactors was a direct measure of the effect of the carbon. After each experiment the resulting solutions were analyzed for pH and sulfate content using a barium turbidity test (Cofer et al., 1980).

Results and Discussion

The results of several bubbler experiments are presented graphically in Figure 2. The yield of sulfate under identical exposure conditions is much larger in the presence of the carbon black. The carbon acts as a catalyst for the oxidation of SO_2 by both air and air with 100 ppm NO_2 added. Apparently the carbon provides sites for the SO_2 oxidation to occur, because as the blank curves indicate, neither air nor NO_2 has any

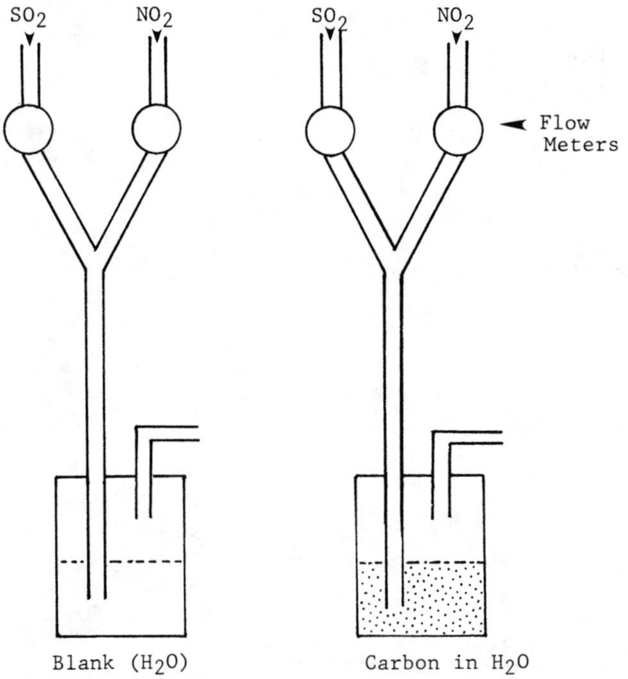

Fig. 1. Apparatus used to investigate SO_2 oxidation in presence of aqueous carbon suspension.

significant oxidation effect in water alone. The conversion rate of SO_2 in air at 100 ppm with 1 percent C is 9.3%/hr, and increases to 58%/hr when 100 ppm NO_2 is added. These results are in agreement with those obtained in previous experiments in our laboratory (Cofer et al., 1980) in which NO_2 significantly enhanced the oxidation of SO_2 on carbon particles. Some enhancement of the oxidation of SO_2 by NO_2 has also been observed on dry soot by Britton and Clarke (1980) and on dry V_2O_5 by Barbaray et al. (1978).

To distinguish the oxidizing effect of NO_2 from that of the O_2 in air, the same experiment was performed with the SO_2 and NO_2 mixed in N_2. The results are shown in Figure 3. Here only the oxidizing capacity of the NO_2 is measured. The experiments in N_2 indicate that NO_2 acts as an oxidizing agent independent of the O_2 in air, and comparison of results from Figures 2 and 3 shows that the effects of O_2 and NO_2 are additive. The intercept of 0.64 mg sulfate in Figures 2 and 3 was found to be due to sulfate present on the carbon surface as received from the manufacturer.

Several 20-hr runs were made at SO_2 and NO_2 concentrations of 100 ppm and flow rates of 100 cm^3/min for each gas. The sulfate yields were the same as would be predicted by linear extrapolation of the data in Figure 2 (30 mg). This indicates that SO_2 was converted to sulfate with no ob-

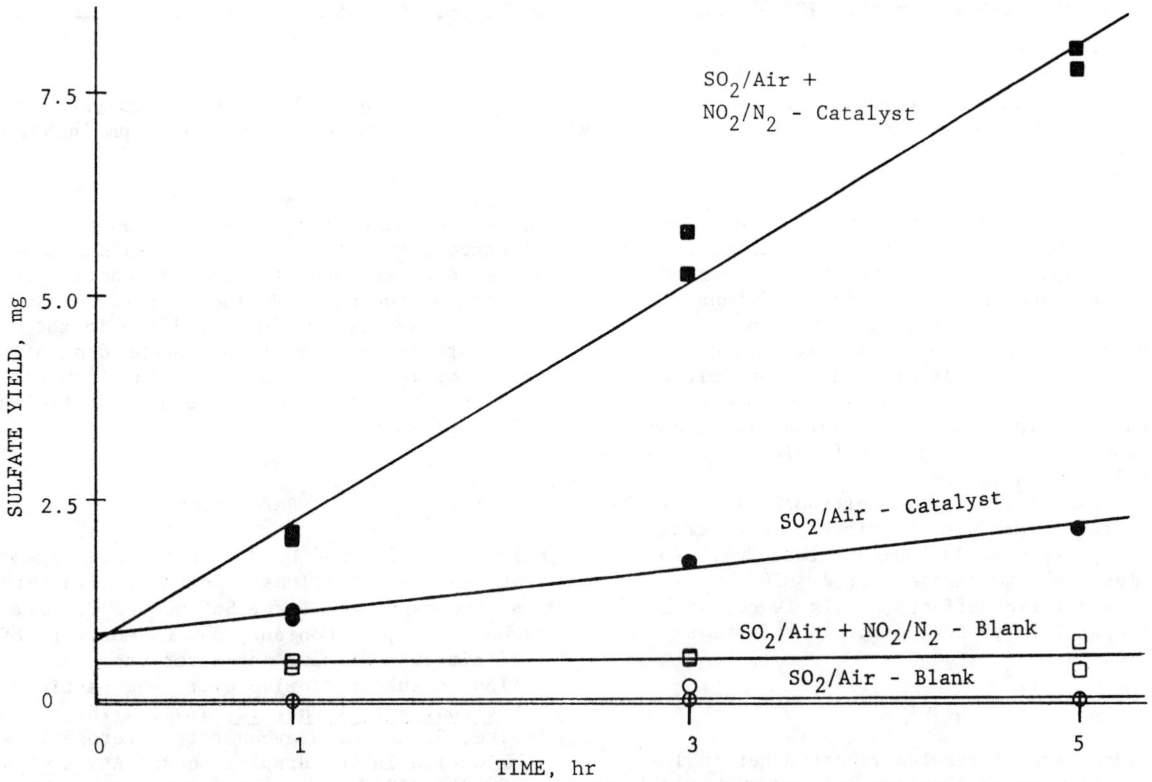

Fig. 2. Effect of NO_2 and catalyst on sulfate yield in air. Blank: 10 ml H_2O; catalyst: 100 mg carbon black in 10 ml H_2O; flow rates: 100 cm^3/min, each gas; concentrations: 100 ppm, each gas.

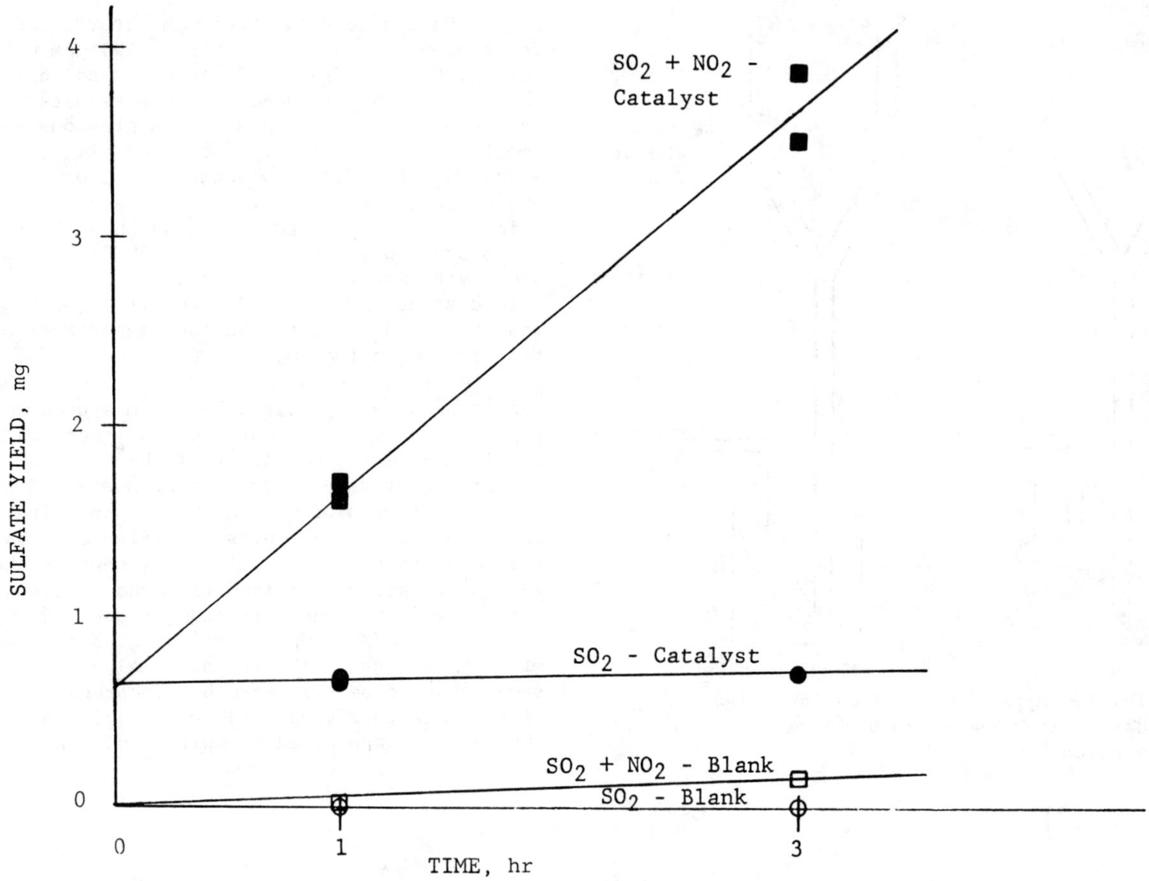

Fig. 3. Effect of NO_2 and catalyst on sulfate yield in N_2. Blank: 10 ml H_2O; catalyst: 100 mg carbon black in 10 ml H_2O; flow rates: 100 cm^3/min, each gas; concentrations: 100 ppm in N_2, each gas.

servable saturation effect for these long exposures. The pH of the solutions for these runs was as low as 1.5, indicating that the conversion takes place under highly acidic conditions. Saturation effects have consistently been observed for SO_2 on dry particles by others (Britton and Clarke, 1980; Baldwin and Golden, 1979; Judeikis et al., 1978) but we have observed no saturation in either the gravimetric or bubbler experiments as long as sufficient H_2O vapor or liquid H_2O is present.

If we take the liberty of extrapolating the 20-hr runs to typical atmospheric SO_2 concentrations of 0.01 ppm this would indicate that "wet" soot particles could be exposed for 2×10^6 hours without saturation effects. This is certainly much longer than typical particle lifetimes in the atmosphere.

Conclusions

The experimental results reported herein lead to the following conclusions. (1) Carbon particles significantly catalyze the oxidation of SO_2 to sulfate by air and/or NO_2. (2) NO_2 is an effective oxidizer for SO_2 in aqueous suspensions of carbon, and its effect is independent of the presence or absence of air; the reaction is controlled by the NO_2, provided sufficient SO_2 is present. (3) The yield of sulfate in the reactions studied appears to be independent of pH at values as low as 1.5, and no saturation occurs for long exposure times so long as sufficient H_2O is present.

References

Baldwin, A. C., and D. M. Golden, Heterogeneous atmospheric reactions: Sulfuric acid aerosols as tropospheric sinks, Science, 206, 562, 1979.

Barbaray, B., J. Contour, and G. Mouvier, Effects of nitrogen dioxide and water vapor on oxidation of sulfur dioxide over V_2O_5 particles, Environ. Sci. Tech., 12, 1294, 1978.

Bielke, S. and G. Gravenhorst, Heterogeneous SO_2 oxidation in the droplet phase, Atmos. Environ. 12, 231, 1978.

Britton, L. G., and A. G. Clarke, Heterogeneous reactions of sulphur dioxide and SO_2/NO_2 mix-

tures with a carbon soot aerosol, Atmos. Environ., 14, 829, 1980.
Calvert, J., F. Su, J. Bottenheim, and O. Strausz, Mechanism of the homogeneous oxidation of sulfur dioxide in the troposphere, Atmos. Environ., 12, 197, 1978.
Chang, S., and T. Novakov, Soot-catalyzed oxidation of sulfur dioxide, in Man's Impact on the Troposphere, edited by J. S. Levine and D. R. Schryer, NASA RP-1022, pp. 349-369, 1978.
Cofer, W. R. III, D. R. Schryer, and R. S. Rogowski, The enhanced oxidation of SO_2 by NO_2 on carbon particlates, Atmos. Environ., 14, 571, 1980.
Hegg, D., and P. Hobbs, Cloud water chemistry and the production of sulfates in clouds, Atmos. Environ., 15, 1597, 1981.
Judeikis, H. S., B. T. Stewart, and A. G. Wren, Laboratory studies of heterogeneous reactions of SO_2, Atmos. Environ., 12, 1633, 1978.
Novakov, T., S. G. Chang, and A. B. Harker, Sulfates as pollution particulates: Catalytic formation on carbon (soot) particles, Science, 186, 259, 1974.
Penkett, S. A., B. M. R. Jones, K. A. Brice, and A. E. J. Eggleton, The importance of atmospheric ozone and hydrogen peroxide in oxidizing sulphur dioxide in cloud and rainwater, Atmos. Environ., 13, 123, 1979.
Rosen, H., T. Novakov, and B. A. Bodhaine, Soot in the arctic, Atmos. Environ., 15, 1371, 1981.

SULFUR DIOXIDE ABSORPTION, OXIDATION, AND OXIDATION INHIBITION IN FALLING DROPS: AN EXPERIMENTAL/MODELING APPROACH

Elmar R. Altwicker and Clement Kleinstreuer

Department of Chemical Engineering and Environmental Engineering,
Rensselaer Polytechnic Institute, Troy, New York 12181

Abstract. Results are reported on the absorption of sulfur dioxide into falling water drops, the extent of oxidation of S(IV) species in the liquid phase, and the inhibition of this oxidation. Experimental results obtained in a laboratory "string-of-falling-drops" system for 0.36-second drop fall time have been compared to calculations based on several absorption models found in the literature. Some of the differences between the experimental results and those from the model calculations emphasize the need for more detailed understanding of initial and non-steady-state behavior. Our own modeling approach, which consists primarily of a hydroaerodynamic and a mass-transfer-with-chemical-reaction submodel, is proposed. Initial comparisons between model and experiment show reasonable agreement, but also point to the need for more detailed understanding of rate processes at the interface. The significance of the experimental/modeling approach lies in its applicability to non-steady-state processes, such as initial rates of SO_2 absorption and reaction in cloud and raindrops, plume washout, and interaction of droplet sprays with waste gases.

Introduction

Removal of trace gases from polluted atmospheres by water drops is one of the important scavenging mechanisms in clouds, rain, and wet scrubbers. The mass transfer of sulfur dioxide into drops, the oxidation of the absorbed sulfur species, and the inhibition of such oxidation continue to be important topics in atmospheric and control technology research (Husar et al., 1978; Beilke et al., 1975; Altwicker, 1976, 1979). The first step in this sequence, the mass transfer of sulfur dioxide from the bulk gas phase to the bulk liquid phase, has been the subject of numerous theoretical and experimental investigations. Experimental systems have ranged from quiescent bulk liquid-phase studies to investigations which utilized suspended drops and laboratory simulations of rain showers as well as field studies (Johnstone and Williams, 1939; Garner and Lane, 1959; Terraglio and Manganelli, 1967; Beilke and Georgii, 1968; Hales, 1972; Miller and dePena, 1972; Matteson and Giardina, 1974; Dana et al., 1975; Barrie and Georgii, 1976; Barrie, 1978; Hikita et al., 1978; Carmichael and Peters, 1979; Roberts and Friedlander, 1980). Comprehensive modeling approaches to wet removal of sulfur compounds from the atmosphere have been taken by Hales (1978), Adamovicz and Hill (1977), Overton et al. (1979), Baboolal et al. (1981), and Walcek et al. (1981). There exists a particular need to further examine initial rates under realistic conditions. This is particularly important since the absorption of sulfur dioxide is accompanied by oxidation of dissolved S(IV) species, a process which can be catalyzed by trace metal ions, peroxides, ozone, and other oxidants, and can also be inhibited to varying degrees by several classes of organic and inorganic compounds. Further, some of the experimental systems cited make extrapolations to processes based on moving drops difficult. Some of these features have promoted our laboratory/modeling approach, which permits the investigation of initial rates of absorption and reaction under a variety of conditions; a thorough understanding of these should eventually lead to an understanding of combined gas and particle removal in such systems.

The experimental system we have employed can be termed the "string-of-falling-drops" generator. This system exhibits several advantages over other methods mentioned previously. Among these are an ability to generate one or more monodisperse drop streams which can be exposed to pollutants at realistic concentrations; at the end of the exposure time the falling drops can be collected and analysis carried out in bulk. The principles and practice of this system have been described by several investigators (Dabora, 1967; Arrowsmith and Foster, 1973; Rajagopalan and Tien, 1973) based on the work of Lord Rayleigh; monodisperse drops are created by forcing water under pressure through an orifice (20-gauge hypodermic needle, blunt edge) which is vibrated axially via audio frequency. The phenomenon of the jet and its subsequent breakup into drops have been discussed

TABLE 1. Experimental Parameters

Drop Characteristics	Equivalent[a] diameter, D	0.84 mm
	Surface area[b] for absorption, A	1.11 mm^2
	Volume, V	0.31 mm^3
	Fall distance	101.8 cm
	Exposure time	0.36 sec
	Jet breakup and drop formation time	0.004 sec
	Initial velocity, V_i	120 cm s^{-1}
	Terminal velocity, V_t	364 cm s^{-1}
Experimental Conditions	Reactor temperature	18.5^o - 21^oC
	Reactor SO_2 concentration	1.1 ±0.25 ppm
	Initial drop pH	5.45 - 5.71

[a]Based on V = 4/3 ab^2, where a is the minor axis length of an oblate spheroid and b is its major length.

[b]One-half of geometric drop surface area.

[c]TCM (tetrachloromercurate) and H_2O_2 solutions were used to collect the drops.

in terms of both capillary instability and mechanical resonance (Scarlett and Parkins, 1977). In our present arrangement two streams of 0.84-mm (equivalent diameter) drops fall through air containing ppm levels of sulfur dioxide confined to a 10.8-cm-diameter tube, and exit through a small opening into absorbing solution (Table 1) to quench subsequent reactions. (S(IV) analysis was performed via the West-Gaeke method, and total S analysis was carried out on an ion chromatograph). Key operating parameters are shown in Table 1. A flow sheet is shown in Figure 1.

Results (Experimental)

Representative results from the laboratory system (Altwicker and Chapman, 1981; E. R. Altwicker and K. Nass, unpublished data, 1981) are listed in Table 2 and compared to species concentrations calculated from several absorption models. The terms [S(IV)] and [total S] designate the concentrations of these species in the liquid phase at the end of the lifetime of the drops (0.36 sec); S(IV) denotes SO_2 plus HSO_3^- plus $SO_3^=$ (although under our conditions HSO_3^- was the dominant species). The ranges and standard deviations listed are for at least eight separate experimental runs each at sulfur dioxide concentrations of 1.1 ±0.25 ppm. Six to eight separate bulk samples were collected during each run for analysis. Inspection of the table shows that a portion of the absorbed sulfur dioxide was oxidized, although the liquid phase contained no added catalyst. It is worthwhile to emphasize that the initial oxygen to S(IV) ratio (as well as the ratio of any potentially catalytic impurity to S(IV)) in the liquid phase is essentially infinity; i.e., the initial condition is very different from that typically used in bulk and stop-flow experiments.

These experimental results on "uncatalyzed" absorption have been compared to those calculated from several recent models by the application of the respective governing equations (Table 2). The various assumptions that were made to arrive at the governing equations are well known and have been summarized by Altwicker and Chapman (1981). Where concentration ranges are given in Table 2 they reflect the range of experimental sulfur dioxide concentrations (Table 1) and, in the case of the Hales and Barrie models, the fact that two different Reynolds numbers were used in the calculations (to take into account the initial, V_i, and the terminal drop velocity, V_T). These models and the Beilke-Gravenhorst model overpredict the observed values, while the Carmichael-Peters model underpredicts. The model presented by Barrie explicitly includes the parameters δ_ℓ and δ_g, i.e., the liquid and gas film thickness; however, the calculated values differ little from those of Hales and Beilke and

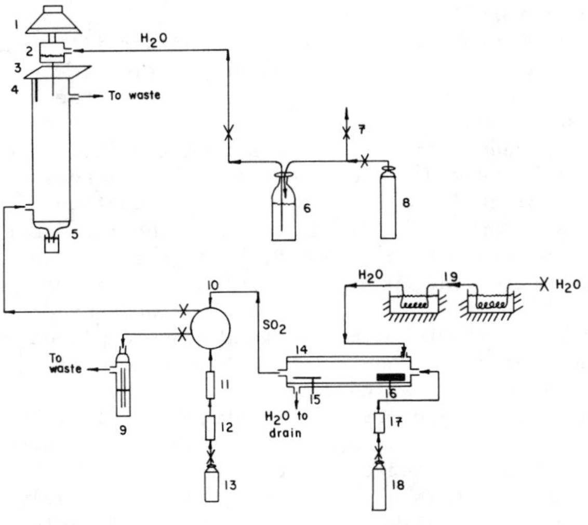

1. Speaker
2. Reservoir and Needle
3. Reaction Chamber Seal
4. Thermometer
5. Collection Funnel
6. Water Supply Reservoir
7. Pressure Release Valve
8. Nitrogen Air Cylinder
9. Impinger
10. Mixing Bulb
11. Rotameter
12. Silica Drying Column
13. Mixing Gas Tank (N_2 or Zero Grade Air)
14. Condenser
15. Thermometer
16. Permeation Tube
17. Rotameter
18. N_2 Carrier Gas Tank
19. Insulated Water Baths for Cooling Water Coils

Fig. 1. Experimental analysis.

TABLE 2. Comparison of Experimental SO_2 Absorption (and Liquid-Phase Oxidation) Quantities With Results Computed From Several Models

Reference	Governing Equation	S(IV)[a] μmol/l	[Total S][b] μmol/l
Altwicker & Chapman (1981); E. R. Altwicker & K. Nass (unpublished data, 1981)	-	1.43 ±0.12	1.94 ±0.3
Hales (1972, 1978)	$\frac{\Delta M}{\Delta t} = \frac{D_g A}{D} C_{SO_2} [2 + 0.6\ Re^{1/2} Sc^{1/3}]$	5.6 - 10.6	
Barrie (1978)	$\frac{\Delta M}{\Delta t} = \frac{D_\ell A}{\delta_\ell} \{[HSO_3^-]_\ell^I - [HSO_3^-]_\ell\}$	3.9 - 5.7	
Beilke & Gravenhorst (1978)	$M_t(g) = 4\pi D_g r\, C_{SO_2}\left[t + \frac{2r\sqrt{t}}{\sqrt{\pi D_g}}\right]$	3.8 - 4.7	
Carmichael & Peters (1979)[c]	$\frac{\Delta M}{\Delta t} = 2A\left(\frac{D_\ell}{\pi t}\right)^{0.5} \alpha^{0.5}(\Phi_i - \Phi_b)$	0.12 - 0.15	
Danckwerts (1970)	$A^* = \frac{q}{4(D_\ell V_\ell h)^{0.5}};\ q = \frac{\Delta M}{\Delta t}$	0.62 - 1.17	

[a]As HSO_3^-.
[b]As $SO_4^{=}$.
[c]$\Phi_i = H[SO_2]_g + [HSO_3^-]_\ell$ where H = Henry's law constant; $\Phi_b = 0$.

Gravenhorst. In their more comprehensive analysis, Carmichael and Peters (1979) pointed out that the simplified treatment reflected in the governing equations for the Hales and Barrie models substantially overestimates sulfur dioxide absorption. However, their model (subject to the assumptions made in the calculations, cf. notation) underestimates the observed results. The Danckwerts model presented in Table 2 suggests an equivalent "laminar" jet. The most obvious reason for using this approach is the fact that the "string of falling drops" appears to have jet properties (Arrowsmith and Foster, 1973); we have put "laminar" in quotation marks since the drop cannot be laminar in the exact sense employed by Danckwerts. An equivalent jet length, h, for the dropstream can be calculated from $h = 4V\ t/\pi D^2$, where V_ℓ = volumetric flow rate, t = fall time, and D = equivalent drop diameter. The calculated A* can be equated to Φ_i (cf. Table 2). The results from this calculation agree more closely with the experimental observations. To avoid any misunderstanding it should be emphasized that these results were obtained by evaluation of the difference, $\Delta M/\Delta t$, between the initiation of the experiment, t = 0, and the quenching of the drops, t = 0.36 sec, and that it was assumed that Δt = drop lifetime = 0.36 sec (cf. Table 1).

Although great care has been taken in the selection of reagents and preparation of solutions (Chapman, 1980), the fact that oxidation has been observed may be due to the fact that at the time of drop formation any unmeasured impurities (for example, $Fe^{+3} < 10^{-6}$ M) as well as oxygen are present in the liquid phase in large excess relative to the initial instantaneous S(IV). Experiments with initial known S(IV) concentrations are planned.

The effect of added catalysts (Co, Fe, Mn) on absorption and oxidation is clearly measurable in our system. Representative results are given in Tables 3 and 4. Inhibition of the oxidation under both catalyzed and "uncatalyzed" conditions has also been observed (Table 4). The inhibitors used are known to function as chainstoppers (diethyl hydroxylamine (DEHA) and ascorbic acid). Even in the "uncatalyzed" case nearly complete absence of oxidation could be demonstrated by 10^{-4} M DEHA.

Previous work (Schroeter, 1960; Altwicker and Sekulic, 1974; Altwicker, 1979) has demonstrated induction periods for the oxidation on the order of seconds or longer and a dependency of the length of the induction period on inhibitor con-

TABLE 3. Experimental Observation on the Effect of Metal Catalyst on S(IV) Oxidation in the Falling Drops (concentrations in μmol/l)

Catalysts	S(IV)	Total S	% Oxidation
Uncatalyzed	1.43 ±0.12	1.94 ±0.3	26
Mn^{+2} 10	0.84	3.67	77.1
Mn^{+2} 50	1.73	3.48	50.3
Mn^{+2} 100	2.69	4.32	37.8
Co^{+2} 10	1.31	2.24	41.5
Fe^{+3} 10	1.33	4.17	68.2

TABLE 4. Effect of Inhibitors (concentrations in μmol/l)

Catalyst	Inhibitor	% Oxidation
None	DEHA, 10^{-4}	~0
Mn, 10^{-4}	None	38
Mn, 10^{-4}	DEHA, 10^{-4}	30
Co, 10^{-5}	None	41.5
Co, 10^{-5}	Ascorbic acid, 10^{-5}	~0

centration, a finding consistent with a radical chain initiation. The results reported here with catalysts essentially showed the absence of an induction period. Since it is known that inhibitors of the type used function usually by lengthening the induction period, the mechanism of the inhibition (Altwicker, 1980) is not clear. Perhaps a complexation step is involved; this would be plausible in view of the very large initial catalyst to S(IV) ratio in the drops (cf. following section).

Modeling Framework

The "string-of-falling-drops" system exhibits certain properties which have not been detailed in the model comparisons shown in Table 2; therefore, exact agreement between these models and the experimental results could hardly be expected. Among such factors are: (1) the drops accelerate over part of their fall distance; (2) convective diffusion may be required to maintain the SO_2 concentration between drops near the bulk gas-phase concentrations; (3) a surrounding gas-phase velocity gradient is induced by the falling drops and a changing SO_2 concentration occurs through this gradient; and (4) the effective area for absorption is not necessarily equal to one-half the geometric surface area, due to wake effects and slight distortion from sphericity.

In our modeling approach the removal of sulfur dioxide by drops is conceptualized as a sequence of fluid spheres which are falling at dynamic (drop) Reynolds numbers in the range of 0.1 to 250 through a gaseous environment.

The generalized as well as the problem-oriented equations are discussed with respect to the submodels of the modeling framework given in Figure 2. For specific solutions, variables such as the ambient temperature gradient, trace gas concentration profile, and humidity distribution are model inputs which would have to be obtained from experiments, existing data bases, or outputs of specific, separate transport models.

Hydroaerodynamic Submodel

The generalized equations representing the dynamics of liquid particles falling through a gaseous semi-infinite environment are the Navier-Stokes equations coupled with a balance of forces (and torques) exerted on the drops. To solve for this two-phase flow field outside and inside the drop is a formidable task (Baboolal et al., 1981). Two simplified sets of modeling equations have been programmed. The first submodel version considers a solitary spherical drop falling through stagnant air, neglecting wake formation, entrainment, drop interaction, changes in drop shape, etc.

$$m\frac{D\vec{v}}{Dt} \approx mv\frac{dv}{dz} = F_{grav} - F_{buoy} - F_{drag} \qquad (1a)$$

For particle Reynolds numbers $Re_p < 300$ drops can be regarded as rigid spheres (Clift et al., 1978), so that

$$\frac{dv}{dz} = \frac{g}{v}\left(1 - \frac{\rho_a}{\rho_f}\right) - \frac{3C_D(v)\, v|v|}{8vr_p}\frac{\rho_a}{\rho_f} \qquad (1b)$$

The drag coefficient $C_D(v)$ is obtained from White (1974):

$$C_D = \frac{24}{Re_p} + \frac{6}{1 + \sqrt{Re_p}} + 0.4 \text{ for } 0 \le Re_p \le 10^5 \qquad (1c)$$

Clearly, such a restricted model is inadequate for the detailed description of the experimental system. The second version for the hydroaerodynamic submodel takes into account a stream of individual drops which form a nonlinear conical (wake) corridor into which ambient air plus trace gases are constantly entrained. Hence the principal variables in addition to the drop velocity are the "radius" of influence (wake corridor) and the wake velocity which possibly reduces the effective drag. The force balance now yields

$$\frac{dv}{dz} = \frac{g}{v}\left(1 - \frac{\rho_a}{\rho_f}\right) + \frac{3C_D\rho_a(\bar{u}_w - v)^2}{8vr_p\rho_f}\frac{(\bar{u}_w - v)}{|\bar{u}_w - v|} \qquad (2a)$$

The second equation is a balance of the change of z-momentum per unit length of the drop stream due

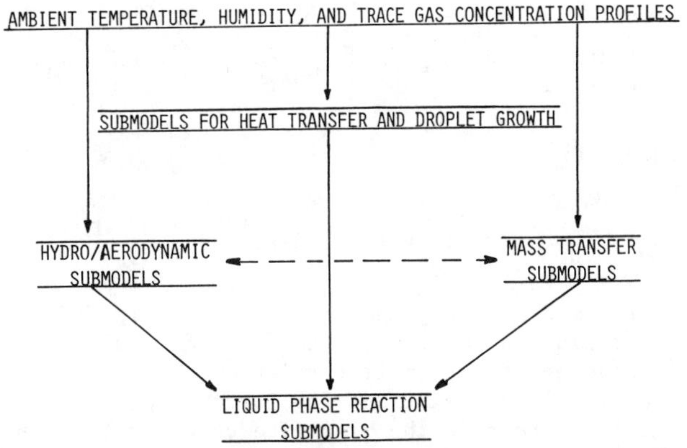

Fig. 2. Submodels of simulator for wet removal of atmospheric trace gases.

to drag and the momentum change in the wake (Arrowsmith and Foster, 1973):

$$\rho_a \frac{\partial}{\partial z} \int_{r=0}^{r} ru[u-u(R)]dr = u\left[r\frac{\partial u}{\partial r}\right]_{r=0}^{r} \quad (2b)$$

For $r \to 0$ the right-hand side of the equation becomes the momentum depletion due to drag, and for $r \to \infty$ the presence of the stream of drops is negligible so that this side is equal to zero. Hence, integration of equation (2b) from r = 0 to r = ∞ delivers

$$\left[r\frac{\partial u}{\partial r}\right]_{r=0}^{r=\infty} = \frac{1}{2\pi}\left[\frac{\dot{m}g}{v}\left(1 - \frac{\rho_a}{\rho_f}\right) - m\frac{dv}{dz}\right] \quad (2c)$$

which is a z-momentum balance on the whole cross section of the air "jet" (Schlichting, 1979). Integration of equation (2a) across the central core, i.e., from r = 0 to r = b, yields an equation for b(z), the radius of half maximum air velocity inside the induced wake cone. The parameter $\dot{m} = m \cdot f$ represents the mass flow rate of drops.

Submodels for Mass Transfer and Chemical Reactions

The generalized equation for mass transfer with chemical reactions is the transient convection-diffusion equation for trace gases with appropriate sink and source terms. Following Barrie (1978) the problem is conceptualized via the double-layer model for the air-drop interface. Ambient SO_2 diffuses through an external gas resistance layer and forms bisulfite instantaneously at the interface

$$SO_2 + H_2O \rightleftarrows HSO_3^- + H^+$$

The bisulfite $(HSO_3^-)^I$ diffuses through the (controlling) liquid film while undergoing some conversion ($SO_4^=$). Hence, the modeling equations are

$$\left\{\begin{matrix}\text{Flux through}\\ \text{gas film}\end{matrix}\right\} = \frac{D_g}{\delta_g}\left[(SO_2)_g - (SO_2)_g^I\right] \quad (3a)$$

where

$$\delta_g = \frac{2r_p}{Sh} = 2r_p\left[1.56 + 0.616\, Re^{1/2} Sc^{1/3}\right]^{-1} \quad (3b)$$

as proposed by Pruppacher and Beard (1971) for a drop radius range of 0.002 to 0.06 cm. Hales (1972) suggested a similar relationship (Frössling's correlation) for large drops.

At the interface the bisulfite concentration is

$$[HSO_3^-]^I = \left\{K_1 K_H [SO_2]_g^I\right\}^{1/2} \quad (3c)$$

with K_1, K_H being equilibrium constants as given by Barrie (1978). For the diffusion of HSO_3^- through the liquid film, which can be taken as $\delta_\ell = 0.1\, r_p$ (Barrie, 1978), the equation by Danckwerts (1970) was employed:

$$\left\{\begin{matrix}\text{absorption rate}\\ \text{of } HSO_3^- \text{ per}\\ \text{unit area}\end{matrix}\right\} = \frac{D_\ell}{\delta_\ell}\left\{[HSO_3^-]^I - \frac{[HSO_3^-]^B}{\cos h\sqrt{M}}\right\}\frac{\sqrt{M}}{\tan h\sqrt{M}} \quad (3d)$$

where $M = \frac{k^1\delta_\ell^2}{D_\ell}$ is the Damkohler number, I = interface, B = bulk, and k^1 can be single or composite rate constant; in the simplest case, $SO_2 + H_2O \rightleftarrows H^+ + HSO_3^-$, the forward rate constant. Most of the HSO_3^- diffuses into the core of the drop. This process is again simulated using Danckwerts' relationship:

$$\left\{\begin{matrix}\text{mass transfer rate}\\ \text{of } HSO_3^- \text{ into bulk}\\ \text{per unit area}\end{matrix}\right\} = \frac{D_\ell k^1}{\sin h\sqrt{M}}\left\{[HSO_3^-]^I - [HSO_3^-]^B \cos h\, \delta_\ell \frac{k^1}{D_\ell}\right\} \quad (3e)$$

Metal-catalyzed conversion of bisulfite (HSO_3^-) to sulfate ($SO_4^=$) may take place simultaneously with the mass transfer process. A first-order rate expression with an empirical constant k[Fe(III)], tabulated by Moller (1980), is used:

$$\frac{d[SO_4^=]}{dt} = k_{ox}[HSO_3^-]$$

$$\text{where } S(IV) \sim HSO_3^- \text{ and } k_{ox} = k[Fe(III)] \quad (3f)$$

Equations (3a) through (3f) are combined to solve for the formation of sulfur species in the liquid phase.

Solution Method and Results

The semi-infinite flow domain has been vertically discretized into n layers so that $n \cdot \Delta z = L$, the total fall height for the given system. Starting from the point of drop releases as z = 0 (initial conditions), the spatial increments Δz can be mapped into time intervals Δt, and vice versa, via the local drop velocity. Assemblage of all submodels into an integrated math model yields a set of coupled ordinary differential equations as well as algebraic equations which are solved simultaneously at each z_i with an explicit marching routine (Euler's method).

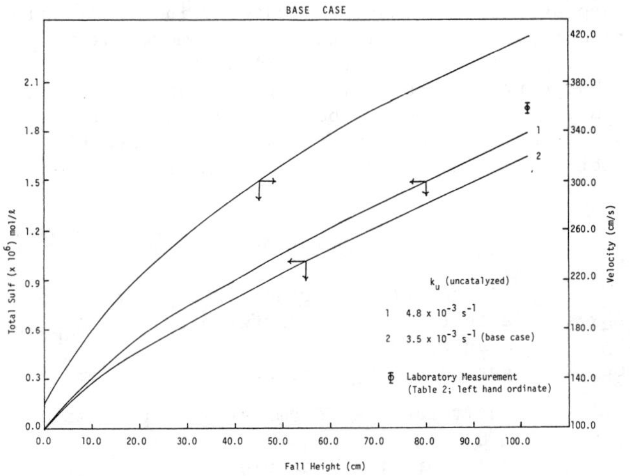

Fig. 3. Prediction of cumulative total sulfur species concentration and drop Reynolds number in test column.

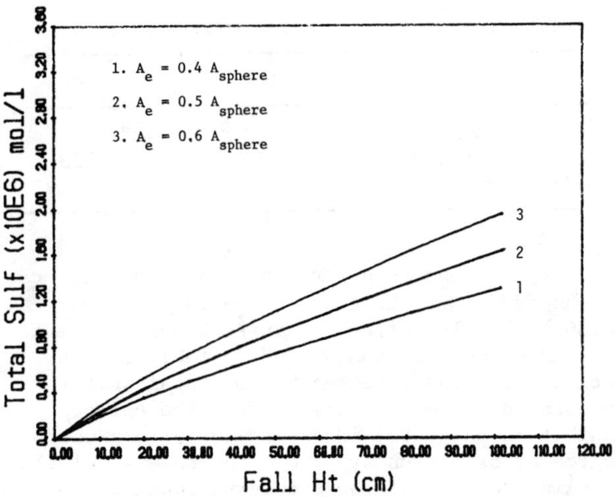

Fig. 5. Computer simulation for various effective mass transfer areas.

The computer simulation model was tested using laboratory data from Altwicker and Chapman (1981) (see Table 2). It can be seen from Figure 3 that the drops do not reach terminal velocity within the relatively short test section. (The calculations in Table 2 were based on initial and terminal velocity of single drops.) The nonlinear accumulation of sulfur species (SO_2, HSO_3^-) in the drops reaches a concentration of 1.626 μmol/l at the end of the fall distance of 100 cm (or exposure time of 0.36 sec). This value compares well with the measured concentration for S_{total} of 1.94 μmol/l (Table 2); it was computed using equations (3a) to (3e). The results of a sensitivity analysis for crucial system parameters are shown in Figures 4 to 6. Although the trends look reasonable, it is evident that changes in the effective mass transfer area A_e and the (rate-controlling) liquid film thickness δ_ℓ are not as remarkable as anticipated. One explanation is that A_e is directly and δ_ℓ is inversely proportional to the total concentration of sulfur species absorbed. Since both parameters are assumed to be only a function of the drop radius r_p, changes in one parameter are to a certain extent counterbalanced by the other. In mathematical terms

$$S_{total} \propto \frac{1}{\delta_\ell},\ A_e \quad \text{or} \quad S_{total} \propto \frac{1}{r_p},\ r_p^2$$

A rate constant of $k_u = 3.5 \times 10^{-3}\ s^{-1}$ for the uncatalyzed oxidation was used in the solution of the base case (Figure 3).

Since it is well known that the oxidation of S(IV) species in solution can be catalyzed and inhibited, we have commenced the investigation of

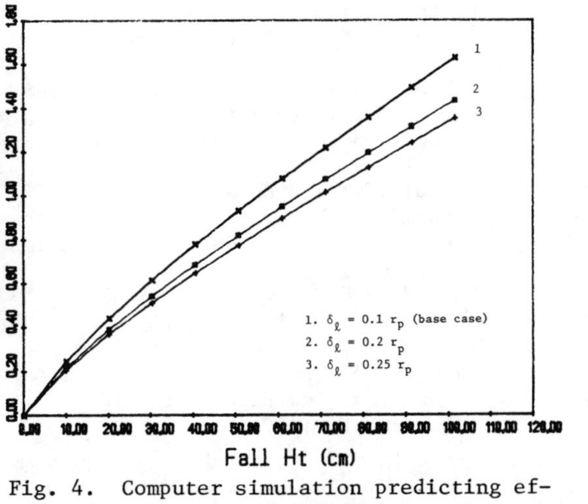

Fig. 4. Computer simulation predicting effects of various liquid film thicknesses.

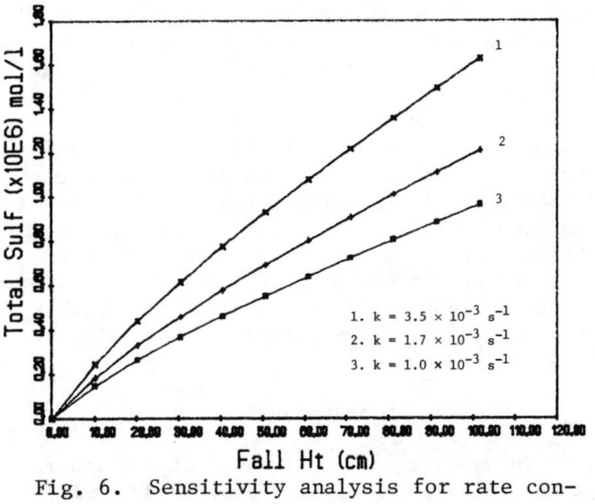

Fig. 6. Sensitivity analysis for rate constant of uncatalyzed reaction.

TABLE 5. Fe^{+3}-Catalyzed Oxidation

Species	Experimental	Model
$[Fe^{+3}]$, mol/l	1×10^{-5}	1×10^{-5}
Total S, μmol/l	4.17	2.3
S(IV)/total S	0.32	0.19

these conditions to determine if such effects can be observed over the time period in question (0.36 sec). The effect of three transition metal cations present is summarized in Table 3. In each case, a large percentage of total sulfur appears in the +6 valency state. The effects of inhibitors and metal ions plus inhibitors on the degree of oxidation are shown in Table 4. The modeling results of these observations will be very much a function of the numerical values adopted for particular parameters (Table 5). Thus, better agreement between experiment and model could have been obtained if a value substantially greater than 3.5×10^{-3} sec^{-1} had been chosen for k_{ox}. The experimental results reported here and previously (Altwicker and Chapman, 1981) as well as recent analyses of earlier experimental findings (Freiberg and Schwartz, 1981) indicate that instantaneous values up to 1 s^{-1} are not unreasonable for the rate coefficient. The catalyzed experiments begin with catalyst S(IV) ratio in the drop near infinity at the time of drop formation, and at the termination of the experiment the ratio is still substantially greater than is used in the "typical" bulk experiments that do not incorporate mass transfer (Martin et al., 1981). It is possible, therefore, that complexation of S(IV) species at the liquid interface may play a role in the mechanism of transport into the bulk and oxidation in both the film and the bulk. Experimental results with Mn (Table 3) may possibly reflect the presence of such an equilibrium reaction, i.e., S(IV) + $Mn^{+2} \rightleftarrows$ complex, and the decrease in percent oxidized with increasing Mn concentration may reflect the rate of oxidation proceeding through the complex. Recent evidence (Stauff and Jaeschke, 1978) suggests the possible role of such complexes in the oxidation.

As was mentioned earlier, the complete inhibition of the "uncatalyzed" and catalyzed reactions by free radical inhibitors (Table 4) is not really consistent with induction periods ranging from several seconds to several minutes. Therefore no single mechanism can be postulated. The presence of unmeasured impurities cannot be ruled out absolutely; further experimentation clearly is required.

Concluding Remarks

The fundamental attractiveness of the laboratory system is its potential to generate drop streams of varying drop velocities, numbers, and sizes encompassing the range encountered in clouds and raindrops. Our aim is to give a comprehensive description of the system to justify its use as a laboratory tool for initial rate studies and the study of other gaseous pollutants, especially pollutant mixtures, and to extend the investigations to simultaneous gas and particle scavenging.

Notation

A	1/2 geometric surface area of drop
A_e	effective mass transfer area, 0.5 A_{sphere}
C_D	drag coefficient
C_{SO_2}	gas-phase SO_2 concentration, mol/cm^3
d	equivalent jet diameter = drop diameter, D
D	diameter of the droplet
D_g	diffusivity of SO_2 in air (0.12 $cm^2\ s^{-1}$)
D_ℓ	diffusivity of HSO_3^- in liquid
F	force component in vertical direction
f	drop frequency
g	acceleration due to gravity
h	equivalent jet length = $\frac{4V_\ell t}{\pi D^2}$, cm
k	$2A^*\left(\frac{D_\ell}{\pi t}\right)^{0.5}\alpha^{0.5}$; $A^* = \Phi_i$ was assumed (see Table 2)
k_g	mass transfer coefficient
K_1, K_H	equilibrium constants (Barrie, 1978)
L	fall height
m	mass of the drop
$\dot{m}$	mass flow rate of stream of drops, m·f
$M_t(g)$	SO_2 transferred after time, t, μmol
$\frac{\Delta M}{\Delta T}$	change in moles per unit time between initiation of experiment and quenching of drops
q	total rate of absorption = $\frac{\pi DhQ(t)}{t}$
$Q(t)$	amount absorbed/unit surface area during t
r	b, the radius of half maximum air velocity inside the induced wake cone
R	specified radius of the air corridor; hence, u(R) is a kind of reference velocity in this (air) momentum balance, whereas u = r(r) is the local variable
r_p	radius of the droplet
Re	Reynolds number v D/ν_a; Re_i = 65 and Re_t = 200 were used in the calculation (Table 2)
Sc	Schmidt number, ν_a/D_g; $Sc(SO_2)$ = 1.19
t	residence time in reactor
u_w	velocity of air in wake corridor
v	velocity of droplet in vertical direction
V_ℓ	volumetric flow rate of liquid
α	enhancement coefficient

$$= \left(1 + \frac{[HSO_3^-]}{[SO_2]}\frac{K_1}{[H^+]} + \frac{[SO_3^=]}{[SO_2]}\frac{K_1K_2}{[H^+]^2}\right)$$

$$\left(1 + \frac{K_1}{[H^+]} + \frac{K_1 K_2}{[H^+]^2}\right)^{-1}$$

δ film or boundary-layer thickness

δ_ℓ liquid film thickness, taken as 10% of drop radius in Table 2 calculations

ρ density

ν kinematic velocity

Acknowledgement. The authors would like to thank E. Chapman, S. Ramachandran, I. Gujral, and K. Nass for valuable and thorough assistance and the National Science Foundation for partial support.

References

Adamovicz, R. F., and F. B. Hill, A model for the reversible washout of sulfur dioxide, ammonia, and carbon dioxide from a polluted atmosphere, Atmos. Environ., 11, 917-927, 1977.

Altwicker, E. R., and T. Sekulic, Inhibition of sulfur dioxide oxidation during absorption, Environ. Lett., 7, 125, 1974.

Altwicker, E. R., Absorption, oxidation and oxidation/inhibition of sulfur dioxide in aqueous solution, DECHEMA-Monographien, 80, Part 2, 343-363, 1976.

Altwicker, E. R., Oxidation/inhibition of sulfite ion in aqueous solution, AIChE Symposium 188, Vol. 75, 145-150, 1979.

Altwicker, E. R., Oxidation and oxidation/inhibition of sulfur dioxide, Adv. Environ. Sci. Engr., 3, 80-91, 1980.

Altwicker, E. R., and E. Chapman, Mass transfer of sulfur dioxide into falling drops: A comparison of experimental data with absorption models, Atmos. Environ., 15, 297-300, 1981.

Arrowsmith, A., and P. J. Foster, The motion of a stream of monosized liquid drops in air, Chem. Engr. J., 5, 243-250, 1973.

Baboolal, L. B., H. R. Pruppacher, and J. H. Topalian, A sensitivity study of a theoretical model of SO_2 scavenging by water drops in air, J. Atmos. Sci., 38, 856-870, 1981.

Barrie, L. A., An improved model of reversible SO_2 washout by rain, Atmos. Environ., 12, 407-412, 1978.

Barrie, L. A., and H. W. Georgii, An experimental investigation of the absorption of sulfur dioxide by water drops containing heavy metal ions, Atmos. Environ., 10, 743-749, 1976.

Beilke, S., and H. W. Georgii, Investigation on the incorporation of sulfur dioxide into fog and rain droplets, Tellus, 20, 435-441, 1968.

Beilke, S., and G. Gravenhorst, Heterogeneous SO_2 oxidation in the droplet phase, Atmos. Environ., 12, 231-239, 1978.

Beilke, S., D. Lamb, and J. Miller, On the uncatalyzed oxidation of atmospheric SO_2 by oxygen in aqueous systems, Atmos. Environ., 9, 1083-1090, 1975.

Carmichael, G. R., and L. K. Peters, Some aspects of SO_2 absorption by water-generalized treatment, Atmos. Environ., 13, 1505-1515, 1979.

Chapman, E., A laboratory investigation of sulfur dioxide removal by falling water drops, M.E. Thesis, Rensselaer Polytechnic Institute, Troy, N.Y., 1980.

Clift, R., H. R. Grace, and M. E. Weber, Bubbles, Drops and Particles, Academic Press, New York, 1978.

Dabora, E. K., Production of monodisperse sprays, Rev. Sci. Instr., 38, 502-506, 1967.

Dana, M. T., J. M. Hales, and M. A. Wolf, Rain scavenging of SO_2 and sulfate from power plant plumes, J. Geophys. Res., 80, 4119-4129, 1975.

Danckwerts, P. B., Gas-Liquid Reactions, McGraw-Hill, New York, 1970.

Freiberg, E., and S. E. Schwartz, Oxidation of SO_2 in aqueous droplets: Mass transport limitation in laboratory studies and the ambient atmosphere, Atmos. Environ., 15, 1145-1154, 1981.

Garner, F. H., and J. J. Lane, Mass transfer to drops of liquid suspended in a gas stream, Trans. Inst. Chem. Engrs., 37, 162-172, 1959.

Hales, J. M., Fundamentals of the theory of gas scavenging by rain, Atmos. Environ., 6, 635-659, 1972.

Hales, J. M., Wet removal of sulfur compounds from the atmosphere, Atmos. Environ., 12, 389-399, 1978.

Hikita, H., S. Asai, and H. Nose, Absorption of sulfur dioxide into water, AIChE J., 24, 147-149, 1978.

Husar, R. B., J. P. Lodge, and D. J. Moore, Sulfur in the atmosphere, Proceedings of the International Symposium held in Dubrovnik, Yugoslavia, September 7-14, Pergamon Press Ltd., N.Y. and London, 1977.

Johnstone, H. F., and G. C. Williams, Absorption of gases by liquid droplets, Ind. Eng. Chem., 31, 993, 1939.

Martin, L. R., D. E. Damschen, and H. S. Judeikis, Sulfur dioxide oxidation reactions in aqueous solutions, EPA 600/7-81-085, Environmental Protection Agency, Washington, DC, May 1981.

Matteson, M. J., and P. J. Giardina, Mass transfer of SO_2 to growing droplets: Role of surface electric properties, Environ. Sci. Tech., 8, 50-55, 1974.

Miller, J. M., and R. G. dePena, Contribution of scavenging sulfur dioxide to the sulfate content of rain water, J. Geophys. Res., 77, 5805-5916, 1972.

Moeller, D., Kinetic model of atmosphere SO_2 oxidation based on published data, Atmos. Environ., 14, 1067-1076, 1980.

Overton, J. H., V. P. Aneja, and J. L. Durham, Production of sulfate in rain and raindrops in polluted atmospheres, Atmos. Environ., 13, 355-367, 1979.

Pruppacher, H. R., and K. V. Beard, A wind tunnel investigation of the rate of evaporation of small water drops falling at terminal velocity, J. Atmos. Sci., 28, 1455-1464, 1971.

Rajagopalan, R., and C. Tien, Production of

monodispersed drops by forced vibration of a liquid jet, Calif. J. Chem. Engrg., 51, 272-279, 1973.

Roberts, D. L., and S. K. Friedlander, Sulfur dioxide transport through aqueous solutions: Part I. Theory, AIChE J., 26, 593-610, 1980.

Scarlett, B., and C. S. Parkins, Droplet production by controlled jet break-up, Chem. Engr. J., 13, 127-141, 1977.

Schlichting, H., Boundary Layer Theory, McGraw-Hill, New York, 1979.

Schroeter, L. L., Sulfur Dioxide, Pergamon Press, N.Y. and London, 1966.

Stauff, J., and W. Jaeschke, Chemilumineszenz der SO_2-Oxidation, Z. Naturforsch, 33b, 293-299, 1978.

Terraglio, F. P., and R. M. Manganelli, The absorption of atmospheric sulfur dioxide by water solutions, J. Air Poll. Control Assoc., 17, 403-406, 1967.

Walcek, C., P. K. Wang, J. H. Topalian, S. K. Mitra, and H. R. Pruppacher, An experimental test of a theoretical model to determine the rate at which freely falling water drops scavenge SO_2 in air, J. Atmos. Sci., 38, 871-876, 1981.

White, F. M., Viscous Fluid Flow, McGraw-Hill, New York, 1974.

SULFUR-DIOXIDE/WATER EQUILIBRIA BETWEEN 0° AND 50°C. AN EXAMINATION OF DATA AT LOW CONCENTRATIONS

Howard G. Maahs

National Aeronautics and Space Administration, Langley Research Center, Hampton, Virginia 23665

Abstract. A discrepancy exists in the values currently in use for the first dissociation constant of SO_2 in water. To resolve this discrepancy, experimental data reported in the literature have been examined, and the correlation

$$\log K_1 = \frac{853.0}{T} - 4.740, \quad K_1 \pm 0.0009 \text{ M}$$

for the first thermodynamic dissociation constant is proposed between 0°C and 50°C. At 25°C, K_1 = 0.0132 M. Data for the second thermodynamic dissociation constant have also been examined, and the correlation

$$\log K_2 = \frac{621.9}{T} - 9.278, \quad K_2 \pm 0.38 \times 10^{-8} \text{ M}$$

is proposed. At 25°C, $K_2 = 6.42 \times 10^{-8}$ M. For Henry's law constant, the correlation

$$\log K_H = \frac{1376.1}{T} - 4.521, \quad K_H \pm 0.028 \text{ M/atm}$$

is recommended between 0°C and 50°C, and for SO_2 partial pressures from about 2×10^{-4} to 1.3 atm. At 25°C, K_H = 1.242 M/atm. At partial pressures below about 2×10^{-4} atm, the proper value for K_H is uncertain.

Introduction

In studies involving SO_2 dissolved in cloud water or in other aqueous systems of SO_2 and its ionic forms, accurate information concerning the vapor-liquid equilibria of SO_2 with its aqueous solution and the dissociation of SO_2 in solution is often important. For example, an accurate estimate of the concentration of total sulfur(IV) in equilibrium solution in cloud droplets requires accurate values for both Henry's law constant K_H and the first dissociation constant of sulfur dioxide K_1. In particular, for a typical atmo sphere containing 325 ppm CO_2 and between 1 ppb and 10 ppm SO_2 it can be shown that, to a good approximation, the equilibrium concentration of total sulfur(IV) in solution is

$$[S(IV)]_t \sim \left(K_1 K_H P_{SO_2}\right)^{1/2} \qquad (1)$$

where P_{SO_2} is the SO_2 partial pressure in atm. But an uncertainty exists regarding the best value for K_1, and a variety of values appear in common use. At 25°C, these range from 0.0127 M to 0.0174 M, with typical values usually lying near either extreme. Along with this is the curious inconsistency that there is but one value in common use for the thermodynamically related Henry's law constant (1.23 M/atm). To resolve these inconsistencies, the pertinent literature reporting experimental determinations of K_1 and K_H has been critically examined, and preferred values are recommended for both. Experimental values reported for the second dissociation constant K_2 are also examined, and a preferred value is recommended. Correlations of the best data for all three equilibrium constants are developed over the temperature range 0°C to 50°C. Also, ionic activity corrections required in more concentrated solutions are briefly discussed, and recommendations are made for estimating the mean activity coefficients of the ionic species in solution.

Background

The first dissociation constant of SO_2 in water

$$SO_2\ (aq) \overset{K_1}{\rightleftarrows} H^+ + HSO_3^-$$

is defined as

$$K_1 = \frac{a_{H^+}\, a_{HSO_3^-}}{a_u} = \frac{[H^+][HSO_3^-]}{[S(IV)]_u} \frac{\gamma_{H^+}\, \gamma_{HSO_3^-}}{\gamma_u}$$

$$= K_{1c} \frac{\gamma_{H^+}\, \gamma_{HSO_3^-}}{\gamma_u} = K_{1c}\, \gamma^2_{HHSO_3} \qquad (2)$$

where a_i is the activity of species i, [i] is the solution concentration of species expressed either as c_i mol/l (M, molarity) or as m_i mol/kg H_2O (*M*, molality), γ_i is the individual activity coefficient for species i in solution, γ_{HHSO_3} is the mean activity coefficient for the $H^+HSO_3^-$ ionic pair, and K_{1c} is a concentration-based dissociation constant. The symbol u refers to that portion of valence-(IV) sulfur, S(IV), which is not ionized in solution, and includes all undissociated SO_2 in solution, whether unhydrated or hydrated as either $SO_2 \cdot nH_2O$ or molecular H_2SO_3. Although an impressive number of studies have failed to reveal any direct evidence supporting the existence of H_2SO_3 (Falk and Giguère, 1958; Jones and McLaren, 1958; Davis and Chatterjee, 1975; Wang and Himmelblau, 1964), some studies argue for the existence of H_2SO_3 based on indirect evidence (Flis et al., 1965; Guthrie, 1979). Since the activity coefficient for a nonionized species in reasonably dilute solution is virtually unity, γ_u has been set equal to 1.

The second dissociation constant of SO_2 in water

$$HSO_3^- \overset{K_2}{\rightleftarrows} H^+ + SO_3^=$$

is similarly defined as

$$K_2 = \frac{a_H + a_{SO_3^=}}{a_{HSO_3^-}} = \frac{[H^+][SO_3^=]}{[HSO_3^-]}\,\frac{\gamma_H + \gamma_{SO_3^=}}{\gamma_{HSO_3^-}}$$

$$= K_{2c}\,\frac{\gamma_H + \gamma_{SO_3^=}}{\gamma_{HSO_3^-}} \qquad (3)$$

In equations (2) and (3), the K's are true thermodynamic dissociation constants, whereas the K_c's, by their very definitions, are concentration-dependent. The K's have units of concentration, although for most practical purposes they are numerically the same whether in molar or molal units. Clearly, however, the units for the K_c's and the concentration base for the γ_i's must be specified, particularly in more concentrated solutions where molarity and molality differ appreciably.

Henry's law, a quantitative relationship between the solubility of a gas and its pressure at equilibrium, has been expressed in several (mutually exclusive) ways (Gerrard, 1972; Gerrard, 1980; Glasstone, 1946). Of these, particularly useful for presenting SO_2 solubility data in water is

$$K_H = \frac{[S(IV)]_u}{p_{SO_2}} \qquad (4)$$

where p_{SO_2} is the partial pressure of SO_2 in equilibrium with the aqueous solution. (Strictly speaking, $SO_2(g)$ in the vapor phase should be in Henry's law equilibrium only with dissolved $SO_2(aq)$, not total undissociated $S(IV)_u$. However, if H_2SO_3 exists and is in equilibrium with $SO_2(aq)$ (as Flis et al. (1965) propose), then it follows that equation (4) is still a valid definition for K_H.) Common practice is to express $[S(IV)]_u$ in equation (4) in units of molality, although molarity is also used. This distinction can often be ignored, however, the difference being less than 1 percent even at concentrations $[S(IV)]_u$ as high as 0.1 *M* (Johnstone and Leppla, 1934; Morgan and Maass, 1931). Typically, Henry's law is obeyed only in dilute solution; however, the concentration range over which it is actually obeyed depends on the particular solute and solvent.

The First Dissociation Constant

From the foregoing discussion it is clear that K_1 is a true dissociation constant which, unlike K_{1c}, is applicable at all solution concentrations. Some of the confusion in the literature regarding the proper value for the dissociation constant has apparently arisen because this distinction has not always been clearly made. For example, Campbell and Maass (1930) and Morgan and Maass (1931), ignoring activity coefficient corrections, derived K_{1c} = 0.0174 M at 25°C from their extensive conductivity data on solutions of SO_2 in water, but implied their result to be a thermodynamic dissociation constant. Later, their data were reanalyzed by Johnstone and Leppla (1934) who obtained K_1 = 0.013 M by including corrections for ionic activity.

Many experimental determinations of both K_1 and K_{1c} are reported in the literature. Compilations of the earlier literature are given by Plummer (1950) and Sillén and Martell (1964). To avoid the problem of dealing with a concentration-dependent dissociation constant, in the present paper we concentrate solely on measurements of K_1. Deviations from thermodynamic ideality encountered in nonideal solutions of moderate ionic strength can be treated in terms of the mean activity coefficient γ_{HHSO_3}, discussed briefly later. Values for K_1 obtained from the literature are shown in Figure 1 plotted as pK_1 (= $-\log K_1$) as a function of 1/T. The experimental methods associated with these determinations are summarized in Table 1.

The K_1 value of Sherrill and Noyes (1926) is from their analysis of the conductivity data of Kerp and Baur (1907) and Lindner (1912); the values of Johnstone and Leppla (1934) are based on the conductivity data of Maass and coworkers (Campbell and Maass, 1930; Morgan and Maass, 1931). In both of these studies the mean activity coefficient had to be estimated, there being no experimental data available. The three values for K_1 reported by Yui (1940) are from three separate experiments, all based on pH measurements in solution, with γ_{HHSO_3} being calculated by the method of Kielland (1937).

Tartar and Garretson's (1941) value is from

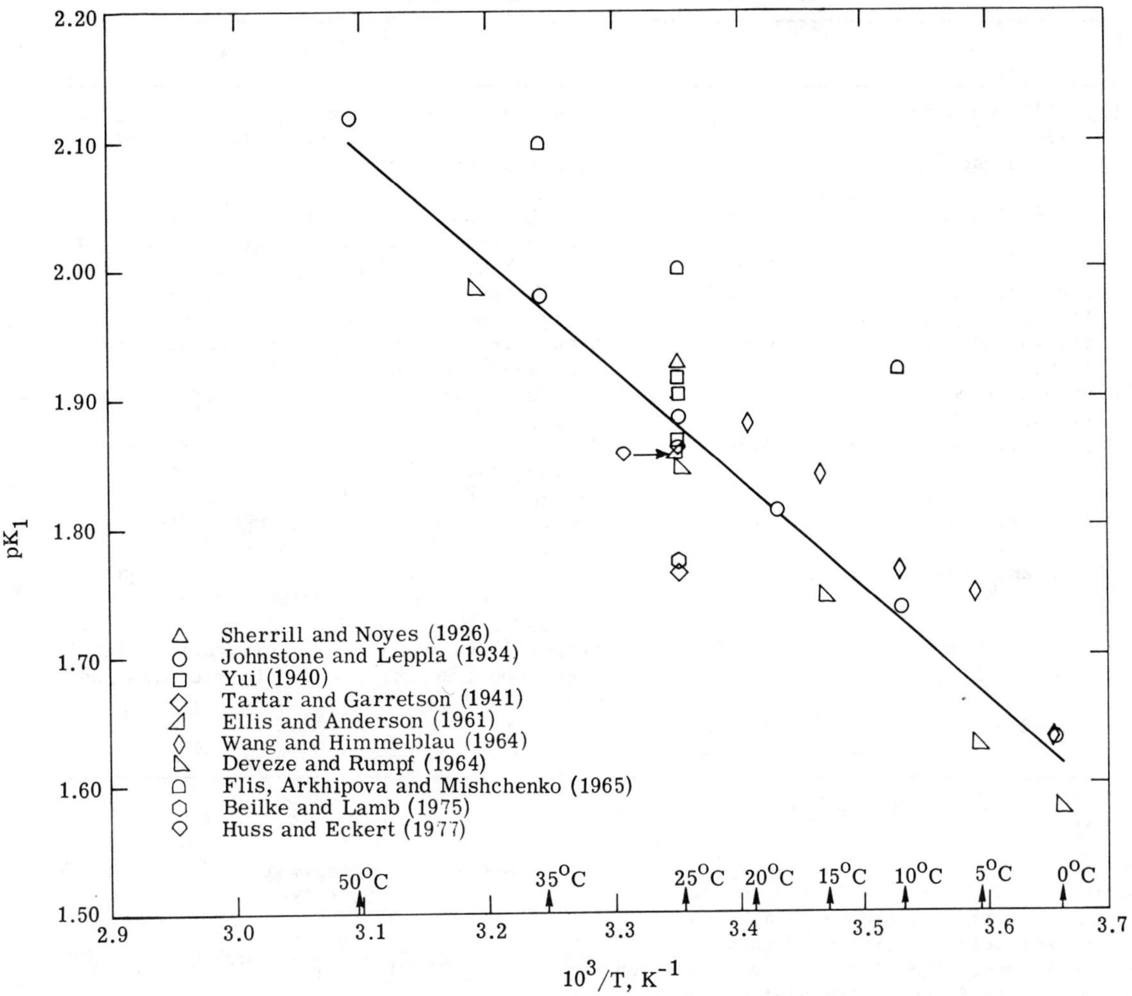

Fig. 1. The first dissociation constant of sulfur dioxide in aqueous solution.

measurements in electrochemical cells. To derive their value, they employed an iterative technique assuming a working value for K_1 and making estimates of the activity coefficients. Their final value for K_1 was obtained by extrapolation to zero ionic strength. The sensitivity of their final value to their initial working value was examined by repeating their calculations starting with a different working value. Since this leads to a diverging solution, it is concluded that their method of analysis does produce a unique value for K_1.

The value of K_1 reported by Ellis and Anderson (1961) is from their conductance measurements of SO_2 solutions in water, using estimates of γ_{HHSO_3} calculated from the Debye-Hückel equation including correction for ionic size. Wang and Himmelblau's (1964) values derive from a kinetic study of the reaction

$$SO_2 + H_2O \rightleftharpoons H^+ + HSO_3^-$$

and not from a direct determination of the dissociation constant; also, they involve large activity corrections because the measurements were made at relatively high solution concentrations ($[HSO_3^-]$ as high as 0.8 M). Because of this and the inherent difficulties in making precise measurements in kinetic studies, their values may not be as accurate as those based on more direct (equilibrium) methods at lower concentrations. Deveze and Rumpf (1964) employed a spectrophotometric method to study solutions of SO_2 in various acids. Activity corrections were made depending on the nature of the particular acid solutions.

The values from the work of Flis et al. (1965) are based on material balance differences of fairly close concentrations determined spectrophotometrically and iodometrically, and are possibly not as precise as those obtained by other methods not involving small differences. The value of Beilke and Lamb (1975) is based on an assumed correct value for K_H of 1.24 M/atm along with pH measurements to obtain $[H^+]$ and $[HSO_3^-]$.

TABLE 1. The First Dissociation Constant

Source	K_1 at 25°C, M	Method
Sherrill and Noyes (1926)	0.012	From conductivity data; γ_{H^+} taken as γ_{HCl}, $\gamma_{HSO_3^-}$ from Debye-Hückel limiting law.
Johnstone and Leppla (1934)	0.013	From conductivity data; γ_{HHSO_3} taken as γ_{HCl}.
Yui (1940)	0.0136 0.0125 0.0121	Titration and pH measurements (three separate experiments); γ_{HHSO_3} calculated from Kielland (1937).
Tartar and Garretson (1941)	0.0172	Electrochemical cells. Extrapolation to zero ionic strength.
Ellis and Anderson (1961)	0.0139	Conductivity data; γ_{HHSO_3} calculated from Debye-Hückel equation.
Wang and Himmelblau (1964)	0.012[a]	From a kinetic study of HSO_3^- formation, γ_{HHSO_3} estimated several ways.
Deveze and Rumpf (1964)	0.0143	Spectrophotometric method in several acid solutions; γ_{H^+} taken as that of the acid, $\gamma_{HSO_3^-}$ taken as γ_{H^+}.
Flis et al. (1965)	0.010	Spectrophotometric study; γ_{HHSO_3} from Debye-Hückel limiting law.
Beilke and Lamb (1975)	0.0171	From a kinetic study of HSO_3^- formation; based on an assumed value for K_H = 1.24 M/atm.
Huss and Eckert (1977)	0.0139 0.0138	Conductivity study with minimization technique. Spectrophotometric study with minimization technique.
Adjusted average	0.0132 ±0.0009	See text

[a]From extrapolation of their data.

It is not stated whether corrections were made for ionic activity. However, examination of the individual data reveals a decrease in K_1 with decreasing concentration. Extrapolation of these data to zero concentration (plotted as a function of the square root of ionic strength) yields the much lower value for K_1 of about 0.0154 M. Huss and Eckert's (1977) values are from both conductivity and spectrophotometric measurements, with activity coefficients obtained by a minimization technique.

On the basis of the preceding discussion, the values from Wang and Himmelblau (1964), Flis et al. (1965), and Beilke and Lamb (1975) are considered somewhat less certain than the remainder. Of these, the value reported by Tartar and Garretson (1941) is seen to differ substantially from the others; the reason for this is not known. It cannot be attributed to uncertainties in the activity coefficients because Tartar and Garretson's value is based on extrapolation to zero ionic strength. Omitting this one extreme value, however, and taking as the best estimate for K_1 the average of the nine remaining values, we get K_1 = 0.0132 M at 25°C. The temperature dependence for K_1 may reasonably be taken to be given by the data of Johnstone and Leppla (1934) and Deveze and Rumpf (1964) as shown in Figure 1. Taking the best slope to be the mean for these two data sets and forcing the curve through 0.0132 M at 25°C, a reasonable representation of the data for K_1 between 0°C and 50°C is

$$\log K_1 = \frac{853.0}{T} - 4.740, \quad K_1 \pm 0.0009 \text{ M} \qquad (5)$$

This line is shown on Figure 1.

The Second Dissociation Constant

In cloud water at the lower pH levels and up to about 5.5 or 6, $[SO_3^=]$ can justifiably be neglected relative to $[HSO_3^-]$. At higher pH's, however, the contribution of $[SO_3^=]$ to $[S(IV)]_T$ becomes increasingly important, necessitating an accurate value for the second dissociation constant K_2. Experimental values for K_2 obtained from the literature are listed in Table 2. With the exception of the value of Salomaa et al. (1969), agreement is very good. The temperature dependence of K_2 can be obtained from the data of Hayon et al. (1972), who report values of K_2 for several additional temperatures over the range 5° to 50°C. Using this temperature dependence and the average value for K_2 at 25°C from Table 2, a reasonable representation of the data is

$$\log K_2 = \frac{621.9}{T} - 9.278, \quad K_2 \pm 0.38 \times 10^{-8} \text{ M} \qquad (6)$$

TABLE 2. The Second Dissociation Constant

Source	K_2 at 25°C, M	Method
Yui (1940)	6.31×10^{-8} 6.16×10^{-8}	Titration and pH measurements (two separate experiments); γ_i calculated from Kielland (1937).
Tartar and Garretson (1941)	6.24×10^{-8}	Electrochemical cells. Extrapolation to zero ionic strength.
Cuta et al. (1957)	7.1×10^{-8}	pH and spectrophotometric measurements. Extrapolation to zero ionic strength.
Salomaa et al. (1969)	3.63×10^{-8}	pH measurements and titration in light and heavy water.
Hayon et al. (1972)	6.31×10^{-8}	Spectrophotometric study; γ_i calculated from Güntelberg approximation.
Adjusted average	6.42×10^{-8} $\pm 0.38 \times 10^{-8}$	The value of Salomaa et al. has been omitted.

Henry's Law Constant

Experimental determination of Henry's law constant K_H (equation (4)) requires the concentration of the unionized solute to be known. This concentration can be obtained either from the degree of dissociation obtained from conductivity measurements, or from the species balance

$$[S(IV)]_u = [S(IV)]_t - [HSO_3^-] = [S(IV)]_t - \left(\frac{[S(IV)]_u K_1}{\gamma_H + \gamma_{HSO_3^-}}\right)^{1/2} \quad (7)$$

where $[S(IV)]_t$ is the concentration of total sulfur(IV) in solution, and where use is made of the fact that $[SO_3^=] \ll [HSO_3^-]$. Thus, to a very good approximation, $[H^+] = [HSO_3^-]$. Since K_H and K_1 are thusly related, the value for K_H derived from any set of experimental data depends on the value selected for K_1.

A large number of measurements of equilibrium solubility of SO_2 in water are reported in the literature. Most of these are for relatively high partial pressures of SO_2, high temperatures, or are in solutions containing a variety of added ionic species. However, three studies reporting experimental measurements for gas-phase SO_2 con-

TABLE 3. Henry's Law Constant

Source	Average K_H at 25°C, M/atm or *M*/atm[a]	Based on K_1 at 25°C, M	Partial Pressure of SO_2, atm: Low	High
Hales and Sutter (1973)	1.39 (M)[b]	0.0132	4.4×10^{-8}[c]	2.6×10^{-6}[c]
Terraglio and Manganelli (1967)	1.07 (M)[d]	0.017	0.31×10^{-6}	3.3×10^{-6}
Recalculation	1.39 (M)[d]	0.0132	"	"
Beilke and Lamb (1975)	1.61 (M)[e]	0.0132	0.52×10^{-6}	12.6×10^{-6}
Johnstone and Leppla (1934)	1.23 (M)	0.013	2.6×10^{-4}	1.3
Saito and Yui (1959)	1.26 (*M*)[f]	0.0125	2.6×10^{-4}	0.15
Flis et al. (1965); Arkhipova et al. (1968)	1.40 (M)	0.010[g]	2.6×10^{-3}	9.2×10^{-3}
Recalculation	1.25 (M)	0.0132	"	"
Vosolsobe et al. (1965)	1.27 (*M*)[f]	0.013	0.02	0.17
Rabe and Harris (1963)	1.20 (*M*)[f]	0.013	0.05	1.0
Adjusted average (see text)	1.242 ±0.028			

[a] M (mol/l) or *M* (mol/kg H_2O). A distinction is not necessary at low concentrations.
[b] Calculated from their data for solution in pure water.
[c] The entries in column 3 of Hales and Sutter's Table 1 should be multiplied by 10^{-6}.
[d] Within the temperature range 23° to 27°C.
[e] Calculated from data given.
[f] Calculated from correlating expression.
[g] Implicit in their data, though not explicitly used.

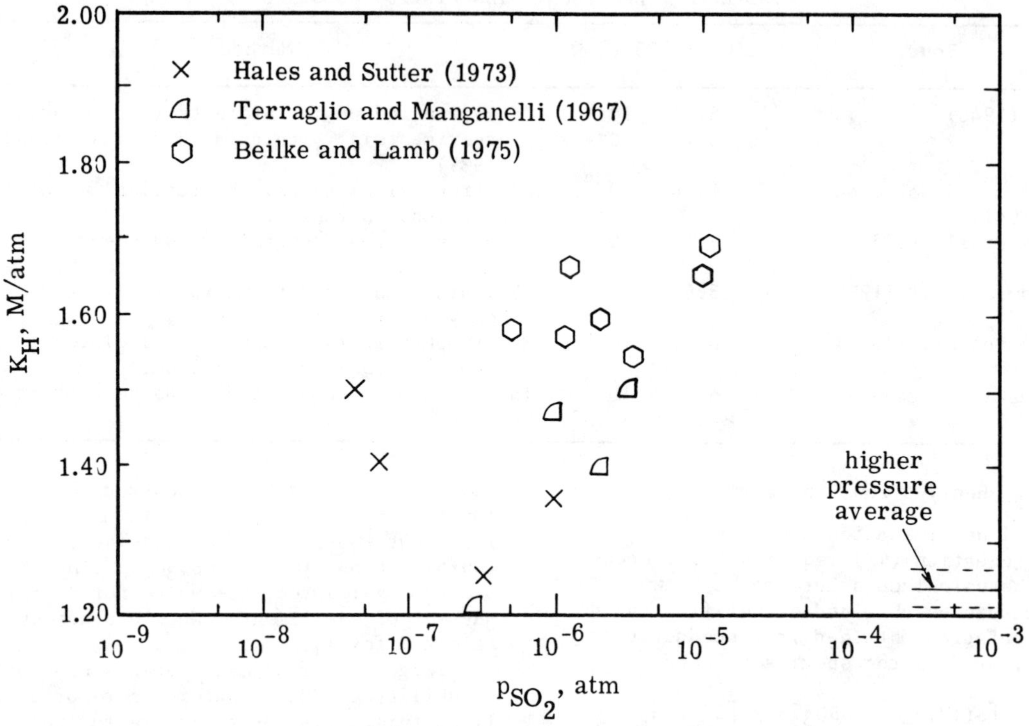

Fig. 2. Dependence of Henry's law constant on SO_2 partial pressure, at 25°C.

centrations below about 10 ppm have been found. These are listed in Table 3 along with several other experimental studies at higher partial pressures. The K_H values for these latter studies all include appropriate corrections for ionic activity. The K_H values reported by Terraglio and Manganelli (1967), Arkhipova et al. (1968), and Flis et al. (1965) are based on selected values for K_1 of 0.017 M and 0.010 M, respectively, and have been recalculated using the recommended value for K_1 of 0.0132 M. For the data of Terraglio and Manganelli, the resulting large increase in K_H occasioned by this recalculation (from 1.07 M/atm to 1.39 M/atm) is a result of the fact that in dilute solution the degree of ionization approaches unity, and the value of K_H becomes very sensitive to the value used for K_1. This illustrates that in studies at low concentrations, use of an accurate value of K_1 is particularly important. Beilke and Lamb (1975) do not specifically report an experimental value for K_H, but one has been calculated from their data using $K_1 = 0.0132$ M.

The K_H values in Table 3 based on low SO_2 partial pressures (the first three data sources) are not values from single determinations, but represent averages of several determinations over a range of partial pressures. If the individual data from these three studies are plotted as a function of partial pressure (see Figure 2), there is observed to be a dependence of K_H on partial pressure. This dependence is most clear in the data of Hales and Sutter (1973) and Terraglio and Manganelli (1967). In Figure 2, an average value for K_H from the five higher pressure data sources (discussed later) is shown for comparison. The lack of a clear and consistent trend of the low pressure data makes their interpretation difficult. **The trend of Hales and Sutter's data possibly suggests a quite high value for K_H at $p_{SO_2} = 10^{-9}$ atm, but this is not certain. What happens between 10^{-5} atm and 2.6×10^{-4} atm is likewise uncertain. The reason that the lower** pressure data are, on the whole, higher than the higher pressure data is also not known. But the fact that they are higher, and by as much as 30 to 40 percent, is cause for concern since the partial pressure of SO_2 in tropospheric cloud systems is very low, on the order of 10^{-9} atm. Clearly, additional studies are needed to resolve these ambiguities.

The variation of average K_H with temperature for all data sources in Table 3 is shown in Figure 3 as a function of 1/T. The curves are plots of data correlations developed by the various authors. Each curve has been drawn to span that temperature range for which supporting experimental data exist. The close agreement among the several correlations and data (corrected to $K_1 = 0.0132$ M) for the five higher pressure data sources ($p_{SO_2} > 2.6 \times 10^{-4}$ atm) is striking. At 25°C, an average K_H for these data is 1.242 M/atm, and a reasonable straight-line fit through this value over the temperature range 0°C to 50° is given by

$$\log K_H = \frac{1376.1}{T} - 4.521, \quad K_H \pm 0.028 \text{ M/atm} \qquad (8)$$

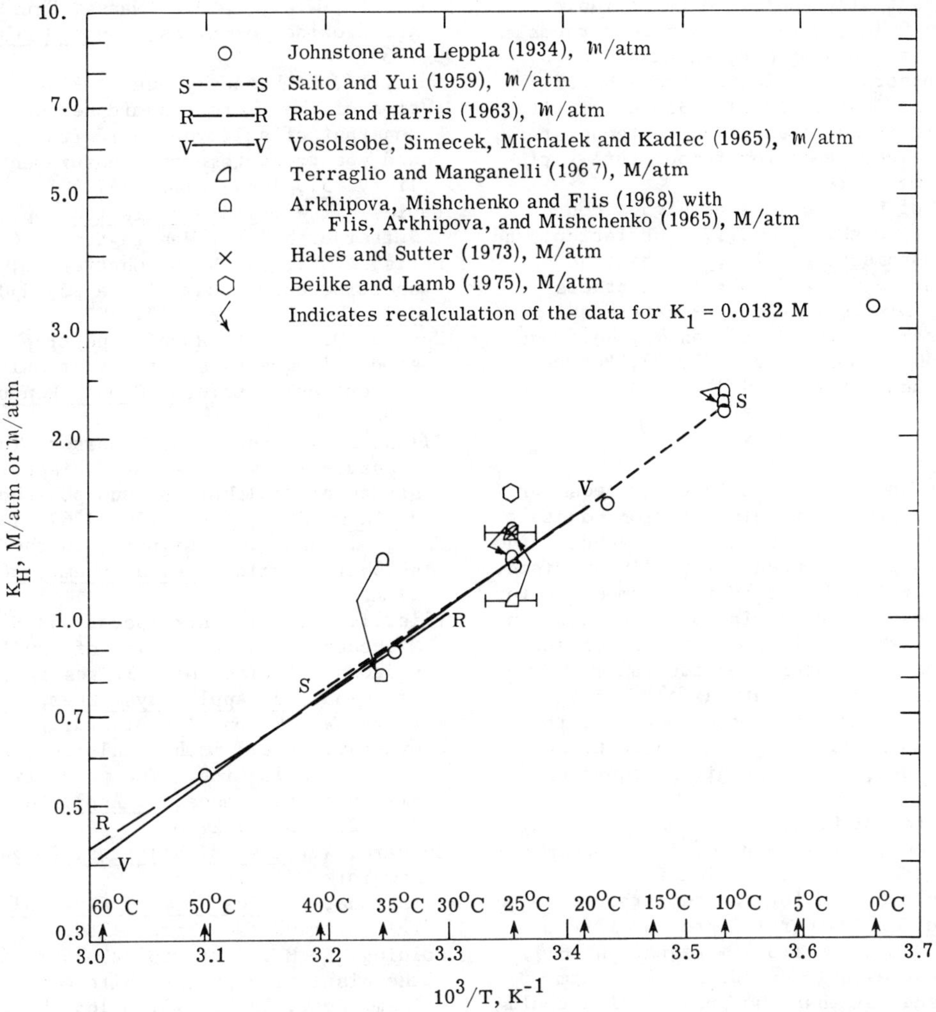

Fig. 3. Henry's law constant as a function of temperature.

For the sake of clarity, this line has not been drawn on Figure 3. The average value of 1.242 M/atm for K_H at 25°C is the value plotted in Figure 2 for $p_{SO_2} > 2.6 \times 10^{-4}$ atm.

Ionic Activity

At very low concentrations, as is often the situation for SO_2 dissolved in cloud droplets or raindrops, the ionic activity of a species may be justifiably taken simply as its concentration. At higher concentrations, however, such as might be encountered in evaporating cloud droplets, corrections for ionic activity can become important, and accurate values of the activity coefficients are needed if effective use is to be made of the thermodynamic dissociation constants. Experimental values for the mean activity coefficient γ_{HHSO_3} are reported by Ratkowsky and McCarthy (1962) and Huss and Eckert (1977). Those of Ratkowsky and McCarthy (which, incidentally, are based on $K_1 = 0.0172$ M) are nontypical of other 1-1 electrolytes in that they exhibit an anomalous downward curvature with increasing ionic strength, while those of Huss and Eckert (based on $K_1 = 0.0138$ M) are more as expected. Moreover, a reanalysis of Ratkowsky and McCarthy's data by Huss and Eckert yields γ_{HHSO_3} values identical to their own. A comparison of these experimental data with values calculated for γ_{HHSO_3} using Kielland's (1937) method shows excellent agreement up to an ionic strength of 0.06 M (the limit of Huss and Eckert's data). Agreement is also excellent with experimental data reported for γ_{HCl} (Harned and Owen, 1958). Hence, a reasonable approach for correcting for ionic activity in the SO_2-water system would be either to use $\gamma_{HHSO_3} = \gamma_{HCl}$, or, if more convenient, to calculate γ_{HHSO_3} from Kielland's method (valid up to an ionic strength of about 0.1 M) and use this method to estimate other activity coefficients which may be required. (A convenient summary of Kielland's method can be found in Lee and Sillén (1959).)

For estimating activity coefficients at ionic strengths above 0.1 M, more approximate methods must be used. One such method, which has been used by some authors up to ionic strengths of 0.5 M, is that of Davies (1962). Since activity coefficients are quite insensitive to temperature, they need not be corrected for temperature variations in the atmosphere.

In solutions of S(IV) oxides at higher concentrations there is a shift in S(IV) speciation, and a variety of tautomeric species, dimers (the pyrosulfite ions $HS_2O_5^-$ and $S_2O_5^=$) and others begin to form. Discussions of the nature and equilibria of some of these species may be found in Golding (1960), Hayon et al. (1972), Bourne et al. (1974), and Guthrie (1979).

Summary

A discrepancy has been noted in the values currently in use for the first dissociation constant of SO_2 in water. To resolve this discrepancy, the pertinent literature has been critically examined, and a value of 0.0132 M at 25°C is recommended for the true thermodynamic dissociation constant. An expression correlating this dissociation constant as a function of temperature for the range 0°C to 50°C has been developed (equation (5)). A value of 6.42×10^{-8} M at 25°C is recommended for the second thermodynamic dissociation constant, and a correlating expression is also given (equation (6)).

Literature data for Henry's law constant have also been examined, and a value of 1.242 M/atm at 25°C is recommended. A correlation for Henry's law constant is proposed for the temperature range 0°C to 50°C and for SO_2 partial pressures from about 2×10^{-4} atm up to 1.3 atm (equation (8)). At partial pressures below about 2×10^{-4} atm existing data suggest that the proper value could possibly be as much as 30 to 40 percent higher. This uncertainty is of particular concern because in many tropospheric cloud systems of interest, the partial pressures of SO_2 are very low, on the order of 10^{-9} atm. Additional studies are clearly needed to resolve these uncertainties.

Ionic activity corrections and speciation in more concentrated solutions are briefly discussed. and recommendations are made for estimating the activity coefficients of the various ionic species in solutions of moderate ionic strength.

References

Arkhipova, G. P., K. P. Mishchenko, and I. E. Flis, Equilibrium in dissolution of gaseous SO_2 in water at 10°-35°C, J. Appl. Chem. USSR, 41, 1069-1071, 1968.

Beilke, S., and D. Lamb, Remarks on the rate of formation of bisulfite ions in aqueous solution, Am. Inst. Chem. Eng. J., 21, 402-404, 1975.

Bourne, D. W. A., T. Higuchi, and I. H. Pitman, Chemical equilibriums in solutions of bisulfite salts, J. Pharm. Sci., 63, 865-868, 1974.

Campbell, W. B., and O. Maass, Equilibria in sulfur dioxide solutions, Can. J. Res., 2, 42-64, 1930.

Cuta, F., E. Beranek, and J. Pisecky, Determination of the thermodynamic second dissociation constant of sulfurous acid from potentiometric and spectrophotometric measurements, Chem. listy, 51, 1614-1617, 1957.

Davies, C. W., Ion Association, pp. 39-41, Butterworth Inc., Washington, D.C., 1962.

Davis, A. R., and R. M. Chatterjee, A vibrational-spectroscopic study of the SO_2-H_2O system, J. Sol. Chem., 4, 399-412, 1975.

Deveze, D., and P. Rumpf, Spectrophotometric study of aqueous solutions of sulfur dioxide in various acid buffers, Compt. Rendu., 258, 6135-6138, 1964.

Ellis, A. J., and D. W. Anderson, The effect of pressure on the first acid dissociation constants of "sulphurous" and phosphoric acids, J. Chem. Soc., 1961, 1765-1767, 1961.

Falk, M., and P. A. Giguère, On the nature of sulphurous acid, Can. J. Chem., 36, 1121-1125, 1958.

Flis, I. E., G. P. Arkhipova, and K. P. Mishchenko, Investigation of equilibria in aqueous solutions of sulfites at temperatures of 10-35°, J. Appl. Phys. Chem, 38, 1466, 1965.

Gerrard, W., Solubility of hydrogen sulphide, dimethyl ether, methyl chloride, and sulfur dioxide in liquids. The prediction of solubility of all gases, J. Appl. Chem. Biotechnol., 22, 623-650, 1972.

Gerrard, W., Gas Solubilities, Pergamon Press, New York, 1980.

Glasstone, S., Textbook of Physical Chemistry, Van Nostrand, New York, 1980.

Golding, R. M., Ultraviolet absorption studies of the bisulphite-pyrosulphite equilibrium, J. Chem. Soc., 1960, 3711-3716, 1960.

Guthrie, J. P., Tautomeric equilibria and pK_a values for "sulfurous acid" in aqueous soluution: A thermodynamic analysis, Can. J. Chem., 57, 454-457, 1979.

Hales, J. M., and S. L. Sutter, Solubility of sulfur dioxide in water at low concentrations, Atmos. Environ., 7, 997-1001, 1973.

Harned, H. S., and B. B. Owen, The Physical Chemistry of Electrolytic Solutions, pp. 716, 725, Reinhold Publishing Corp., New York, 1958.

Hayon, E., A. Treinin, and J. Wilf, Electronic spectra, photochemistry, and autooxidation mechanism of the sulfite-bisulfite-pyrosulfite systems. The SO_2^-, SO_3^-, SO_4^-, and SO_5^- radicals, J. Am. Chem. Soc., 94, 47-57, 1972.

Huss, A., Jr., and C. A. Eckert, Equilibria and ion activities in aqueous sulfur dioxide solutions, J. Phys. Chem., 81, 2268-2270, 1977.

Johnstone, H. F., and P. W. Leppla, The solubility of sulfur dioxide at low partial pressures. The ionization constant and heat of ionization of sulfurous acid, J. Am. Chem. Soc., 56, 2233-2238, 1934.

Jones, L. H., and E. McLaren, Infrared absorption

spectra of SO_2 and CO_2 in aqueous solution, J. Chem. Phys., 28, 995, 1958.

Kerp, W., and E. Baur, The ionization constant of sulphurous acid, Arb. kaiserl. Gesundheitsamt., 26, 297-300, 1907.

Kielland, J., Individual activity coefficients of ions in aqueous solutions, J. Am. Chem. Soc., 59, 1675-1678, 1937.

Lee, T. S., and L. G. Sillén, Chemical Equilibrium in Analytical Chemistry, pp. 239-243, Interscience Publishers, New York, 1959.

Lindner, J., Electrolytic dissociation of sulfurous acid, Monatsh., 33, 613-672, 1912.

Morgan, O. M., and O. Maass, An investigation of the equilibria existing in gas-water systems forming electrolytes, Can. J. Res., 5, 162-199, 1931.

Plummer, A. W., Thermodynamic data for SO_2-H_2O. Bibliography and critical analysis, Chem. Eng. Progress, 46, 369-374, 1950.

Rabe, A. E., and J. F. Harris, Vapor liquid equilibrium data for the binary system, sulfur dioxide and water, J. Chem. Eng. Data, 8, 333-336, 1963.

Ratkowsky, D. A., and J. L. McCarthy, Spectrophotometric evaluation of activity coefficients in aqueous solutions of sulfur dioxide, J. Phys. Chem., 66, 516-519, 1962.

Saito, S., and N. Yui, The distribution coefficient and its applications. V. Vapor-liquid equilibriums of sulfur dioxide, Nippon Kagaku Zasshi, 80, 139-141, 1959.

Salomaa, P., R. Hakala, S. Vesala, and T. Aalto, Solvent deuterium isotope effects on acid-base reactions. III. Relative acidity constants of inorganic oxyacids in light and heavy water. Kinetic applications, Acta. Chem. Scand., 23, 2116-2126, 1969.

Sherrill, M. S., and A. A. Noyes, The interionic attraction theory of ionized solutes. VI. The ionization and ionization constants of moderately ionized acids, J. Am. Chem. Soc., 48, 1861-1873, 1926.

Sillén, L. G., and A. E. Martell, Stability Constants of Metal-Ion Complexes, pp. 229-232, Special Publ. No. 17, The Chemical Society, London, 1964.

Tartar, H. V., and H. H. Garretson, The thermodynamic ionization constants of sulfurous acid at 25°C, J. Am. Chem. Soc., 63, 808-816, 1941.

Terraglio, F. P., and R. M. Manganelli, The absorption of atmospheric sulfur dioxide by water solutions, J. Air Pollut. Control Assoc., 17, 403-406, 1967.

Vosolsobe, J., A. Simecek, J. Michalek, and B. Kadlec, Solubility of sulfur dioxide in water, Chem. Prumysl., 15, 401-404, 1965.

Wang, J. C., and D. M. Himmelblau, A kinetic study of sulfur dioxide in aqueous solution with radioactive tracers, Am. Inst. Chem. Eng. J., 10, 574-580, 1964.

Yui, N., Dissociation constants of sulfurous acid, Bull. Inst. Phys. Chem. Research (Toyko), 19, 1229-1236, 1940.

THEORETICAL LIMITATIONS ON HETEROGENEOUS CATALYSIS BY TRANSITION METALS IN AQUEOUS ATMOSPHERIC AEROSOLS

Charles J. Weschler

Bell Telephone Laboratories, Inc., Holmdel, New Jersey 07733

T. E. Graedel

Bell Telephone Laboratories, Inc., Murray Hill, New Jersey 07974

Abstract. The potential for transition metals commonly found in atmospheric aerosols to function as heterogeneous catalysts has been evaluated by assessing their abundance, chemical form, stable oxidation states, and bonding properties. These factors indicate that iron, manganese, and perhaps copper (depending on its atmospheric abundance) are most likely to function in this way. For an atmospheric reaction heterogeneously catalyzed by a transition metal, the rate-determining step is likely either reaction at the catalyst surface or permeation of the reactant through the organic film that is sometimes observed on aqueous atmospheric aerosols. The former is expected to be rate limiting and zero order in reactant concentration at high atmospheric concentrations of the reacting species (greater than 10 ppb). The latter becomes rate limiting and first order in reactant at lower reactant concentrations. Heterogeneous catalysis by transition metals is likely to be of little consequence for species with rapid gas-phase reaction pathways, but may be significant for slower processes such as the oxidation of sulfur dioxide.

Introduction

Transition metals are known constituents of aqueous atmospheric aerosols and participate in numerous chemical processes occurring within these systems. Such processes include redox (oxidation-reduction) reactions in which the transition metal serves as a heterogeneous catalyst. In this paper we assess the potential importance of transition metals in this capacity.

A schematic picture of an aqueous atmospheric aerosol facilitates the discussion. A summary of the physical picture of an atmospheric aerosol at moderate to high humidities (Graedel and Weschler, 1981) is presented in Figure 1. Several aspects of the diagram deserve mention. The first is the commonly present insoluble core. Compounds that contain transition metals may occur on the surface of this core (Linton et al., 1976; Keyser et al., 1978), and we are addressing the ability of these compounds to catalyze reactions within the aerosol. Second, the core is generally enveloped in an aqueous solution that typically accounts for 30 percent of the total weight (Ho et al., 1974). This solution contains significant amounts of dissolved inorganic and organic compounds (Ketseridis et al., 1976; Graedel, 1978). Third, the solution is covered on at least some occasions with a film of organic material (Husar and Shu, 1975); any polar portions of the organic molecules are likely to be preferentially oriented toward the core as a result of solution bonding. This organic film presumably coalesces and thickens as the particles reside in the atmosphere (Grosjean and Friedlander, 1975; Husar and Shu, 1975). The features shown in Figure 1 are also somewhat applicable to raindrops. Fly-ash particles have less in common with typical aerosols and raindrops; they are generally too fresh to have thick water shells or organic surface films.

In speaking of transition metals as heterogeneous catalysts we mean that the compounds that contain the transition metals are in a different phase than the reactants. This paper is concerned only with transition metal containing compounds that exist as solids on the surface of the core, and reactants (gases, liquids, or salts) that are dissolved in the aqueous solution that envelops the core. We are not speaking of transition metals dissolved in the liquid phase, but of heterogeneously catalyzed liquid-phase oxidations.

Selected Considerations

Several important factors must be considered in evaluating transition metals as potential heterogeneous catalysts: (1) the abundance of the transition metal in atmospheric aerosols; (2) the chemical form of the metal in the aerosol; (3) the stable oxidation states that are available to the metal; (4) the ability of the metal to reversibly

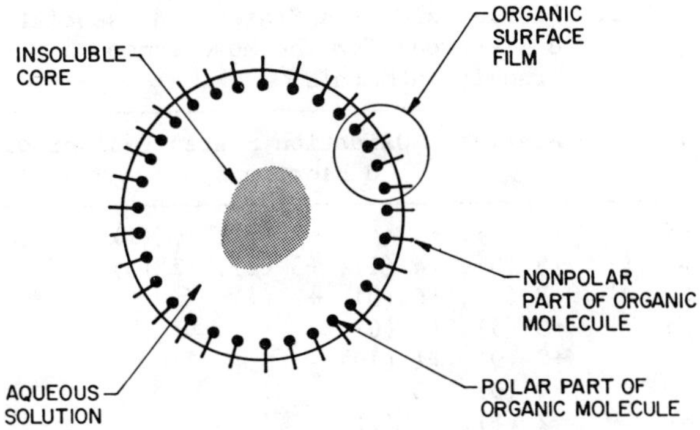

Fig. 1. Schematic diagram of the possible structure of aged atmospheric aerosols (Graedel and Weschler, 1981).

bond the adsorbate (i.e., to bond strongly enough to activate the adsorbate, but not so strongly that a stable complex is formed); and (5) the rate of the heterogeneously catalyzed pathway compared with the rate of the noncatalyzed pathway and with the lifetime of the aerosol.

Abundance of Transition Metals in Aerosols

Of the transition metals, the first-row transition elements make by far the largest contribution to the atmospheric aerosol. Typical ranges reported for the atmospheric concentrations of the most abundant transition elements are listed in Table 1, and these elements are subclassified into four groups with respect to these concentrations. The concentration range in each group differs by a factor of 10 from that of an adjacent group. The concentrations of Co, Sc, Hg, and Ag are so low that these species are unlikely to be significant as heterogeneous catalysts.

In this paper we will not go into a detailed discussion of analogous catalytic processes in fly ash or raindrops. However, to permit comparisons, also listed in Table 1 are concentration ranges for transition metals in fly ash and precipitation. The ratio of fly ash to gaseous effluents in power plant plumes is significantly larger than typical ratios of atmospheric aerosols to atmospheric gases. Catalysis on fly ash may thus be more important than catalysis on aerosols. Of special interest here, however, are the relative differences in elemental abundance between the two systems. In particular, the high relative concentrations of vanadium and nickel deserve mention. In precipitation, the ordering of transition metals by concentration is generally similar to that in aerosols. Cloud water is of average density ~1 g/m^3 (Mason, 1971); it is of interest that the solution concentrations of columns 5 and 6 thus convert to volume densities similar to those in aerosols (columns 2 to 4).

Chemical Forms of the Transition Metals

The majority of the publications containing atmospheric concentrations for transition metals simply report elemental analyses and, for obvious reasons of analytical complexity, do not identify the chemical forms in which these elements are present. To a large extent the chemical form of a transition metal in the aerosol depends on the origins of the metal. The element can come pri-

TABLE 1. Abundance of Transition Metals in Atmospheric Systems

Element	Aerosol Concentration Range, μg/m^3			Fly-ash Concentration Range, ppm[d]		Precipitation Concentration Range, μg/l[e]	
	Urban[a]	Nonurban[a]	Global[b]	Oil-Fired	Coal-Fired	Urban	Rural
Fe	1.1 -2.1	0.19 -0.51	0.5 -2.0	(4-6)×10^3	(4-14)×10^4	12-1800	1-2800
Cu	0.15 -0.26[c]	0.16 -0.36[c]	0.01 -0.80[c]	100-500	30-400	8-120	0.4 -150
Zn	-	-	0.10 -0.80	40-400	20-1200	18-180	1-310
Ti	0.03 -0.05	0.009 -0.028	0.10 -0.08	300-700	100-7000	36	-
Mn	0.04 -0.08	0.004 -0.018	0.01 -0.12	200-500	50-500	1.9-26	0.2 -84
V	0.016-0.052	0.004 -0.008	0.01 -0.10	(1-13)×10^4	200-500	16-68	0.13-23
Ni	0.014-0.017	0.002 -0.011	0.008 -0.05	(4-23)×10^3	200-600	2.4-110	0.1 -48
Cr	0.006-0.009	0.002 -0.004	0.005 -0.02	100-1000	100-1000	0.15-15	0.01-30
Cd	0.001-0.004	0.0001-0.0005	0.001 -0.010	1-4	2-2200	0.5-2.3	0.08-46
Co	0.001	0.002	0.0003-0.003	50-500	5-70	1.8	0.01
Sc	-	-	0.0001	<1-5	10-50	-	-
Hg	-	-	0.0003-0.002	<1	<1	-	-
Ag	-	-	0.001 -0.003	<1	<1-20	-	-

[a]Arithmetic means, 1970-1976, from EPA (1979).
[b]Rahn (1976).
[c]The Cu values are likely high due to contamination from wearing of the commutator in the hi-vol motor (King and Toma, 1975).
[d]Henry and Knapp (1980).
[e]Thorton (1980).

marily from natural sources, from anthropogenic sources, or from some combination of the two. The contributions from these sources can be estimated by comparing the concentration of an element in an atmospheric aerosol to its concentration in the Earth's crust or the sea. In Table 2 the eight most abundant transition metals are classified as enriched, intermediate, or nonenriched based on such comparisons. Transition metals that have predominantly natural sources (nonenriched) are likely to be present in the aerosol as weathered soil, clay, or mineral particles. Transition metals that are enriched are likely to have entered the atmosphere as a result of smelting, refining, combustion, incineration, wear, corrosion, etc., and are present primarily as oxides, hydroxides, and carbonates. Compounds that contain enriched metals such as Cu and Zn have more recent origins and are likely, as a first approximation, to be more reactive than compounds that contain nonenriched metals.

Available Oxidation States

The stable oxidation states of the most abundant transition metals are listed in Table 3 together with the associated d electrons. A key feature of a good redox catalyst is the availability of surface sites that are associated with single valency changes of the cation (Morrison, 1977) (e.g., Mn^{+2}/Mn^{+3}, Fe^{+2}/Fe^{+3}). Such redox sites are considered highly active in electron exchange. (In terms of solid-state chemistry, the concept of a redox system controlling electron exchange can be viewed as the Fermi energy at the surface of the solid controlling the exchange of unpaired electrons (Morrison, 1977).) Thus Zn, with only one stable oxidation state, is unlikely to function as an efficient redox catalyst. Two adjacent oxidation states are available to Cr, Fe, Cu, Ti, and Ni, while multiple states characterize Mn and V.

Bonding Tendency

If a compound containing a transition metal is to function as an efficient catalyst, it should be capable of bonding to the reacting species (the absorbates). However, the catalyst cannot bond the reactants (or products) too strongly. If this were to occur, the species would not vacate the catalytic surface and soon all the active sites would be occupied. As early as 1950 Dowden noted that the formation of coordinate bonds appropriate for a heterogeneous catalyst appears favored on transition metal compounds possessing partially occupied "atomic" d orbitals. The data presented in Figure 2, adapted from Giordano (1969), offer support for such a correlation. While this simple approach permits a rough estimate of bonding tendencies, it must be emphasized that transition metal bond strengths depend in a complex fashion on both the nature of the coordinating species (in this case, the reactants or products) and the nature of the transition metal compound (see, for example, Basolo and Pearson (1968)). For example, despite the implications of the above discussion, V_2O_5 is a common industrial catalyst for the oxidation of sulfur dioxide in the contact process. In fact, catalysis does not necessarily require a chemical bond between the reacting species and the metal; outersphere electron transfer can initiate a catalytic process (Basolo and Johnson, 1964; Basolo and Pearson, 1968).

TABLE 2. Chemical Forms of the More Common Transition Metals

Element	Chemical Form
Cu, Zn	Enriched - primarily oxides, hydroxides, and carbonates
Fe, Mn, V, Ni, Cr	Intermediate - minerals/ oxides, hydroxides, and carbonates
Ti	Nonenriched - primarily rutile (TiO_2) and ilmenite ($FeTiO_3$)

TABLE 3. Stable Oxidation States and Associated d Electrons for the More Common Transition Metals[a]

Element	Available Oxidation States (Number of d Electrons)
MN	+2 (5), +3 (4), +4 (3), +6 (1), +7 (0)
V	+5 (0), +4 (1), +3 (2), +2 (3)
Cr	+3 (3), +6 (0), +2 (4)
Fe	+3 (5), +2 (6)
Cu	+2 (9), +1 (10)
Ti	+4 (0), +3 (1)
Ni	+2 (8), +3 (7)
Zn	+2 (10)

[a]First listed value is most stable state.

Likely Transition Metal Catalysts

Evaluation of the four previously mentioned factors suggests that the transition metals potentially important as heterogeneous catalysts in atmospheric aerosols are (in order of decreasing likelihood):

$$Fe > Mn \sim Cu(?) > Ti > V \sim Ni \sim Cr > Zn$$

This ordering is based on the following arguments. Although a fraction of the aerosol iron is present in minerals (e.g., ferrous silicates, ferric aluminosilicates, and pyrites), the total atmospheric abundance of Fe is so large that iron oxides and hydroxides dominate the atmospheric aerosols. Mn and Cu are more likely to function as efficient catalysts than Zn or Ti for reasons of available oxidation states (Zn) or bonding tendencies (Ti). However, it should be noted that

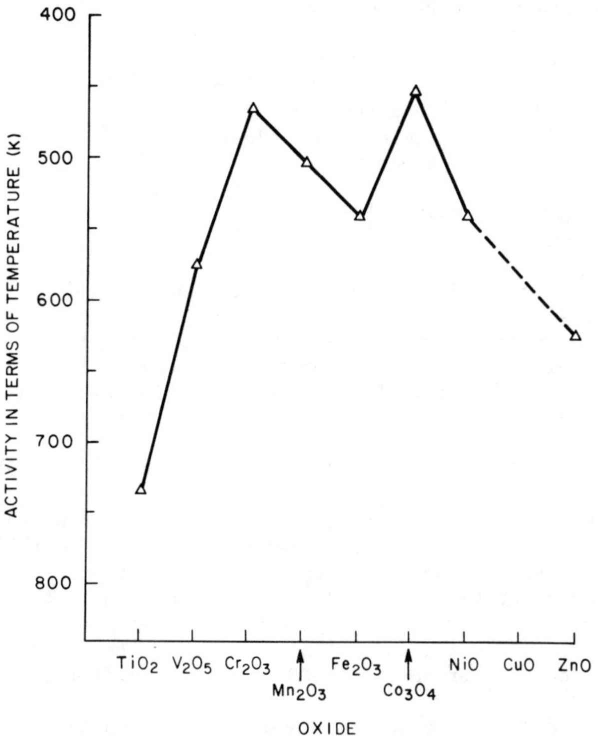

Fig. 2. Catalytic activity of the more common transition metal oxides for the oxidation of NH_3 (expressed as the temperature required to oxidize 5 percent of the NH_3)(after Giordano, 1969).

the reported abundance of Cu in the atmospheric aerosol may be artificially high due to reentrained copper particles that result from wearing of the commutator in the hi-vol motor frequently used in air sampling networks (King and Toma, 1975). Cr, Ni, and V are limited as catalysts in aerosols primarily by their concentrations.

Rate of Heterogeneously Catalyzed Pathway

Having identified transition metals likely to serve as heterogeneous catalysts, we now examine the rate of the catalyzed reaction. The reaction of atmospheric species catalyzed by a transition metal compound on the surface of an insoluble aerosol core and the possible return of reaction products to the atmosphere can be envisioned as occurring in seven steps: (1) transport of the reactant(s) to the gas-liquid interface, (2) transport of reactant(s) across the interface and through the liquid to the particle surface, (3) adsorption of reactant(s) at the particle surface, (4) reaction on the surface, (5) desorption of product(s) from the particle surface, (6) transport of product(s) through the liquid and across the gas-liquid interface, and (7) transport of product(s) from the interface into the bulk gas phase.

The potential importance of transition metals as heterogeneous catalysts depends on the rates of the catalyzed and noncatalyzed reactions. To make such comparisons it is useful to estimate an upper limit for the rate of the former reaction. The rate-determining step in a heterogeneous reaction involving a gas and a transition metal is most likely to be either step 1 (the rate of collision), step 2 (the rate of transport across the organic skin and through the liquid), or step 4 (the rate of reaction at the surface). The other steps should be relatively fast assuming that a sufficient number of active sites are available and that the liquid is not saturated with reaction product. As an example of an order of magnitude assessment of catalytic potential, the following paragraphs estimate upper limit values for steps 1, 2, and 4.

The rate of collision between a gas molecule and an aerosol (Present, 1958) is given by

$$\text{rate of collision} = \left|\frac{8RT}{\pi M}\right|^{\frac{1}{2}} \left|\frac{(A)}{4}\right| [G] \qquad (1)$$

where R is the gas constant, T is the temperature, M is the molecular weight, A is the surface area of aerosol per unit volume of space, and [G] is the concentration of the gas molecule. To evaluate equation (1), consider that in a typical particle size distribution, the total surface area of the aerosol per unit volume, A, is approximately 0.0014 m^2/m^3 if the aerosol concentration is 100 $\mu g/m^3$ and the mean density of the particles is 2.2 g/cm^3 (Judeikis and Siegel, 1973). A reasonable value for the molecular weight of the gas is 64 atomic mass units (amu) or 0.064 kg/mol. These estimates yield:

$$\text{rate of collision} = (315\ m/s)(3.5 \times 10^{-4}\ m^2/m^3)[G]$$
$$= 0.11[G]\ sec^{-1} \qquad (2)$$

An interesting feature of the schematic diagram of the atmospheric aerosol (Figure 1) is the organic surface film. The electron microscope studies of Husar and Shu (1975) demonstrated the existence of such films; the generality of their presence is not known. However, the circumstantial evidence that coated particles are common is substantial. First, alkanes, fatty acids, and other organic compounds are abundant in aerosols (Ketseridis et al., 1976; Simoneit and Mazurek, 1981). Second, compositional patterns in the gas and aerosol phases suggest that the presence of lipids on aerosols results in part from gas to particle deposition (Marty et al., 1979; Hahn, 1980). Third, the low aqueous solubilities of the lipids suggest that they will be largely retained on the surfaces of the particles (Hahn, 1980). If so, molecules with water-soluble end groups will be oriented approximately perpendicular to the surface (Adam, 1968), as shown in Figure 1.

The presence of an organic film on an aqueous

TABLE 4. Selected Permeability Coefficients for Trace Atmospheric Gases

Gas	Membrane	Temp (°C)	P(barrers)	Reference
CH_2CH_2	Poly(ethylene)	25	6.6	Li and Long (1969)
H_2S	Cellulose acetate	30	20	Heilman et al. (1956)
NH_3	High-density poly(ethylene)	23	7.6	Lee and Hart (1980)
SO_2	Ethyl cellulose	25	260	Hsieh (1963)

droplet is known to retard the rate of droplet evaporation (Archer and LaMer, 1955; Snead and Zung, 1968; May, 1972; Strathdee and Given, 1977; Chang and Hill, 1980). Since water permeation is inhibited, the permeation of reactive atmospheric molecules into the droplet will be inhibited as well. Data specifically applicable to this problem are not available, but an initial estimate may be made by considering the organic surface film to be a reasonable simulation of a polymeric membrane. The expression for the rate of transport of a gaseous species across an organic membrane is:

$$\text{rate of transport} = \left|\frac{DS}{L}(p_1 - p_2)\right|(A) \qquad (3)$$

where D and S are, respectively, the diffusion and solubility coefficients for a given gas/polymer combination, p_1 and p_2 are the partial pressures of the gas on either side of the membrane, L is the thickness of the membrane, and A is the total surface area of the aerosol per unit volume. The product of D and S is referred to as the permeability coefficient (P).

We choose values of $A = 14\ cm^2/m^3$ (see equation (1)) and $L = 5$ Å (i.e., a monolayer, although the film is often likely to be considerably thicker (cf. Husar and Shu, 1975); note that a monolayer organic film of unit density that envelops a particle of 1 µm diameter represents only about 0.3 percent of the total particle mass). Then if the concentration of the permeating gas is assumed to be zero within the aerosol and if we substitute P = DS, equation (3) becomes

$$\text{rate of transport} = 2.1 \times 10^{-6}\ P[G]\ sec^{-1} \qquad (4)$$

where P is expressed in barrers (1 barrer $\equiv 10^{-10}$ cm^3 permeating gas STP·cm thickness/(cm^2 area·cm pressure·sec)), a unit common in the permeation literature.

What are appropriate values for P? Permeability coefficients have seldom been measured for the atmospheric trace gases, but a few values chosen from the literature are given in Table 4. The coefficient is very sensitive to the membrane through which permeation occurs (e.g., Felder et al. (1975) quote values for SO_2 that cover the range 10^{-7} to 10^4 barrers). Aerosols are known to contain both aliphatic and aromatic alcohols and acids (Schuetzle et al., 1975; Cronn et al., 1977; Grosjean et al., 1978); these are structurally similar to the cellulose acetate and ethyl cellulose for which data are available. Consequently, a liberal estimate of the permeability coefficient for trace atmospheric gases is approximately 150 barrers. Equation (4) then becomes

$$\text{rate of transport} = 3.2 \times 10^{-4}\ [G]\ sec^{-1} \qquad (5)$$

This estimate for the rate of transport across the organic film should be a reasonable upper limit for step 2. Any contribution to the rate from transport through the liquid would only make the value smaller. Normally, however, transport through the liquid is fast relative to transport across the organic film (Pruppacher and Klett, 1978).

The next step in the evaluation process is to derive a rate for reaction at the transition metal surface (step 4) to compare with the rates derived for steps 1 and 2. For an industrial redox process a reasonable upper limit on the catalytic activity, the rate per unit area at which molecules react, is 10^{14} molec/sec-cm^2 (see, for example, Sleight (1980) and references therein). (If the catalyst contains an active site every 40 Å, this is roughly a "turnover rate" of 1000 molec/min.) To estimate the catalytic surface area of the transition metal, assume the metal is present in the atmosphere at a concentration of 1 µg/m^3 (approximately the atmospheric concentration of iron), has a density of 5 g/cm^3 (similar to Fe_2O_3), and exists as 0.1-µm-diameter spheres (an assumption that will give an exaggeratedly high value for the surface area, consistent with an upper limit estimate). Using these values, the total catalytic surface area of the transition metal would be approximately 0.1 cm^2 per m^3 of air. The product of this value and the catalytic activity yields an upper limit for step 4:

TABLE 5. Upper Limit Rate Comparisons (molec/sec-m^3)

Gas Concentration, ppb	Step 1	Step 2	Step 4
0.01	3×10^{13}	9×10^{10}	1×10^{13}
1	3×10^{15}	9×10^{12}	1×10^{13}
100	3×10^{17}	9×10^{14}	1×10^{13}

rate of reaction at the transition metal surface

$$= 10^{13} \text{ molec sec}^{-1} \text{ m}^{-3} \quad (6)$$

Rates for steps 1, 2, and 4 at selected gas concentrations are compared in Table 5. At low atmospheric concentrations of the reacting species (<0.1 ppb), permeation across the organic surface film is likely to be rate determining and a reasonable upper limit on the heterogeneously catalyzed pathway is 3.2×10^{-4} [G] molec sec^{-1} m^{-3} where the units for [G] are molec/m^3. (Note, however, that this limit will be inapplicable for a freshly-generated aerosol that has not yet acquired an organic shell and will be an upper limit to the rate for an aerosol with a shell more than a monolayer thick. It appears from ambient data (Grosjean and Friedlander, 1975; Schuetzle et al., 1975) that organic deposition to aerosols varies on a scale of a few hours and is correlated with the degree of photochemical activity; the organic aerosol surface is thus in a state of constant evolution.) At high atmospheric concentrations of the reacting species (greater than 10 ppb), reaction at the transition metal surface is likely to be rate determining and a reasonable upper limit on the rate is 1×10^{13} molec sec^{-1} m^{-3}. At moderate reactant concentrations (in the range of 1 ppb) either step 2 or step 4 is likely to determine the rate. Under no conditions is step 1 rate determining, and under no conditions is the rate of the heterogeneously catalyzed path expected to be faster than 10^{13} molec sec^{-1} m^{-3}.

To decide whether a pathway heterogeneously catalyzed by a transition metal is potentially significant for a given reaction, the upper limit for the metal-catalyzed rate at the appropriate reactant concentration is compared to the homogeneous rate for the reaction in question. For example, consider the oxidation of tropospheric NO. A typical rural concentration for NO is approximately 0.1 ppb (Kley et al., 1981); this translates to an upper limit for the rate of the metal-catalyzed path of approximately 10^{12} molec/sec-m^3. The rate at which O_3 oxidizes NO at this rural concentration is approximately 4×10^{13} molec/sec-m^3. The comparison indicates that heterogeneous transition metal catalysis is of little consequence to the tropospheric oxidation of NO. Homogeneous oxidation rates for selected atmospheric processes are listed in Table 6. The type of comparison just outlined, incorporating the appropriate atmospheric concentrations for the reactants in question, indicates that heterogeneous catalysis by transition metals contained within aerosols is capable of contributing to the oxidation of H_2S, SO_2, HCHO, CH_4, and CO, and to the decomposition of H_2O_2. However, it is important to remember that the rates in Table 5 are upper limit estimates. The actual rate for the catalyzed process is likely to be several orders of magnitude slower. (The organic surface film may be considerably thicker than one monolayer, the permeation coefficient may be smaller, the catalytic surface area may be less than the generous estimate used above, the catalytic activity may be closer to 10^{12} molec/sec-cm^2, etc.) Comparisons between rates in Table 5 and homogeneous rates can demonstrate that for some reactions heterogeneous catalysis by metals is of no significance. However, when such a comparison fails to eliminate the heterogeneous path, that path is not necessarily important. More accurately, it is potentially significant.

TABLE 6. Initial Homogeneous Gas-Phase Reaction Rates for Selected Atmospheric Gases[a]

Species	Gas-Phase Reactant	Urban $-d[G]/dt$, molec/$s^{-1}m^{-3}$	Rural $-d[G]/dt$, molec/$s^{-1}m^{-3}$
H_2O_2	hν	5×10^{12}	2×10^{11}
CO	HO·	3×10^{13}	4×10^{11}
NO	O_3	4×10^{15}	4×10^{13}
NO_2	hν	1×10^{16}	2×10^{13}
H_2S	HO·	1×10^{11}	5×10^{9}
SO_2	HO·	6×10^{11}	1×10^{10}
CH_4	HO·	3×10^{11}	1×10^{11}
$CH_3CH_2CH_3$	HO·	2×10^{12}	2×10^{11}
CH_2CHCH_3	HO·	2×10^{13}	3×10^{11}
$CH_2CHCH_2CH_3$	HO·	2×10^{12}	3×10^{11}
HCHO	hν	1×10^{13}	4×10^{11}
CH_3CHO	hν	3×10^{12}	6×10^{10}

[a]Rate constants from Graedel (1978). Concentrations of atmospheric gases from Graedel and Weschler (1981), and Graedel (1978).

Conclusions

Of the transition metals commonly found in atmospheric aerosols, Fe, Mn, and perhaps Cu (depending on its true concentration) are the most likely to make a significant contribution as heterogeneous catalysts. At low atmospheric concentrations of the reacting species, the rate for the heterogeneously catalyzed reaction occurring within the aerosol may often be determined by the rate at which the reactant crosses the organic film that coats at least some aqueous atmospheric surfaces. This process is first order in reactant concentration. At high reactant concentration, reaction at the catalyst surface is more probably rate determining and is zero order in reactant concentration. Upper limit rate estimates for heterogeneously catalyzed processes indicate that metal catalysis within atmospheric aerosols is of little consequence for fast tropospheric processes such as the oxidation of NO. However, for slower processes such as the oxidation of SO_2 or H_2S, heterogeneous catalysis by transition metals contained within atmospheric particles is potentially significant.

References

Adam, N. K., The Physics and Chemistry of Surfaces, Dover, New York, p. 24ff, 1968.

Archer, R. J., and V. K. LaMer, The rate of evaporation of water through fatty acid monolayers, J. Phys. Chem., 59, 200-208, 1955

Basolo, F., and R. C. Johnson, Coordination Chemistry, pp. 164-169, W. A. Benjamin, Menlo Park, CA, 1964.

Basolo, F., and R. G. Pearson, Mechanisms of Inorganic Reactions, 2nd ed., pp. 475-476, John Wiley & Sons, New York, 1968.

Chang, D. P. Y., and R. C. Hill, Retardation of aqueous droplet evaporation by air pollutants, Atmos. Environ., 14, 803-807, 1980.

Cronn, D. R., R. J. Charlson, R. L. Knights, A. L. Crittenden, and B. R. Appel, A survey of the molecular nature of primary and secondary components of particles in urban air by high-resolution mass spectroscopy, Atmos. Environ., 11, 929-937, 1977.

Dowden, D. A., Heterogeneous catalysis, Part I. Theoretical basis, J. Chem. Soc., 242-265, 1950.

Environmental Protection Agency, Air Quality Data for Metals from the National Air Surveillance Network, EPA-600/4-79-054, August 1979.

Felder, R. M., R. D. Spence, and J. K. Ferrell, Permeation of sulfur dioxide through polymers, J. Chem. Eng. Data, 20, 235-242, 1975.

Giordano, N., Choice and evaluation of heterogeneous catalysts, Chim. L'Industria, 51, 1189-1199, 1969.

Graedel, T. E., Chemical Compounds in the Atmosphere, Academic Press, New York, 1978.

Graedel, T. E., and C. J. Weschler, Chemistry in aqueous atmospheric aerosols and raindrops, Rev. Geophys. Space Phys., 19, 505-539, 1981.

Grosjean, D., and S. K. Friedlander, Gas-particle distribution factors for organic and other pollutants in the Los Angeles atmosphere, J. Air Poll. Contr. Assoc., 25, 1038-1044, 1975.

Grosjean, D., K. Van Cauwenberghe, J. P. Schmid, P. E. Kelley, and J. N. Pitts, Jr., Identification of C_3-C_{10} aliphatic dicarboxylic acids in airborne particulate matter, Environ. Sci. Tech., 12, 313-317, 1978.

Hahn, J., Organic constituents of natural aerosols, Ann. N.Y. Acad. Sci., 338, 359-376, 1980.

Heilman, W., V. Tammela, J. A. Meyer, V. Stannett, and M. Szware, Permeability of polymer films to hydrogen sulfide gas, Ind. Eng. Chem., 48, 821-824, 1956.

Henry, W. M., and K. T. Knapp, Compound forms of fossil fuel fly ash emissions, Environ. Sci. Tech., 14, 450-456, 1980.

Ho, W., G. M. Hidy, and R. M. Govan, Microwave measurements of the liquid water content of atmospheric aerosols, J. Appl. Met., 13, 871-879, 1974.

Hsieh, P. Y., Diffusibility and solubility of gases in ethyl-cellulose and nitrocellulose, J. Appl. Polym. Sci, 7, 1743-1756, 1963.

Husar, R. B., and W. R. Shu, Thermal analysis of the Los Angeles smog aerosol, J. Appl. Met., 14, 1558-1565, 1975.

Judeikis, H. S., and S. Siegel, Particle-catalyzed oxidation of atmospheric pollutants, Atmos. Environ., 7, 619-631, 1973.

Ketseridis, G., J. Hahn, R. Jaenicke, and C. Junge, The organic constituents of atmospheric particulate matter, Atmos. Environ., 10, 603-610, 1976.

Keyser, T. R., D. F. S. Natusch, C. A. Evans, Jr., and R. W. Linton, Characterizing the surfaces of environmental particles, Environ. Sci. Tech., 12, 768-773, 1978.

King, R. B., and J. Toma, Copper emissions from a high volume air sampler, NASA TMX-71693, Lewis Research Center, Cleveland, Ohio, March 1975.

Kley, D., J. W. Drummond, M. McFarland, and S. C. Liu, Tropospheric profiles of NO_x, J. Geophys. Res., 86, 3153-3161, 1981.

Lee, C. O., and G. K. Hart, The permeability of polyethylene to ammonia, J. Appl. Polym. Sci., 25, 955-957, 1980.

Li, N. N., and R. B. Long, Permeation through plastic films, AIChE J., 15, 73-80, 1969.

Linton, R. W., A. Loh, D. F. S. Natusch, C. A. Evans, Jr., and P. Williams, Surface predominance of trace elements in airborne particles, Science, 191, 852-854, 1976.

Marty, J. C., A. Saliot, P. Buat-Menard, R. Chesselet, and K. A. Hunter, Relationship between the lipid compositions of marine aerosols, the sea surface microlayer, and subsurface water, J. Geophys. Res., 84, 5707-5716, 1979.

Mason, B. J., The Physics of Clouds, 2nd ed., Clarendon Press, Oxford, 1971.

May, K. R., Comments on "Retardation of water drop evaporation with monomolecular surface films," J. Atmos. Sci., 29, 784-785, 1972.

Morrison, S. R., The Chemical Physics of Surfaces, Plenum Press, New York, 1977.

Present, R. D., Kinetic Theory of Gases, p. 21, McGraw-Hill, New York, 1958.

Pruppacher, H. R., and J. D. Klett, Microphysics of Clouds and Precipitation, p. 305ff, D. Reidel, Dordrecht, 1978.

Rahn, K. A., The Chemical Composition of the Atmospheric Aerosol, Technical Report, Grad. School of Oceanography, Univ. of Rhode Island, Kingston, 1976.

Schuetzle, D., D. Cronn, A. L. Crittenden, and R. J. Charlson, Molecular composition of secondary aerosol and its possible origin, Environ. Sci. Tech., 9, 838-845, 1975.

Simoneit, B. R. T., and M. A. Mazurek, Air pollution: The organic components, CRC Crit. Rev. Environ. Control, 11(3), 219-276, 1981.

Sleight, A. W., Heterogeneous catalysts, Science, 208, 895-900, 1980.

Snead, C. C., and J. T. Zung, The effects of insoluble films upon the evaporation kinetics of liquid droplets, J. Coll. Inter. Sci., 27, 25-31, 1968.

Strathdee, G. G., and R. M. Given, Coadsorption of n-alkyl alcohols and hydrogen sulfide at the aqueous solution interface, J. Phys. Chem., 81, 327-332, 1977.

Thorton, J. D., The metal and strong acid composition of rain and snow in Minnesota, M. S. Thesis, Univ. of Minnesota, Minneapolis, 1980.

Evidence Regarding Heterogeneous Reactions in the Atmosphere

EVIDENCE FOR HETEROGENEOUS REACTIONS IN THE ATMOSPHERE

George M. Hidy

Environmental Research & Technology, Inc.,
2625 Townsgate Road, Suite 360, Westlake Village, California 91361

Abstract. Verification of heterogeneous reactions in the atmosphere through observations has remained a difficult, perhaps unachievable task because of the diversity and complexity of simultaneous chemical interactions which are suspected to occur. However, recent measurements combined with data analysis show promise for supplying both direct and indirect evidence of heterogeneous sulfur oxide and nitrogen oxide chemistry. Examples of useful methods are provided, which include (1) direct interpretation of observations in and near clouds and inference from thermodynamic properties, (2) inspection of combinations of aerometric data, (3) inference from statistical analysis, (4) comparison of observations with a validated air quality model, and (5) differences in particle size/composition distributions. An example involving thermodynamics is dry ammonium nitrate undergoing equilibrium transformation to the vapor phase, which is very sensitive to temperature. The other sample results presented suggest that heterogeneous oxidation of SO_2 to sulfate may occur in the presence of suspended liquid water, particularly in winter, either through media buffered by absorbed ammonia or via suspended soot in droplets. No observational evidence has been found supporting metal-oxide- or ion-catalyzed reactions of sulfur dioxide or nitrogen oxides under atmospheric conditions.

Introduction

The hypothesis has existed for many years that heterogeneous chemical reactions are active in the formation or modification of atmospheric aerosol particles. Perhaps best known in the early literature is the attack of sodium chloride in sea salt by acidic gases to form sodium sulfate or sodium nitrate, with release of hydrogen chloride (Robbins et al., 1959). Indirect evidence of this reaction has been reported by many investigators, based on the depletion of chloride relative to sodium in samples taken from marine air (e.g., Miller et al., 1972).

A variety of other heterogeneous reactions involve sulfur dioxide (SO_2) and nitrogen oxide (NO_x) oxidation to sulfate and nitrate products. The SO_2 reactions have been more extensively studied than the NO_x reactions. The SO_2 reactions may take place in liquid water droplets or at particulate substrates of metal oxides or carbon (soot) (e.g., Bielke and Gravenhorst, 1978; S. G. Chang, private communication, 1980). The reactions may be thermodynamic in nature, involving absorption and solution of gases, or they may be surface- or volume-catalyzed by trace constituents. Examples of thermodynamically dominated systems include (1) oxidation of SO_2 dissolved in water, enhanced by acid buffering from ammonium ion (e.g., Scott and Hobbs, 1967), (2) condensation or volatilization of "dry" ammonium nitrate interacting with temperature change (e.g., Stelson et al., 1979) and (3) acidic gas interactions directly with a particle substrate. Examples of the metal-salt-catalyzed SO_2 oxidation reactions in water droplets include those identified with iron or manganese salts. An example of soot-catalyzed SO_2 oxidation in droplets was reported by S. G. Chang (private communication, 1980). Metal-oxide- or soot-particle-catalyzed SO_2 reactions have been reported by Judeikis et al. (1978) and Novakov et al. (1974), respectively. Little is known about analogous heterogeneous nitrogen oxide reactions, but they undoubtedly are possible. Although specific heterogeneous reactions can be identified, it remains to be proven whether or not any of them are actually important in the atmosphere.

For many years, the principal mechanism for atmospheric sulfate formation in the troposphere was thought to occur through a heterogeneous pathway. However, since the early 1970's workers have concentrated on homogeneous gas-phase reactions to account for sulfate behavior at least in summer conditions. Likewise, the nitrate aerosol is believed to be linked mainly with nitric acid vapor formation by homogeneous reactions in photochemically driven processes. With expanding knowledge of aerosol systems, an increasing inventory of experimental observations now exists which cannot be accounted for adequately by homogeneous reaction pathways. Therefore, heterogeneous chemistry is again being con-

TABLE 1. Some Sulfate Production Rates and Associated Parameters Measured in Wave Clouds Over Western Washington (from Hegg and Hobbs, 1981)

Date of Sample	Temperature (K)	Ambient Concentrations of Ozone (ppb)	Time for Sulfate Production in Cloud (min)	Ambient Concentrations of SO_2 ($\mu g\ m^{-3}$)	Cloud Water pH	Ambient Concentrations of Ammonia (ppb)	Production of Sulfate in Cloud ($\mu g\ m^{-3}$)[a]	Sulfur Production Rate[b] (%/hr)
May 15, 1980	266	40	9	7.8	5.9	--	4.2	300
July 13, 1979	263	--	5.8	7.0	4.4	--	0.46	70
June 9, 1980	274	40	31	14.8	4.3	--	0.31	4
September 3, 1980	258	30	3	5.8	5.1	~1	0.63	200

[a]Downwind concentration of $SO_4^{=}$ minus the upwind concentration.
[b]Expressed as a percentage of the SO_2 in the ambient air.

sidered as an alternate means for interpreting data. Although considerable progress has been made, very little data currently exist which allow unambiguous interpretation of the dominance of heterogeneous processes in the troposphere. In this paper some examples of recent evidence for such reactions are discussed which are derived from macroscopic observations of particles in the atmosphere. The reported observations are examined first as direct evidence, then as indirect evidence. The latter grouping remains the largest collection of available studies.

Direct Evidence for Heterogeneous Processes

Evidence for the heterogeneous chemistry of sulfur dioxide is difficult to separate out because of the confounding influences of simultaneous transport and mixing processes, deposition processes, and homogeneous gas-phase transformation. Because of this difficulty, virtually no cases of heterogeneous reactions have been isolated until very recently. Two studies have emerged which relied on aircraft observations outside of clouds and collection of cloud water nearby to establish directly the activity of clouds for producing sulfate.

In the first study, Lazrus et al. (1981) reported such observations in air overrunning a surface warm front over the Ohio River Valley. Preexisting acidity from sulfur oxides and nitrogen oxides in the clear air upwind could not account for the acidifying sulfate and nitrate sampled in cloud water downwind. Therefore, the investigators concluded that heterogeneous processes were responsible for the added anion concentrations in cloud water.

In the second study, Hegg and Hobbs (1981) reported experiments sampling air in and near stratiform clouds over Los Angeles and wave clouds over western Washington state. They found evidence for sulfate production in such clouds in some cases, but not in every case. The measurements in lenticular wave clouds are particularly convincing because these types are considered in "steady state." That is, the air upwind rises through the condensation zone and leaves downwind

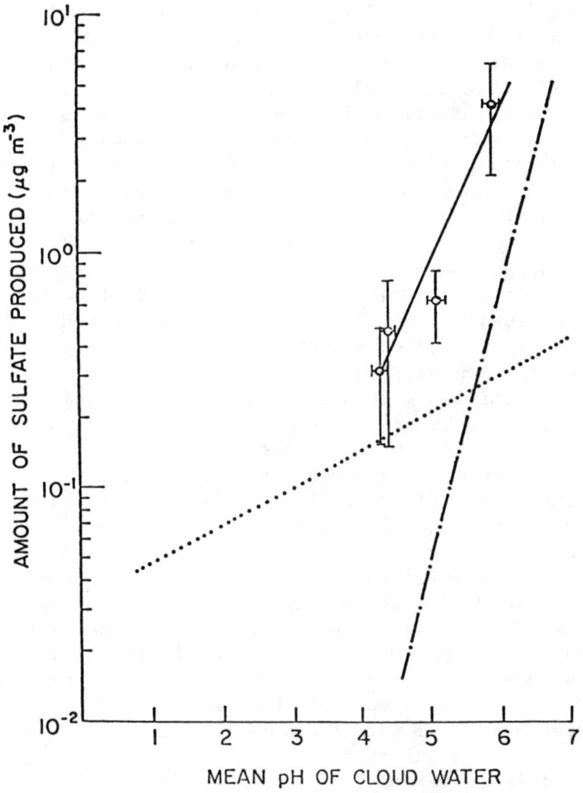

Fig. 1. Sulfate produced in wave clouds in western Washington state as a function of pH. A theoretical dependence for O_2 and O_3 oxidation in water droplets is shown by (— • —). This mechanism enhanced by H_2O_2 oxidation is given by the dotted line (••••) (Hegg and Hobbs, 1981).

TABLE 2. Distribution of Nitrate Samples Collected at Low Humidity (below deliquescent point of NH_4NO_3 at sampling temperature) (from Grosjean, 1981)

$HONO_2$, equil	Number of Samples	Diurnal Distribution	NO_3^- ($\mu g/m^3$)	$HONO_2$ Measured ($\mu g/m^3$)	$HONO_2$, equil ($\mu g/m^3$)	T °C	RH %
Achieved	5	2 night 3 early a.m.	high, 6-28	9-30	3-11	14-22	35-50
Close (within 5 $\mu g/m^3$	7	daytime	low, <5	12-37	14-37	24-32	31-48
Not achieved	5	daytime	low, <3	4-20	30-44	30-33	16-46

by sinking over the crest of the wave. The cloud is maintained by condensation of water vapor at the "crest" of the wave. Thus the cloudy air reactor remains stationary so that repeated sampling of incoming air can be achieved, with repeated collection of cloud water under approximately the same conditions.

An example of Hegg and Hobb's (1981) results is shown in Table 1. They indicate a significant production of sulfate within the wave clouds sampled on different occasions. The calculated rate of oxidation is droplet-pH dependent as the aqueous theory predicts, but the apparent oxidation rate was very high (>100 percent SO_2/hr) for short time periods on two occasions. This rate is higher than would be expected from available models, though it has been suggested that such rates could be attributed to H_2O_2 mechanism (R. Martin, personal communication, 1981). The rate of production in the Hegg and Hobbs (1981) work decreased substantially over long time periods, evidently due to suppression of absorption by acidification of droplets.

Some additional results of Hegg and Hobbs (1981) are given in Figure 1, showing the measured pH dependence. The broken lines in the figure show two different heterogeneous oxidation rate schemes. The first involves a combined dissolved oxygen and ozone rate expression; the second includes this combination plus the rate for dissolved H_2O_2 oxidation. The latter is much less sensitive to droplet acidity than the other two. The sulfate produced in the wave clouds is larger than either theory estimates. The reasons for this have not as yet been explained. The authors did not state assumptions about metal ion catalysts, but the level of ammonia was taken as 0.1 ppb in the calculations.

A different example of direct evidence is derived from recent nitric acid observations taken in the summer of 1980 at Claremont, CA. Grosjean (1981) has analyzed data obtained for NO, NO_2, $HONO_2$, and particulate nitrate. He has reported that the observed behavior of particulate nitrate sampled on inert Teflon filters in relatively dry air as a function of temperature is consistent with the very high volatility of ammonium nitrate projected from equilibrium considerations for the reaction $NH_4NO_3(s) \rightleftarrows HONO_2(g) + NH_3(g)$ (e.g., Stelson et al., 1979). At ambient humidities lower than the deliquescent point of NH_4NO_3, the stability of NH_4NO_3 in ambient air is given by this equilibrium expression, and thus depends only on the precursor concentration and the temperature. At humidities higher than the deliquescent point of NH_4NO_3, the equilibrium no longer applies, and aqueous chemistry must be considered. In the latter case, the temperature dependence of NH_4NO_3 stability in aqueous aerosols is still unknown. In the case of solid NH_4NO_3, the equilibrium constant varies strongly with temperature.

Formation of NH_4NO_3 at the high daytime summer temperatures typically encountered in southern California requires the photochemical production of significant amounts of nitric acid. For example, if the ambient temperature is 35°C and the humidity is below the deliquescent point of NH_4NO_3 at that temperature (e.g., less than 50 percent RH), ~53 $\mu g/m^3$ of nitric acid (~21 ppb) would be required (not to mention an equimolar amount of ammonia) to satisfy the vapor-solid equilibrium. In this case negligible amounts of particulate nitrate would be formed even though significant amounts of nitric acid, up to 50 $\mu g/m^3$, may be produced during a smog episode.

Grosjean (1981) has categorized the data according to temperature and humidity averaged over the experiments at Claremont, CA. The resulting breakdown indicated that 17 samples were collected at humidities below the deliquescent point of NH_4NO_3 at the reported sampling temperature (Table 2). For these, comparisons of the measured nitric acid concentrations with those calculated from the Stelson et al. (1979) model indicated that the $HONO_2$ concentrations required to satisfy equilibrium were achieved in only five cases, for which high particulate nitrate concentrations were observed, in agreement with theory. In the 12 remaining cases in which measured $HONO_2$ levels were too low to satisfy the equilibrium condition, the measured particulate nitrate was always low, again in agreement with the theory. As expected from the strong temperature dependence, the five high-nitrate cases corresponded to nighttime and early morning samples, while the low nitrate cases were all daytime samples, even though appreciable levels of nitric acid, up to 44 μg/m^3, were present during the photochemical smog episodes studied.

Indirect Evidence for Heterogeneous Processes

By far the most extensive evidence for heterogeneous chemistry in the atmosphere comes from indirect inference based on observational data. Since the measurements of nitrate in the air are generally viewed with suspicion because of sampling artifacts, their interpretation for an atmospheric effect is considered highly uncertain, with a few exceptions as described previously. In view of this, let us concentrate on atmospheric sulfate behavior for the remainder of the paper.

The dependence of sulfate concentrations on relative humidity has been cited by many authors as evidence which is consistent with heterogeneous oxidation. Relative humidity is associated with liquid water content. Hence, it follows that the higher the relative humidity, the more moisture present in hygroscopic particles, within which reactions may take place. A sulfate production dependence on ozone can arise either from ozone as an indicator of homogeneous reactions, or in heterogeneous processes from the absorption of ozone in water drops also containing SO_2. Thus an ozone-sulfate dependence in itself does not verify homogeneous gas-phase reactions.

An example of the association between ozone, relative humidity, and sulfate production was derived from recent observations in Los Angeles (S. L. Heisler, private communication, 1980). In this experiment, particulate samples were obtained at 2-hour intervals over days with different smog categories at Anaheim and Azusa. Anaheim is a few miles inland from the Pacific Ocean, and near large sources of SO_2. Azusa is twenty miles further inland, and is on a path downwind from the SO_2 sources much of the time. Thus, the combination qualitatively should yield results showing change with reaction time if the winds always blow inland.

The Los Angeles sampling experiment was conducted in the fall and winter of 1977 and the spring of 1978 to obtain a series of measurements on selected days corresponding to low ozone (<20 ppb, hourly maximum) and low sulfate (<20 μg/m^3, 24-hour average), high ozone (>20 ppb) and low sulfate, low ozone and high sulfate (>20 μg/m^3), and high ozone and high sulfate days. The results are shown in Figure 2 for midday maximum $SO_4^=$ samples at the daily ozone maximum as a function of relative humidity. Although the results are quite scattered, there is a definite separation in the clusters of observations for different ozone and humidity conditions. Sulfate accumulation at both sites was highest on high

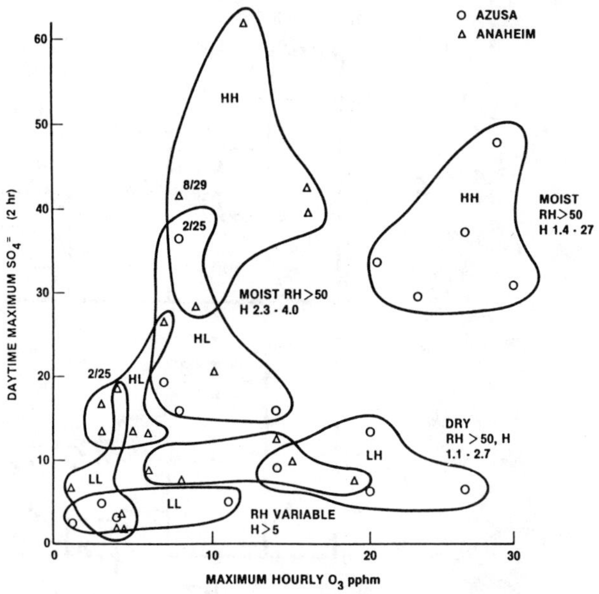

Fig. 2. Daytime maximum 2-hour average sulfate versus hourly maximum ozone concentrations for category days at Azusa and Anaheim. RH is relative humidity in percent and H is mixing height in thousands of feet. HH, HL, and LL indicate $SO_4^=/O_3$ categories (S. L. Heisler, private communication 1980).

ozone/high relative humidity days, and lowest during low ozone/low humidity days. However, moderately high daily maximum sulfate concentrations were obtained without high ozone concentrations but with relative humidity larger than 50 percent. These days generally included fog and/or low stratus cloud cover. Low sulfate levels also were observed with relative humidity less than 50 percent even with elevated ozone concentrations. This kind of evidence suggests qualitatively that aqueous droplet reactions may be active in the Los Angeles environment, since relative humidity above 50 percent appears to be a persistent requirement for elevated sulfate concentrations. Again, the ozone dependence cannot be interpreted unambiguously as a measure of homogeneous or heterogeneous processes.

There has been much said about the propensity for Los Angeles air to produce particulate sulfate and nitrate. One possible reason why particulate nitrate is found especially in Los Angeles air has been attributed to relatively high levels of ambient ammonia. Ammonia can react with nitric acid vapor or in water droplets to form ammonium nitrate. If ammonia concentrations are high enough, not only can nitrate be explained stoichiometrically as an ammonium salt, but sulfate also can be accounted for as an ammonium salt.

As a part of the study done by S. L. Heisler (private communication, 1980), an attempt was made to verify the abundance of ammonia in Los Angeles air. This was done by a dual filter

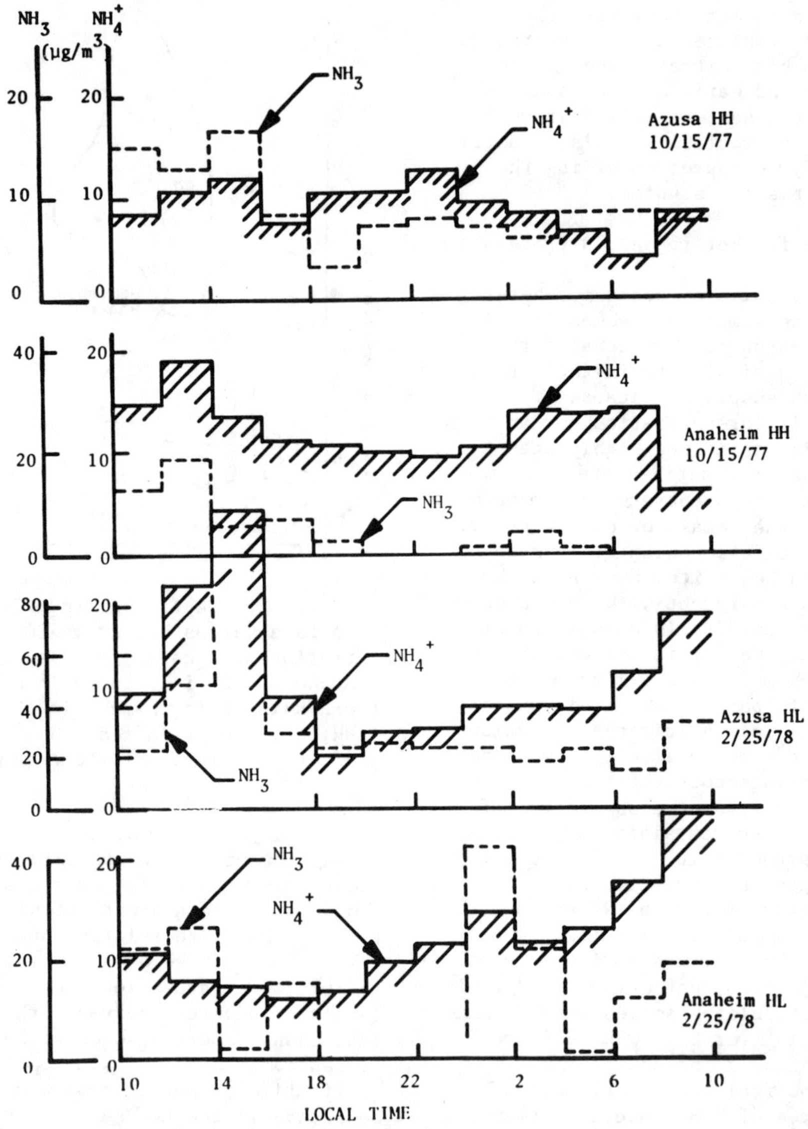

Fig. 3. Diurnal changes in ammonia and ammonium ion on October 15, 1977 (HH = high sulfate - high ozone) and February 25, 1978 (HL = high sulfate, low ozone) at Azusa and Anaheim (S. L. Heisler, private communication, 1980).

method, where aerosol was removed in the first filter and ammonia was absorbed on a backup filter impregnated with potassium bisulfate. Sample results for the presence of ammonium in the aerosol and ammonia in the backup filter are shown in Figure 3. In general there was an excess of ammonia in Los Angeles air relative to the collected particles. The excess appears to be greater on the 2 days shown in Azusa (inland, east of Los Angeles) compared with Anaheim nearer the ocean. The excess ammonia provides an ample reservoir for buffering acidic reactions in moist aerosol particles or in fog in the Los Angeles area. However, such data in themselves do not prove that ammonia plays an active role in particulate sulfate or nitrate production.

S. L. Heisler (private communication, 1980) also attempted to explore the presence of the valence state of metals in the filter-collected aerosol particles at Azusa and Anaheim. Photoelectron spectroscopy was used to investigate the valence states of iron and manganese since this is known to be a factor in the catalytic activity of metals for SO_2 oxidation. Insufficient material was collected to obtain information on this question. Attempts to correlate iron, manganese, or vanadium concentrations with sulfate behavior also failed to give any significant relationships.

Thus, this work, like other past experiments, did not provide any evidence for metal-catalyzed oxidation reactions in the air.

Another approach to rationalize atmospheric chemical processes has used patterns of diurnal change. Although this method has provided considerable insight into smog chemistry, hour-by-hour regularity of changes (e.g., Hidy and Mueller, 1980) may have been overinterpreted. Large variabilities in diurnal changes in sulfate and nitrate at different stations in Los Angeles were obtained in the fall, winter, and spring observations of S. L. Heisler (private communication, 1980). These showed that air transport and mixing processes combined with chemical reactions yield a complex picture which requires interpretation on a case-by-case basis. Nevertheless, this method gives valuable descriptions of persistent or prevalent chemical factors in selected areas.

Contrasts in sulfate chemistry at nine rural sites selected to represent regional processes (scales of 100 to 1000 km in geographical extent) over the greater northeastern United States were obtained from daily 3-hour averaged observations of the Sulfate Regional Experiment (SURE) (P. K. Mueller, unpublished data, 1981). Nine (Class I) stations extending eastward from Rockport, IN to Montague, MA were operated continuously for over a year between August 1977 and December 1978.

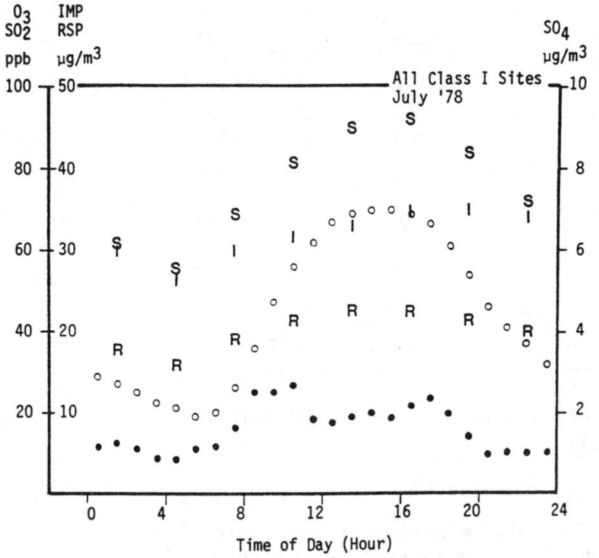

Fig. 4. Diurnal distribution of air quality parameters observed during the SURE in the eastern United States. Values are geometric means for particulates and arithmetic means for gases. I and R represent the particulate mass concentration of inhalable particulate matter (IPM; <15 µm diameter) and respirable particulate matter (RPM; <2.5 µm diameter), respectively; ● = SO_2, ○ = O_3 (P. K. Mueller, unpublished data, 1981).

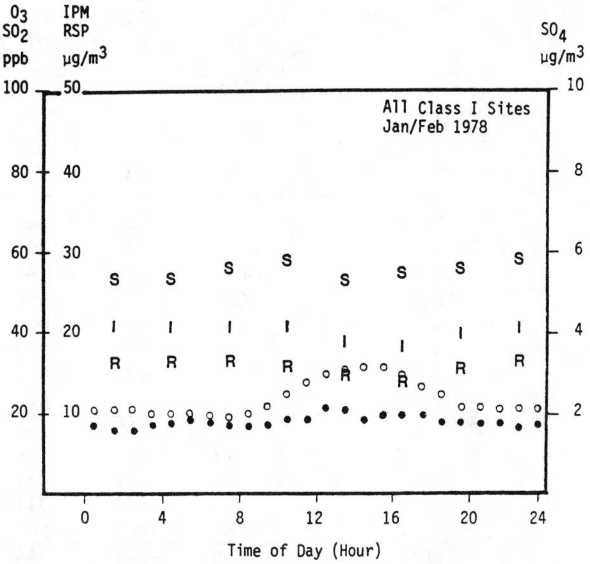

Fig. 5. Diurnal distribution of air quality parameters observed during the SURE in the eastern United States. Values are geometric means for particulates and arithmetic means for gases. I and R represent the particulate mass concentration of inhalable particulate matter (IPM; <15 µm diameter) and respirable particulate matter (RPM; <2.5 µm diameter), respectively; ● = SO_2, ○ = O_3 (P. K. Mueller, unpublished data, 1981).

Summer (July 1978) and winter (January-February 1978) differences are shown in Figures 4 and 5 for the nine station means over the two 30-day periods. The strong average diurnal correspondence in summer (Figure 4) between sulfate and ozone can be interpreted as evidence for acceleration of homogeneous gas-phase oxidation with photochemical processes, since relative humidity has a strong inverse diurnal dependence. However, this correspondence again does not necessarily rule out heterogeneous oxidation in moist air in summer, considering the midday increase in particle concentration. Probably both reaction paths occur, but such data do not permit interpretation of their relative importance without confirmation from analysis using a diagnostic model.

In winter, the diurnal change in sulfate is less dramatic than in summer, and it does not follow the daily ozone concentration (Figure 5). Instead, sulfate peaks in midmorning, earlier than ozone, and again at night. Sulfate also follows particulate mass more closely in winter than in summer. The average SO_2 concentration in winter is relatively steady in concentration and has a 24-hour mean which is larger than in summer. This suggests a lower rate (or efficiency) of SO_2 oxidation to sulfate. The average diurnal sulfate and ozone variation is much smaller in winter, but the mean sulfate level is similar.

TABLE 3. Component Loadings for the Photochemical Principal Component[a] of the St. Louis RAPS Data (values × 1000) (from Henry and Hidy, 1981)

Variable[b]	Site 105	108	112	115	118	120	122	124
TSPM	300	511	441	500	408	288	294	429
TS	-088	-116	-097	-	-251	-	-	-283
SO_2	-084	024	-	-104	-	042	-066	-
H_2S	003	059	-	-	-	-	-	-
NO	-117	008	-052	-115	-388	-146	061	-501
NO_2	-039	179	307	149	340	339	015	-414
O_3MX	713	797	756	734	743	584	804	749
AVOX	683	657	641	662	757	649	753	783
THC	215	289	-	007	-	254	025	-
NMHC	272	169	-	-	-	-	-	-
TMIN	840	842	871	849	892	836	914	886
TMAX	901	901	911	900	950	891	948	938
TAVG	888	889	901	883	933	874	951	933
RHAV	-306	-241	-244	-324	-311	-278	-137	-377
IRH	-246	-229	-191	-307	-279	-181	-142	-261
AHUM	694	689	724	707	746	702	803	754
WS00	-568	-551	-566	-649	-324	-504	-178	-072
WS12	-357	-481	-384	-492	-092	-456	-202	014
WSAV	-543	-577	-515	-617	-202	-502	-234	-035
WDQ1	057	085	071	164	-050	178	006	-071
WDQ2	385	091	200	191	335	330	175	134
WDQ3	235	028	227	-112	170	156	211	368
WDQ4	-627	-442	-461	-444	-518	-622	-327	-395
PCIP	-225	-065	-018	-249	036	009	045	-027
PRES	-284	-245	-220	-260	-323	-339	-381	-406
DELP	-170	-172	-244	-141	-227	-328	-010	-165
MXAM	-631	-637	-617	-637	-562	-661	-477	-507
STAM	362	282	412	276	214	379	084	190
MXPM	397	503	382	432	456	428	445	564
STPM	-333	-321	-244	-330	-350	-301	-467	-470
VENT	054	022	014	-045	-445	030	-100	-095
Sulfate average	13	14	12	11	11	10	12	11
and std. deviations	7.7	9.0	8.3	6.5	7.6	6.7	8.7	7.8
Percentage sulfate variability	25.1	25.8	23.4	26.4	12.4	24.1	24.7	10.9

[a]Consistently large values are boxed.

[b]Variables included in the St. Louis principal component analysis (daily averages used are from midnight to midnight):

TSPM	Total suspended particulates minus 110% of sulfate and nitrate ($\mu g/m^3$)
TS	Total sulfur gas (ppm)
SO_2	Sulfur dioxide (ppm)
H_2S	Hydrogen sulfide (ppm)
NO	Nitric oxide (ppm)
NO_2	Nitrogen dioxide (ppm)
O_3MX	Daily maximum 1-hour oxidant (ppm)
AVOX	Daily average oxidant
THC	Total hydrocarbon gas (ppm)
NMHC	Nonmethane hydrocarbon gas (ppm)
TMIN	Daily minimum temperature (°C)
TMAX	Daily maximum temperature (°C)
TAVG	Daily average temperature (°C)
RHAV	Daily average relative humidity (%)
IRH	$(1\text{-RHAV})^{-1}$
AHUM	Absolute humidity (mb of H_2O)
WS00	Wind speed at midnight (m/s)
WS12	Wind speed at noon (m/s)
WSAV	Daily average wind speed (m/s)
WDQ1	Frequency of wind direction from 0° - 90° (%) (0° = North)
WDQ2	Frequency of wind direction from 90° - 180° (%)

WDQ3	Frequency of wind direction from 180° - 270° (%)
WDQ4	Frequency of wind direction from 270° - 360° (%)
PCIP	Precipitation (in. of H_2O/day)
PRES	Station pressure (in. of Hg)
DELP	Pressure at 0 hours minus pressure at 2300 hours
MXAM	Mixing height at 6 a.m. CST (m)
STAM	Temperature difference across morning inversion (°C)
MXPM	Mixing height at 6 p.m. CST (m)
STPM	Temperature difference across afternoon inversion (°C)
VENT	$[1/4\ (MXAM + MXPM)\ (WSAM + WSPM)]^{-1}$ (s/m^2) where WSAM + WSPM is wind speed in the center of the mixing layer
WSAM	Morning wind speed
WSPM	Afternoon wind speed

In winter the relative humidity at stations remains high in much of the data, but the absolute humidity drops dramatically compared with summer. This difference in the behavior of sulfate and ozone between summer and winter may be an indication of differences in the predominance of homogeneous and heterogeneous processes in the two seasons. Other results suggest that ozone and absolute humidity must be high for sulfate formation by homogeneous reactions (e.g., Henry and Hidy, 1979). The winter result from the SURE then suggests intensification of a heterogeneous contribution, but again does not preclude the homogeneous gas-phase pathway.

The ambiguities in interpretation of data where effects of transport, mixing, chemistry, and removal processes occur together are considerable. To avoid these, attempts have been made to separate out such processes by factor analysis (e.g., Henry and Hidy, 1979). This method basically sorts large bodies of aerometric data into groups, called principal components, which are independent of one another but explain most of the variability of the total data set. Regression of these components on sulfate gives the independent groupings of physical variables which account for variability in sulfate. The principal components have been found to be consistent with hypothesized atmospheric physical and chemical processes.

A component associated with homogeneous gas-phase photochemical oxidation has been identified in data taken in three different cities, Los Angeles, New York, and St. Louis. This component accounts for a significant part of the sulfate variability in these sets. An example for St. Louis is shown in Table 3 for the data obtained from the EPA Regional Air Pollution Study (RAPS) (Henry and Hidy, 1981). This component is nearly uniform in its character for eight different (numbered) stations. Association with photochemistry comes from high ozone values, high temperature contributions, high absolute humidity, and association with low morning mixing height (negative MXAM in Table 3). With collections of data it is often not possible to separate relative humidity effects from absolute humidity variation. Thus, results shown in Figure 2, for example, may as well be absolute-humidity related. An important feature of the principal component analysis is its ability to separate clearly the influences of these variables.

Factor analysis for data taken in Salt Lake City revealed a different interpretation from the St. Louis condition. Sulfate concentrations in Los Angeles, New York, and St. Louis are found to be maximum in summer, but they are highest in winter in Salt Lake City. The component accounting for the largest variability in sulfate in the Salt Lake City area is listed in Table 4. This component shows a strong positive association with relative humidity, a negative association with absolute humidity, and a modest positive association with particulate matter other than sulfate (TSPM). In addition, a positive association with SO_2 is found along with a strong negative association with ozone, temperature dispersion parameters of negative mixing height, and ventilation. We interpret this pattern to be associated with heterogeneous reactions for sulfate as compared with homogeneous photochemical reactions (see also Henry and Hidy (1981)). Thus, this result is taken as evidence for the significance of heterogeneous chemistry in Salt Lake City, as compared to its smaller significance in the other cities investigated.

As another example of indirect evidence of the role of heterogeneous chemistry, let us return to sulfate events in the Los Angeles area. For some time workers have found difficulties in explaining the relatively high concentrations of sulfate on the western side of the Los Angeles Basin. Some workers have postulated primary sulfate emissions as a factor; others have postulated the sea-land breeze multiday carryover as a factor. The first explanation has largely been eliminated by diagnostic modeling, while the second requires knowledge for testing of offshore pollution transport, which generally is not available. Testing with air quality models also could be done, but requires multiday simulation with confidence in reliability. This has not yet been achieved. A third explantion suggests that additional sulfate can be produced effectively by an oxidation reaction which is nearly zero order in SO_2, giving a virtual source on the west side of the metropolitan area. Such a reaction has been found through soot-catalyzed oxidation of absorbed SO_2 in water or fog droplets (S. G. Chang, private communication, 1980).

The patterns of sulfur oxide behavior for Los Angeles calculated with a large numerical model are shown in Figure 6. The model accounts for transport, mixing, SO_2 oxidation as a function of photochemical activity, and dry deposition (for details and validation see Henry et al. (1980)). We note that major SO_2 source areas are located to the southwest, northwest, and northeast. The

TABLE 4. Annual Data Set: First Principal Component[a] for the Salt Lake City Area (values × 1000)(from Henry and Hidy, 1981)

Variable[b]	Salt Lake City	Ogden	Kearns	Magna
TSPM	307	204	029	193
NO_3	615	602	504	426
SO_2	563	407	445	496
AVOX	-656	-586	-672	-687
MXOX	-471	-533	-573	-599
TMAX	-894	-901	-923	-922
TMIN	-894	-908	-916	-918
TAVG	-913	-923	-943	-943
RH12	806	801	834	836
RHAV	851	841	876	873
IRHN	611	573	616	598
AH12	-593	-594	-600	-601
AHAV	-671	-669	-673	-672
WS00	-349	-304	-286	-266
WS12	-311	-245	-196	195
WSAV	-491	-419	-397	-366
WDQ1	320	276	246	299
WDQ2	-269	-277	-208	-194
WDQ3	-074	005	-031	-029
WDQ4	201	210	155	140
PPTM	-040	-005	025	011
PPTA	-006	072	094	086
PRES	228	228	201	205
PRTD	016	025	073	080
MXPM	-732	-741	-755	-730
DTPM	556	594	476	463
MXAM	080	114	178	182
DTAM	069	042	-110	-129
VENT	568	593	523	486
Sulfate var.(%)	40	27	24	21
Sulfate regression coefficient	4.0072	2.123	4.051	8.22

[a]Consistently large values are boxed.

[b]List of variables and abbreviations for Salt Lake City (all daily averages are from 11:00 a.m., except oxidant, which is midnight to midnight):

SO_4	Sulfate
TSPM	Total suspended particulate minus 110% of sulfate and nitrate
NO_3	Particulate nitrate (NO_3 is affected by filter artifact from NO_2 absorption; NO_3 is then taken as a NO_2 surrogate)
SO_2	Sulfur dioxide
AVOX	Daily average oxidant
MXOX	Maximum hourly oxidant
TMAX	Maximum temperature
TMIN	Minimum temperature
TAVG	Average temperature
RH12	Noon relative humidity
RHAV	Daily average relative humidity
IRHN	$(1\text{-RH12})^{-1}$
AH12	Noon absolute humidity
AHAV	Average absolute humidity
WS00	Midnight wind speed
WS12	Noon wind speed
WSAV	Average wind speed
WDQ1	Frequency of winds from 0° - 90° (0° = North)
WDQ2	Frequency of winds from 90° - 180°
WDQ3	Frequency of winds from 180° - 270°
WDQ4	Frequency of winds from 270° - 360°
PPTM	Maximum hourly precipitation
PPTA	Average precipitation
PRES	Average station pressure
PRTD	Change in station pressure
MXPM	Afternoon mixing height
DTPM	Temperature difference across afternoon inversion
MXAM	Morning mixing height
DTAM	Temperature difference across morning inversion
VENT	Inverse of average of wind speed in center of mixing layer × average inversion height

calculations are based on a well-validated numerical model capable of simulating 24-hour average sulfate concentrations within 60 percent or less of measurements. The calculations in Figure 6 illustrate the pattern of sulfate and sulfur dioxide estimated for a moderate photochemical smog day on August 17, 1973. This day was selected for availability of short-duration observations of various pollutants, and for particularly detailed wind data. In the figure, observed values are shown at the top for comparison with the model estimates in the map. The results shown are for an empirical surrogate for homogeneous chemistry based on ozone behavior. The comparison between measurement and calculation for the homogeneous oxidation baseline is shown in the figure. Sulfate concentrations calculated by three selected mechanisms are compared in Table 5. These mechanisms are (1) homogeneous oxidation baseline, (2) a first-order reaction in SO_2 with constant rate coefficient, and (3) a homogeneous component with soot-catalyzed oxidation (zero order in SO_2) added. The latter was calculated, relating the soot concentration distribution to coefficient of haze (COH) values measured on that day at several sites in the area by the South Coast Air Quality Management District. The COH values are empirically correlated with soot in the Los Angeles air (Henry et al., 1980). The authors noted further that fog was present through midday on the case day on the west side of the basin, so that suspended liquid droplets were available for the reaction medium. To complete the calculation, the fog droplets were assumed to have a pH of 4.5 for the modeling.

Comparison of the results in Table 5 suggests that the aqueous soot-catalyzed oxidation process does indeed enhance sulfate levels on the west side (Anaheim and downtown Los Angeles). How-

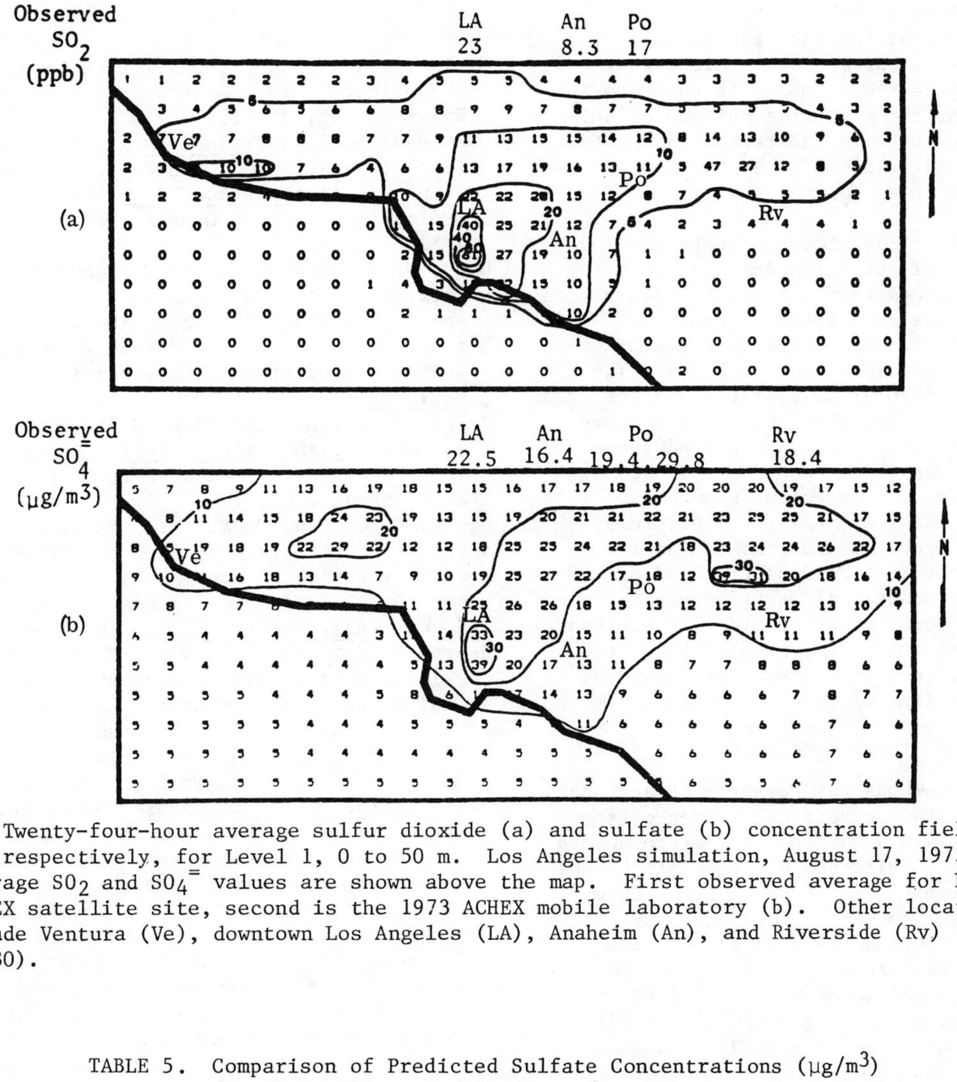

Fig. 6. Twenty-four-hour average sulfur dioxide (a) and sulfate (b) concentration fields, in ppb and $\mu g/m^3$, respectively, for Level 1, 0 to 50 m. Los Angeles simulation, August 17, 1973. Observed average SO_2 and $SO_4^=$ values are shown above the map. First observed average for Pomona (Po) is the ACHEX satellite site, second is the 1973 ACHEX mobile laboratory (b). Other locations indicated include Ventura (Ve), downtown Los Angeles (LA), Anaheim (An), and Riverside (Rv) (from Henry et al., 1980).

TABLE 5. Comparison of Predicted Sulfate Concentrations ($\mu g/m^3$) for Three Chemical Mechanisms (from Henry et al., 1980)

Time	Anaheim			Downtown Los Angeles			Pomona			Riverside		
	Homo.[a]	H. + Car.[b]	Const. Rate[c]	Homo.	H. + Car.	Const. Rate	Homo.	H. + Car.	Const. Rate	Homo.	H. + Car.	Const. Rate
2-3	12	14	14	22	27	26	14	14	17	16	16	18
5-6	14	19	18	30	39	45	16	16	23	17	17	21
8-9	19	26	25	69	83	111	16	16	24	18	18	23
11-12	23	28	27	39	48	50	12	14	16	9	9	10
14-15	6	8	5	6	8	5	15	16	15	6	7	6
17-18	6	7	6	6	8	6	18	19	16	7	8	6
20-21	13	16	17	8	11	8	5	6	6	4	3	3
23-24	19	23	29	8	14	10	10	10	10	3	3	3
24-hr avg.	13	17	17	25	31	35	13	13	15	11	11	12
Observed 24-hr avg.		16			23			19			18	

[a]Homogeneous empirical chemistry.
[b]Homogeneous plus carbon-based heterogeneous chemistry.
[c]Constant first-order rate.

ever, it does not appreciably influence the sulfate levels calculated for the eastern part of the basin (Pomona and Riverside). Although the contribution is relatively small, it does provide an explanation for the west-side buildup of sulfate. The results provide indirect evidence for the potential significance of the soot-based heterogeneous mechanism for the conditions simulated.

A final example of recent experiments showing indirect evidence of heterogeneous processes in sulfate behavior comes from detailed investigation of particle size distributions. Hering and Friedlander (1981) have reported that two distinct types of sulfate distributions are found. The first peaks at 0.54 ±0.07 μm diameter and the second at 0.20 ±0.02 μm. From analysis of particle growth laws, they have interpreted these differences in terms of the dominance of SO_2 oxidation inside particles (a volume growth law) in the former case. In the latter, the particle growth fits a homogeneous gas-phase oxidation process, with subsequent condensation and diffusional (surface-dominated) particle growth. In Los Angeles, the heterogeneous oxidation process appears to be dominant in the samples examined, while the reverse is found in samples taken in dry, desert conditions of the Southwest.

Summary and Conclusions

With improving methods for measurement and analysis of atmospheric observations, evidence is emerging which suggests the role of heterogeneous processes in sulfur oxide and nitrogen oxide chemistry. Direct evidence for heterogeneous chemistry is emerging from sampling in and near clouds, and from thermodynamic analysis. Indirect evidence comes from several methods ranging from inspection of large bodies of data and statistical inference to diagnostic air quality modeling. Homogeneous gas-phase SO_2 oxidation continues to be capable of explaining much of the available urban and rural sulfur oxide observations. Yet there is an increasing suggestion that heterogeneous processes also play a role, perhaps in winter and in cases where high condensed water content is present as a reaction medium. Although the aqueous soot-catalyzed reaction can now be tested with available data for its significance in SO_2 oxidation, similar tests for metal-catalyzed reactions have not been attempted. Future experiments combined with knowledge from theory and laboratory experiments will continue to reveal elements of these elusive heterogeneous reactions.

Acknowledgements. I am indebted to my associates Drs. R. C. Henry, S. L. Heisler, P. K. Mueller, D. Grosjean, and S. K. Friedlander for their contributions to experimental studies in the atmosphere. Our work at ERT could not have been done without the sponsorship of the Electric Power Research Institute, the American Petroleum Institute, and the Southern California Edison Company.

References

Beilke, S., and G. Gravenhorst, Heterogeneous SO_2 oxidation in the droplet phase, Atmos. Environ., 12, 231-239, 1978.

Grosjean, D. G., Critical Evaluation and Comparison of Measurement Methods for Nitrogeneous Compounds in the Atmosphere, Chapter 5, Coordinating Research Council, Atlanta, Georgia, 1981.

Hegg, D. A., and P. V. Hobbs, Cloud water chemistry and the production of sulfates in clouds, Atmos. Environ., 15, 1597, 1981.

Hering, S., and S. K. Friedlander, Origins of sulfur size distributions, submitted to Atmos. Environ., 1981.

Henry, R. C., D. G. Godden, G. M. Hidy, and N. J. Lordi, Simulation of Sulfur Oxide Behavior in Urban Areas, American Petroleum Institute, Washington, D.C., 1980.

Henry, R. C., and G. M. Hidy, Multivariate analysis of particulate sulfate and other air quality variables by principal components - Part I. Annual data for Los Angeles and New York, Atmos. Environ., 13, 1581-1596, 1981.

Henry, R. C., and G. M. Hidy, Multivariate analysis of particulate sulfate and other air quality variables by principal components - Part II. Salt Lake City, Utah, and St. Louis, Mo., Atmos. Environ., in press, 1982.

Hidy, G. M., and P. K. Mueller (Eds.), Origins and Character of Los Angeles Smog, pp. 395 ff., Wiley-Interscience, N.Y., 1980.

Judeikis, H. S., B. T. Stewart, and A. G. Wren, Laboratory studies of heterogeneous reactions of SO_2, Atmos. Environ., 12, 1633-1642, 1978.

Lazrus, A., E. Likens, V. Mohnen, and P. Haagtensen, Acidity in air and water in a case of warm frontal precipitation, paper presented at American Meteorological Society/Canadian Meteorological and Oceanographic Society Conference on the Long Range Transport of Airborne Pollutants and Acid Rain, Albany, N.Y., April 1981.

Miller, M. S., S. K. Friedlander, and G. M. Hidy, A chemical element balance for the Pasadena aerosol, in Aerosols and Atmospheric Chemistry, edited by G. M. Hidy, p. 301, Academic Press, N.Y., 1972.

Novakov, T., S. G. Chang, and A. B. Harker, Sulfates as pollution particulates. Catalytic formation on carbon (soot) particles, Science, 186, 259-261, 1974.

Robbins, R. C., R. D. Kaadle, and D. L. Eckhardt, The conversion of sodium chloride to hydrogen chloride in the atmosphere, J. Meteorol., 16, 53-56, 1959.

Scott, W. D., and P. V. Hobbs, The formation of sulfate in water droplets, J. Atmos. Sci., 24, 54-57, 1967.

Stelson, A. W., S. K. Friedlander, and J. H. Seinfeld, A note on the equilibrium relationship between ammonia and nitric acid and particulate ammonium nitrate, Atmos. Environ., 13, 369, 1979.

SOOT-CATALYZED ATMOSPHERIC REACTIONS

T. Novakov

Lawrence Berkeley Laboratory, University of California, Berkeley, California 94720

Abstract. This paper reviews the work of the Atmospheric Aerosol Research group at Lawrence Berkeley Laboratory on soot-catalyzed atmospheric reactions. Data are presented on the soot concentrations in various geographic regions of the United States. The experimental results for SO_2 oxidation on both dry and wet soot particles are reviewed.

Introduction

Atmospheric aerosol particles are classified as primary or secondary, depending on their origin. Primary particles are produced by sources such as combustion devices and are introduced into the atmosphere in particulate form. Secondary particles are formed in the atmosphere by chemical reactions among primary and secondary gaseous species, primary particles, and gaseous and liquid water.

Depending on the phases of the reactants, atmospheric reactions can be homogeneous or heterogeneous. Homogeneous reactions involve only gaseous species, while heterogeneous processes may involve gases and solid particles, gases and liquid droplets, or three-phase systems with gases, liquid droplets, and solid materials occluded in these droplets. These heterogeneous processes may be catalytic or stoichiometric, and may proceed in the bulk of a droplet, on the gas-solid interface, or on the solid-liquid interface.

In this paper we shall discuss the last two categories, specifically in conjunction with the role of combustion-generated soot particles in the oxidation of sulfur dioxide. Soot is synonymous with primary carbonaceous particulate material. It is a chemically complex material consisting of an organic component and a component variously referred to as elemental, graphitic, or black carbon, but for consistency we shall use the term black carbon here. Soot and smoke were the first air pollutants to be recognized and controlled. The word "smog" is a contraction of "smoke" and "fog." This kind of smog has become known as London-type smog.

Soot in the atmosphere not only contributes to the total particulate concentration in ambient air but also may serve as an efficient catalyst for atmospheric reactions such as the oxidation of SO_2 to sulfate. Soot has properties similar to those of activated carbon, which is well known to be a catalytically and surface-chemically active material.

While at one time the presence of soot in the atmosphere of industrial cities was obvious, it has become less obvious in more recent times. Improvements in combustion technology and the use of better-grade fuels have led to the virtual elimination of visible smoke emissions. The emphasis of air pollution control thus shifted away from primary particulate emissions toward controlling gaseous emissions, especially in view of the newer concept of Los Angeles-type photochemical smog, which was believed to contain neither smoke nor fog. According to such a view, the haze over the Los Angeles Air Basin on polluted days is due almost entirely to the photochemical conversion of certain invisible gases to light-scattering particles consisting of sulfates, nitrates, and secondary organics, but almost no soot.

The assessment of the chemical role of soot in the atmosphere in general, and in photochemical environments such as Los Angeles in particular, had to start with an empirical assessment of the soot concentrations. The results of these studies, as shown herein, have clearly demonstrated that soot is ubiquitous not only in urban atmospheres but also in remote regions such as the Arctic (Rosen et al., 1981). Therefore Los Angeles, with its abundant coastal fog, contains both components of London-type fog: smoke (or soot) and fog.

Soot in the Atmosphere

The first indication of the presence of soot in the Los Angeles atmosphere came from photoelectron spectroscopy (ESCA) of ambient samples (Novakov, 1973). These spectra showed that most of the particulate carbon was in a neutral chemical state compatible with combustion-generated soot. According to the photochemical hypothesis, the secondary organics should be oxygenated, resulting in an easily detectable chemical shift of the ESCA carbon peak. In our spectra, only a very small fraction of the carbon peak of the ambient samples was found to be chemically shifted. With ESCA we also found that the ambient carbonaceous material was relatively nonvolatile when the samples were heated to 350°C in vacuum. Such measurements, combined with visual observation of the filter samples (all of which appeared black or grey), led us to postulate that a substantial fraction of the total particulate carbon in the Los Angeles basin is in the form of soot. In recent years, we initiated a field measurement program to identify and quantify the amount of soot in many locations across the United States.

The principal approach used in our laboratory relies on the use of black carbon as a tracer for primary carbonaceous material (Rosen et al., 1980; Novakov, 1981) because black carbon can be produced only in a combustion process and is therefore definitely primary. The methodology that we adopted involved systematic measurements of the ratio of black carbon to total carbon for a large number of samples collected directly from sources, source-dominated environments, and well-aged ambient air (24-hour samples) (Hansen et al., 1980). The ambient samples were collected in areas with widely differing atmospheric chemical characteristics (e.g., degree of photochemical activity, source composition, geographic location). Measurements of this ratio from a number of source samples give insights into the relative black-to-total-carbon ratio of primary emissions and the source variabilities. Secondary material will not contain the black component but will increase the total mass of carbon and therefore reduce the black to total carbon fraction. That is, under high photochemical conditions one would expect this ratio to be significantly smaller than under conditions obviously heavily influenced by sources.

Because of the large number of samples that had to be analyzed, a fast-throughput optical attenuation method (Rosen et al., 1980) was used for determining black carbon. The validity of the optical attenuation method was checked by performing Raman spectroscopic (Rosen et al., 1978) and optoacoustic (Yasa et al., 1978) measurements on some of the ambient and source samples. Total particulate carbon was determined by a combustion method.

The optical attenuation method (Rosen et al., 1980) compares the transmission of a 633-nm He-Ne laser beam through a loaded filter relative to that of a blank filter. The relationship between the optical attenuation ATN and the black carbon content $[C_{black}]$ (μg black carbon per cm^2 of filter) can be written as

$$[C_{black}] = ATN/K \tag{1}$$

where $ATN = -100\ \ell n(I/I_0)$, I and I_0 are the transmitted light intensities for the loaded filter and for the filter blank, and K is the proportionality constant.

Besides the black carbon, particulate material also contains organic material which is not optically absorbing. The total amount of particulate carbon $[C_{tot}]$ (in μg/cm^2) is then

$$[C_{tot}] = [C_{black}] + [C_{org}] \tag{2}$$

We define specific attenuation (σ) as the attenuation per unit mass of total carbon:

$$\sigma \equiv \frac{ATN}{[C_{tot}]} = K[C_{black}]/[C_{tot}] \tag{3}$$

The determination of specific attenuation therefore gives an estimate of black carbon as a fraction of total carbon.

The proportionality constant K, which is equal to the specific attenuation of black carbon alone, was recently shown to have an average value of 20 (Hansen et al., 1980). In principle the percentage of soot (by definition soot is equivalent to primary carbonaceous material) in ambient particles can be determined from the ratio of ambient specific attenuation to an average specific attenuation of major primary sources (Novakov, 1981):

$$[soot]/[C_{tot}] = \sigma_{ambient}/\sigma_{source} \tag{4}$$

The average and extreme values of specific attenuation are listed in Table 1 along with the black carbon fraction of a number of source samples.

The percentage of soot in ambient carbonaceous particulates can be estimated by comparing the specific attenuation (σ) of sources with that of ambient samples. The fraction of soot is given in equation (4). The mean specific attenuations of ambient samples (weekends excluded) are listed in Table 2 in order of decreasing σ and soot fractions obtained by using equation (4) and $\sigma_{source} = 5.85$.

Based on this estimate, the New York City carbonaceous aerosol is essentially primary soot. A

TABLE 1. Specific Attenuation (σ) and Black Carbon (BC) (% of total C) of Source Samples

Source	No. Samples	Average		Highest		Lowest	
		σ	% BC	σ	% BC	σ	% BC
Parking garage	12	5.4	27	7.7	39	2.25	11
Diesel	6	5.6	28	5.7	29	3.5	18
Scooter	9	5.1	26	6.1	31	4.2	21
Tunnel	63	6.3	32	12.5	63	3.7	19
Natural gas	6	2.6	13	3.3	17	1.9	10
Garage and tunnel		5.85	29				

TABLE 2. Mean Specific Attenuation of Ambient Samples

Site	No. Samples	$\bar{\sigma}$	SDEV	Soot (%)
New York	211	5.69	1.34	97
Gaithersburg	155	4.72	1.51	81
Argonne	221	4.35	1.64	74
Berkeley	513	4.28	1.47	73
Anaheim	444	3.99	1.71	68
Fremont	461	3.74	1.25	64
Denver	42	3.47	1.49	59

different value of σ_{source} would certainly change the estimated soot percentage. However, New York City's average soot content would nevertheless remain the highest, irrespective of the actual numerical value of σ_{source}. It is logical that samples from this location have the highest soot content because the site represents a heavily traveled street canyon. Fremont and Anaheim samples have the smallest soot content on the average, as might be expected, because both sites represent receptor sites.

These results demonstrate that soot is certainly a major fraction of ambient particulate carbon at all locations studied. These findings also suggest that in the atmosphere there is a catalytically active material present in high concentrations, so that the assessment of its role in heterogeneous atmospheric chemistry is warranted.

Soot-Catalyzed SO_2 Oxidation

In this section we review our laboratory results on heterogeneous oxidation on soot particles in air and present results of numerical calculations which suggest that soot-catalyzed oxidation can be an important mechanism for sulfate formation in the atmosphere.

Novakov and coworkers (Novakov, 1973; Novakov et al., 1974) used photoelectron spectroscopy (ESCA) to study the oxidation of SO_2 on soot particles produced by a propane flame. It was found that under some conditions, a significant amount of sulfate can be produced by the catalytic action of soot particles. Although these early experiments were qualitative, it was nevertheless possible to conclude the following: (1) the reaction product is in a 6^+ oxidation state (i.e., sulfate), (2) soot-catalyzed oxidation of SO_2 is more efficient at high relative humidity, (3) the oxygen in air plays an important role in SO_2 oxidation, (4) soot-catalyzed oxidation exhibits a saturation effect, and (5) SO_2 can be oxidized on other types of graphitic carbonaceous particles, such as ground graphite particles and activated carbon. Results from the experiments with combustion-produced soot particles are essentially similar to those obtained for activated carbon by Davtyan and Tkach (1961) and Siedlewski (1965).

Soot-catalyzed SO_2 oxidation can proceed by two mechanisms: a "dry" mechanism, in the presence of water, and a "wet" mechanism, when the soot particles are covered by a liquid water layer. Soot particles can acquire a liquid water layer by condensation or by hygroscopic action of, for example, small amounts of sulfuric acid or sulfate salts. The experiments mentioned (Novakov et al., 1974; Davtyan and Tkach, 1961; Siedlowski, 1965) involved the dry mechanism. The wet mechanism is much more efficient than the dry and is applicable to situations in plumes, clouds, fogs, and the ambient atmosphere when the aerosol particles are covered with a liquid water layer. The dry mechanism is expected to operate in stacks or under conditions of low relative humidity.

A description of the dry mechanism was given by Yamamoto et al. (1972), who studied the reaction

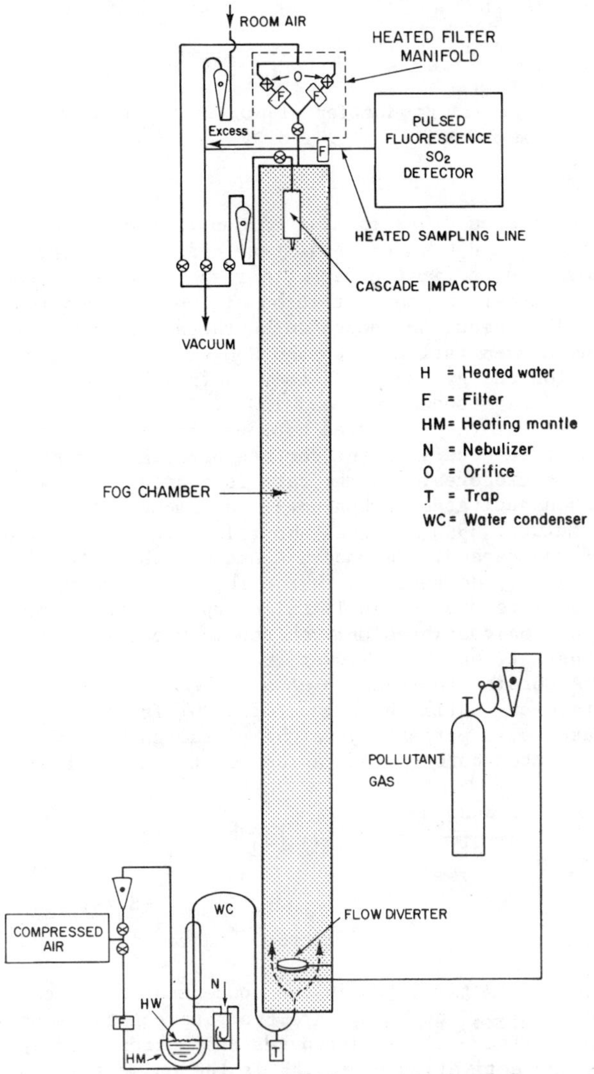

Fig. 1. Laboratory fog chamber and associated equipment (from Benner et al., 1981).

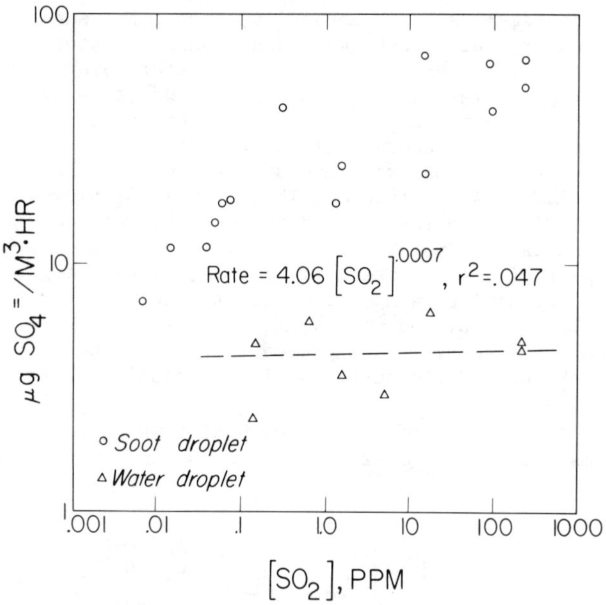

Fig. 2. Production of SO$_4$ in fog droplets (from Benner et al., 1981).

kinetics on dry activated carbon in the presence of O$_2$ and H$_2$O vapor. The rate of reaction was found to be first order with respect to SO$_2$, provided that the concentration of SO$_2$ was less than 0.01 percent, and depended on the square root of the concentration of O$_2$ and H$_2$O vapor. The activation energy was found to vary from -4 to -7 kcal/mol between 70°C and 150°C, depending on the origin of the activated carbon. Initially the reaction occurs on the surface of both micropores and macropores, and the rate is constant for a given activated carbon until the amount of accumulated H$_2$SO$_4$ reaches about 10 percent by weight of the carbon. Beyond that amount, the rate gradually decreases with the reaction time until the micropore volume is filled up by H$_2$SO$_4$. The reaction continues only on the macropores at a constant, but much slower, rate. According to Yamamoto et al. (1972), a rate expression (until the amount of H$_2$SO$_4$ formed reaches 10 percent by weight of the carbon) for activated carbon used can be written as follows:

$$\frac{d[H_2SO_4]}{dt} = [C][SO_2][O_2]^{0.5}[H_2O]^{0.5}$$

$$\times (k_{micro} + k_{macro})^{-E_a/RT} \qquad (5)$$

where t is time, [C] is the concentration of carbon, k_{micro} and k_{macro} are the rate constants on the surface of the micropores and macropores, E_a is the activation energy, R is the universal gas constant, and T is absolute temperature.

The dry mechanism is relatively inefficient

because the reaction product remains on the carbon surface and acts as the catalyst poison. The situation is entirely different when soot (or another carbon) is covered with a layer of liquid water and the catalytic oxidation occurs at the solid-liquid interface. In this case there is constant regeneration of active sites because the reaction product is soluble in water and therefore leaves the soot surface.

Such reactions were studied in detail by Chang et al. (1979) and Brodzinsky et al. (1980), who used both combustion soots and activated carbons. These studies used suspensions of activated carbon in water to which different concentrations of sulfurous acid were added. The results of these studies can be summarized as follows: (1) the reaction rate is first order and 0.69th order with respect to the concentration of carbon and dissolved oxygen, respectively; (2) the reaction rate is effectively pH independent (pH < 7.6); (3) the activation energy of the reaction is 11.7 kcal/mol; (4) there is a mass balance between the consumption of sulfurous acid and the production of sulfuric acid; and (5) the reaction rate has a complex dependence on the concentration of H$_2$SO$_3$, ranging between a second and zeroth order reaction. Other workers have also studied similar

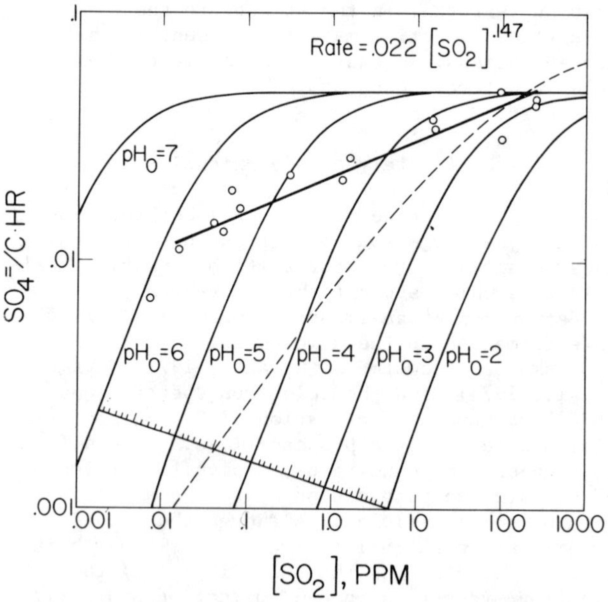

Fig. 3. Normalized rate of SO$_4$ formation in wet soot droplets versus SO$_2$. The open circles and the solid line drawn through the circles were obtained from fog chamber data. The family of curves labeled pH$_0$ = 2 through pH$_0$ = 7 was calculated from equation (6), where pH$_0$ is the selected pH at which the droplets were buffered before exposure to SO$_2$ and CO$_2$. The dashed line shows the rate of SO$_4$ formation for wet nonbuffered soot droplets that were equilibrated with SO$_2$ and 320 ppm CO$_2$ (from Benner et al., 1982).

systems. For example, Cofer et al. (1980) studied the effect of NO_2 on soot-catalyzed SO_2 oxidation and found that significant enhancement of oxidation occurs when NO_2 is present.

The oxidation of S(IV) to S(VI) can be expressed simply by the symbolic net reaction

$$2S(IV) + O_2 \rightarrow 2S(VI) \qquad (a)$$

(For this and following reactions, let C = carbon content (assuming it to be proportional to active surface area); S(IV) = $H_2O \cdot SO_2$, HSO_3^-, and $SO_3^=$; S(VI) = HSO_4^-, and $SO_4^=$. The experimental results of Chang et al. (1979) and Brodzinsky et al. (1980) yield the following empirical rate law for this reaction:

$$\frac{d[S(VI)]}{dt} = k[C][O_2]^{0.69} \frac{\alpha[S(IV)]^2}{1 + \beta[S(IV)] + \alpha[S(IV)]^2} \qquad (6)$$

where k = 1.69×10^{-5} $mol^{0.31} \cdot l^{0.69}/g \cdot sec$, $\alpha = 1.50 \times 10^{12}$ l^2/mol^2, $\beta = 3.06 \times 10^6$ l/mol, [C] = g carbon/l, $[O_2]$ = mol dissolved oxygen/l, [S(IV)] = total mol S(IV)/l, and [S(VI)] = total mol S(VI)/l. Using the Arrhenius equation, the rate constant may be expressed as

$$k = Ae^{-E_a/RT} \qquad (7)$$

where E_a = 11.7 kcal/mol and A = 9.04×10^3 $mol^{0.31} \cdot l^{0.69}/g \cdot sec$.

The fact that the reaction rate is first order with respect to the activated carbon catalyst is representative of surface catalysis. The reaction will then proceed via the adsorption of the reaction species onto a catalytically active site. A series of adsorption steps can explain the fractional and varying order of reaction with respect to O_2 and S(IV) and are proposed here in the following four-step reaction:

$$C + O_2 \underset{k_{-1}}{\overset{k_1}{\rightleftharpoons}} C \cdot O_2 \qquad (b)$$

$$C \cdot O_2 + S(IV) \underset{k_{-2}}{\overset{k_2}{\rightleftharpoons}} C \cdot O_2 \cdot S(IV) \qquad (c)$$

$$C \cdot O_2 \cdot S(IV) + S(IV) \underset{k_{-3}}{\overset{k_3}{\rightleftharpoons}} C \cdot O_2 \cdot 2S(IV) \qquad (d)$$

$$C \cdot O_2 \cdot 2S(IV) \xrightarrow{k_4} C + 2S(VI) \qquad (e)$$

Benner et al. (1982) extended the above research by laboratory fog chamber studies on dispersed water droplets that contain soot particles. In these experiments, particles of Nuchar-SN carbon, preextracted to remove $SO_4^=$, were resuspended by nebulization, and the resultant mist was passed over heated water and subsequently cooled to cause the nebulized particles to grow into larger droplets (Figure 1). These droplets were exposed to SO_2 which was introduced into the bottom of the fog chamber. Particulate samples were collected at the top of the chamber and analyzed for $SO_4^=$ by liquid chromatography and for carbon by combustion.

When pure water droplets were exposed to SO_2 in the fog chamber, the rate of $SO_4^=$ formation (μg $SO_4^=/m^3 \cdot h$) equals 4.06 (ppm $SO_2)^{0.0007}$ (Figure 2). If the droplets contained Nuchar-SN particles, the rate of $SO_4^=$ formation was found to be significantly faster than for pure water droplets. For example, wet soot particles exposed to 0.007 ppm SO_2 produced $SO_4^=$ faster than pure water droplets exposed to 222 ppm SO_2. The data for $SO_4^=$ formation by wet soot particles are plotted with open circles in Figure 2. When the soot/droplet data in Figure 2 are normalized to carbon concentration, the carbon-normalized rate of $SO_4^=$ formation can be plotted as shown in Figure 3 (open circles). The equation which best fits the carbon-normalized soot droplet data is $SO_4^=/C \cdot h$ = $0.222(SO_2)^{0.147}$.

In Figure 3 a family of curves shows the rate of $SO_4^=$ formation, normalized to C, as a function of $[SO_2]$ for various initial $[H^+]$'s. This family of curves was calculated from the rate expression in equation (6). The pH effect is related to the absorption of SO_2 by the droplet. The droplet with the lower initial pH will have a lower equilibrium [S(IV)], and the oxidation rate will be similarly slower. The family of curves was cal-

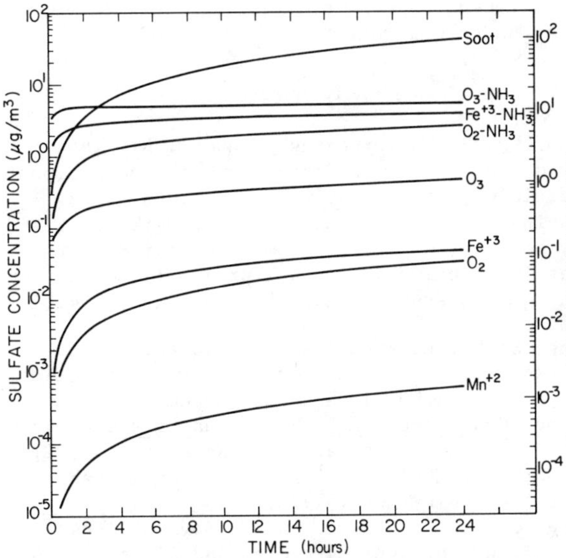

Fig. 4. Comparison of the relative significance of various SO_2 conversion processes in aqueous droplets (from Chang et al., 1981).

culated assuming that the droplet had a specified initial pH and equilibrated with SO_2 and 320 ppm CO_2 to produce a certain equilibrium [S(IV)] before the oxidation started.

The dashed line in Figure 3 shows the S(IV) oxidation rate expected for wet soot particles in equilibrium with SO_2. The difference between the fog chamber data and this dashed line is due to the alkaline nature of Nuchar-SN. Neutralization increased the oxidation rate because it brought more $SO_2 \cdot H_2O$ and HSO_3^- into the droplet.

Chang et al. (1981) have carried out box-type calculations to compare the relative importance of sulfate production mechanisms by soot particles and other mechanisms involving liquid water. The following initial conditions were used in the calculation: liquid water, 0.05 g/m^3; SO_2, 0.01 ppm; O_3, 0.05 ppm; NH_3, 5 ppb; and CO_2, 0.000311 atm. Concentrations of particulate Fe and Mn of 250 ng/m^3 amd 20 ng/m^3, respectively, were assumed. However, only 0.13 percent of the total iron and 0.25 percent of the Mn are water soluble, according to Gordon et al. (1975). The concentration of soot was taken as 10 μg/m^3.

The results of this calculation, shown in Figure 4, indicate that O_3 and soot can be important mechanisms for sulfate aerosol formation. In general the O_3 mechanism is more important under high pH and/or photoactivity conditions when the concentration of O_3 is high, whereas the soot process is more important when the lifetime of fog or clouds is long and the pH of the droplets is low.

Acknowledgment. This work was supported by the Assistant Secretary for the Environment, Office of Health and Environmental Research, Pollutant Characterization and Safety Research Division of the U.S. Department of Energy under Contract No. W-7405-ENG-48, and by the National Science Foundation under Contract No. ATM 80-13707.

References

Benner, W. H., R. Brodzinsky, and T. Novakov, Oxidation of SO_2 in droplets which contain soot particles, Atmos. Environ., in press, 1982.

Brodzinsky, R., S.-G. Chang, S. S. Markowitz, and T. Novakov, Kinetics and mechanism for the catalytic oxidation of sulfur dioxide on carbon in aqueous suspensions, J. Phys. Chem., 84, 3354, 1980.

Chang, S.-G., R. Brodzinsky, R. Toossi, S. S. Markowitz, and T. Novakov, Catalytic oxidation of SO_2 on carbon in aqueous suspensions, in Proc. Conference on Carbonaceous Particles in the Atmosphere, edited by T. Novakov, p. 122, Report No. LBL-9037, Lawrence Berkeley Laboratory, Berkeley, Ca., 1979.

Chang, S.-G., R. Toossi, and T. Novakov, The importance of soot particles and nitrous acid in oxidizing SO_2 in atmospheric aqueous droplets, Atmos. Environ., 15, 1287, 1981.

Cofer, W. R. III, David R. Schryer, and Robert S. Rogowski, The enhanced oxidation of SO_2 by NO_2 on carbon particulates, Atmos. Environ., 14, 571-575, 1980.

Davtyan, O. K., and Yu. A. Tkach, The mechanism of oxidation, hydrogenation, and electrochemical oxidation on solid catalysts. II. The catalytic activity of surface 'oxides' on carbon, Russian J. Phys. Chem., 35, 486, 1961.

Gordon, G. E., D. D. Davis, G. W. Israel, H. E. Landsberg, and T. C. O'Haver, Atmospheric Impact of Major Sources and Consumers of Energy, Report NSF/RA/E-75/189, National Science Foundation, 1975. (Available from NTIS as PB 262 574.)

Hansen, A. D. A., et al., The use of an optical attenuation technique to estimate the carbonaceous component of urban aerosols, in Atmospheric Aerosol Research Annual Report FY-1979, p. 8, Report No. LBL-10735, Lawrence Berkeley Laboratory, Berkeley, Ca., 1980.

Novakov, T., Chemical characterization of atmospheric pollution particulates by photoelectron spectroscopy, in Proceedings, 2nd Joint Conference on Sensing of Environmental Pollutants, p. 197, Instrument Society of America, Pittsburgh, 1973.

Novakov, T., Microchemical characterization of aerosols, in Nature, Aim and Methods of Microchemistry, edited by H. Malissa, M. Grasserbauer, and R. Belchers, p. 141, Springer-Verlag, Vienna, 1981.

Novakov, T., S.-G. Chang, and A. B. Harker, Sulfates as pollution particulates: Catalytic formation on carbon (soot) particles, Science, 186, 259, 1974.

Rosen, H., A. D. A. Hansen, R. L. Dod, and T. Novakov, Soot in urban atmospheres: Determination by an optical absorption technique, Science, 208, 741, 1980.

Rosen, H., A. D. A. Hansen, L. Gundel, and T. Novakov, Identification of the optically absorbing component of urban aerosols, Appl. Opt., 17, 3859, 1978.

Rosen, H., T. Novakov, and B. A. Bodhaine, Soot in the Arctic, Atmos. Environ., 15, 1371, 1981.

Siedlewski, J., The mechanism of catalytic oxidation on activated carbon. The influence of free carbon radicals on the adsorption of SO_2, Int. Chem. Eng., 5, 297, 1965.

Yamamoto, K., M. Seki, and K. Kawazoe, Absorption of sulfur dioxide on activated carbon in the flue gas desulfurization process. II. Rate of oxidation of sulfur dioxide on activated carbon surfaces, Nippon Kagaku Kaishi, 6, 1046, 1972.

Yasa, Z., N. Amer, H. Rosen, A. D. A. Hansen, and T. Novakov, Photoacoustic investigation of urban aerosol particles, Appl. Opt., 18, 2528, 1978.

THE RELATIVE IMPORTANCE OF VARIOUS URBAN SULFATE AEROSOL PRODUCTION MECHANISMS - A THEORETICAL COMPARISON

Paulette Middleton

National Center for Atmospheric Research*, P.O. Box 3000, Boulder, Colorado 80307

C. S. Kiang

School of Geophysical Sciences, Georgia Institute of Technology, Atlanta, Georgia 30322

Volker A. Mohnen

Atmospheric Science Research Center, State University of New York at Albany, Albany, New York 12222

Abstract. Theoretical estimates have been made to demonstrate the relative importance of various pathways for the production of sulfate aerosols in an urban atmosphere away from the stationary sources under different atmospheric conditions. We have incorporated photochemical reactions, vapor condensation, and catalytic and noncatalytic oxidation on a wetted aerosol surface into our theoretical consideration. From our calculations it is found that under daytime conditions, with photochemical reactions, sulfuric acid vapor condensation and liquid-phase oxidation by H_2O_2 can be the dominant sulfate aerosol production mechanisms. Gas to particle conversion is expected to be an even more important pathway to sulfate aerosol formation under daytime conditions, since reactions involving radical clusters such as $HSO_3 \cdot H_2O$, $HSO_5 \cdot H_2O$, and $SO_3 \cdot H_2O$ are approximated by H_2SO_4 condensation in our estimates. Under nighttime conditions, without photochemical reactions, sulfate aerosol production in general is lower than under daytime conditions, and catalytic and noncatalytic oxidation mechanisms on the wetted aerosols become important pathways for SO_2-to-sulfate conversion.

Introduction

Although atmospheric suspended sulfates have been identified as a potential environmental problem for many years, the necessity for an increase in the use of sulfur-containing fuels has focused added attention on the possible adverse effects associated with airborne particulate sulfates. These suspended sulfates are generally in the submicrometer size range and are principally derived from atmospheric reactions of gaseous precursors. However, the observed sulfate concentrations are not always highly correlated with SO_2 concentrations. This suggests that factors other than local SO_2 emissions may also control the rate of sulfate formation. Thus, besides the possible local adverse health effects and visibility degradation, the presence of such amounts of particulate acidic sulfate, which is highly hygroscopic, may also alter cloud formation, precipitation, and albedo over a much larger region.

In order to gain an understanding of the nature and behavior of these sulfate particles suspended in the atmosphere, one must consider the combined effects of source input, transport, mixing, and removal, as well as microphysical and chemical transformations. In this paper we limit our discussion to the relative importance of various transformation processes involved in urban sulfate aerosol production under different atmospheric conditions.

The relative importance of the processes is investigated by comparing theoretical estimates of the rates of sulfate aerosol formation due to simultaneous condensation of SO_2 gas-phase oxidation products and SO_2 oxidation reactions on the wetted aerosol surface. The list of sulfate production mechanisms included in this study should not be assumed to be a complete representation of reality for several reasons: (1) possibly important nonlinear reactions are not described, (2) synergisms between sulfate, nitrate, and organic aerosol production mechanisms are not included, and (3) the behavior of the H_2O_2 reaction at very low pH values as described in the laboratory may not be realistic in the atmosphere.

Different urban atmospheric conditions are

*The National Center for Atmospheric Research is sponsored by the National Science Foundation.

studied by varying temperature, initial particle composition with respect to particle size, and initial gas concentrations. Since uncertainties in rate constants and concentrations of intermediate reactive species are taken into account in our study, the contribution of each reaction to the total sulfate production is expressed as a range. The transformation processes considered, the calculation schemes used, and the results of our study are presented in the following sections.

Chemical and Microphysical Processes

The overall rate of urban sulfate aerosol production is a complex function of gas-phase photochemical oxidation reactions leading to condensible vapors, the conversion of these vapors into sulfate particles, and the catalytic and noncatalytic oxidation of SO_2 on wetted aerosol surfaces.

Photochemical Reactions in the Gas Phase

Possible atmospheric homogeneous SO_2 oxidation processes in the gas phase are listed in Table 1. From the conversion rates calculated for urban conditions by Calvert et al. (1978), it is shown that the SO_2 + OH radical reaction is probably the major SO_2 gas-phase oxidation pathway in the urban environment.

Gas to Particle Conversion

The detailed kinetics describing the transformations of SO_2 gas-phase oxidation products into sulfate aerosols, either by mixing of these vapors to form tiny sulfate aerosols, and/or by condensation of these gaseous oxidation products or vapor clusters of these vapors onto preexisting aerosols, are not well known. Several pathways leading from oxidation products HSO_3 and SO_3 to sulfate aerosols have been discussed recently (Castleman et al., 1975; Friend and Vasta, 1980; Davis et al., 1979). In the absence of rate constant data for all of the individual kinetic steps leading from single molecule radicals to large stable clusters, equilibrium theories based on thermodynamic properties of molecular clusters have been developed (e.g., Kiang et al., 1973). Since thermodynamic data are available for only H_2SO_4 and H_2O mixtures, these systems are used as approximations to radical cluster systems. Estimates using the H_2SO_4-H_2O system as an approximation will probably underestimate gas to particle conversion rates (Davis et al., 1979; Friend and Vasta, 1980).

It has been demonstrated using this approximation that new particle formation can occur under atmospheric conditions of high H_2SO_4 concentrations (>10^{-5} ppm) and high relative humidity (>50 percent) even in the presence of preexisting particles (Middleton and Kiang, 1978). These tiny new particles coagulate with each other and with larger particles so rapidly that a significant increase in sulfate levels for the larger particles (>0.1 μm) can be obtained as a result of this new particle formation. At lower H_2SO_4 concentrations most of the H_2SO_4 vapor condenses onto preexisting particles (Middleton and Kiang, 1978). Estimated vapor condensation rates are even faster when scavenging of molecular clusters by preexisting particles is explicitly considered (Gelbard and Seinfeld, 1979).

Away from stationary sources, HSO_3 concentrations and therefore, by our approximation, H_2SO_4 concentrations are estimated to be less than 10^{-5} ppm (Graedel et al., 1976). In this concentration range, which we have considered in this investigation, condensation is the dominant gas to particle conversion process. Uncertainty in the rate of production of condensible vapor H_2SO_4 and the rate of condensation of the vapor is taken into account in our estimates.

Catalytic and Noncatalytic Oxidation Reactions in Wetted Aerosols

Sulfate can also be produced by various catalytic and noncatalytic oxidation reactions in wetted aerosols. SO_2 conversion to sulfate in clean water is a slow process, but is known to be accelerated in the presence of ammonia (Junge and Ryan, 1958; Van Den Heuvel and Mason, 1963; Scott and Hobbs, 1967; Johnstone and Coughanowr, 1958), metallic contaminations such as iron (Brimblecombe and Spedding, 1974; Neytzell de Wilde and

TABLE 1. Possible Atmospheric Gas-Phase SO_2 Oxidation Processes[a]

	Initial Steps
(1)	$SO_2 + h\nu(2400\text{-}3400\ \text{Å}) \rightarrow SO_2^*$
(2)	$SO_2^* + O_2 \rightarrow (SO_4)$
(3)	$SO_2 + O + M \rightarrow SO_3 + M$
(4)	$SO_2 + O_3 \rightarrow SO_3 + O_2$
(5)	$SO_2 + NO_3 \rightarrow SO_3 + NO_2$
(6)	$SO_2 + N_2O_5 \rightarrow SO_3 + N_2O_4$
(7)	$SO_2 + CH_3O_2 \rightarrow SO_3 + CH_3O$
(8)	$SO_2 + HO_2 \rightarrow SO_3 + OH$
(9)	$SO_2 + OH + M \rightarrow HSO_3 + M$
	Possible Subsequent Reactions
(10)	$SO_3 + H_2O \rightarrow SO_3{\cdot}H_2O \rightarrow H_2SO_4$
(11)	$SO_3^{\cdot} + H_2O \rightarrow$ nucleus
(12)	$HSO_3^{\cdot} + OH \rightarrow H_2SO_4$
(13)	$HSO_3^{\cdot} + OH \rightarrow H_2O + SO_3$
(14)	$HSO_3^{\cdot} + HSO_3^{\cdot} \rightarrow H_2S_2O_6$
(15)	$HSO_3^{\cdot} + H_2O \rightarrow HSO_3^{\cdot}(H_2O)$
(16)	$HSO_3^{\cdot} + O_2 \rightarrow HSO_5$
(17)	$HSO_5^{\cdot} + H_2O \rightleftarrows HSO_5^{\cdot}(H_2O)$

[a]For details see Calvert et al. (1978) for reactions (1) to (9), Castleman et al. (1975) for reaction (10), Friend and Vasta (1980) for reaction (11), and Davis et al. (1979) for reactions (12) to (17).

TABLE 2. Chemical Equilibrium Constants

Equilibrium Constant Expression[a]	Value of Equilibrium Constant 25°C	15°C	5°C
$K_w = [H^+][OH^-]$	1.008×10^{-14}	4.25×10^{-15}	1.93×10^{-15}
$K_{HS} = [SO_2 \cdot H_2O]/P_{SO_2}$	1.25	1.83	2.78
$K_{1S} = [H^+][HSO_3^-]/[SO_2 \cdot H_2O]$	0.0174	0.0219	0.0393
$K_{2S} = [H^+][SO_3^=]/[HSO_3]$	6.3×10^{-8}	7.9×10^{-8}	1.02×10^{-7}
$K_{HC} = [CO_2 \cdot H_2O]/P_{CO_2}$	0.034	0.043	
$K_{1C} = [H^+][HCO_3^-]/[CO_2 \cdot H_2O]$	4.45×10^{-7}	3.78×10^{-7}	
$K_{2C} = [H^+][CO_3^=]/[HCO_3^-]$	4.68×10^{-11}		
$K_{HA} = [NH_3 \cdot H_2O]/P_{NH_3}$	57.0	104.0	198.0
$K_{1A} = [NH_4^+][OH^-]/[NH_3 \cdot H_2O]$	1.774×10^{-5}	1.65×10^{-5}	1.53×10^{-5}
$K_{HO_3} = [O_3 \cdot H_2O]/P_{O_3}$	0.0123	0.016	0.022
$K_{HH} = [H_2O_2 \cdot H_2O]/P_{H_2O_2}$	7×10^4	3.94×10^5	2.51×10^6
$K_{HO_2} = [O_2 \cdot H_2O]/P_{O_2}$	1.08×10^{-3}	1.28×10^{-3}	1.68×10^{-3}

[a]Concentrations in mol l^{-1}, pressures in atmospheres. Values obtained from the following data sources: McKay (1971) K_w, H_{HS}, K_{1S}, K_{2S}, K_{HA}, K_{1A}, K_{HC}, K_{1C}, K_{2C}; Mellor (1964) K_{HO_3}; Scatchard et al. (1952) K_{HH}; International Critical Tables (1928) K_{HO_2}.

Taverner, 1958) and manganese ions (Matteson et al., 1969; Barrie and Georgii, 1976), soot (Chang et al., 1979; Novakov et al., 1974), or strong oxidizing agents such as ozone and hydrogen peroxide (Penkett et al., 1979).

The rate of uncatalyzed oxidation in clean water has been studied by many workers. (See McKay (1971) and Beilke et al. (1975) for detailed discussion.) In our study we use the rate expression derived by McKay (1971) from the measurements of Fuller and Crist (1941). The possible order of magnitude overestimate of sulfate production by uncatalyzed oxidation obtained using this expression (Beilke et al., 1975) is taken into account in our analysis.

The rate expressions for metal-catalyzed oxidation of SO_2 in the atmosphere are still a matter of discussion (Beilke and Gravenhorst, 1978; Hegg and Hobbs, 1978). The apparently large differences in rate expressions for both the iron-catalyzed reaction (Freiberg, 1974; Overton et al., 1979) and the manganese-catalyzed reaction (Matteson et al., 1969; Barrie and Georgii, 1976) are taken into account in our analysis. The catalytic oxidation by soot in aqueous solution used in our study was derived by Chang et al. (1979).

Penkett et al. (1979) reported their experimental studies of the rate of SO_2 conversion by ozone and hydrogen peroxide. From their activation energy data, the temperature and pH dependence of these reactions can be deduced. Due to the controversy over these reaction rates, the alternative rates proposed by Erickson et al. (1977) for the O_3 reaction and Martin and Damschen (1981) for the H_2O_2 reaction are also considered.

Ammonia in solution acts as a buffer, thereby decreasing particle acidity. Since all of the wetted aerosol oxidation reaction rates considered, except for the soot reaction rate, depend on particle acidity, ammonia is included in our estimates.

Calculation Scheme

The calculation scheme developed to determine the rates of sulfate production and corresponding gas depletion is based on the concept that vapor transfer to the aerosols and vapor conversion within the aerosols are coupled kinetic processes. Previous studies have assumed continuous equilibrium between the gases and aerosols. In our scheme the rates are calculated in a sequence of kinetic and equilibrium steps.

Initially the gases are assumed to be in equilibrium with the aerosols (step 1). In step 2, the formation of sulfate occurs due to time-dependent oxidation processes on the wetted aerosol surface. Simultaneously, in step 3, sulfuric acid vapor condenses onto the aerosol. This increase in sulfate due to both condensation and oxidation processes has disturbed the initial gas and aerosol equilibrium. To reestablish equilibrium (step 4) more gas must dissolve. The amount of a particular gas available for going into solution, however, depends on the rate at which that gas is being transferred to the aerosols. Thus, the rate of change of aerosol composition due to time-dependent sulfate oxidation must be compared with the rate of gas transfer to determine which is the rate-limiting process (step 5). If the solution kinetics are occurring at a faster rate than the gas-phase kinetics, then the amount of change in aerosol composition is determined by the amount of gas reaching the aerosol. Similarly, if the gas kinetics are occurring at a faster rate, then the amount of gas actually going into solution is determined by the rate of change of aerosol composition. Once the rate-limiting process has been determined, the new

TABLE 3. Liquid-Phase Oxidation Kinetic Rate Constants

Rate Expression Values	Value of Rate Constant		
	25°C	15°C	5°C
Oxygen			
McKay (1971)			
K'_{O_2}	0.013	0.0048	0.004
K''_{O_2}	59.0	20.0	6.56
a	1/2	1/2	1/2
Beilke et al. (1975)			
K'_{O_2}	0	0	0
K''_{O_2}	$1/2 \times 10^{-4}$	2.74×10^{-4}	6.61×10^{-4}
a	-0.16	-0.16	-0.16
Iron			
Freiberg (1974)			
K'_{Fe}	3.4×10^{2}*	1.87×10^{2}	1.07×10^{-2}
K''_{Fe}	0	0	0
b	2	2	2
c	1	1	1
Overton et al. (1979)			
K'_{Fe}	64.21		
K''_{Fe}	4700		
b	1	1	1
c	0	0	0
Manganese			
Matteson et al. (1969)			
K'_{Mn}	3.7×10^{2}		
K_s	5×10^{-5}		
K''_{Mn}	0	0	0
d	1	1	1
Barrie and Georgii (1976)			
K'_{Mn}	0	0	0
K_s	0	0	0
K''_{Mn}	$9.4 \times 10^{2}\ s^{-1}$		
d	0	0	0
Soot			
Chang et al. (1979); S. G. Chang (private communication, 1978)			
K_{soot}	2.55×10^{-4}†	1.26×10^{-4}	5.90×10^{-5}
Ozone & Hydrogen Peroxide**			
Penkett et al. (1979)			
K_{O_3}	$K_{O_3} = A_{O_3}\ \exp(-E_{O_3}/RT)$		
$K_{H_2O_2}$	$K_{H_2O_2} = A_{H_2O_2}{}^{*}\ \exp(-E_{H_2O_2}/RT)$		

*$(1\ mol^{-1}\ s^{-1})$

†$(1\ g^{-1}\ s^{-1})$

**K_{O_3} and $K_{H_2O_2}$ are a function of pH and temperature. Values for Arrheenius constants and activation energies are as follows:

sulfate concentration, the aerosol acidity (pH), and the corresponding depletion of each gas are determined for the time step t, which is 1 second in these calculations (step 6). This "pseudo-kinetic" sequence is repeated until the dominant sulfate transformation mechanism has been identified. In our calculations we have assumed that the liquid-phase oxidation rates are not affected by the rates of diffusion of reactive species in solution. Details of this calculation scheme are discussed in Middleton and Kiang (1979).

In order to illustrate the variation in sulfate production with respect to particle size, we have divided up the particle range of interest (0.1 to 10.0 μm in diameter) into 10 categories. Thus, the total concentration of sulfate produced in each category after time t is given as the sum of sulfate produced by condensation of H_2SO_4 vapor and sulfate produced by oxidation of SO_2 on the wetted aerosol surface.

The rate of sulfate production due to H_2SO_4 condensation in the particle radius range r_j to $r_j + dr$ is given as

$$\frac{d[H_2SO_4]_j}{dt} = 4\pi r_j n(r_j) D_{H_2SO_4} \alpha_{H_2SO_4} G_{H_2SO_4} \qquad (1)$$

where the diffusion coefficient of H_2SO_4 in air, $D_{H_2SO_4}$, and the accommodation coefficient, $\alpha_{H_2SO_4}$, are given as 0.1 and 1.0 $cm^2\ s^{-1}$ respectively, $G_{H_2SO_4}$ is the concentration of H_2SO_4 vapor, and $n(r_j)$ is the concentration of particles of radius r_j.

The rate of sulfate production due to SO_2 oxidation in the wetted aerosols of radius r_j to $r_j + dr$ is expressed as

$$\begin{aligned}\frac{d[SO_4^{=}]_j}{dt} = \Bigg(&\left\{K'_{O_2} + K''_{O_2}[H^+]_j^{\,a}\right\}[SO_3^{=}]_j \\ &+ [Fe^{+3}]_j\left\{\frac{K'_{Fe}[HSO_3^-]_j^{\,b}}{[H^+]_j^{\,c}} + K''_{Fe}[SO_3^{=}]_j\right\} \\ &+ [Mn^{+2}]_j^{\,d}\left\{\frac{K'_{Mn}[SO_2\cdot H_2O]_j}{[SO_2\cdot H_2O]_j + [Mn^{+2}]_j + K_s} + K''_{Mn}[SO_3^{=}]_j\right\} \\ &+ K_{soot}[C_x]_j[O_2\cdot H_2O]_j^{0.75} \\ &+ K_{H_2O_2}[H_2O_2\cdot H_2O]_j[HSO_3^-]_j \\ &+ K_{O_3}[O_3\cdot H_2O]_j[HSO_3^-]_j\Bigg) \\ &\times \frac{4\pi}{3} r_j^3 n(r_j)\beta_j \end{aligned} \qquad (2)$$

For our calculations the aerosols are assumed to be liquid water droplets which contain various amounts of iron, manganese, and soot. These concentrations are given as $[Fe^{+3}]_j$, $[Mn^{+2}]_j$, and $[C]_j$, respectively; $[SO_2\cdot H_2O]$, $[O_2\cdot H_2O]$, $[O_3\cdot H_2O]$, and $[H_2O_2\cdot H_2O]$ are the concentrations of the dissolved gases SO_2, O_2, O_3, and H_2O_2, respectively; and $[SO_3^{=}]$, $[HSO_3^-]$, and $[SO_4^{=}]$ are the concentrations of sulfite, bisulfite, and sulfate in the aerosol, respectively. Chemical equilibrium constants used to calculate these concentrations are given in Table 2. Concentration factors a, b, c, and d, and rate constants, indicated by K, are given in Table 3. The alternative rate expressions used for the O_3 and H_2O_2 reactions are given at the end of Table 3. If solution kinetics are rate limiting, $\beta_j = 1$; if gas kinetics are rate limiting, $\beta_j < 1$.

Results

The relative importance of sulfate production mechanisms in different urban environments is studied by varying the initial gas concentrations, temperature, and particle composition with respect

TABLE 3. (continued)

Parameters	pH 1	2	3	4	4.6	5	6.6	8.2
E_{O_3} (kcal mol^{-1})	7.25	11.13	6.45	4.49		6.53		
$A_{O_3} = 10^{\alpha}:\alpha$	10.17	13.56	10.58	9.37		11.16		
$E_{H_2O_2}$ (kcal mol^{-1})					7.27		9.23	20.69
$A_{H_2O_2} = 10^{\alpha}:\alpha$					8.87		8.96	14.56

Alternative rate expression used for O_3 from Erickson et al. (1977) as interpreted by Overton et al. (1979) is:

$$[O_3\cdot H_2O]\{5.9 \times 10^2[SO_2\cdot H_2O] + 2.2 \times 10^9[SO_3^{=}] + 3.1 \times 10^5[HSO_3^-]\}$$

Alternative rate expression used for H_2O_2 from Martin and Damschen (1981) is:

$$[H_2O_2\cdot H_2O][SO_2\cdot H_2O]/(0.1 + [H^+])$$

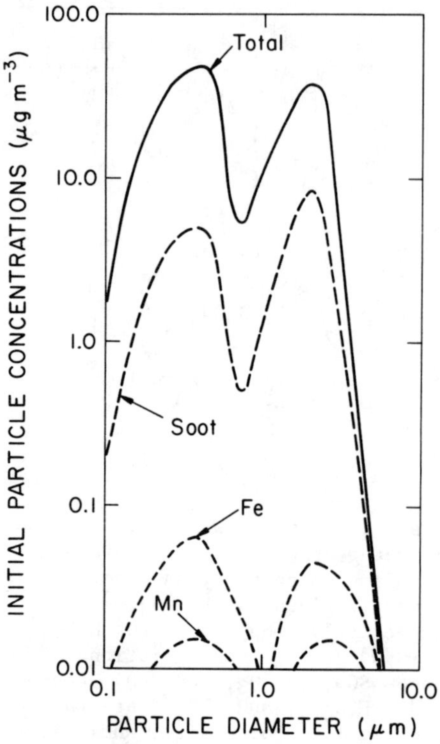

Fig. 1. Initial particle composition with respect to particle sizes.

to particle size. The initial particle composition with respect to particle size is illustrated in Figure 1 and listed in Table 4. The relative amounts of soot in the particles are determined by assuming soot is approximately 80 percent of the total carbon (Novakov et al., 1974) measured under urban conditions (Appel et al., 1978). Iron and manganese are taken to be a factor of 100 and a factor of 1000 less than the soot, respectively. It is assumed that 10 to 100 percent of the soot is an effective catalyst and 0.1 to 1.0 percent of the iron and manganese are effective. These ranges are also given in Table 4.

Initial SO_2, NO_3, and O_3 concentrations were obtained from measurements representative of urban environments (Appel et al., 1978; Spicer, 1977). Estimates of representative urban daytime H_2O_2 concentrations are also taken from measurements (Kok et al., 1978) and photochemical model estimates (Graedel et al., 1976). The H_2SO_4 concentrations are inferred from the assumed production rate for H_2SO_4 vapor over the time period of the calculation. This rate can be calculated from reaction (9) in Table 1 using a rate constant in the range 0.8 to 7.2 × 10^{-31} cm^6 s^{-1} and OH concentration in the range 10^6 to 10^7 molec cm^{-3} (Middleton and Kiang, 1978). The ranges of gas concentrations and temperatures used in the study are listed in Table 5.

Average sulfate aerosol production with respect to particle size under different atmospheric conditions is illustrated in Figure 2. The range of contribution of each sulfate transformation mech-

TABLE 4. Initial Particle Concentrations for Daytime Conditions[a]

Diameter (μm)	Mixing Ratio	Concentration (μg m^{-3})[b]			
		Total	Soot	Fe	Mn
0.1	2.1×10^{-12}	2.1	0.2	2×10^{-3}	2×10^{-4}
0.2	2.4×10^{-11}	23.7	2.4	2×10^{-2}	2×10^{-3}
0.3	4.3×10^{-11}	42.8	4.3	4×10^{-2}	4×10^{-3}
0.4	5.0×10^{-11}	49.5	5.0	5×10^{-2}	5×10^{-3}
0.5	3.3×10^{-11}	33.0	3.3	3×10^{-2}	3×10^{-3}
0.7	5.7×10^{-12}	5.7	0.6	6×10^{-3}	6×10^{-4}
1.0	1.3×10^{-11}	13.1	1.3	1×10^{-2}	1×10^{-3}
2.0	4.2×10^{-11}	42.4	8.5	8×10^{-2}	8×10^{-3}
3.0	5.9×10^{-12}	5.9	1.2	1×10^{-2}	1×10^{-3}
4.0	6.3×10^{-13}	0.6	0.1	1×10^{-3}	1×10^{-4}

[a]For nighttime conditions, the values are three-fourths of the daytime conditions. Particle size distribution from Middleton and Brock (1976).

[b]Ranges of effective catalysts in "solution" (moles of catalyst per liter of liquid in the aerosol) are as follows:

Diameter (μm)	Soot	Fe	Mn
0.1 - 1.0	0.8 - 8.0	1.8×10^{-5} to 1.8×10^{-4}	1.8×10^{-6} to 1.8×10^{-5}
1.0 - 4.0	1.6 - 16.0	3.6×10^{-5} to 3.6×10^{-4}	3.7×10^{-6} to 3.7×10^{-5}

TABLE 5. Atmospheric Conditions and Initial Gas Concentrations

Gas Concentrations (ppm)	Summer Day (25°C)	Winter Day (5°)	Summer Night (15°C)
SO_2	0.01	0.01	0.01
CO_2	332.0	332.0	332.0
NH_3	0.01	0.01	0.005
O_3	0.1	0.05	0.01
H_2O_2	10^{-4}-10^{-3}	10^{-5}-10^{-3}	10^{-5}-10^{-4}
H_2SO_4	10^{-7}-10^{-6}	10^{-8}-10^{-7}	10^{-12}-10^{-11}

anism to the total sulfate aerosol production over the calculation time period for different atmospheric conditions is given in Figure 3. Ranges of particle acidity and sulfate aerosol production in the 0.1 to 1.0 and 1.0 to 10.0 μm size categories for different conditions are summarized in Figure 4. A typical rate of change of sulfate production due to each mechanism with respect to time is illustrated in Figure 5.

Discussion

It is important to stress that the purpose of this investigation is to study the relative importance of various possible pathways for sulfate

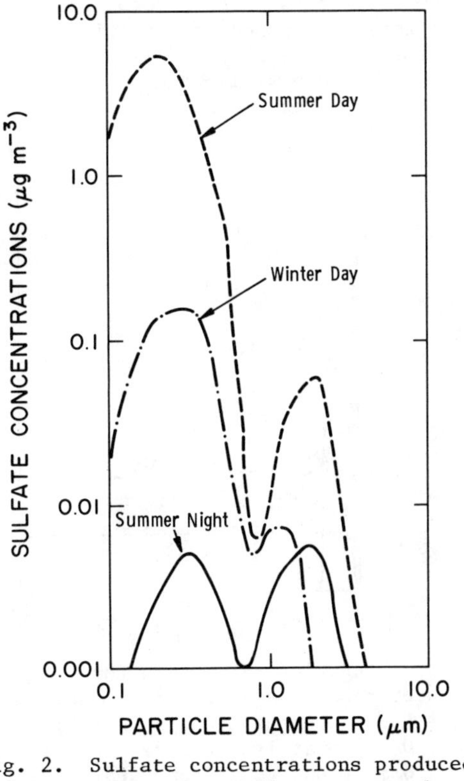

Fig. 2. Sulfate concentrations produced after 5 minutes under different atmospheric conditions.

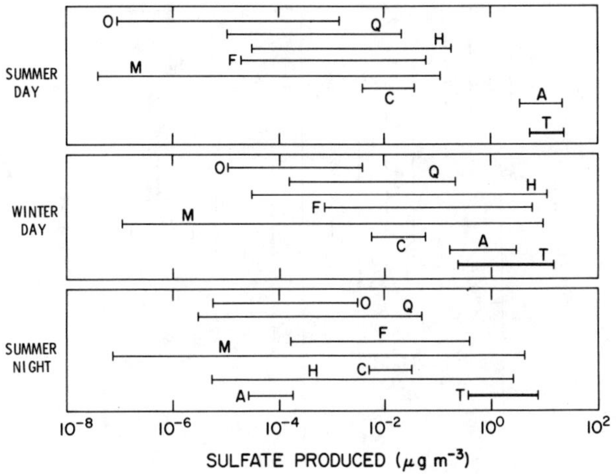

Fig. 3. Range of contributions of each mechanism to the total sulfate production after 5 minutes for different atmospheric conditions. T = total sulfate, A = H_2SO_4 condensation, H = H_2O_2 oxidation, O = uncatalyzed oxygen oxidation, Q = O_3 oxidation, F = iron-catalyzed oxidation, M = manganese-catalyzed oxidation, and C = soot-catalyzed oxidation.

aerosol production under different atmospheric conditions. In order to assess the contribution of each formation mechanism to observed urban sulfate levels, other factors such as primary source input, removal, transport, and mixing must be coupled together with the short-time-interval chemical and microphysical transformation calculation used in this study.

From our investigation it is found that the total sulfate aerosol production changes noticeably under different atmospheric conditions. These changes are the result of variations in the relative importance of the different sulfate production mechanisms. Factors influencing these changes are the concentration of photochemically produced gaseous species, temperature, and aerosol acidity and size distribution.

The noticeable variation in sulfate aerosol production under atmospheric conditions representative of daytime and nighttime and summer and winter daytime are illustrated in Figure 2. It is found that the rate of sulfate aerosol formation is higher under summer daytime conditions than under winter daytime or nighttime conditions.

The variation of total sulfate aerosol production is the result of changes in the individual sulfate aerosol formation reaction rates under different atmospheric conditions. In Figure 3, the range of contributions of each reaction rate to the total sulfate aerosol production is given for each case studied. The ranges reflect the ranges of gas concentrations used, the amount of effective catalysts in solution, and the variation in different rate expressions for the uncatalyzed oxidation and the iron- and manganese-

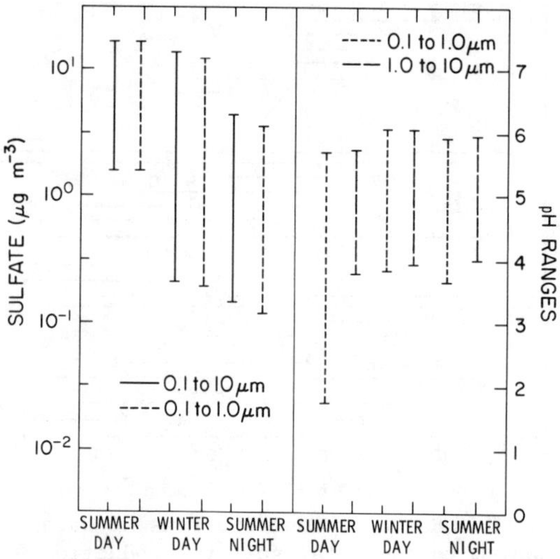

Fig. 4. Range of sulfate produced and particle acidity (pH) for particle size ranges and atmospheric conditions indicated.

catalyzed oxidations. It is found that, in general, H_2SO_4 condensation and H_2O_2 oxidation on the wetter aerosol can be dominant pathways to sulfate aerosol formation under daytime conditions. Under summer daytime conditions sulfate formation is higher, and H_2SO_4 condensation can be the more important mechanism. Under winter daytime conditions sulfate formation is lower, and H_2O_2 oxidation can be the more important mechanism. Under nighttime conditions, the rate of sulfate aerosol formation is lower and catalytic and noncatalytic oxidation mechanisms on the wetted aerosol surface become the major sulfate formation mechanisms. In particular, catalytic oxidation by soot, iron, and manganese appears to be the most important pathway under nighttime conditions.

The changes in the relative importance of sulfate aerosol formation mechanisms under different atmospheric conditions are influenced, to varying degrees, by the concentration of photochemically produced gaseous species, temperature, and aerosol acidity and size distribution. As is illustrated in Figure 3, a major factor appears to be the concentration of photochemically produced gases. Temperature changes can influence the relative importance of H_2SO_4 condensation and H_2O_2 oxidation rates, since the ratio of H_2O_2 in solution to H_2O_2 in the vapor is more sensitive to temperature than the H_2SO_4 condensation rate. Thus H_2O_2 oxidation can become the more important sulfate aerosol formation pathway under winter daytime conditions when temperatures are lower.

The effects of changes in particle acidity which are independent of sulfate concentration are observed by studying the changes in individual sulfate formation rates under conditions of different NH_3 concentrations. When the NH_3 concentration is lower, less ammonium is present in solution and the particle acidity increases. As a result, sulfate aerosol formation rates which are inversely proportional to particle acidity (the O_2, O_3, iron, and manganese reactions) all decrease noticeably.

Variations in the soot concentrations are studied in particular, since the soot mechanism appears to be the major sulfate aerosol formation mechanism which is not directly or indirectly related to photochemical activity. The increase in soot alone, however, does not alter the total sulfate aerosol formation under summer daytime conditions with photochemical reactions. It should be noted that under conditions of higher particle acidity (lower NH_3) and reduced photochemical activity (lower H_2SO_4 and H_2O_2), soot catalysis could become a competitive pathway to sulfate formation even under daytime conditions.

The relationship of particle size and sulfate aerosol production is summarized in Figure 4. It is found that for all of the cases studied, most of the sulfate aerosol production occurs in the 0.1- to 1.0-μm range. Essentially all of the H_2SO_4 condensation and over half of the SO_2 oxidation to sulfate on the wetted aerosol surface occur in this size range. Thus, most of the sulfate aerosol will be produced in the smaller particles where the aerosol volume mixing ratio (Table 4) is higher.

It should be stressed again that these short-time-scale calculations have been designed to study the rates of sulfate aerosol formation under different atmospheric conditions. Over a longer time period - for example, over an hour - each mechanism rate will change noticeably, and extrapolations of our short-time-scale sulfate

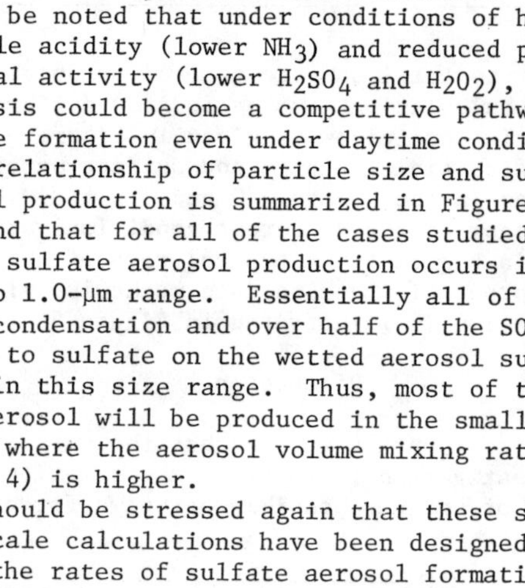

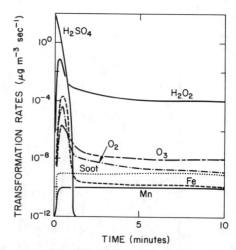

Fig. 5. Reaction rates as a function of time under summer daytime conditions, case 1. H_2SO_4 = H_2SO_4 condensation, H_2O_2 = H_2O_2 oxidation, O_2 = uncatalyzed oxygen oxidation, O_3 = O_3 oxidation, Fe = iron-catalyzed oxidation, Mn = manganese-catalyzed oxidation, and soot = soot-catalyzed oxidation.

aerosol formation rates to longer time scales will be incorrect. To illustrate this point, each sulfate aerosol formation mechanism rate is examined as a function of time over a longer time period. It is shown that under daytime conditions (Figure 5) the rate of sulfate aerosol production due to H_2SO_4 condensation reaches a maximum within the first minute. After a minute, most of the initial H_2SO_4 vapor has been consumed. As the other gases are slowly depleted and the particle acidity increases, the sulfate production rates on the wetted aerosol surface also decrease. Only the soot catalysis mechanism rate remains constant since it is not influenced by changes in gas concentration or particle acidity. Since H_2SO_4 condensation is the dominant mechanism in this case, the total rate of sulfate aerosol production over an hour is determined within the first few minutes. The correct rate of sulfate aerosol production is 5.6 g m^{-3} h^{-1}, whereas a rate linearly extrapolated from 3.2 gm^{-3} for 5 minutes would be incorrectly given as 38.4 g m^{-3} h^{-1}. Only for cases where the dominant sulfate production mechanism rate is constant in time should a linear extrapolation of the short-time-scale calculation result be considered as a reasonable estimate.

Conclusions

From these calculations it is concluded that under daytime conditions with photochemical reactions, H_2SO_4 vapor condensation and liquid-phase oxidation by H_2O_2 can be the dominant sulfate production mechanisms. Gas-to-particle conversion is expected to be an even more important pathway to sulfate aerosol formation since reactions involving radical clusters are approximated by H_2SO_4 condensation in our estimates. Under nighttime conditions sulfate production rates are lower, and other catalytic and noncatalytic oxidation mechanisms on the wetted aerosol surface become important conversion mechanisms. The variations in temperature, particle acidity, and soot composition studied here do not significantly change the total sulfate aerosol production under daytime conditions. However, the relative importance of the different individual reaction rates is sensitive to these changes. In particular H_2O_2 oxidation is more rapid at lower temperatures. Also, soot catalysis becomes a more competitive pathway to sulfate aerosol formation under conditions of higher particle acidity, since this process is the only oxidation mechanism on the wetted aerosol surface which is not directly influenced by particle acidity. From this study is also concluded that most of the sulfate aerosol is produced in the 0.1- to 1.0-μm size range.

Several recommendations for future research are made as a result of this model study. Equilibrium and kinetic constants used in these calculations are taken from experiments in which the solutions are close to ideal and the gas and catalyst concentrations used are often higher than typical atmospheric concentrations. Since urban aerosols are typically not dilute aqueous solutions, but rather complex solutions and/or solid particles coated with thin films, a more accurate description of condensation of low-concentration gases onto such multicomponent systems is needed. Laboratory and/or theoretical studies testing the effects of deviations of real aerosols from the ideal solutions usually assumed in model studies must be performed before rigorous assessment of aerosol and gas interactions can be made. Variations of kinetic rate constants with respect to temperature are needed for several oxidation mechanisms considered herein.

It is also clear from this study that proper interpretations of observed sulfate levels in urban aerosols require simultaneous measurement of major gaseous species and both size and composition of aerosols. In particular, particle acidity and sulfate concentration with respect to particle size should be measured.

The calculations presented here qualitatively illustrate the competition between sulfate production mechanisms. Reliable quantitative assessments for specific urban environments would require consideration of simultaneous combined effects of gas and aerosol transformations, primary source input, removal, transport, and mixing. For such a complex model to be feasible, it will be necessary to develop extensive parameterization of many of the microphysical and chemical processes. Before we can have confidence in such parameterizations, a thorough understanding of these processes is essential. The results of the study presented in this paper are a step toward this understanding.

References

Appel, B. R., E. L. Kothny, E. M. Hoffer, G. M. Hidy, and J. J. Wesolowski, Sulfate and nitrate data from the California aerosol characterization experiment (ACHEX), Environ. Sci. Technol., 12, 418-425, 1978.

Barrie, L. A., and H. W. Georgii, An experimental investigation of the absorption of sulfur dioxide by water drops containing heavy metal ions, Atmos. Environ., 10, 743-749, 1976.

Beilke, S., and G. Gravenhorst, Heterogeneous SO_2 oxidation in the droplet phases, Atmos. Environ., 12, 231-239, 1978.

Beilke, S., D. Lamb, and J. Muller, On the uncatalyzed oxidation of atmospheric SO_2 by oxygen in aqueous systems, Atmos. Environ., 9, 1083-1090, 1975.

Brimblecombe, P., and D. J. Spedding, The catalytic oxidation of micromolar aqueous sulfur dioxide - I. Oxidation in solutions containing iron(III), Atmos. Environ., 8, 937-945, 1974.

Calvert, J. G., F. Su, J. W. Bottenheim, and O. P. Strausz, Mechanism of the homogeneous oxidation of sulfur dioxide in the troposphere, Atmos. Environ., 12, 197-226, 1978.

Castleman, A. W., R. E. Davis, H. R. Munkelwitz, I. N. Tang, and W. P. Wood, Kinetics of association reactions pertaining to H_2SO_4 aerosol formation, Int. J. Chem. Kinet., Symp. No. 1, 629-640, 1975.

Chang, S. G., R. Brodzinsky, R. Toosi, S. S. Markowita, and T. Novakov, Catalytic oxidation of SO_2 on carbon in aqueous suspensions, in Proc. Conference on Carbonaceous Particles in the Atmosphere, edited by T. Novakov, p. 122, Report No. LBL-9037, Lawrence Berkeley Laboratory, Berkeley, CA, 1979.

Davis, D. D., A. R. Ravishankara, and S. Discher, SO_2 oxidation via the hydroxyl radical: Atmospheric fate of HSO_x radicals, Geophys. Res. Lett., 6, 113-116, 1979.

Erickson, R. E., L. M. Yates, R. L. Clark, and D. McEwen, The reaction of sulfur dioxide with ozone in water and its possible atmospheric significance, Atmos. Environ., 11, 813-817, 1977.

Freiberg, J., Effects of relative humidity and temperature on iron-catalyzed oxidation of SO_2 in atmospheric aerosols, Environ. Sci. Technol., 8, 731-734, 1974.

Friend, J. P., and R. Vasta, Nucleation by free radicals from the photooxidation of sulfur dioxide in air, J. Phys. Chem., 84, 2423, 1980.

Fuller, E. C., and R. H. Crist, The rate of oxidation of sulfite ions by oxygen, J. Am. Chem. Soc., 63, 1644-1650, 1941.

Gelbard, F., and J. H. Seinfeld, The general dynamic equation for aerosols - Theory and application to aerosol formation and growth, J. Collid Inter. Sci., 68, 363-383, 1979.

Graedel, T. E., L. A. Farrow, and T. A. Weber, Kinetic studies of the photochemistry of the urban troposphere, Atmos. Environ., 10, 1095-1116, 1976.

Hegg, D. A., and P. V. Hobbs, Oxidation of sulfur dioxide in aqueous systems with particular reference to the atmosphere, Atmos. Environ., 12, 241-253, 1978.

International Critical Tables, Vol. 3, p. 257, McGraw-Hill, N.Y., 1928.

Johnstone, H. F., and D. R. Coughanowr, Absorption of SO_2 from air, Ind. Engn. Chem., 50, 1169-1172, 1958.

Junge, C. E., and T. G. Ryan, Study of the SO_2 oxidation in solution and its role in atmospheric chemistry, Q. J. Roy. Meteorol. Soc., 84, 46-55, 1958.

Kiang, C. S., D. Stauffer, V. A. Mohnen, J. Bricard, and D. Vigla, Heteromolecular nucleation theory applied to gas-to-particle conversion, Atmos. Environ., 7, 1279-1283, 1973.

Kok, G. L., K. R. Darnall, A. M. Winer, J. N. Pitts, and B. W. Gay, Ambient air measurements of hydrogen peroxide in the California south coast air basin, Environ. Sci. Technol., 12, 1077-1080, 1978.

Martin, R. L., and D. E. Damschen, Aqueous oxidation of sulfur dioxide by hydrogen peroxide at low pH, Atmos. Environ., 15, 1615-1621, 1981.

Matteson, M. J., W. Stoeber, and H. Luther, Kinetics of the oxidation of sulfur dioxide by aerosol of manganese sulfate, Ind. Engng. Chem. Fund., 8, 677-687, 1969.

McKay, H. A. C., The atmospheric oxidation of sulfur dioxide in water droplets in presence of ammonia, Atmos. Environ., 5, 7-24, 1971.

Middleton, P., and J. R. Brock, Simulation of aerosol kinetics, J. Colloid Inter. Sci., 54, 249-264, 1976.

Middleton, P., and C. S. Kiang, A kinetic aerosol model for the formation and growth of secondary sulfuric acid particles, J. Aerosol Sci., 9, 359-385, 1978.

Middleton, P., and C. S. Kiang, Relative importance of nitrate and sulfate aerosol production mechanisms in urban atmospheres, in Nitrogeneous Air Pollutants, edited by D. Grosjean, pp. 269-288, Ann Arbor Science Publ., Inc., Ann Arbor, Michigan, 1979.

Neytzell de Wilde, F. G., and L. Taverner, Experiments relating to the possible production of an oxidizing acid leach liquor by autooxidation for the extraction of uranium, Proc. 2nd U.N. Int. Conf. on the Peaceful Uses of Atomic Energy, Vol 3, pp. 271-275, 1958.

Novakov, T., S. G. Chang, and A. B. Harker, Sulfates as pollution particulates: Catalytic formation on carbon (soot) particles, Science, 186, 259-261, 1974.

Overton, J. H., V. P. Aneja, and J. L. Durham, Production of sulfate in rain and raindrops in polluted atmospheres, Atmos. Environ., 13, 355-367, 1979.

Penkett, S. A., B. M. R. Jones, K. A. Brice, and A. E. J. Eggleton, The importance of atmospheric ozone and hydrogen peroxide in oxidizing sulfur dioxide in cloud and rainwater, Atmos. Environ., 13, 123-137, 1979.

Scatchard, G., G. M. Kavanagh, and L. B. Ticknor, Vapor-liquid equilibrium. VIII. Hydrogen peroxide-water mixtures, J. Am. Chem. Soc., 74, 3715-3720, 1952.

Scott, W. D., and P. V. Hobbs, The formation of sulfate in water droplets, J. Atmos. Sci., 24, 54-57, 1967.

Spicer, C. W., Photochemical atmospheric pollutants derived from nitrogen oxides, Atmos. Environ., 11, 1089-1095, 1977.

Van Den Heuvel, A. P., and B. J. Mason, The formation of ammonium sulfate in water droplets exposed to gaseous sulfur dioxide and ammonia, Q. J. Roy. Meteorol. Soc., 89, 271-275, 1963.

IMPORTANCE OF HETEROGENEOUS PROCESSES TO TROPOSPHERIC CHEMISTRY: STUDIES WITH A ONE-DIMENSIONAL MODEL

R. P. Turco

R & D Associates, Marina del Rey, California 90291

O. B. Toon, R. C. Whitten, and R. G. Keesee

NASA Ames Research Center, Moffett Field, California 94035

P. Hamill

Systems and Applied Sciences Corp., Palo Alto, California 94306

Abstract. The chemistry of the troposphere is affected by processes which involve the interactions between gases, aerosols and cloud droplets. Such "heterogeneous" processes are not usually considered in studies of the global tropospheric chemical cycles. (In the present context, the term "heterogeneous processes" refers to multiphase processes, processes involving particles and particle interactions, surface phenomena, and chemistry in solution, all as opposed to "homogeneous" gas-phase processes.) We have developed a one-dimensional model of tropospheric air composition which incorporates a number of heterogeneous physical and chemical processes. Gases, aerosols, and hydrometeors interact through the physical mechanisms of nucleation, condensation, evaporation, coagulation, coalescence, and deliquescence. Material is removed from the atmosphere by precipitation, sedimentation, and dry deposition. Chemical transformations occur both in the vapor and condensed (aerosol, raindrop) phases. The model also accounts for the sources and vertical diffusion of gases and particles, and for the changes in solar intensity caused by light-scattering from aerosols and clouds. We describe the structure of the model and compare preliminary computational results with other simulations and field data to demonstrate the accuracy of the model. It is shown that rainout and washout processes strongly influence the distributions of tropospheric gases and aerosols under certain conditions.

Introduction

In recent years it has become apparent that man's activities can influence the global chemical cycles of the Earth's atmosphere. Yet our understanding of these cycles, and our ability to predict man's impact on them, has remained limited. Current concern centers around possible anthropogenic perturbations of the nitrogen, carbon, and sulfur cycles, all of which interact to some degree, and all of which involve heterogeneous chemical and physical processes. Most previous theoretical studies have concentrated on the homogeneous (gas-phase) chemistry of these cycles, with heterogeneous processes afforded only a highly parameterized treatment (e.g., Logan et al., 1981). Here, a model is introduced which includes gases, aerosols, and hydrometeors (cloud and rain drops) as coequal, interactive model elements. The heterogeneous processes considered in constructing the model are listed in Table 1.

Model Description

The present model is a one-dimensional, time-dependent simulation of tropospheric gases and particles. Fifteen height levels are included, extending from the ground to 14 km in 1-km intervals. Boundary conditions are established at the surface and at the tropopause near 14 km. The model has three major elements: gases, aerosols, and clouds. The interactions between these elements are depicted schematically in Figure 1. The basic structure of the model is summarized in Table 2.

Chemistry

The gas-phase chemistry incorporates 32 "active" species in the following families: odd oxygen (O, O_3), hydrogen (OH, HO_2, H_2O_2), nitrogen (NO, NO_2, NO_3, N_2O_5, HNO_2, HNO_3, HO_2NO_2, NH_2, NH_3), carbon (CO, CH_2O, CH_3O, CH_3O_2, CH_3OOH, $CH_3O_2NO_2$), and sulfur (S, SO, SO_2, SO_3, HS, H_2S, HSO_3, H_2SO_4, OCS, CS, CS_2, $(CH_3)_2S$). In addi-

TABLE 1. Heterogeneous Processes Contributing to Tropospheric Chemistry

Processes	Key Species
Condensation of gases on aerosols	H_2SO_4, H_2O, HNO_3
Dissolution of gases in aerosols	H_2O_2, SO_2, O_3, NH_3
Chemical reaction in aerosol solution	SO_2, NO_x
Reactions of gases on solid surfaces	SO_2, NO_2
Vapor nucleation	H_2SO_4/H_2O, C_xH_y
Dry deposition of gases	SO_2, HNO_3, O_3
Dry deposition of aerosols	$SO_4^=$, C_x
Scavenging of gases by cloud drops	SO_x, NO_x
Precipitation removal of soluble vapors	SO_x, NO_x
Collection of aerosols by hydrometeors	$SO_4^=$, NO_3^-, C_x
Nucleation of aerosols into cloud drops	$SO_4^=$, NO_3^-
Aqueous chemistry in cloud and rain drops	$SO_4^=$, NO_3^-, $CO_3^=$, HO_x
Natural and anthropogenic emission and absorption of gases and particles from land and oceans	-

tion, a few reactive species (H, $O(^1D)$, $S(^1D)$, $SO_2(^{1,3}B_1)$, CS_2^*, CH_3, CHO) are treated using photochemical equilibrium relations, and the concentrations of several relatively inert species (N_2, O_2, H_2O, CH_4, N_2O) are held fixed. The major chemical groups, and many of the important interaction pathways between the groups, are illustrated in Figure 2.

The gaseous photochemical scheme includes about 75 chemical reactions and 16 photolytic processes. The reaction set for the O-H-N-C constituents is the same as that of Logan et al. (1981); the sulfur reaction set is adopted from Turco et al. (1982). Reaction rate coefficients and photolysis cross sections and quantum yields are taken from the recent recommendations of DeMore et al. (1981). Photodissociation rates are computed at arbitrary solar zenith angles using a two-stream radiation transport model (Joseph et al., 1976; Meador and Weaver, 1980), taking into account the effects of O_3 and NO_2 absorption, Rayleigh scattering, aerosol and cloud turbidity, and surface albedo. The solar fluxes compiled by Delaboudiniere et al. (1978) are used. The vertical transport of gaseous constituents is accomplished by the use of an empirical eddy diffusion coefficient (Bauer, 1974); the coefficient has a value of 1×10^5 cm^2/sec between 0 and 10 km and decreases logarithmically to 5×10^3 cm^2/sec at 14 km.

Sources for gases include emissions at the ground (NO, CO, SO_2) and subsidence from the

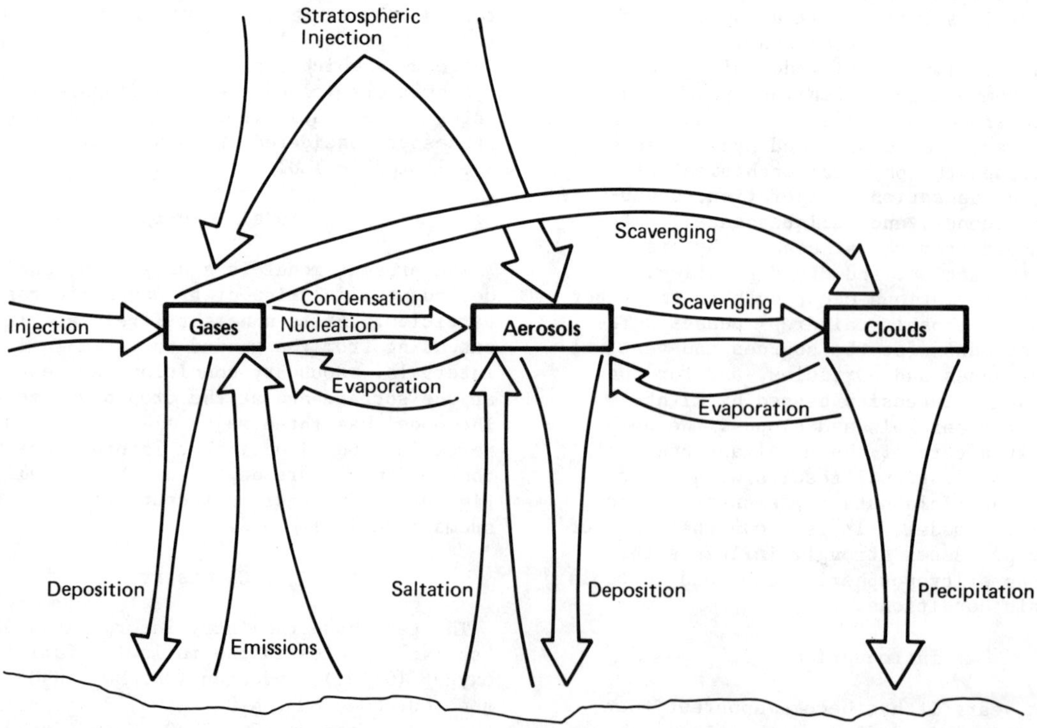

Fig. 1. Physical interactions between tropospheric gases, aerosols, and clouds, as simulated in a one-dimensional model.

stratosphere (NO, NO_2, HNO_3, O_3). Sinks include deposition at the surface, with nominal deposition velocities of 0.5 cm/sec for gases such as HNO_3, H_2O_2, H_2SO_4, etc. (Sehmel, 1980), and scavenging by aerosols and hydrometeors (Hales, 1972). The latter process is reversible in that dissolved gases can evaporate from solutions that are supersaturated (Fuchs and Sutugin, 1971).

The scheme for solving the gas continuity equations is described by Turco and Whitten (1977) and includes a treatment of "diurnal averaging" of photochemical rates, as proposed by Turco and Whitten (1978).

Aerosol and cloud-water chemical reactions are currently being added to the model. The chemical species and interactions considered for aerosols are depicted in Figure 3, and the rainwater constituents treated in the model are listed in Table 3. Specific aqueous reaction mechanisms have been adopted from previous studies (e.g., Freiberg, 1974; Larson and Harrison, 1977; Peterson and Seinfeld, 1979; Möller, 1980; Tang, 1980; Graedel and Weschler, 1981).

Microphysics

The microphysical interactions of a multicomponent aerosol system are analyzed over a size dispersion extending from 0.001 to ~10 μm radius. The aerosols are basically sulfuric-acid/water droplets containing up to six additional "core" components, identified as ammonium sulfate, ammonium nitrate, silicates, sea salt, soot, and metal oxides; the last four components are considered as generic aerosol constituents.

In addition to the liquid (aqueous) aerosols, "dry" particles, or condensation nuclei (cn), are also treated. The dry particles are composed of the same six materials which are imbedded as cores in the sulfuric acid aerosols. Dry particulates are introduced into the atmosphere at the surface by natural processes (e.g., wind saltation and bursting bubbles) and anthropogenic processes (e.g., industrial effluents and agricultural operations). Above the ground, particles also have both natural sources (e.g., in situ nucleation and stratospheric aerosol subsidence) and anthropogenic sources (e.g., aircraft and rocket emissions). Dry aerosols can be nucleated into wet aerosols when the local supersaturations of H_2SO_4 and/or water vapor are sufficiently large. The dry and wet aerosols also interact through coagulation. As a result of nucleation and coagulation, material collects within the aerosols as cores. The cores continue to grow by chemical reaction. For example, ammonia gas absorbed by sulfuric acid aerosols produces ammonium sulfate core material (which itself may be fluid).

When aerosols evaporate, the cores are left behind as dry particles. To simulate accurately the resultant size dispersion of the dust, the size distribution of the cores in aerosols of each size is approximated using the first two volume moments of the cores. (See Turco et al. (1979a, b) for a description of the core volume moment equations used here).

TABLE 2. Structure of a One-Dimensional Time-Dependent Tropospheric Model

Major Elements	Physical & Chemical Processes
Gases	42 O-N-H-C-S species
	75 homogeneous reactions
	16 photolytic processes
	Two-stream radiation transport
	Vertical mixing
	Surface deposition
	Particle scavenging
Aerosols	40 sizes: 0.001 - 10 μm
	8 chemical components
	Dry and wet aerosols
	Microphysics: nucleation, condensation/evaporation, coagulation, sedimentation, diffusion, deposition
	Cloud scavenging: Brownian/turbulent diffusion, ventilated convection, thermophoresis and diffusiophoresis
	Chemistry: gas absorption, soot catalysis
	Light scattering
Hydrometeors	Empirical drop size and number versus height and time
	24 dissolved components: $SO_x^=$, NO_x^-, $CO_x^=$, NH_4^+, HO_x, O_3, C_xH_y, metals, soot, sand
	Microphysics: gravitational fall, gas and aerosol scavenging, aerosol nucleation
	Chemistry: aqueous photochemistry, ion equilibrium

Several aerosol nucleation processes are considered in the model. H_2SO_4/H_2O binary homogeneous and ion nucleation rates are calculated taking into account preexisting aerosols and the gas kinetic limitations to these processes that apply at very low H_2SO_4 vapor concentrations. Heterogeneous heteromolecular nucleation of dry dust into sulfuric acid aerosols is also accounted for using the scheme of Hamill et al. (1982).

All of the particles are subject to coagulation (Fuchs, 1964; Hidy and Brock, 1970), gravitational sedimentation (Kasten, 1968) and vertical diffusion. The aerosols and dust are also scavenged by cloud and rain drops through the processes of Brownian and turbulent coagulation, thermophoresis and diffusiophoresis, and inertial impaction (Pruppacher and Klett, 1978). The liquid aerosols grow (shrink) by absorbing (evaporating) gaseous species, mainly H_2SO_4 and H_2O. Particles are, in addition, subject to deposition

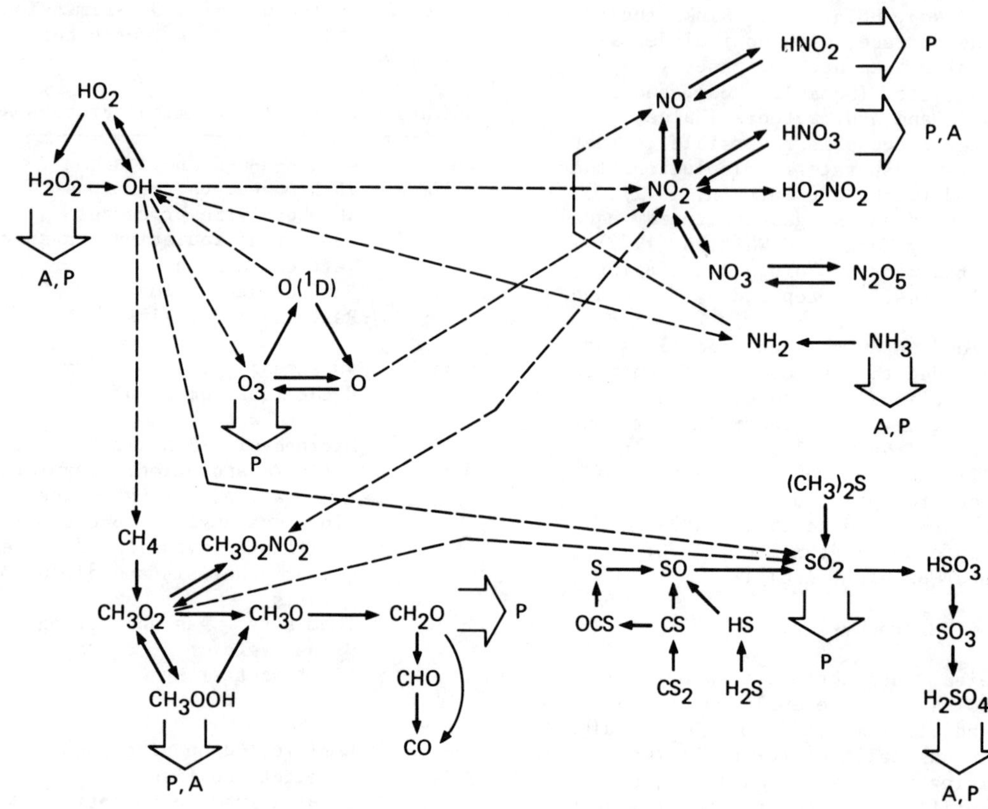

Fig. 2. Gas-phase constituents and major reaction pathways (solid lines). Interactions between chemical families are indicated by dashed lines. Heavy (double) arrows show key heterogeneous pathways involving aerosols (A) and precipitation (P).

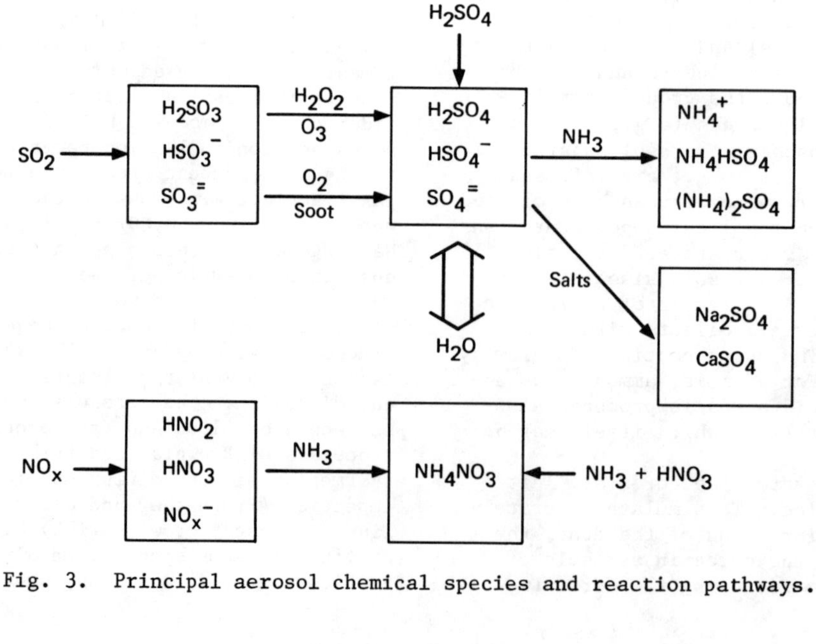

Fig. 3. Principal aerosol chemical species and reaction pathways.

TABLE 3. Aqueous Constituents

Constituents	Major Sources
Carbonates (H_2CO_3, HCO_3^-, $CO_3^=$)	CO_2 equilibrium
Sulfites (H_2SO_3, HSO_3^-, $SO_3^=$)	SO_2 dissolution
Sulfates (HSO_4^-, $SO_4^=$)	Sulfite oxidation by dissolved O_2, O_3, H_2O_2, and metal catalysts Condensation of gaseous H_2SO_4 Scavenging of aerosol H_2SO_4, $(NH_4)_2SO_4$, and Na_2SO_4
Nitrites (HNO_2, NO_2^-)	Dissolution of NO, NO_2, and HNO_2
Nitrates (NO_3^-)	Nitrite oxidation by dissolved O_3 Dissolution of gaseous NO_2 and HNO_3 Scavenging of aerosol HNO_3, NH_4NO_3, and $NaNO_3$
Ammonium (NH_4OH, NH_4^+)	Dissolution of NH_3 Scavenging of aerosol $(NH_4)_2SO_4$ and NH_4NO_3
Active hydrogen (OH, HO_2, H_2O_2)	Dissolution of H_2O_2 O_3 photochemical reactions in solution
Other cations (H^+, Na^+, Ca^{++}, Fe^{+3})	Aerosol scavenging
Other anions (OH^-)	Water dissociation
Solids (soot, sand)	Particle scavenging

at the ground (Slinn, 1977). The coupled aerosol/dust/gas continuity equations employed in the model take into consideration all of the physicochemical processes mentioned. The basic mathematical scheme for obtaining solutions of the equations is described by Turco et al. (1979a) and Toon et al. (1979).

Cloud and rain drops are simply treated as a monodispersed collection of hydrometeors which can change size with height and time in a predetermined way. For example, variations in droplet size are related to rainfall rates through empirical correlations (Markowitz, 1976; Best, 1950). The raindrop continuity equation describes the production of drops in the cloud and their fall to the ground using the fall velocities defined by Beard (1976).

Aerosols and gases scavenged by cloud and rain drops affect precipitation chemistry. Aerosols may also be nucleated into cloud drops when the local humidity exceeds the Köhler barrier, as determined by the aerosol composition (Hänel, 1976). As a result of scavenging and nucleation processes, water drops can accumulate up to 24 compounds, as indicated in Table 3.

Simulations

When these results were first presented, in 1981, only preliminary calculations had been carried out with the model. These were made to test the accuracy of the numerical algorithms for gas-phase chemistry, radiation transport, multicomponent aerosol coagulation, and raindrop scavenging of gases and aerosols, and to establish the raindrop trace constituent continuity equations and the mass conservation of aerosol material transferred to raindrops. The calculations were not "tuned" to fit any particular geophysical conditions. Temperature and humidity profiles were taken from the U.S. Standard Atmosphere (1976). Some of the key model boundary conditions are summarized in Table 4. Typical predicted concentrations for several of the gas species are shown in Figure 4. Comparison of these curves with those presented by Logan et al. (1981) shows that for roughly the same lower boundary conditions and gas sources the calculated abundances are quite similar. More detailed comparisons are impractical because of numerous small differences between the models.

Our simulations indicate that the stratospheric source of NO_x is significant in the upper troposphere, as pointed out by Liu et al. (1980). Tropopause fluxes of NO, NO_2, HNO_3, and O_3 (Table 4) were taken from the one-dimensional stratospheric model of Turco and Whitten (1977). The total predicted concentration of NO + NO_2 near the tropopause, ~55 pptv (parts per trillion by volume), is significantly smaller than a recent

TABLE 4. Model Boundary Conditions for Gaseous Constituents

Species	Surface Flux[a,b]	Tropopause Flux[a]
NO	3.0×10^9	-1.3×10^7
NO_2	-7.0×10^8	-2.2×10^7
HNO_3	-2.3×10^9	-1.3×10^8
SO_2	-3.6×10^8	0
H_2S	3.9×10^8	0
OCS	-1.4×10^7	0
CS_2	1.0×10^8	0
$(CH_3)_2S$	2.8×10^8	0
CO	5.0×10^{11}	0
O_3	-1.4×10^{11}	-6.0×10^{10}
H_2O_2	-2.0×10^{11}	0

[a]Flux units are molec/cm^2-sec, positive upward.

[b]The net emission or deposition flux at the surface is given.

measurement of ~200 pptv (Kley et al., 1981). The difference may be due partly to a difference in the local odd-nitrogen partition, which is dominated by HNO_3, or partly to a local intrusion of stratospheric air.

Also shown in Figure 4 is the possible effect of a heavy rainfall (~2 cm in 1 hour) on the gas concentrations. The precipitation was allowed to fall uniformly below 5 km. Gases were assumed to have infinite solubilities, which is a reasonable approximation for HNO_3 and H_2O_2 but not for NO, SO_2, and many other constituents. The results in Figure 4, therefore, represent the maximum degree of scavenging of gases. Nonetheless, the large potential depletions, an order of magnitude in one storm, may have important consequences. Tropospheric cleansing by rainfall is, apparently, highly inhomogeneous, quite unlike the uniform "rainout" parameterization employed in most one- and two-dimensional photochemical models. Such localized removal will inevitably lead to spatially irregular gas distributions. According to our preliminary results, washout of HNO_3 may also contribute significantly to rain acidity, as suggested recently by Durham et al. (1981).

Complex photochemical effects occur in the region of raining clouds, as is indicated by the changes in OH concentrations above and below the clouds in Figure 4. Due to light scattering, near-ultraviolet solar fluxes below the clouds are reduced by 60 to 70 percent (Galbally, 1972), and those above the clouds are enhanced by 20 to 30 percent. The changes in light intensities affect molecular photodissociation rates. Species such as NO and OH, moreover, are strongly coupled by chemical reactions. Thus, for example, height variations in OH closely parallel those in NO in Figure 4.

The simulated transfer rates of sulfur and odd-

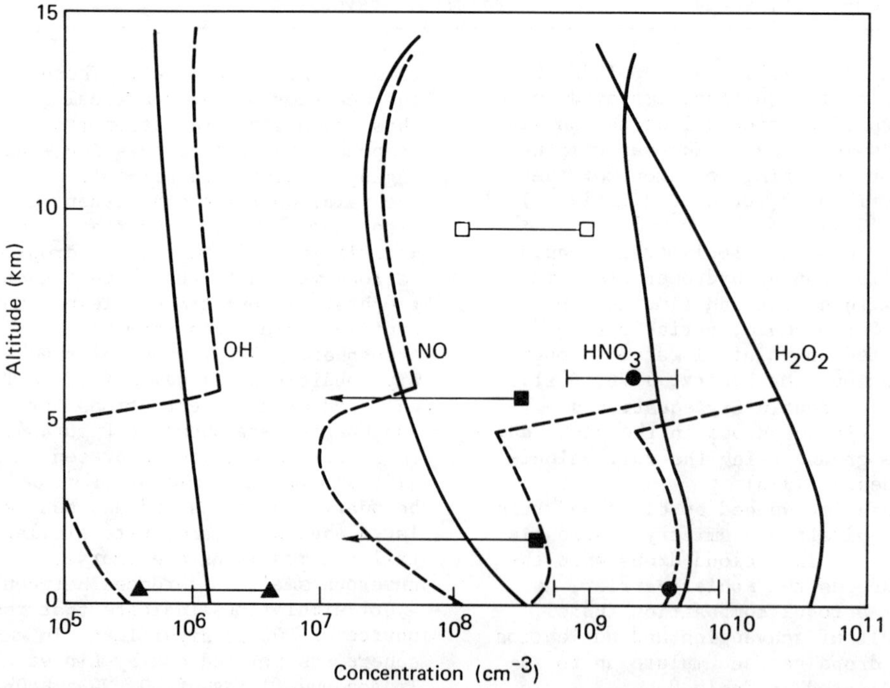

Fig. 4. Predicted tropospheric gas concentrations in the free troposphere at steady state (——) and 1 hour later at the end of a continuous rainfall of ~2 cm/hr (---). The rain occurs below 5 km. Also shown are the following data: acidic nitrate at 70°N to 55°S (●) (Huebert and Lazrus, 1980); NO at 15° to 45°N (■) (Schiff et al., 1979); NO at 41°N taken as one-half of the NO + NO_2 concentration (□) (Kley et al., 1981); and OH at 47°N (▲) (Campbell et al., 1979).

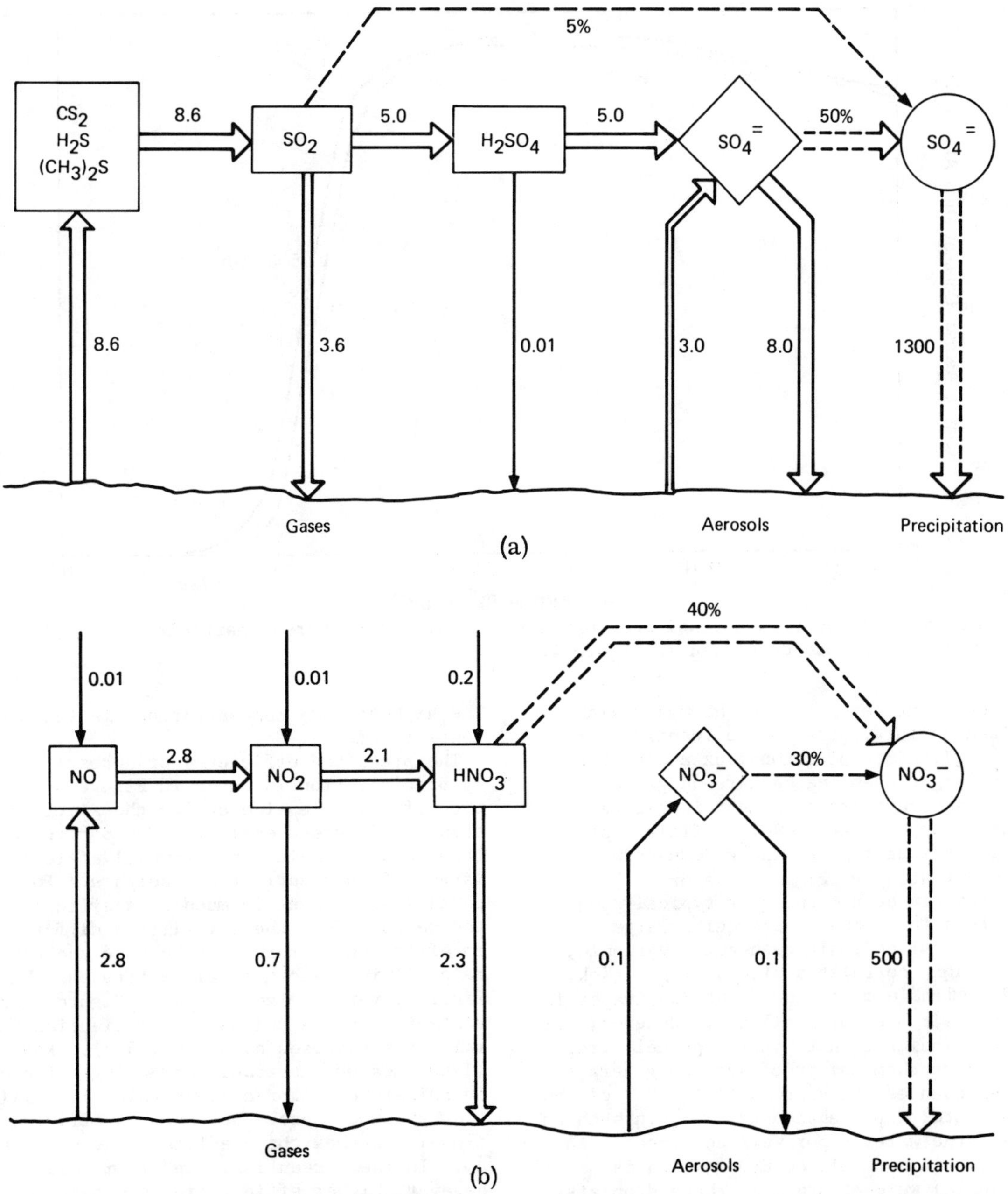

Fig. 5. Calculated sulfur (a) and odd-nitrogen (b) atom fluxes in the presence and absence of rainfall. Fluxes are given in units of 10^8 S atoms/cm^2-sec and 10^9 N atoms/cm^2-sec, respectively, averaged over the globe, for a vertical column of tropospheric air. Solid lines indicate fluxes under cloud-free conditions. Dashed lines indicate average precipitation fluxes corresponding to the rainfall conditions of Figure 4. The percentages of gaseous and particulate material scavenged during the rain storm are presented, rather than the instantaneous fluxes.

nitrogen species through the troposphere are illustrated in Figure 5. Shown are emission fluxes, stratospheric fluxes, gas-to-particle conversion rates, dry deposition fluxes, fractions of gas and aerosol materials scavenged by raindrops, and instantaneous precipitation fluxes. The calculations reveal several important qualitative features. In cloud-free regions of the troposphere, dry deposition is the dominant loss mechanism for both the S and N cycles (as SO_2 and $SO_4^=$ for the sulfur cycle, and as HNO_3 for the odd-nitrogen cycle). Furthermore, in Figure 5(a)

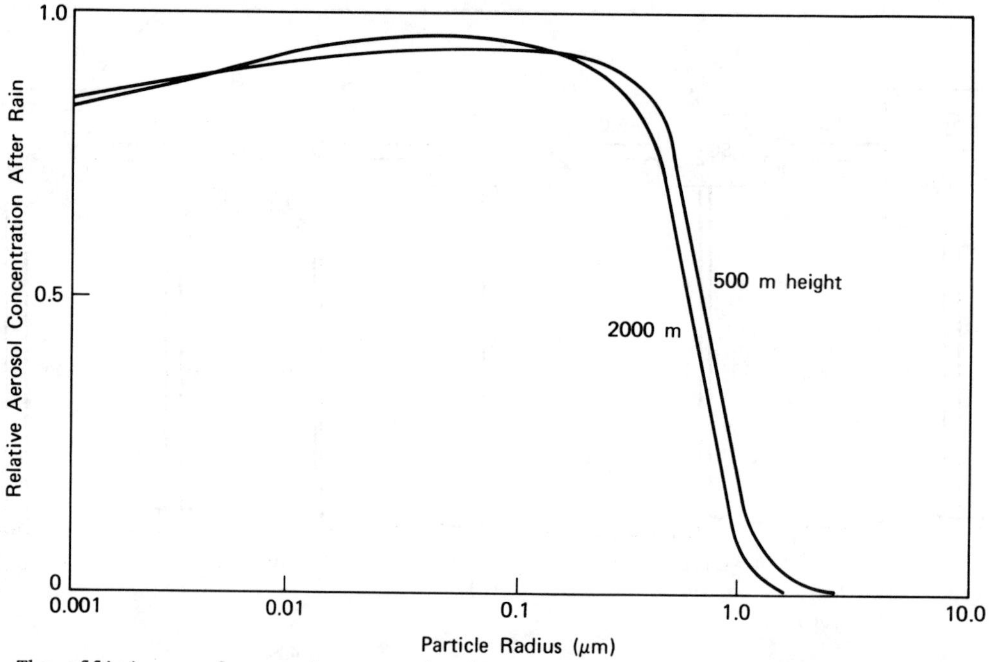

Fig. 6. The efficiency of aerosol scavenging by precipitation versus particle size at two heights for the rainfall episode described in Figure 4.

it is obvious that aerosols are the major sink for oxidized sulfur vapors (predominantly in the form of H_2SO_4). Precipitation fluxes also tend to be much larger than the background gas and particle fluxes, but are transient in nature. Precipitation appears to be very efficient at scavenging particulate $SO_4^=$ and gaseous HNO_3, but less efficient at scavenging SO_2 vapor.

It has recently been found that typical H_2O_2 concentrations in rainwater are quite large (~10^{-6} to 10^{-5} M) and, although quite variable, are not strongly correlated with sunlight (Kok, 1981). No reliable measurements of H_2O_2 vapor in tropospheric air are yet available. However, our calculations indicate that, in air parcels free of clouds, photochemical processes can generate H_2O_2 gas abundances which are 10^{-7} to 10^{-6} of the background water vapor abundance. If, in such an air parcel, the water vapor were condensed into droplets, and if most of the H_2O_2 (which is highly soluble) were absorbed in those droplets, then an aqueous hydrogen peroxide solution in the range of 10^{-6} to 10^{-5} M would result. This photochemical source of aqueous peroxide would not be expected to have a strong diurnal component, inasmuch as the lifetime of H_2O_2 vapor in clean air is longer than a day. In frontal storm systems, the continuous condensation of water vapor in fresh air parcels as cloud and rain water implies a continuous (steady) source of aqueous H_2O_2. In convective storm systems, higher aqueous H_2O_2 concentrations might occur because larger volumes of fresh air are constantly processed through the cloud (Ludlam, 1980), and H_2O_2 vapor could be preferentially absorbed by hydrometeors at the cloud base. Kok (1981) has found the highest H_2O_2 concentrations in the most intense rainfalls.

The predicted efficiency of aerosol scavenging by precipitation is shown in Figure 6 as a function of particle size during the simulated rain episode discussed earlier. For particles larger than ~0.5 μm radius, the removal efficiency is large (>50 percent) at all heights. For smaller particles, washout is much less efficient. In the calculation, there is little differentiation in efficiency with height because the simulated rainfall is uniform at all altitudes. The efficiency versus size curves in Figure 6 are consistent with the collision efficiencies for 1-mm raindrops proposed by Slinn (1977). Radke et al. (1980) measured aerosol scavenging efficiencies in rainstorms. For a heavy rainfall, similar to that presumed here, the observed efficiency closely matches the predicted values above ~0.5 μm. In the "Greenfield gap" from ~0.2 to 0.5 μm, however, higher efficiencies are observed. Radke and coworkers attribute this effect to the swelling of aerosols by absorption of water vapor within humid clouds, leading to enhanced collection cross sections. In the present calculations, the possible expansion and nucleation of the aerosols at high relative humidity in the presence of clouds was neglected. This is an important area for future investigation.

Conclusions

We conclude that accurate analysis and interpretation of global tropospheric chemical cycles will require models which include realistic treatments of heterogeneous chemical and physical processes. Preliminary simulations made with such a model indicate that washout and rainout of gases and aerosols govern their distributions under certain conditions. Likewise, the compounds scavenged by water droplets significantly affect the composition and acidity of cloud and rain water. Comprehensive physicochemical models will soon be available to study in detail the transfer of gaseous and particulate trace materials through the environment.

References

Bauer, E., Dispersion of tracers in the atmosphere and ocean: Survey and comparison of experimental data, J. Geophys. Res., 79, 789-795, 1974.

Beard, K. V., Terminal velocity and shape of cloud and precipitation drops aloft, J. Atmos. Sci., 33, 851-864, 1976.

Best, A. C., The size distribution of raindrops, Quart. J. Roy. Meteor. Soc. London, 76, 16-36, 1950.

Campbell, M. J., J. C. Sheppard, and B. F. Au, Measurement of hydroxyl concentration in boundary layer air by monitoring CO oxidation, Geophys. Res. Lett., 6, 175-178, 1979.

Delaboudiniere, J. P., R. F. Donnelly, H. E. Hinteregger, G. Schmidtke, and P. C. Simon, Intercomparison/Compilation of Relevant Solar Flux Data Related to Aeronomy, COSPAR Manual No. 7, Institut d'Aeronomie Spatiale de Belgique, Belgium, 1978.

DeMore, W. B., L. J. Stief, F. Kaufman, D. M. Golden, R. F. Hampson, M. J. Kurylo, J. J. Margitan, M. J. Molina, and R. T. Watson, Chemical Kinetic and Photochemical Data for Use in Stratospheric Modelling, JPL Publication 81-3, Jet Propulsion Laboratory, Pasadena, Ca., 1981.

Durham, J. L., J. H. Overton, Jr., and V. P. Aneja, Influence of gaseous nitric acid on sulfate production and acidity in rain, Atmos. Environ., 15, 1059-1068, 1981.

Freiberg, J., Effects of relative humidity and temperature on iron-catalyzed oxidation of SO_2 in atmospheric aerosols, Environ. Sci. Technol., 8, 731-734, 1974.

Fuchs, N. A., The Mechanics of Aerosols, translated by R. E. Daisley and M. Fuchs, Pergamon Press, New York, 1964.

Fuchs, N. A., and A. G. Sutugin, Highly dispersed aerosols, in Topics in Current Aerosol Research, 2, edited by G. M. Hidy and J. R. Brock, Pergamon Press, Oxford, pp. 1-60, 1971.

Galbally, I. E., Production of carbon monoxide in rain water, J. Geophys. Res., 77, 7129-7132, 1972.

Graedel, T. E., and C. J. Weschler, Chemistry within aqueous atmospheric aerosols and raindrops, Rev. Geophys. Space Phys., 19, 505-539, 1981.

Hales, J. M., Fundamentals of the theory of gas scavenging by rain, Atmos. Environ., 6, 635-659, 1972.

Hamill, P., R. P. Turco, O. B. Toon, C. S. Kiang, and R. C. Whitten, On the formation of sulfate aerosol particles in the stratosphere, J. Aerosol Sci., in press, 1982.

Hänel, G., The properties of atmospheric aerosol particles as functions of the relative humidity at thermodynamic equilibrium with the surrounding moist air, Adv. Geophys., 19, 73-188, 1976.

Hidy, G. M., and J. R. Brock, The Dynamics of Aerocolloidal Systems, Chapter 10, Pergamon Press, New York, 1970.

Huebert, B. J., and A. L. Lazrus, Tropospheric gas phase and particulate nitrate measurements, J. Geophys. Res., 85, 7322-7328, 1980.

Joseph, J. H., W. J. Wiscombe, and J. A. Weinman, The delta-Eddington approximation for radiative flux transfer, J. Atmos. Sci., 33, 2452-2459, 1976.

Kasten, F., Falling speed of aerosol particles, J. Appl. Meteor., 7, 944-947, 1968.

Kley, D., J. W. Drummond, M. McFarland, and S. C. Liu, Tropospheric profiles of NO_x, J. Geophys. Res., 86, 3153-3161, 1981.

Kok, G. L., Measurements of hydrogen peroxide in rainwater, Trans. Am. Geophys. Union, 62, 884, 1981.

Larson, T. V., and H. Harrison, Acidic sulfate aerosols: Formation from heterogeneous oxidation by O_3 in clouds, Atmos. Environ., 11, 1133-1141, 1977.

Liu, S. C., D. Kley, M. McFarland, J. D. Mahlman, and H. Levy II, On the origin of tropospheric ozone, J. Geophys. Res., 85, 7546-7552, 1980.

Logan, J. A., M. J. Prather, S. C. Wofsy, and M. B. McElroy, Tropospheric chemistry: A global perspective, J. Geophys. Res., 86, 7210-7254, 1981.

Ludlam, F. H., Clouds and Storms: The Behavior and Effect of Water in the Atmosphere, Chapters 7 and 8, The Pennsylvania State University Press, University Park, Penn., 1980.

Markowitz, A. H., Raindrop size distribution expressions, J. Appl. Meteor., 15, 1029-1031, 1976.

Meador, W. E., and W. R. Weaver, Two-stream approximations to radiative transfer in planetary atmospheres: A unified description of existing methods and a new improvement, J. Atmos. Sci., 37, 630-643, 1980.

Möller, D., Kinetic model of atmospheric SO_2 oxidation based on published data, Atmos. Environ., 14, 1067-1076, 1980.

Peterson, T. W., and J. H. Seinfeld, Heterogeneous condensation and chemical reaction in droplets - Application to the heterogeneous atmospheric oxidation of SO_2, Adv. Environ. Sci. Tech., 10, 125-180, 1979.

Pruppacher, H. R., and J. D. Klett, Microphysics of Clouds and Precipitation, D. Reidel, Dordrecht, Holland, 1978.

Radke, L. F., P. V. Hobbs, and M. W. Eltgroth, Scavenging of aerosol particles by precipitation, J. Appl. Meteor., 19, 715-722, 1980.

Schiff, H. I., A. Pepper, and B. A. Ridley,

Tropospheric NO measurements up to 7 km, J. Geophys. Res., 84, 7895-7897, 1979.

Sehmel, G. A., Particle and gas dry deposition: A review, Atmos. Environ., 14, 983-1011, 1980.

Slinn, W. G. N., Some approximations for the wet and dry removal of particles and gases from the atmosphere, Water Air Soil Poll., 7, 513-543, 1977.

Tang, I. N., On the equilibrium partial pressures of nitric acid and ammonia in the atmosphere, Atmos. Environ., 14, 819-828, 1980.

Toon, O. B., R. P. Turco, P. Hamill, C. S. Kiang, and R. C. Whitten, A one-dimensional model describing aerosol formation and evolution in the stratosphere: II. Sensitivity studies and comparison with observations, J. Atmos. Sci., 36, 718-736, 1979.

Turco, R. P., and R. C. Whitten, The NASA Ames Research Center One- and Two-Dimensional Stratospheric Models. Part I: The One-Dimensional Model, NASA TP-1002, 1977.

Turco, R. P., and R. C. Whitten, "A note on the diurnal averaging of aeronomical models," J. Atmos. Terrestr. Phys., 40, 13-20, 1978.

Turco, R. P., P. Hamill, O. B. Toon, R. C. Whitten, and C. S. Kiang, A one-dimensional model describing aerosol formation and evolution in the stratosphere: I. Physical processes and mathematical analogs, J. Atmos. Sci., 36, 699-717, 1979a.

Turco, R. P., P. Hamill, O. B. Toon, R. C. Whitten, and C. S. Kiang, The NASA-Ames Research Center Stratospheric Aerosol Model. I. Physical Processes and Computational Analogs, NASA TP-1362, 1979b.

Turco, R. P., O. B. Toon, and R. C. Whitten, Stratospheric aerosols: Observation and theory, Rev. Geophys. Space Phys., in press, 1982.

U. S. Standard Atmosphere, U. S. Government Printing Office, Washington, DC, 1976.

SULFATE IN THE ATMOSPHERIC BOUNDARY LAYER: CONCENTRATION AND MECHANISMS OF FORMATION

R. F. Pueschel and E. W. Barrett

Office of Weather Research and Modification, NOAA Environmental Research Laboratories, Boulder, Colorado 80303

Abstract. Data on particulate sulfur in a coal-fired power plant plume and on its long-range transport are presented in relation to the total aerosol. The primary plume aerosol that could provide potential catalytic surfaces for sulfate generation is fly ash. Its surface area was determined in situ by Knollenberg optical particle counters. Surface features and the spatial distribution of elements in individual fly ash particles were determined by scanning electron microscopy and X-ray energy dispersive analysis of airborne samples.

The results provide evidence of a selective accumulation of sulfate on the surface on some fly ash particles. The amount of sulfate thus formed, however, is only about 2 percent of the total mass. Hence, fly ash has little catalytic effect on the heterogeneous conversion of SO_2 to $SO_4^=$ above currently accepted rates of 1 percent per hour. Homogeneous nucleation at low rates appears to be equally important for the generation of small sulfate particles that are subject to long-range transport. This finding was confirmed by measurements in east-central Utah, a region of the contiguous U. S. that historically is characterized by excellent atmospheric clarity. Although the total particle mass of the background aerosol in Utah is 2 orders of magnitude less than in the Four Corners area, the small-particle mode is dominated by silicon and sulfur. The spherical shape of these silicon- and sulfur-containing particles indicates that a phase transition was involved in the mechanism of their formation. The increase in sulfur aerosols with height suggests their advection from remote sources. It appears that silicon is as good a tracer for a primary combustion aerosol as is sulfur for a secondary aerosol.

Introduction

Low visibilities, acid rain, and inadvertent cloud modification are some of the effects that have been linked to the presence of anthropogenic sulfur in the atmosphere. Because of the local or, at the most, regional occurrence of these effects, the development of regional sulfur cycles (Rodhe, 1972) becomes necessary in order to predict future trends in environmentally harmful effects of sulfur. The sulfur budget of a region, or of a single source, is determined by source strengths and transport, transformation, and removal processes. Each of these processes is strongly influenced by the mechanisms of homogeneous or heterogeneous conversion of SO_2 to $SO_4^=$.

Our experimental techniques of airborne sampling and subsequent laboratory analysis of aerosols to arrive at mechanisms and rates of sulfate formation have been discussed elsewhere (Mamane and Pueschel, 1980; Pueschel, 1976). A detailed description of the acquisition, retrieval, and processing of aerosol data from Knollenberg optical particle counters and data acquisition systems has been given by Barrett et al. (1979).

Results and Discussion

Sulfate Formation in Power Plant Plumes

Shown in Figures 1 and 2 are examples of the type of information gained on size, shape, and elemental composition of individual particles when they are subjected to the combined techniques of scanning electron microscopy and X-ray energy dispersive analysis (SEM-XEDA). X-ray spectra are shown in Figure 1 for a fly ash particle collected in the plume of the Four Corners Power Plant some 20 km downwind from the stacks. The particle image also shown in Figure 1 results from secondary electron emissions in the SEM mode. Its spherical shape identifies the particle as resulting from a mechanism that involves a phase change from either gas or liquid to solid. In order to generate X-rays in the XEDA mode, the SEM's electron beam is kept stationary. The X-rays, induced by the incidence of electrons on elements inside and on the surface of the particle, are collected for a specified time by a Li-doped Si detector, analyzed in a multichannel analyzer, and displayed on a video screen. The various X-ray spectra shown in Figure 1 correspond to different portions of the particle, as indicated. Typically, fly ash from coal combus-

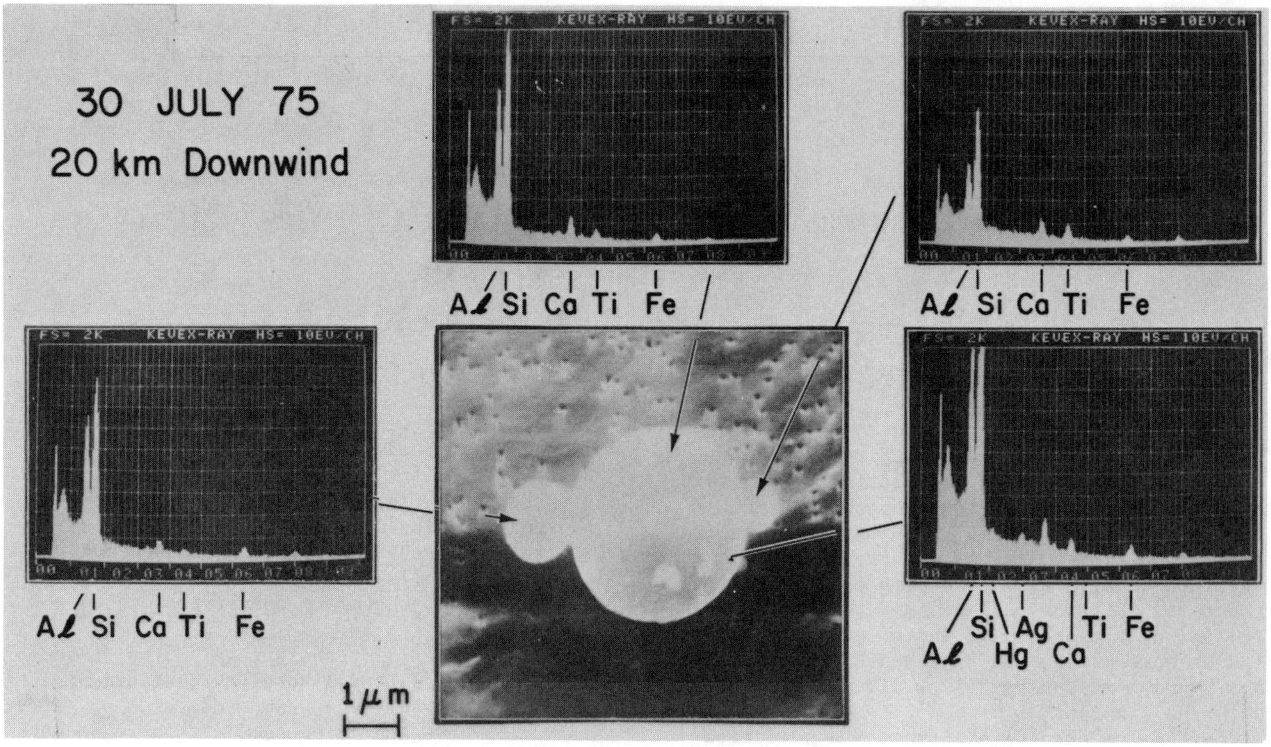

Fig. 1. Typical shape and elemental composition of a fly ash particle.

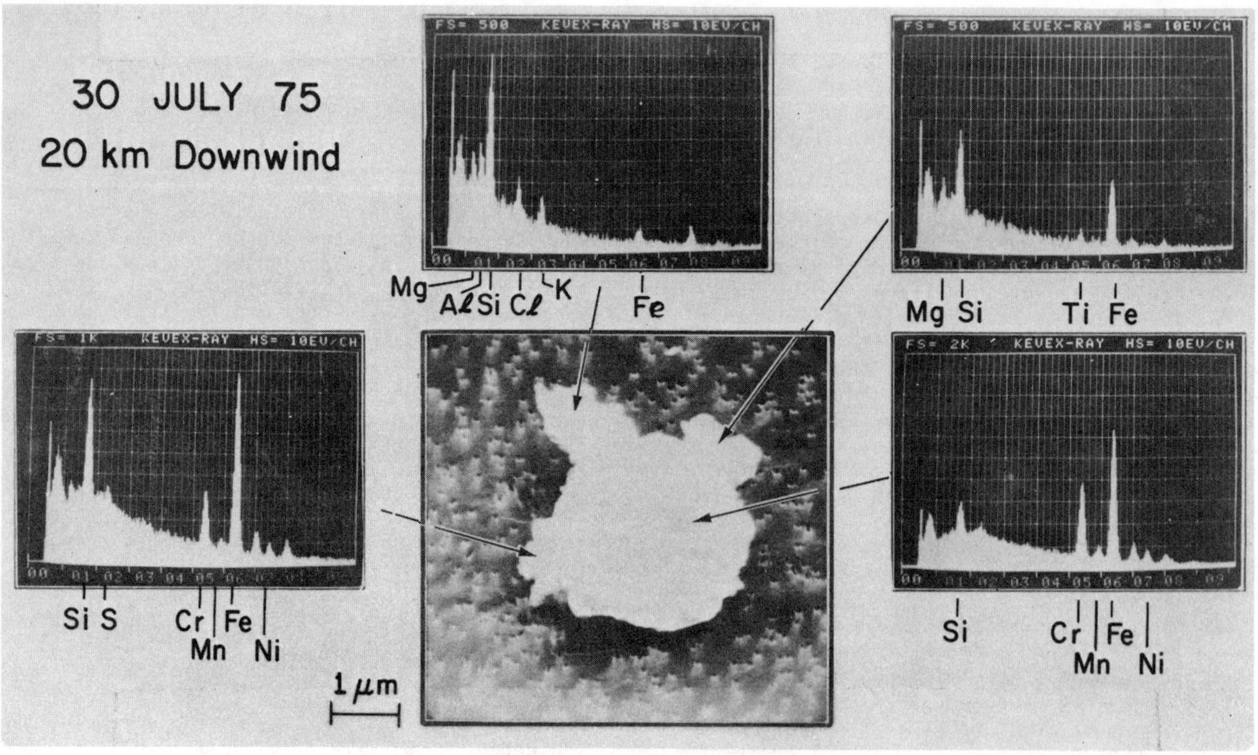

Fig. 2. Typical shape and elemental composition of a soil-derived particle.

tion contains mainly light elements, e.g., Al, Si, K, Ca, and trace amounts of Fe and Ti.

X-ray spectra are shown in Figure 2 for a soil-derived particle collected outside the plume at the same geographic location. In contrast to fly ash, this particle contains heavier elements in higher concentrations. It also follows from Figure 2 that different portions of the same particle contain different elements in varying concentrations, indicating a large degree of chemical inhomogeneity. Physical differences between background and plume aerosols in the Four Corners region are illustrated in Figure 3. Aerosol size distributions are shown in Figure 3(a), as represented by modified gamma distributions (Barrett et al., 1979) that were least-square fitted to Knollenberg probe data collected outside the plume at 0.4, 16.0, and 32 km from the stacks. We note a fair degree of homogeneity in the aerosol content of the portion of the atmosphere that is unaffected by power plant effluents. Differences in the aerosol size distributions from 0.05 μm on up are small if compared with portions of in-plume size distributions, shown in Figure 3(b). While the concentrations in the small-particle mode of the plume aerosol are similar to those found in the ambient air, the number concentrations in the large-particle mode are significantly elevated inside the plume. It is the potential catalytic properties of the surface area associated with this plume aerosol that are of interest for the present discussion. Because the surface area calculated from optical particle counters is based on equivalent spheres, the in-plume data are probably quite realistic because of the spherical shape of fly ash particles.

The statistics for the particle number, surface area, and volume/mass distributions are given in Table 1 for the plume aerosol and in Table 2 for the background aerosol collected at 0.4 km from the stacks. These statistics are defined as follows: the mean radius is that value that divides the area under the curve (of the parameter considered versus particle radius R as a linear (not logarithmic) scale) into two equal

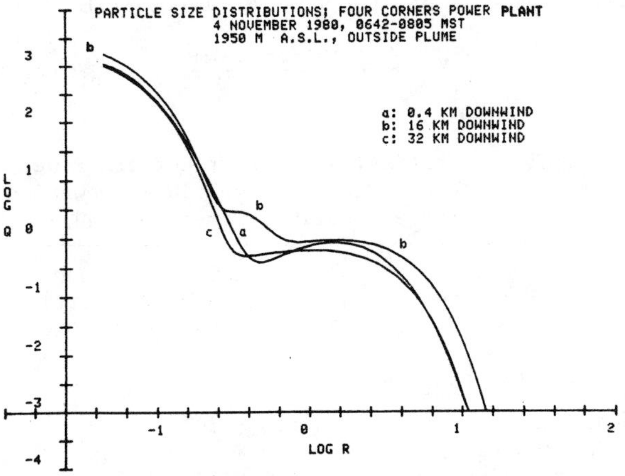

(a)

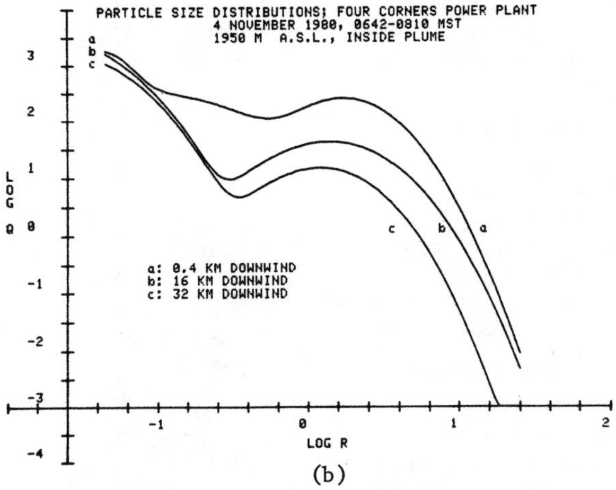

(b)

Fig. 3. Size distributions of aerosol particles in background air (a) and inside the plume (b) at 0.4, 16, and 32 km from the stacks. Log Q = log (dN/d(log R)) where N is the number of aerosol particles per cm^3 and R is in μm.

TABLE 1. Statistics for Four Corners Power Plant Plume Aerosol 0.4 km Downwind (Date: 801101; Time: 082159)

Parameter	Small Particle Mode	Large Particle Mode
Number Distribution		
Mean radius, μm	3.2×10^{-3}	1.6
Modal radius, μm	1.9×10^{-2}	1.2
Median radius, μm	1.0×10^{-2}	1.2
Standard deviation, μm	1.1×10^{-2}	1.4
Skewness parameter	1.7	1.5
Modal ordinate, cm^{-3}	1.9×10^{3}	2.5×10^{2}
Number concentration, cm^{-3}	1.8×10^{4}	2.0×10^{2}
Surface Distribution		
Mean radius, μm	5.8×10^{-2}	4.9
Modal radius, μm	6.8×10^{-2}	3.7
Median radius, μm	5.6×10^{-2}	3.7
Standard deviation, μm	2.6×10^{-2}	4.3
Skewness parameter	0.67	1.5
Modal ordinate, cm^2/cm^3	1.9×10^{-6}	1.4×10^{-4}
Surface area, cm^2/cm^3	1.3×10^{-6}	1.1×10^{-4}
Mass Distribution		
Mean radius, μm	7.0×10^{-2}	8.6
Modal radius, μm	7.8×10^{-2}	6.5
Median radius, μm	6.9×10^{-2}	6.5
Standard deviation, μm	2.5×10^{-2}	7.5
Skewness parameter	0.57	1.5
Modal ordinate, $\mu g/m^3$	2.8	4.5×10^{4}
Mass concentration, $\mu g/m^3$	1.1	3.7×10^{4}

TABLE 2. Statistics for Four Corners Background Aerosol 0.4 km Downwind (Date: 801101; Time: 0821242)

Parameter	Small Particle Mode	Large Particle Mode
Number Distribution		
Mean radius, μm	7.1×10^{-3}	0.46
Modal radius, μm	1.0×10^{-2}	0.55
Median radius, μm	2.0×10^{-4}	6.6×10^{-2}
Standard deviation, μm	1.7×10^{-2}	0.92
Skewness parameter	1.6	1.6
Modal ordinate, cm^{-3}	1.9×10^{3}	2.3
Number concentration, cm^{-3}	8.5×10^{3}	7.0
Surface Distribution		
Mean radius, μm	8.5×10^{-2}	3.9
Modal radius, μm	8.9×10^{-2}	4.0
Median radius, μm	7.5×10^{-2}	4.0
Standard deviation, μm	5.1×10^{-2}	2.5
Skewness parameter	1.0	1.1
Modal ordinate, cm^2/cm^3	5.5×10^{-7}	1.3×10^{-6}
Surface area, cm^2/cm^3	3.7×10^{-7}	9.3×10^{-7}
Mass Distribution		
Mean radius, μm	0.12	5.5
Modal radius, μm	0.12	5.6
Median radius, μm	0.11	5.0
Standard deviation, μm	5.8×10^{-2}	2.8
Skewness parameter	0.95	0.98
Modal ordinate, $\mu g/m^3$	3.8	4.2×10^{2}
Mass concentration, $\mu g/m^3$	2.1	2.4×10^{2}

areas; the modal radius is the abscissa of the curve maximum; the median radius is that which divides the data into two equal populations; the standard deviation is the second moment of the data about the mean radius and measures the broadness or narrowness of the distribution; the skewness parameter is the third moment about the mean divided by the standard deviation, and measures the asymmetry of the curve (on a linear scale) with respect to the mean radius; the modal ordinate is the value of the data at the peak of the curve; and the concentrations (and surface area) are proportional to the total area under the curves. All of these are closed analytical expressions when the data are represented by generalized gamma distributions. Corresponding aerosol statistics are shown in Tables 3 and 4 for samples collected at 16 km downwind from the stacks. It follows from Figure 3 and Tables 1 through 4 that, for the size range 0.05 to 23.5 μm particle radius, both the plume and background aerosols are characterized by two major modes. At 0.4 km from the stacks, the aerosol surface area inside the plume is increased by about 2 orders of magnitude above background. At 16 km from the stacks, this increase is reduced to approximately 1 order of magnitude.

The frequency with which specific elements of atomic number larger than 10 are found in fly ash (Table 5) alludes to the chemical composition of the fly ash surfaces. It can be seen that Si is the dominant element in almost all of the fly ash particles. Al is found in about 80 percent of all particles, and about half contain S. Of the heavier elements, Fe is the most abundant, appearing in about 43 percent of all particles, followed by Ti.

We shall now address the question of the distribution of the sulfur that, according to Table 5, is associated with 50 percent of the particles, in relation to the particle surface. If fly ash surfaces indeed play a catalytic role in the oxidation of SO_2, the resulting sulfate must be found on, rather than inside, the fly ash particles. The X-rays generated by the passage of n electrons of energy e through m grams of an element of atomic number A is (Campbell et al., 1975)

$$X = \frac{6 \times 10^{23} nm\sigma(e)\omega\varepsilon}{A} \tag{1}$$

TABLE 3. Statistics for Four Corners Power Plant Plume Aerosol 16 km Downwind (Date: 801101; Time: 085244)

Parameter	Small Particle Mode	Large Particle Mode
Number Distribution		
Mean radius, μm	4.3×10^{-2}	0.96
Modal radius, μm	4.0×10^{-2}	0.94
Median radius, μm	4.0×10^{-2}	0.67
Standard deviation, μm	1.7×10^{-2}	0.96
Skewness parameter	1.1	1.3
Modal ordinate, cm^{-3}	1.8×10^{3}	35.
Number concentration, cm^{-3}	7.7×10^{2}	40.
Surface Distribution		
Mean radius, μm	5.8×10^{-2}	2.9
Modal radius, μm	5.4×10^{-2}	2.9
Median radius, μm	5.4×10^{-2}	2.6
Standard deviation, μm	2.3×10^{-2}	1.7
Skewness parameter	1.1	1.1
Modal ordinate, cm^2/cm^3	7.7×10^{-7}	1.4×10^{-5}
Surface area, cm^2/cm^3	1.3×10^{-6}	9.4×10^{-6}
Mass Distribution		
Mean radius, μm	6.7×10^{-2}	3.9
Modal radius, μm	6.2×10^{-2}	3.9
Median radius, μm	6.2×10^{-2}	3.6
Standard deviation, μm	2.7×10^{-2}	2.0
Skewness parameter	1.1	1.0
Modal ordinate, $\mu g/m^3$	1.9	3.2×10^{3}
Mass concentration, $\mu g/m^3$	0.79	1.8×10^{3}

where $\sigma(e)$ = cross section for K-shell or L-shell ionization, ω = fluorescence yield for K- or L-shell emission, and ε = detection efficiency. We know from our past research (Pueschel, 1976; Parungo et al., 1978) that under constant operating conditions of the SEM-XEDA unit (electron beam current = 15 μA; electron beam diameter = 0.009 μm; accelerating voltage = 16.5 keV; magnification = 75000×; detector distance = 3 cm) a proportionality is to be expected between X-ray intensity X, emitted by matrix elements from inside a particle, and particle radius R, as long as the particle radius exceeds the electron beam spot radius. This is qualitatively shown in Figure 1, where the intensity of X-ray generation from the elements Al, Si, and Ca is higher the larger the particle. Under these same conditions, the X-ray intensity X from surface-deposited elements will be proportional to the thickness of the surface layer, but independent of particle radius R. Therefore, when the data are least-square fitted to a power-function curve

$$X = aR^b \tag{2}$$

TABLE 4. Statistics for Four Corners Background Aerosol 16 km Downwind (Date: 801101; Time: 085244)

Parameter	Small Particle Mode	Large Particle Mode
Number Distribution		
Mean radius, μm	4.4×10^{-2}	0.76
Modal radius, μm	4.0×10^{-2}	0.85
Median radius, μm	4.0×10^{-2}	0.34
Standard deviation, μm	2.2×10^{-2}	1.1
Skewness parameter	1.2	1.4
Modal ordinate, cm^{-3}	1.1×10^3	1.2
Number concentration, cm^{-3}	5.6×10^2	2.3
Surface Distribution		
Mean radius, μm	6.8×10^{-2}	3.7
Modal radius, μm	6.1×10^{-2}	3.8
Median radius, μm	6.1×10^{-2}	3.3
Standard deviation, μm	3.3×10^{-2}	2.2
Skewness parameter	1.2	1.0
Modal ordinate, cm^2/cm^3	5.3×10^{-7}	7.3×10^{-7}
Surface area, cm^2/cm^3	1.7×10^{-7}	5.0×10^{-7}
Mass Distribution		
Mean radius, μm	8.4×10^{-2}	5.0
Modal radius, μm	7.5×10^{-2}	5.1
Median radius, μm	7.5×10^{-2}	4.6
Standard deviation, μm	4.1×10^{-2}	2.5
Skewness parameter	1.2	0.98
Modal ordinate, $\mu g/m^3$	1.5	2.1×10^2
Mass concentration, $\mu g/m^3$	0.75	1.2×10^2

TABLE 5. Frequency With Which Elements Are Found in Fly Ash Particles Collected in the Plume of the Four Corners Power Plant

Element	Percent of Particles
Si	99.3 ± 1.5
Al	79.8 ± 8.6
Mg	49.3 ±15.1
S	48.3 ± 6.7
Fe	42.5 ±10.9
Ca	40.0 ±10.9
K	23.3 ± 6.4
Ti	16.0 ± 9.2
Cl	12.5 ± 3.3
Zn	5.3 ± 3.5
Cr	<3.0
Mn	<3.0
Cd	<3.0

where a and b are constants, the constant b determines whether an element is part of the particle matrix ($b \gg 0$) or whether it is concentrated on the particle's surface ($b \simeq 0$).

Results are shown in Table 6 for the elements Si, Al, Fe, and S contained in fly ash sampled in the Four Corners Power Plant plume on several occasions over a 3-year period, encompassing a wide range of atmospheric residence times and relative humidities. The proportionality in Table 6 between particle radius R and the X-ray intensities from elements Si, Al, and Fe identifies these elements as volume-distributed. The independence of X-ray intensity emitted from S on particle radius, on the other hand, is proof that sulfur is accumulated on the particle's surface. The results shown in Table 6 are typical for the arid environment of the Four Corners region. We did not encounter a sufficiently high number of days of high humidity to investigate the liquid-phase oxidation of SO_2. The case study of 11 February 1976 (Table 6) indicates that a relative humidity as high as 80 percent has no effect on the amount of sulfate accumulated by fly ash. This is contrary to what has been found in laboratory studies (Haury et al., 1978).

It remains to be shown whether or not this amount of heterogeneously formed sulfur is a significant portion of the total. To do this we have formed, via equations (1) and (2), the ratios of m_S to m_{Si}, m_{Al}, and m_{Fe} for the mass modal radii (from Tables 1 through 4) and the average X-ray intensities (from Table 6). The results are shown in Table 7. The last column in this table gives the mass concentration of sulfur, calculated using the fly ash aerosol statistics of Tables 3 through 5, under the assumption that fly ash resembles glass in its chemical composition, with an average silicon content of 30 percent by mass. The sulfate concentrations of 470.0 and 27.7 $\mu g/m^3$ at 0.4 and 16.0 km, respectively, from the stacks are only 2 percent of the concen-

TABLE 6. Power-Function Curve Fit Between X-Ray Intensity Emitted by Elements Si, Al, Fe, and S, and Particle Radius R

Date	Residence Times (hours)	Relative Humidity (percent)	X_{Si}	X_{Al}	X_{Fe}	X_S
75 07 29	1.0	30	$14.3R^{0.6}$	$8.5R^{0.6}$	----	$2.1R^{0.0}$
75 07 30	1.6	64	$27.5R^{0.5}$	$19.5R^{0.5}$	----	$3.9R^{0.0}$
75 07 30	2.6	54	$41.0R^{0.7}$	$29.4R^{0.8}$	----	$6.3R^{0.1}$
75 07 31	4.0	32	$17.6R^{0.6}$	$13.5R^{0.6}$	----	$2.7R^{0.1}$
76 02 10	1.0	--	$22.5R^{0.6}$	----	----	----
76 02 11	2.0	80	$46.1R^{1.0}$	$23.8R^{0.9}$	$3.1R^{0.5}$	$3.3R^{0.1}$
76 02 19	10.0	--	$16.0R^{0.6}$	$12.1R^{1.5}$	----	$3.1R^{0.1}$
76 10 09	1.0	--	$25.4R^{0.8}$	$15.5R^{0.7}$	$1.2R^{0.4}$	$2.9R^{0.1}$
78 06 06	0.2	--	$18.6R^{1.1}$	----	----	$1.1R^{-0.1}$
78 06 06	2.0	--	$13.3R^{0.9}$	----	----	----
Grand average			$(24.2 \pm 11.2) \times R^{(0.7 \pm 0.2)}$	$(16.3 \pm 7.4) \times R^{(0.8 \pm 0.3)}$	$(2.2 \pm 1.1) \times R^{(0.4 \pm 0.1)}$	$(3.2 \pm 1.5) \times R^{(0.0 \pm 0.1)}$

tration of SO_2 found independently as 2.7×10^4 and 1.5×10^3 μg/m^3 at these same distances (Pueschel and Van Valin, 1978). The fact that we do not pick up a measurable amount of sulfate on fly ash as the plume residence time increases leads us to conclude that heterogeneous oxidation of sulfur dioxide on the surface of fly ash particles must be ruled out as the dominant mechanism of sulfate formation in power plant plumes, in agreement with earlier findings (Mamane and Pueschel, 1979; Parungo et al., 1978). These findings agree with those of laboratory studies of heterogeneous reactions of SO_2 by Judeikis et al. (1978). Most of the sulfate is formed either homogeneously, or heterogeneously on surfaces of particles that cannot be identified by SEM-XEDA techniques. Such particles of submicron size, composed of elements lighter than sodium and of a vapor pressure that is low enough to cause the particles to be stable at 10^{-7} torr (SEM specimen chamber pressure), have been found to exist in the atmosphere and to amount to between 30 and 50 percent of the total aerosol population (Pueschel and Wellman, 1979).

Sulfate in Background Air

The occurrence of long-range transport of submicron particles has been documented by our research in east-central Utah, historically one of the least polluted regions in the contiguous United States (Flowers et al., 1969). Typical particle size distributions, measured in November 1978, are shown in Figure 4. The number, surface area, and volume/mass aerosol statistics, corresponding to curve a in Figure 4, are shown in Table 8. The point to be noted is that the sizes and concentrations of this aerosol are significantly smaller than those of the aerosol found in the vicinity of the Four Corners Power Plant (Figure 3; Tables 1 through 4).

Some physical and chemical parameters of the Utah "background" aerosol, and their variation with altitude, are shown in Figure 5 as a function of particle size. Depicted in Figure 5 for two flight levels is the frequency (ordinate) with which specific elements (abscissa) are found in the total aerosol and in small (diameter less than 0.5 μm) and large (diameter larger than 0.5 μm) aerosol particles. The aerosol population is further subdivided into spherical and nonspherical particles on the premise that spherical particles are generated by a process that involves a phase change, such as combustion, whereas nonspherical particles are produced by mechanical disintegration, such as wind action.

It follows from Figure 5 that on a day of extreme atmospheric clarity (Allee & Pueschel,

TABLE 7. Mass of Sulfur on Fly Ash in the Four Corners Power Plant Plume

Distance from Stacks (km)	Surface Area Modal Radius (μm)	m_S/m_{Si}	m_S/m_{Al}	m_S/m_{Fe}	m_S (μg/m^3)
0.4	6.8×10^{-2}	0.6	1.1	1.2	2.9
0.4	3.7	0.04	0.05	0.16	4.4×10^2
16.0	5.4×10^{-2}	0.8	1.5	1.4	0.25
16.0	2.6	0.05	0.06	0.2	27.

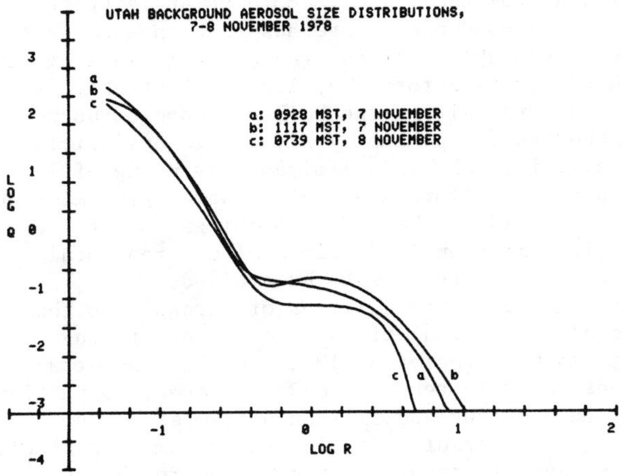

Fig. 4. Size distribution of aerosol in background air in east-central Utah. Log Q = log (dN/d(log R)) where N is the number of aerosol particles per cm^3 and R is in μm.

TABLE 8. Statistics of Utah Background Aerosol Represented by Distribution a in Figure 4

Parameters	Small Particle Mode	Large Particle Mode
Number Distribution		
Mean radius, μm	2.1×10^{-2}	1.4
Modal radius, μm	1.6×10^{-2}	1.1
Median radius, μm	1.6×10^{-2}	1.1
Standard deviation, μm	1.9×10^{-2}	1.1
Skewness parameter	1.5	1.4
Modal ordinate, cm^{-3}	1.1×10^{3}	0.22
Number concentration, cm^{-3}	9.5×10^{2}	0.16
Surface Distribution		
Mean radius, μm	6.9×10^{-2}	3.5
Modal radius, μm	5.1×10^{-2}	2.8
Median radius, μm	5.1×10^{-2}	2.8
Standard deviation, μm	6.2×10^{-2}	2.7
Skewness parameter	1.5	1.4
Modal ordinate, cm^2/cm^3	1.1×10^{-7}	8.2×10^{-8}
Surface area, cm^2/cm^3	9.6×10^{-8}	6.1×10^{-8}
Mass Distribution		
Mean radius, μm	0.12	5.5
Modal radius, μm	9.3×10^{-2}	4.4
Median radius, μm	9.3×10^{-2}	4.4
Standard deviation, μm	0.11	4.2
Skewness parameter	1.5	1.4
Modal ordinate, $\mu g/m^3$	0.53	19.
Mass concentration, $\mu g/m^3$	0.44	15.

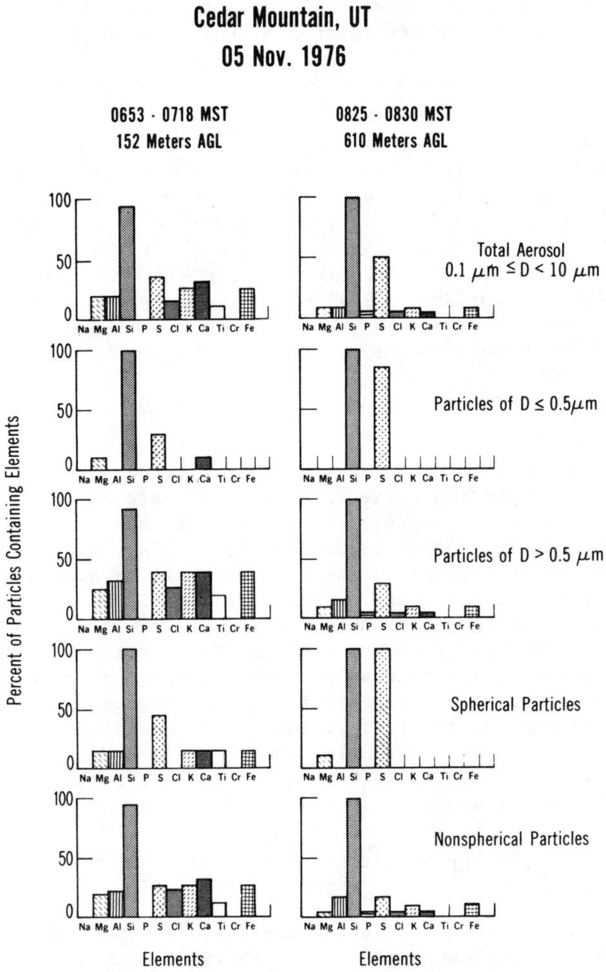

Fig. 5. Frequency of occurrence of elements in aerosols in east-central Utah and variability with particle size, shape, and altitude above ground level (AGL).

1980) sulfur is found predominantly in small particles, the dominant shape of sulfur particles is spherical, and the fraction of sulfur particles increases with altitude. This indicates long-range advection to Utah of a gas-to-particle converted sulfur aerosol from remote sources. All other elements with the exception of Si are found in large nonspherical particles that are concentrated near the ground and hence would be soil derived.

The similarity in shape and size between the sulfur and silicon in small and spherical particles suggests a possible common origin. Since both types of aerosol also have been found in the power plant plume discussed earlier, coal combustion is a likely process for their generation. Hence, submicron silicon spheres may be as good a tracer for a primary coal combustion aerosol (Macias et al., 1980) as are sulfur-containing

TABLE 9. Power-Function Curve Fit Between X-Ray Intensity Emitted by the Elements S and Si, and Particle Radius R[a]

Date	X_{SI}	X_S	X_S/X_{Si}	EF
76 11 05	$27.1R^{0.5}$	$7.7R^{-0.1}$	$0.3R^{-0.6}$	1.7×10^2
76 11 05	$23.1R^{0.8}$	$9.0R^{0.8}$	$0.7R^{-0.6}$	3.7×10^2
76 11 06	$20.9R^{0.4}$	$5.9R^{0.2}$	$0.3R^{-0.2}$	1.7×10^2
76 11 06	$14.3R^{0.7}$	$5.5R^{0.4}$	$0.4R^{-0.3}$	-----
77 09 07	$52.2R^{1.2}$	$7.6R^{0.3}$	$0.2R^{-0.9}$	2.1×10^2
77 08 05	$23.7R^{0.8}$	$10.8R^{0.3}$	$0.5R^{-0.5}$	2.4×10^2

[a]The enrichment factor EF of sulfur is defined in equation (3).

submicron spheres for a secondary aerosol.

Power-function curve fits, $X = aR^b$, are shown in Table 9 for the Utah background aerosol on several occasions. It follows from this table that Si is uniquely identified as a matrix element in both spherical and nonspherical particles. Sulfur, on the other hand, is less uniquely surface-deposited than was the case for fly ash. Apparently, independently formed sulfate has been transported over long distances to interact with a silicon aerosol to form the mixed aerosol dominating the small particle mode.

Also shown in Table 9 is an enrichment factor for sulfur, defined as

$$EF = \frac{[m(S)/m(Si)]aerosol}{[m(S)/m(Si)]crust} \qquad (3)$$

for a m(S)/m(Si) ratio for the Earth's crust of 1.9×10^{-3} (Mason, 1952). There is a large 1-σ rms error associated with EF, reflecting changes in the environmental parameters as well as statistical errors in the analysis. Nevertheless, enrichments of sulfur by a factor of several hundreds in the aerosol can be postulated. A higher enrichment in spherical than in nonspherical particles is indicated. The premise of a simultaneous advection of sulfur and silicon aerosols makes this enrichment a rather conservative estimate.

Summary and Conclusions

In a coal-fired power plant plume the surface area of the primary fly ash aerosol is increased by 2 orders of magnitude above the background aerosol. This surface, composed of aluminum silicates and other light elements, has little catalytic effect on the oxidation of SO_2. Only 2 percent of the total sulfur in the plume is attached to fly ash, and this amount is independent of the residence of the aerosol in the plume from 0 to 10 hours. This suggests the following mechanism of formation. Because of their low vapor pressure, fly ash particles solidify first when exiting from the boiler. More volatile sulfur compounds condense or are adsorbed on some of the large available surface area per unit mass of fly ash particles before they leave the stacks, or even before exiting the boiler. Once airborne, further sulfur pickup by fly ash is negligible. In all, fly ash has a limited scavenging efficiency for sulfur, and control strategies must be concentrated on the reduction of gaseous sulfur dioxide emission if detrimental environmental effects of sulfur are to be avoided.

Background concentrations of aerosols in the vicinity of a coal-fired power plant are increased by a factor of 10 above the concentrations in truly remote areas. However, combustion-produced sulfur and silicon aerosols make up the greater number of particles in the small particle mode, even in remote regions. The morphology of the sulfur-silicon mixed aerosol in remote regions differs from that in power plant plumes in that sulfur has become a matrix element during transport. This suggests that coagulation transforms independently formed sulfur and silicon aerosols during transport.

Acknowledgements. It is a pleasure to thank Dennis Wellman, Farn Parungo, Helen Proulx, Eve Ackerman, Cathy Criley, and Peggy Yotka for their valuable contributions to this phase of the work.

References

Allee, P. A., and R. F. Pueschel, A study of visibilities in Carbon and Emery Counties, Utah, NOAA Data Report ERL ARL-1, National Oceanic and Atmospheric Administration, Air Resources Laboratories, Silver Spring, Maryland, 1980.

Barrett, E. W., F. P. Parungo, and R. F. Pueschel, Cloud modification by industrial pollution: A physical demonstration, Meteorol. Rundsch., 32, 136-149, 1979.

Campbell, J. L., B. H. Orr, A. W. Herman, L. A. McNellis, J. A. Thompson, and W. B. Cook, Trace element analysis of fluids by X-ray fluorescence, Anal. Chem., 47, 1542-1553, 1975.

Flowers, E. C., R. A. McCormick, and K. R. Kurfis, Atmospheric turbidity over the United States, 1961-1966, J. Appl. Meteor., 5, 955-962, 1969.

Haury, G., S. Jordan, and C. Hoffmann, Experimental investigations of the aerosol-catalyzed oxidation of SO_2 under atmospheric conditions, Atmos. Environ., 12, 281-287, 1978.

Judeikis, H. S., T. B. Stewart, and A. G. Wren, Laboratory studies of heterogeneous reactions of SO_2, Atmos. Environ., 12, 1633-1645, 1978.

Macias, E. S., D. L. Blumenthal, J. A. Andersen, and B. K. Cantrell, Size and composition of visibility-reducing aerosols in south-western plumes, in Aerosols: Anthropogenic and Natural - Sources and Transport, edited by Theo J. Knight and Paul J. Lioy, pp. 233-257, New York Academy of Sciences, 1980.

Mamane, Y., and R. F. Pueschel, Oxidation of SO_2

on the surface of fly ash particles, Geophys. Res. Lett., 6, 109-112, 1979.

Mamane, Y., and R. F. Pueschel, Formation of sulfate particles in the plume of the Four Corners Power Plant, J. Appl. Meteor., 19, 779-790, 1980.

Mason, B., Principles of Geochemistry, John Wiley and Sons, New York, 1952.

Parungo, F., E. Ackerman, H. Proulx, and R. Pueschel, Nucleation properties of fly ash in a coal-fired power plant plume, Atmos. Environ., 12, 929-935, 1978.

Pueschel, R. F., Aerosol formation during coal combustion: Formation of sulfates and chlorides on fly ash, Geophys. Res. Lett., 3, 651-653, 1976.

Pueschel, R. F., and C. C. Van Valin, Cloud nucleus formation in a power plant plume, Atmos. Environ., 12, 307-312, 1978.

Pueschel, R. F., and D. L. Wellman, On the nature of the atmospheric background aerosol, paper presented at Fourteenth Conference on Agriculture and Forest Meteorology, Minneapolis, Minn., April 2-6, 1979.

Rodhe, H., A study of the sulfur budget for the atmosphere over northern Europe, Tellus, 24, 128-138, 1972.

WATER VAPOR AND TEMPERATURE DEPENDENCE OF AEROSOL SULFUR CONCENTRATIONS AT FORT WAYNE, INDIANA, OCTOBER 1977

John W. Winchester and Alistair C.D. Leslie

Department of Oceanography, Florida State University, Tallahassee, Florida 32306

Abstract. Variations in 2-hourly aerosol sulfur concentration measurements at EPRI-SURE site 7, Fort Wayne, Indiana, October 1-31, 1977, are found to be significantly correlated with water vapor partial pressure and with temperature, but not with relative humidity. If the variation is due to a variable rate of SO_2 conversion to H_2SO_4 controlled in part by these two parameters, assuming these are not significantly correlated with other causes of aerosol sulfur concentration variation (e.g., SO_2 source strength, atmospheric mixing and dilution, or removal processes), the dependence of the conversion rate on them may be estimated. An analysis indicates the apparent conversion rate of SO_2 to aerosol S is proportional to $P^3_{H_2O}$ with an activation energy E_a = 13 kcal/mol, corresponding at ambient temperatures to a rate doubling in 9°C (cf. saturation P_{H_2O} doubling in 11°C). A dependence of aerosol S concentrations on gaseous SO_2 is observed qualitatively but appears to be weaker, proportional to $P^n_{SO_2}$ with $n \geq 0.4$. The results suggest that oxidation may proceed by equilibrium dissolution of SO_2 in liquid H_2SO_4 aerosol droplets, hydration and ionization to the most readily oxidized species (requiring ca. 3 H_2O molecules in a series of equilibria), and oxidation by a dissolved oxidant at a temperature-dependent rate. The overall dependence on $P^3_{H_2O}$ reflects the corresponding dependence on H_2O activity in solution.

Introduction

Variations in the concentration of aerosol sulfur in polluted nonurban atmospheres of the eastern United States are often considerable, sometimes a factor of ten in a few hours. These variations conceptually must be due to a combination of factors, including SO_2 source strength, physical dilution by cleaner air, the rate of conversion of gaseous sulfur dioxide to liquid sulfuric acid aerosol, and processes for trace gas and aerosol particle removal at the earth surface. We have examined measurements made as air passed over a fixed site at Fort Wayne, Indiana, October 1-31, 1977, in order to identify factors which may influence the rate of conversion of SO_2 to H_2SO_4. Temperature and dew point (a measure of water vapor partial pressure) are found to be correlated with aerosol sulfur concentration. On the assumption that these factors do not vary systematically with the other causes of aerosol S concentration variation, an association with the rate of SO_2 conversion to H_2SO_4 aerosol is implied.

Data Base

Aerosol particles were collected on 0.4 μm Nuclepore by a time sequence "streaker" filter sampler (Nelson, 1977) at site 7, Fort Wayne, Indiana, of the EPRI-SURE network (Mueller et al., 1980; Leslie, 1980). These samples were analyzed for sulfur and other elemental constituents by particle induced X-ray emission (PIXE) (Johansson et al., 1975), and the concentrations of the elements in 2-hourly time increments were computed. SO_2 concentrations were also measured hourly at the site, giving non-zero values at or above the detection limit of 1 ppb about 30 percent of the time. Temperature and dew point observations were made hourly at the Fort Wayne National Weather Service station 10 km from the site.

Figures 1a and 1b display the pattern of variability of aerosol S, specific humidity (from dew point data) raised to the third power for emphasis, and SO_2. Six cold front passages were identified from synoptic weather data and appear to have a significant association with high aerosol S concentrations, although other high aerosol S values were observed as well. If the data are examined carefully in a qualitative sense, they give the impression that maxima in aerosol S are often associated with high values of H_2O, especially if SO_2 is detected. In some instances SO_2 is high but maxima in aerosol S are not observed, perhaps because the general level of H_2O is low. Because of the hint given in Figure 1 that the H_2O vapor content of the atmosphere may be associated with the rate of forming aerosol S from a gaseous SO_2 precursor, we examine the data more closely.

Figure 2 shows a scatter diagram on logarithmic scales of aerosol S and H_2O concentrations plotted for night and day times separately. A significant correlation appears for both subsets

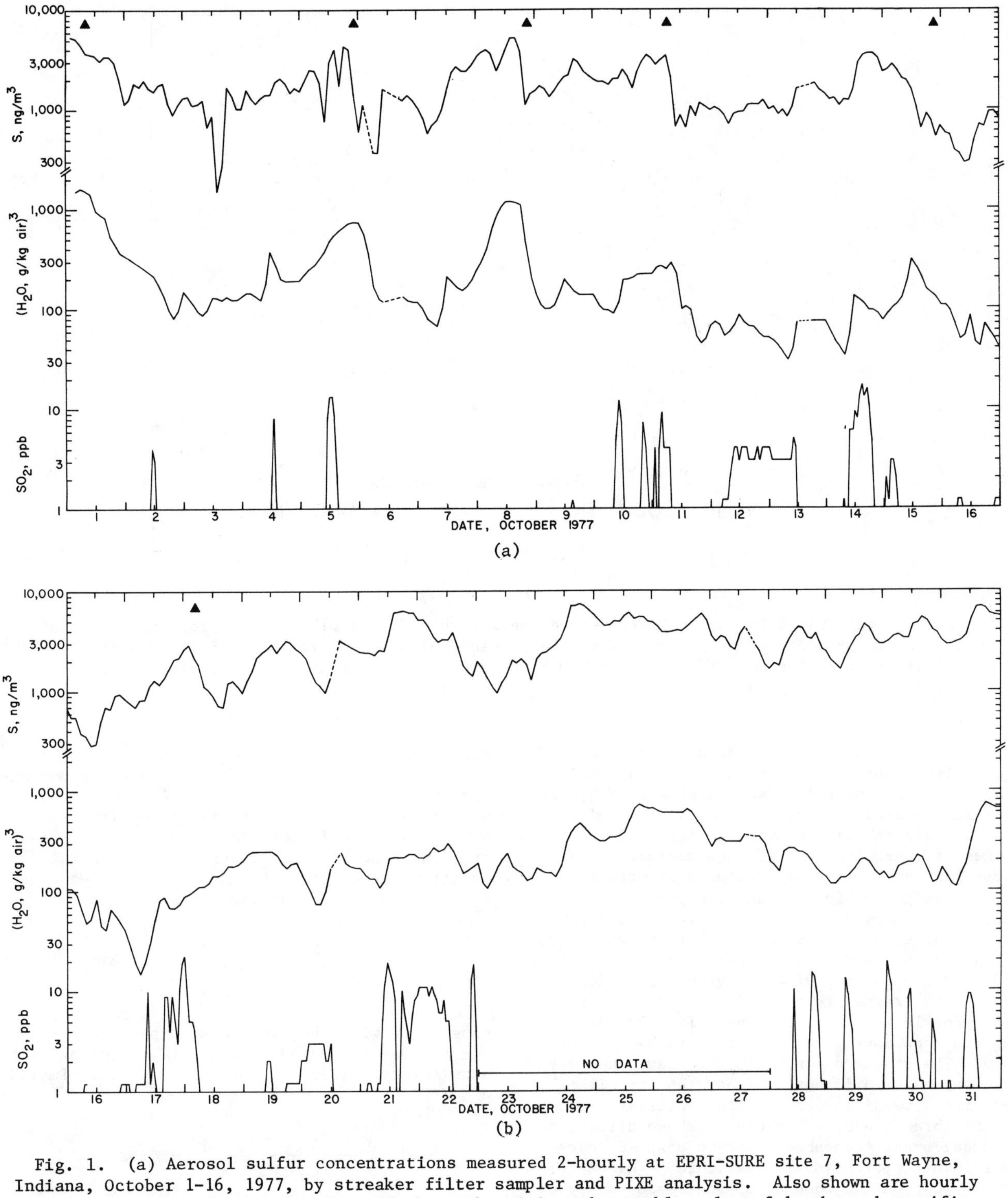

Fig. 1. (a) Aerosol sulfur concentrations measured 2-hourly at EPRI-SURE site 7, Fort Wayne, Indiana, October 1-16, 1977, by streaker filter sampler and PIXE analysis. Also shown are hourly SO_2 concentrations at the site, at and above the minimum detectable value of 1 ppb, and specific humidity, raised to the third power, based on temperatures and dew points reported 3-hourly at the nearby Fort Wayne National Weather Service (NWS) station. Approximate times of cold front passage are shown at the top by triangles. (b) Aerosol sulfur concentrations measured 2-hourly at EPRI-SURE site 7, Fort Wayne, Indiana, October 16-31, 1977, by streaker filter sampler and PIXE analysis.

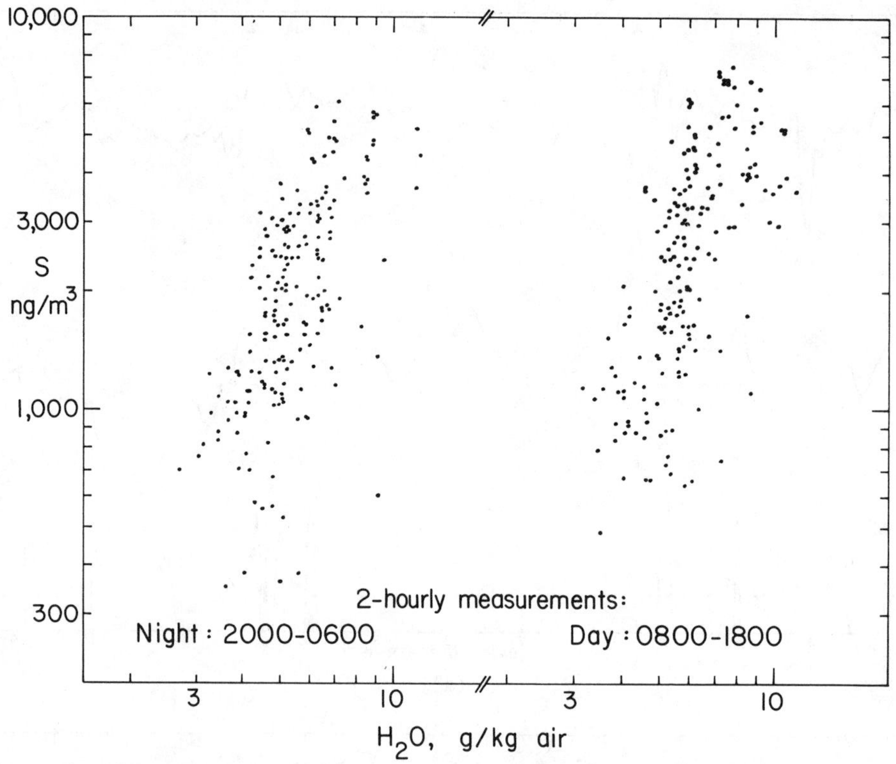

Fig. 2. Aerosol sulfur concentrations and specific humidities of Figure 1 plotted logarithmically for night and day times separately. Regression calculations for each 182 data point subset indicate correlation coefficients $r > 0.6$ ($P \ll 0.001$) and an approximate relationship $S = a(H_2O)^3$.

($r > 0.6$ for each subset of 182 points) and a steep slope approximated by the relationship $S = a(H_2O)^3$. Regression analyses using either H_2O or S as independent variables are consistent with this relationship. The overall night and day ranges of concentrations of each variable are about the same, although daytime concentrations average slightly higher than at night.

Figure 3 shows average diurnal variations of aerosol S, saturation vapor pressure of water (as an equivalent expression for ambient temperature), and partial pressure of water vapor (from dew point data) raised to the third power for emphasis. Hourly aerosol S concentrations by interpolation and hourly temperature data have been used. The standard errors of the means show daytime values significantly higher than at night but with some differences in the detailed forms of the three trends. Because of these differences a causal connection should not be assumed unless additional information is available. In the following section the results of regression analyses indicate a high probability of an influence by both water vapor pressure and temperature on the observed concentration of aerosol S.

Figure 4 presents on a probability plot the extent of variation of the three variables shown in Figure 1. Except for the highest concentrations, all three variables conform well to log normal distributions with the medians and geometric standard deviations shown. This plot can be used to arrive at an approximate sensitivity of aerosol S to variations in H_2O and SO_2 concentrations, under the assumption that a causal relationship exists. For aerosol S and H_2O the ratio of the widths of the distributions is

$$\frac{\ell n\ \sigma_S}{\ell n\ \sigma_{H_2O}} = \frac{\ell n\ 2.12}{\ell n\ 1.276} = 3.08$$

consistent with the functional dependence of $S = a(H_2O)^3$ suggested by Figure 2. Similarly, the sensitivity of aerosol S to SO_2 may be estimated as

$$\frac{\ell n\ \sigma_S}{\ell n\ \sigma_{SO_2}} = \frac{\ell n\ 2.12}{\ell n\ 7.36} = 0.38$$

consistent with a dependence $S = b(SO_2)^{0.4}$ if a causal connection may be assumed. This latter value is made uncertain by the low median concentration of SO_2 of 35 percent of that of aerosol S, the fact that 70 percent of the measurements lie

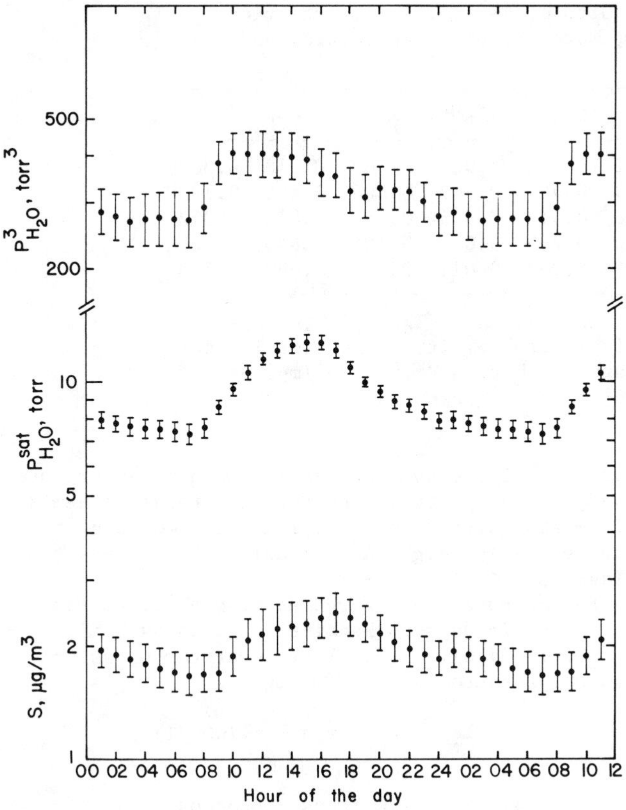

Fig. 3. Average diurnal variations for Fort Wayne site 7, October 1-31, 1977, of aerosol sulfur concentrations (hourly by interpolation of 2-hourly measurements) with nearby NWS hourly temperatures (expressed as corresponding saturation water vapor pressures, $P^{sat}_{H_2O}$) and hourly water vapor partial pressures raised to the third power (corresponding to hourly dew point temperatures), with standard errors of the means of 31 data points.

below the 1 ppb detection limit, and the possibility that variations in SO_2 due to processes other than conversion to H_2SO_4 aerosol are much greater than those due to conversion alone. Especially because of the last possibility, this dependence is an estimated lower limit to the true dependence.

The data base described may be used to test for significant correlations among the parameters to establish grounds for believing that variations in the rate of conversion of SO_2 to H_2SO_4 may be an important cause of the variations in aerosol S concentration and that this rate may depend on atmospheric temperature and water vapor pressure.

Regression Analysis

Variability in aerosol S may in principle be caused by variations in SO_2 source strength, rate of SO_2 conversion to aerosol H_2SO_4, physical mixing with cleaner air, and removal of SO_2 or aerosol. We examine the possibility that significant variation in the conversion rate occurs in response to water vapor partial pressure and ambient temperature. Furthermore, we consider that these variables are not correlated with the other processes contributing to overall aerosol S variability. Therefore, the latter may contribute

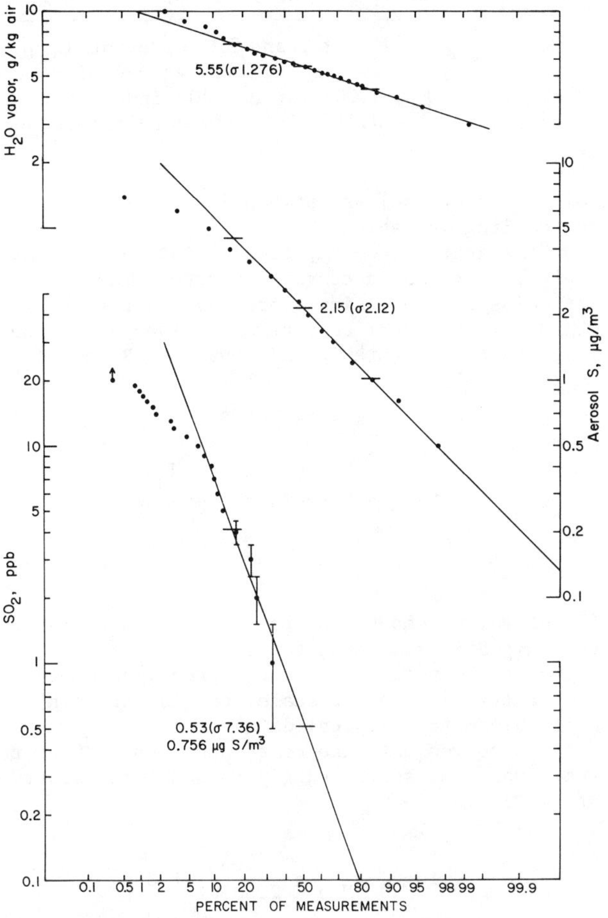

Fig. 4. Percent of measurements in excess of plotted values of 364 2-hourly data for gaseous SO_2, aerosol S, and H_2O vapor concentrations at Fort Wayne site 7, October 1-31, 1977. Log normality over most of the ranges is indicated by linearity; 50 percent concentrations and geometric standard deviations (16-84 percentile ranges) indicate a width of the aerosol S distribution intermediate between those of SO_2 and H_2O. Since only integer values of ppb SO_2 (0, 1, 2, ...) were reported, the assumed ranges of validity at the lowest 4 values are indicated for comparison with the log normal trend. Temperatures were found to be normally distributed with a mean and ±σ range of 51 ± 8°F (10.56 ± 4.44°C).

TABLE 1. R^2 for Fort Wayne, Indiana, October 1-31, 1977, Correlations Among Hourly Aerosol Sulfur, Water Vapor, Temperature, and Relative Humidity[a]

Set	Range	n	*S-W*	*S-T*	*S-W,T*	*S-RH*	*W-T*	*W-RH*	*T-RH*
All	01-24 hr	743	0.410	0.342	0.434	0.003	0.563	0.018	0.303
Day	09-20 hr	372	0.424	0.319	0.461	0.007	0.404	0.178	0.180
Night	21-08 hr	371	0.383	0.456	0.454	0.036	0.854	0.011	0.072
Warm	$t > 50°F$ (10°C)	354	0.242	0.124	0.265	0.064	0.185	0.543	0.185
Cool	$t \leq 50°F$ (10°C)	389	0.297	0.204	0.299	0.012	0.584	0.099	0.133

[a]*S*, *W*, *T*, and *RH* represent $\log_{10}$ S, $1/T_{dp}$, 1/T, and $1/T_{dp} - 1/T$, respectively (see text). Values of $R^2 > 0.10$ imply probability of no correlation $P < 0.001$ for n = 100 independent observations whereas $R^2 < 0.03$ implies $P > 0.1$; 743 hourly data points were used in the regressions.

noise to correlations between S and the rate-controlling variables.

On the average, we may take variations in S to imply variations in conversion rate. The Arrhenius rate equation expresses the rate constant k in terms of an activation energy E_a, the absolute temperature T, and constants R and A:

$$k = Ae^{-E_a/RT}$$

$$\log_{10} k = \log_{10} A - \frac{E_a}{2.303RT}$$

$$\log_{10} k \propto \log_{10} S$$

Therefore, we choose to transform S concentration to $\log_{10}$ S in searching for correlations. We also transform temperature t to reciprocal absolute temperature 1/T (on °F scale, consistent with available meteorological data).

It is convenient also to relate temperatures to water vapor pressures P_{H_2O}. For ambient temperature T,

$$\log_{10} P^{sat}_{H_2O} = \frac{-\lambda_{vap}}{2.303RT} + C$$

where the superscript "sat" indicates saturation, λ_{vap} is the heat of vaporization of water, and C is a constant. For dew point T_{dp},

$$\log_{10} P_{H_2O} = \frac{-\lambda_{vap}}{2.303RT_{dp}} + C$$

For fractional relative humidity RH,

$$\log_{10} RH = \log_{10} \frac{P_{H_2O}}{P^{sat}_{H_2O}} = \frac{-\lambda_{vap}}{2.303R}\left(\frac{1}{T_{dp}} - \frac{1}{T}\right)$$

Therefore, we choose as variables for regression $\log_{10}$ S, 1/T, $1/T_{dp}$, and $(1/T_{dp} - 1/T)$ as measures of aerosol S conversion rate, ambient temperature, water vapor partial pressure, and relative humidity, abbreviated in Table 1 as *S*, *T*, *W*, and *RH*, respectively, e.g., R^2 for column *S-W* refers to $\log_{10} S = C_1(1/T_{dp}) + C_2$.

For computational purposes, the absolute temperature in degrees Rankine is given by T(°R) = t(°F) + 459.67. Equivalent logarithm of water vapor pressure in torr (mm Hg) is

$$\log_{10} P_{H_2O} = a + b(1000/T)$$

$$a = 9.16825 \pm 0.00008$$

$$b = -4.18140 \pm 0.04180$$

in the range 32-86°F (0-30°C). In a regression result for S in ng/m^3 and T in °R,

$$\log_{10} S = a' + b'(1000/T)$$

an equivalent expression for P_{H_2O} in torr (mm Hg) is

$$\log_{10} S = a'' + b'' \log_{10} P_{H_2O}$$

where

$$a'' = a' - ab'/b$$

$$b'' = b'/b$$

In Table 1 we present the results of regression analyses using standard methods (Ryan et al., 1976). These have been tested by lag correlations, where one variable at time t is compared with the other at time t ± n, where lag n = 0,1,2,...,24 hours. The best R^2 values occur for S lagged by 0 or, at most, 1 hour (cf. Leslie, 1980). The values of R^2 adjusted for degrees of freedom indicate for all data points a stronger correlation between aerosol S and water than with temperature, although both are strong. No significant correlation with relative humidity is found. The water content is correlated with temperature although neither water nor temperature individually is very strongly associated with relative humidity. For daytime points similar results are found, although for nighttime points the temperature association is somewhat stronger than that with water. If warm periods are looked at separately from cool, the water and temperature correlations with aerosol S are weaker but with water being the stronger of the two. If both water and temperature are used as two predictors of S in a linear regression equation, usually a somewhat stronger correlation is found than using just one predictor. Dependence of aerosol S on SO_2 cannot be determined reliably due to 70 percent of SO_2 values being less than the detection limit of 1 ppb. Regression of aerosol S against $SO_2 > 1$ ppb gives $R^2 = 0.041$. The low significance is no doubt due in part to disappearances of SO_2, by removal as well as by conversion, as aerosol S is being formed; SO_2 is a necessary precursor. The general finding of Table 1 is that both water and temperature are significantly correlated with aerosol S, generally with water being the better predictor.

Discussion

The foregoing results are consistent with a model in which SO_2 oxidation proceeds at a rate which is regulated both by temperature and by the partial pressure of water vapor in the atmosphere. The approximately cubic dependence on P_{H_2O} suggests some sort of chemical control in which water activity plays a crucial role. The temperature dependence, as will be discussed further below, is not unlike that of many reactions in solution. Therefore, we suggest a mechanism in which gaseous SO_2 dissolves under equilibrium conditions in aqueous sulfuric acid aerosol, and equilibrates in solution in a series of hydration and ionization steps to form a concentration of a readily oxidized species given by the equilibrium expressions which include the activities of the reacting species. If water is a reactant in three successive steps, e.g., hydration of SO_2 and two successive ionizations, then an overall cubic dependence on H_2O would be expected. Water may also participate in equilibria of the oxidizing agent, e.g., H_2O_2 which may compete with H_2O in the hydration of H_2SO_4 (Winchester, 1980).

Atmospheric H_2SO_4 aerosol should contain H_2O corresponding to an equilibrium value of P_{H_2O} in the atmosphere. Figure 5 shows the relationship between P_{H_2O} and three measures of the H_2SO_4/H_2O mixing ratio at 20°C (Washburn, 1928). It is noteworthy that at P_{H_2O} values 10 percent or less of the saturation vapor pressure of water (relative humidity <10 percent), the equilibrium H_2SO_4-H_2O solution still contains up to 40 percent H_2O by weight and 75 percent on a molar basis. However, it is P_{H_2O}, not the weight percent or mole fraction H_2O in solution, which is a measure of water activity in solution and thus the positions of the equilibria which determine the relative abundance of the most readily oxidized species in solution resulting from the hydration and ionization of dissolved SO_2. H_2SO_4 has a great affinity for H_2O and effectively removes it from availability for other reactions even though H_2O appears to be abundant in solution. The "availability" is in fact given by its activity in solution, or, in other words, P_{H_2O}.

The temperature dependence can be estimated by comparing widths of normal distributions, as was done in the discussion of Figure 4. The result of this procedure is consistent with regression computations but does not require that one of the two variables be considered independent and therefore gives a visually realistic trend through the data points (cf. Figure 2 and accompanying discussion). Ambient temperatures were found to be normally distributed with a standard deviation about a mean value given by 51 ± 8°F (10.56 ± 4.44°C). If temperature is a causal factor in

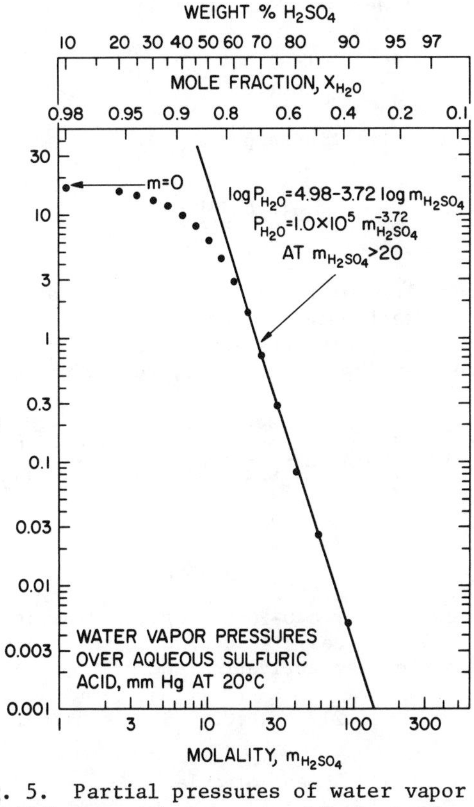

Fig. 5. Partial pressures of water vapor in equilibrium with aqueous sulfuric acid solutions at 20°C (Washburn, 1928). P_{H_2O} is a measure of the chemical potential or activity of dissolved H_2O for any equilibrium process in solution. For $m_{H_2SO_4} > 20$, P_{H_2O} decreases as the 3.72 power of increasing molality. A value of $P_{H_2O} = 3$ mm Hg corresponds to 2.5 g H_2O per kg of air at atmospheric pressure.

the rate of formation of aerosol S, as suggested by the regressions, this information can be used to estimate an activation energy for the oxidation step if we consider the following relationships:

$$\ell n \frac{P_2}{P_1} = -\frac{\lambda_{vap}}{R}\left(\frac{1}{T_2} - \frac{1}{T_1}\right) \qquad \text{Saturation vapor pressure}$$

$$\ell n \frac{k_2}{k_1} = -\frac{E_a}{R}\left(\frac{1}{T_2} - \frac{1}{T_1}\right) \qquad \text{Arrhenius rate equation}$$

If on the average ℓn S is proportional to ℓn k, we use $\ell n\ S_2/S_1$ as being equivalent to $\ell n\ k_2/k_1$ for the relative values at temperatures T_2 and T_1.

On converting the range 51 ± 8°F to reciprocal absolute temperature units, we find a standard deviation σ_T = 1.34373, equal to the standard deviation of the corresponding saturation vapor pressure distribution. The ratio of this width to that of aerosol S is given by

$$\frac{\ell n\ 1.344}{\ell n\ 1.276} = 1.213$$

implying a relationship

$$S = \left(P^{sat}_{H_2O}\right)^{1.213}$$

The formal relationships between the saturation vapor pressure and heat of vaporization of water λ_{vap} and the rate constant k and activation energy E_a are identical. Therefore, we may write

$$\ell n \frac{k_2}{k_1} = n\ \ell n \frac{P^{sat}_2}{P^{sat}_1} = n\left[-\frac{\lambda_{vap}}{R}\left(\frac{1}{T_2} - \frac{1}{T_1}\right)\right]$$

$$E_a = n\lambda_{vap}$$

Numerically, for a 0-20°C range, and relative to λ_{vap} = 590 ± 5 cal/g = 10.62 kcal/mol = 0.46 ev/molecule, the activation energy

$$E_a = 1.213 \times 10.62 = 12.9 \text{ kcal/mol}$$

$$= 0.56 \text{ ev/molecule}$$

This value, comparable to activation energies of many reactions in solution, corresponds to a temperature change for rate doubling of 9°C, which may be compared with that for $P^{sat}_{H_2O}$ doubling of 11°C.

The foregoing discussion has made use only of relationships between aerosol sulfur concentrations, temperature, and dew point, on the assumption that their relationships implied a variability in rate of formation of aerosol S. Times of frontal passage correspond to high values of aerosol S and dew point but otherwise do not appear to be anomalous. Further analysis of the problem making use of several additional aerosol trace elements, spectral analysis of time series data, other EPRI-SURE sites, and time series records much longer than one month will be reported separately.

Acknowledgements. Technical assistance by M.J. Dancy and M. Darzi is gratefully acknowledged, and one of us (J.W.W.) is indebted to the National Center for Atmospheric Research, Boulder, Colorado, and the University of California, Davis, for visiting scientist appointments during sabbatical leave when some of the work was performed. This study was supported in part by the U.S. Environmental Protection Agency.

References

Johansson, T. B., R. E. Van Grieken, J. W. Nelson, and J. W. Winchester, Elemental trace analysis of small samples by proton induced X-ray emission, Anal. Chem., 47, 855-860, 1975.

Leslie, A. C. D., Comments on an aerosol sulfur/water relationship in the eastern United States, in Long Range Aerosol Transport of Metals and Sulfur, edited by H. Lannefors and J. W. Winchester, pp. 46-53, Rep. SNV PM 1337, National Swedish Environment Protection Board, 1980.

Mueller, P. K., G. M. Hidy, K. Warren, T. F. Lavery, and R. L. Baskett, The occurrence of atmospheric aerosols in the northeastern United States, in Aerosols: Anthropogenic and natural, sources and transport, edited by Theo. J. Kneip and Paul J. Lioy, pp. 463-482, Annals of the New York Academy of Sciences, 338, 1980.

Nelson, J. W., Proton-induced aerosol analyses: Methods and samplers, in X-ray Fluorescence Analysis of Environmental Samples, edited by Thomas G. Dzubay, pp. 19-34, Ann Arbor Science Publishers, Ann Arbor, Michigan, 1977.

Ryan, T. A., Jr., B. L. Joiner, and B. F. Ryan, Minitab Student Handbook, Duxbury Press, Boston, Mass., 1976.

Washburn, E. W. (Ed.), International Critical Tables, vols. III and IV, McGraw-Hill, New York, 1928.

Winchester, J. W., A sulfuric acid formation mechanism, in Long Range Aerosol Transport of Metals and Sulfur, edited by H. Lannefors and J. W. Winchester, pp. 54-58, Rep. SNV PM 1337, National Swedish Environment Protection Board, 1980.

EVIDENCE FOR AEROSOL CHLORINE REACTIVITY DURING FILTER SAMPLING

Wang Mingxing* and John W. Winchester

Department of Oceanography, Florida State University, Tallahassee, Florida 32306

Abstract. Concentrations of particulate chlorine are compared with those of other elements as sampled by cascade impactors and filter samplers at rural Xinglong and urban Beijing (cf. Berg and Winchester, 1978; Cicerone, 1981). It is found that a significant Cl deficit for filter as compared with impactor data occurs in the rural area, but in the urban area an apparently much lower deficit is found. The Cl deficit is considered to represent a volatility loss by chemical reactions during filter sampling, but less so during impactor sampling. A possible chemical reaction for Cl loss is acidification and release of HCl. The comparison between urban and rural areas suggests that the urban atmosphere is less acidic, perhaps due to the relatively high abundance of neutralizing particulate matter, including black smoke constituents from coal combustion.

Introduction

The deficit of sea salt Cl in the polluted atmosphere has long been recognized. Particle-to-gas conversion in the marine atmosphere has generally been observed based on the measurement of the Cl/Na ratio in marine aerosols in relation to that of bulk seawater. The variation of this ratio with particle size has also been examined by several investigators, who suggest that the conversion of Cl from particle to gas phase may occur preferentially on small particles. Rahn (1971) observed a general absence of Cl in the smallest particles (less than 1.0 μm radius) collected by an Andersen cascade impactor at inland sites in northern Canada. Wilkniss and Bressan (1972) have reported measurements over the Atlantic and Pacific Oceans by filtrations showing that the mean Cl/Na ratio for particles larger than 2 μm in radius is 1.88 ±0.34, while this value is 1.4 ±0.26 for all the particles greater than 0.2 μm in radius, representing a 22-percent deficit below the seawater ratio Cl/Na = 1.80. Martens et al. (1973) have reported large Cl losses from the smallest sea salt particles collected by Andersen impactor in ambient air from the San Francisco Bay area and from Puerto Rico, based on the Cl/Na ratio in comparison with that of seawater. Increasing loss of Cl with decreasing particle size was observed, with over 90 percent of the Cl lost from the smallest particles (radii 0.2 to 0.4 μm), and less than 10 percent of the Cl lost from the large particles (radii greater than 5 μm). They found a linear correlation between measured gaseous NO_2 concentration and total Cl loss in the San Francisco samples. More recently, Meinert and Winchester (1977) reported a complete absence of particulate Cl in the 0.1- to 0.5-μm-radius range in the samples of the North Atlantic marine aerosol taken by cascade impactor in Bermuda. A similar result has been reported by Johansson et al. (1974).

Such deficits have been generally found in marine atmospheres which are exposed to air pollution, for instance over the North Atlantic, but not in clean atmospheres, such as over the South Pole or central Pacific. Gordon et al. (1977) have reported that the Cl/Na ratio between 75 and 2000 m above the Bahamas is 1.83 ±0.04, which indicates no appreciable deficit of Cl over that area. These earlier observations suggest that the Cl deficit represents a conversion of Cl from particle to gas phase and is caused by reaction with air pollutants, possibly acid-forming pollutants which lead to the release of gaseous HCl.

Relatively few measurements exist for investigating the chemical reactions in the sampling devices. Berg and Winchester (1977) measured the gaseous Cl and particulate Cl simultaneously with two different particle-collecting devices. The combined totals of inorganic gaseous Cl and particulate Cl concentrations measured by different devices appear to agree for all their samples, while the ratios of particulate to gaseous Cl collected by filters and impactors differ greatly, sometimes by a factor of 10.

Chemical reactions for conversion of particulate to gaseous Cl apparently can occur in sampling devices, although the exact mechanism is still not well understood. Moreover, previous studies have

*On leave from the Institute of Atmospheric Physics, Chinese Academy of Sciences, Beijing, as a U.S. National Academy of Sciences Distinguished Scholar and courtesy professor at Florida State University.

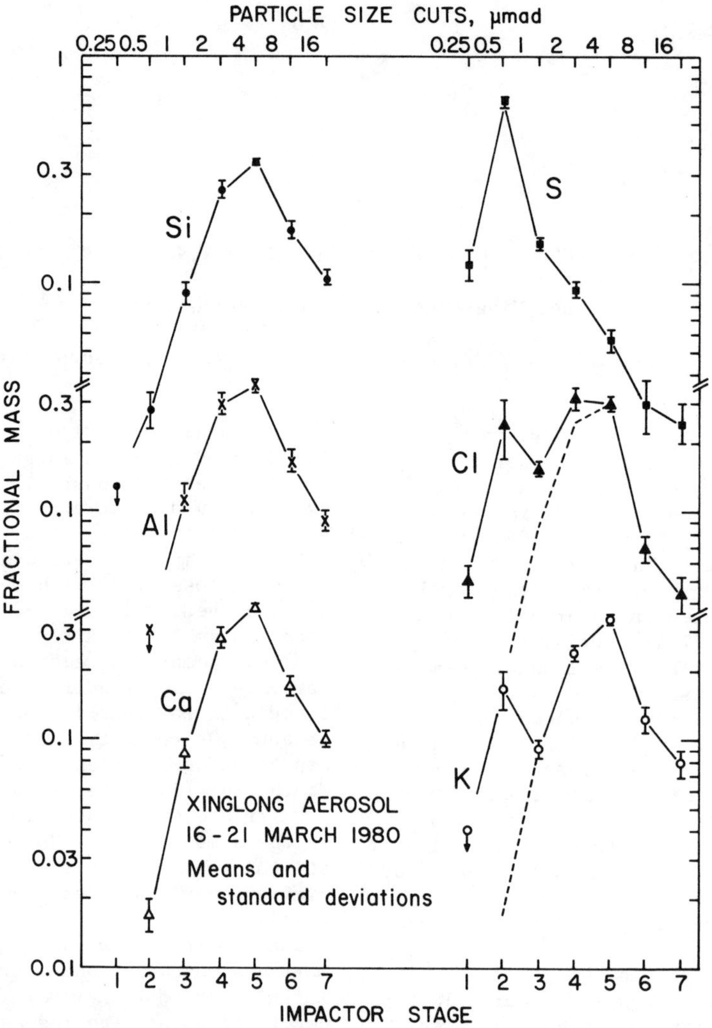

Fig. 1. Size distribution of two groups of elements at Xinglong: Si, Al, Ca, which exhibit only a coarse mode, and S, Cl, and K, which exhibit a fine mode. Dashed line represents estimated coarse-mode component similar to Ca. Particle size cuts in μm aerodynamic diameter (μmad) between impactor stages are given at top.

emphasized sea salt Cl, overlooking aerosol Cl of pollution origin, e.g. from fossil fuel combustion. The present study has been carried out in a location where atmospheric Cl is believed to be largely derived from pollution sources, including coal combustion. The results show that a particulate Cl deficit occurs preferentially on filters as compared to cascade impactors.

Experimental

A comprehensive aerosol sampling program was carried out by a joint American/Chinese research group at an astronomical observing station of the Chinese Academy of Sciences located in a rural area near Xinglong approximately 100 km northeast of Beijing. The program aimed to determine the composition of polluted aerosol due to coal combustion (Wang et al., 1981b; Winchester et al., 1981b, c). During the cold season, from November to March, this rural area provides an excellent opportunity to determine the elemental composition of fine particulate matter released into the atmosphere by coal combustion and aged during transport. The area is downwind of Beijing, the capital of the People's Republic of China, Tianjin, one of the largest industrial cities of the country, and the densely populated North China plain. In this region, large amounts of coal are consumed for space heating and industry, while vehicular and industrial emissions from petroleum and metallurgical dusts are relatively less significant.

Two kinds of aerosol samplers, which are compatible with elemental analysis by PIXE (proton-induced X-ray emission) were operated concurrently

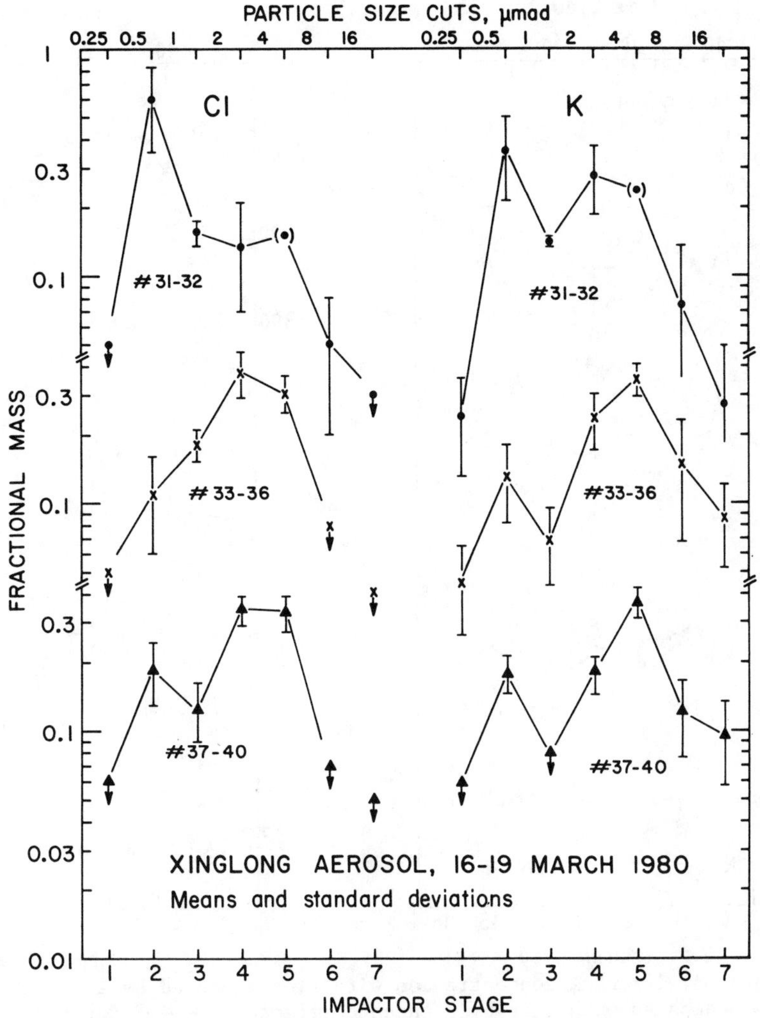

Fig. 2. Size distribution of Cl and K averaged for different time periods characterized by different weather systems: mostly polluted (top), mostly clean marine (center), and mixed (bottom). Numbers refer to specific cascade impactor samples (Winchester et al., 1981c).

at the Xinglong site, side by side on the roof of the central research building of the astronomical station (Wang et al., 1981a). At the same time a filter sampler was operated at the tower facility of the Institute of Atmospheric Physics, located at the north edge of the city proper of Beijing (Ren et al., 1982). The samples collected were analyzed by PIXE at the Florida State University and the University of Lund, Sweden. Concentrations of up to 17 elements were determined. The results show that for rural Xinglong a deficit of particulate Cl is likely to be found on the filter sampler relative to the impactor. Even more interesting is that in the urban area of Beijing, where large amounts of particulate matter, including black coal smoke, are generally observed, the Cl concentration by filter sampler is relatively higher than at Xinglong, indicating a lower deficit. This result may be compared with that of Overein (1972), where rainwater pH was observed to be less acidic close to a combustion source than it was some distance downwind. The evidence for and the possible chemical processes causing the aerosol Cl deficit at Xinglong and its apparent inhibition by high particulate concentrations in Beijing will be discussed further.

Results and Discussion

In Figure 1 the size distribution of various elements is shown as measured by cascade impactor. The elements S, Cl, and K show a significant fine mode, which indicates that a large fraction of these elements is probably due to a high-temperature pollutant source, while the other elements, Si, Ca, and Fe, are found in a coarse mode, indicating an essentially terrestrial dust source. The size distribution of Cl, in particular,

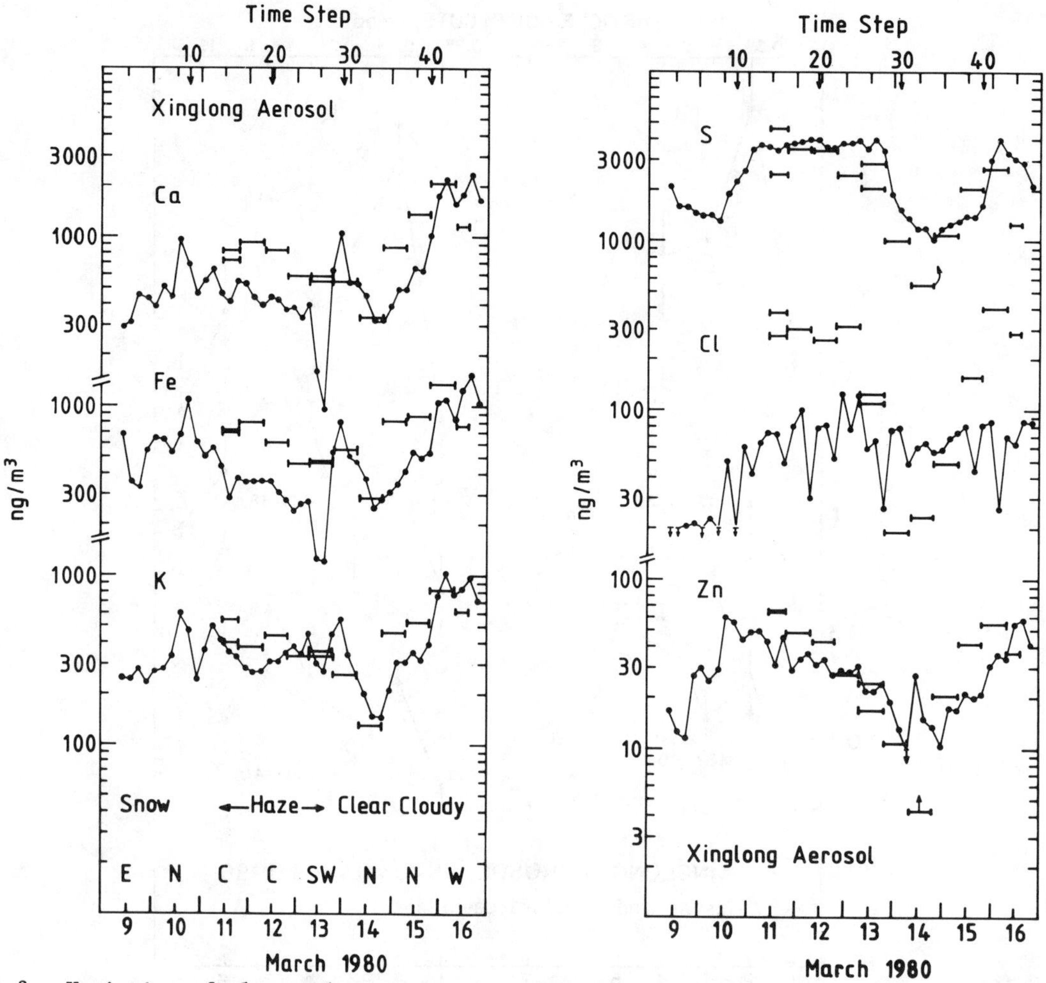

Fig. 3. Variation of elemental concentration with time measured by a streaker Nuclepore filter sampler (•——•) compared with that by a cascade impactor (├——┤) for the first week at Xinglong (Winchester et al., 1981b).

indicates that a substantial portion of the particulate Cl sampled in this work is released from a pollution source, especially coal combustion. This is in contrast to the marine aerosol observations mentioned previously, where virtually a complete absence of particulate Cl in fine particles with radius less than 1 μm is often observed.

In Figure 2 the size distributions of Cl and K are shown averaged for different time periods. As can be seen from the figure, the size distribution of Cl for impactor samples 33 to 36 (corresponding to the time period March 17-18) is similar to that often found in the marine atmosphere, which indicates that marine air had entered into the sampling area. Indeed, a meteorological analysis shows that the air flow starting on the night of March 16 is easterly, which may bring marine air to the sampling site from the Bohai Bay, some 200 km southeast (Winchester et al., 1981c). For all the other samples Cl exhibits a fine mode concentration and is associated with air flow from the south and west, so that we believe the fine particulate Cl sampled in Xinglong is most likely a pollutant, especially from coal combustion.

In Figures 3 and 4 elemental concentrations sampled by the different sampling devices are compared. They seem to be in approximate agreement for all elements except Cl. For elements found mostly in a coarse mode the concentrations sampled by the filter sampler are actually somewhat lower than those sampled by impactor. This is judged to be due to a greater sampling efficiency for coarse particles by the impactor. For typically fine mode elements a closer agreement is found between the two types of samplers. For Cl, however, the concentrations on the filter are very much lower, sometimes by a factor of 10, than those determined by the impactor sampler. It is quite apparent that a considerable amount of

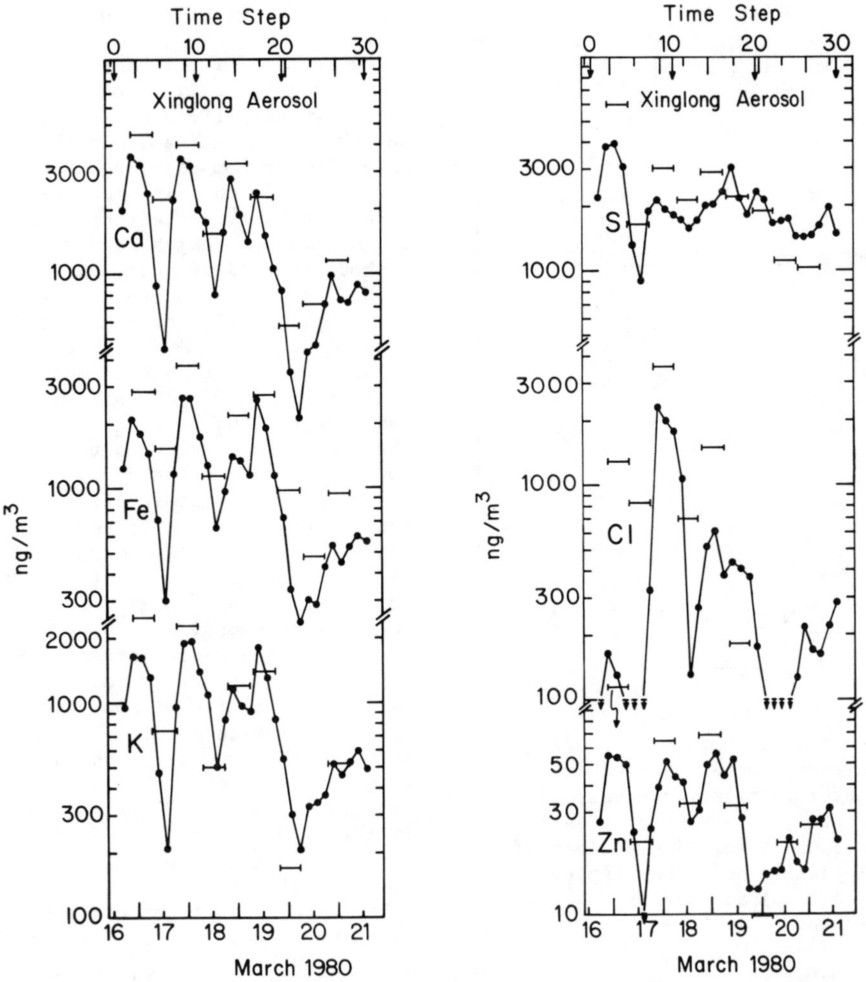

Fig. 4. Variation of elemental concentration with time measured by a streaker Nuclepore filter sampler (•——•) compared with that by a cascade impactor (⊢——⊣) for the second week at Xinglong. Impactor data are averages of duplicate samples.

particulate Cl must have been lost from the filter during sampling. This suggests that chemical reactions involving the release of gaseous Cl, e.g. HCl, can take place on a Nuclepore filter during sampling but not to such a degree on impaction surfaces.

In Figure 5 a comparison is made between the S concentrations averaged over time periods of similar meteorological and air flow conditions and the corresponding Cl deficits indicated by the ratios of averaged concentrations based on impactor samples to those on filter samples. The strong similarity between the two suggests that the chemical reactions for Cl loss must involve reactive substances associated with air pollution sulfur. Because acidic substances are probably abundant in such air, release of gaseous HCl is expected to occur during sampling, although oxidants may also react to release Cl_2.

Acidification by both H_2SO_4 and HNO_3 may lead to HCl evolution, either by acidic aerosol or vapor or by gaseous precursors which are oxidized to acids during sampling. Because oxidation reactions are expected to be slower than acidifications, direct reaction between chloride and preformed acids drawn through the sampler would seem more likely.

Possible simplified chemical reactions for release of gaseous HCl are:

$$2Cl^- + H_2SO_4 \rightarrow SO_4^= + 2HCl \quad (1)$$

$$Cl^- + HNO_3 \rightarrow NO_3^- + HCl \quad (2)$$

On the basis of present evidence we cannot be sure which reaction type is necessarily dominant at the Xinglong site. From aerodynamic considerations, a reactive gas may have a greater chance of reacting with a predeposited aerosol while being drawn through a Nuclepore filter than it would around a cascade impactor surface. In Figures 3 and 4

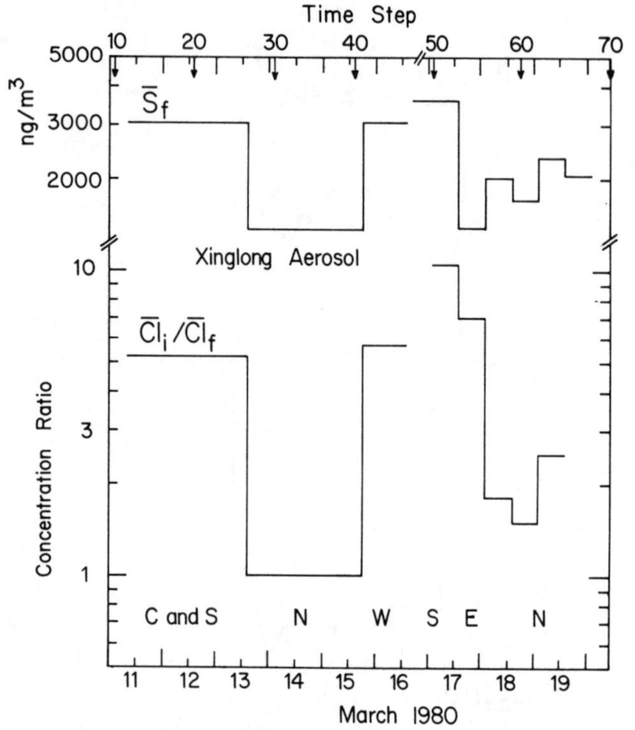

Fig. 5. Correspondence between S concentration by filter sampler, S_f, and Cl deficit, measured by impactor to filter concentration ratio for Cl, Cl_i/Cl_f, as averages of data shown in Figures 3 and 4 for time periods of similar meteorological conditions. Approximate periods of calm (C) and different wind directions are shown.

reasonably good agreement is shown between aerosol S concentrations measured by filter and impactor samplers, implying no significant SO_2 uptake and retention by the filter during sampling. Nitric acid vapor could cause the observed Cl deficit preferentially on the filter, but sulfuric acid aerosol, whose concentration relative to sulfate salts could vary according to the general level of pollutants in the air, may also account for the observed variability in Cl deficit.

It is interesting to see what happens closer to the polluted city where the air pollution problem is quite visible. Filter samples collected during the same time period at the edge of Beijing apparently show no such deficit according to the following reasoning. Average elemental weight ratios for rural Xinglong and urban Beijing aerosols are listed in Table 1. All of them except for S and Cl are similar, although the absolute concentrations for the two locations differ by a factor of at least 4. The Cl/Fe ratio is much lower in Xinglong than in Beijing based on streaker filter samples. If the impactor data were used for particulate Cl, these ratios are comparable. This is an indication that the Cl deficit is much less on the Beijing filter.

The reason for this result cannot yet be determined unambiguously. The comparison of the elemental weight ratios between the two locations has shown that aerosols in these two locations have similar composition. A possibly important difference is that in urban Beijing there was much more black smoke containing elemental and organic carbon, which is not analyzed by the PIXE method, as well as metal oxide fly ash constituents. This particulate matter may be responsible for the inhibition of possible Cl loss mechanisms, e.g.: (1) elemental carbon or metal oxide constituents in the coal smoke may reduce the atmospheric acidity by absorbing acid pollutant gases, e.g., SO_2 or HNO_3, or they could reduce the speed of reaction on the filter, (2) carbon soot collected on the filter may absorb HCl released due to chemical reaction, thus inhibiting the Cl loss from the filter, or (3) the heavy loading of particulate matter on Beijing's filter sample itself may reduce the speed of HCl release to the air, thus reducing the extent of the Cl loss during sampling.

Whatever the exact mechanism, it seems that to improve air quality by restricting the emission of particulate matter without controlling gaseous SO_2 and NO_x emission may lead to an undesirable result. The removal of the suspended particles from the atmosphere may cause an increase in the acidity of the air downwind, and eventually increase acidic aerosol fallout and acid rain. The same idea has been emphasized by Likens and Bormann (1974).

The S/Fe ratio is also measurably greater in Xinglong (3.06) than in Beijing (2.15). This may

TABLE 1. Elemental Weight Ratios in Beijing and Xinglong Aerosols by Streaker Filter Sampler[a]

Element/Fe	Beijing	Xinglong
Al/Fe	2.15	2.04
Si/Fe	4.51	4.53
K/Fe	0.62	0.77
Ca/Fe	1.32	1.84
Ti/Fe	0.13	0.15
Mn/Fe	0.040	0.049
Zn/Fe	0.076	0.039
Pb/Fe	0.059	0.039
Sr/Fe	0.024	0.024
S/Fe	2.15	3.06
Cl/Fe	0.69	0.19
Cl/Fe (impactor)[b]	-	0.52[b]

[a]Averages for March 6-20 at Beijing and March 16-21 at Xinglong sites.

[b]Impactors were operated concurrently with streakers at Xinglong, but not at Beijing. Element ratios for impactors were in approximate agreement with those for streakers in all cases except Cl/Fe (cf. Figures 3 and 4).

reflect a more complete conversion of SO_2 gas to particulate S, e.g. H_2SO_4, thus accounting for part of the greater apparent acidity at Xinglong. These ratios in Table 1 may also be compared with those of exceptionally clean northerly air sampled by impactor near Beijing (Winchester et al., 1981a) where fine-mode S/Fe = 40 and Cl/Fe = 0.8.

Acknowledgements. We are grateful to Dr. Alistair C. D. Leslie for his assistance in manipulating data on computer. This study was supported in part by the U.S. National Academy of Sciences Committee on Scholarly Communication with the People's Republic of China and the China Association of Science and Technology. Additional support from the U.S. Environmental Protection Agency is acknowledged for assistance in obtaining elemental composition data used in the study.

References

Berg, W. W., and J. W. Winchester, Organic and inorganic gaseous chloride in the marine aerosol, J. Geophys. Res., 82, 5945-5953, 1977.

Berg, W. W., and J. W. Winchester, Aerosol chemistry of the marine atmosphere, in Chemical Oceanography, vol. 7, 2nd ed., edited by J. P. Riley and R. Chester, Academic Press, New York, 1978.

Cicerone, R. J., Halogens in the atmosphere, Rev. Geophys. Space Phys., 19, 123-139, 1981.

Gordon, C. M., E. C. Jones, and R. E. Larson, The vertical distribution of particulate Na and Cl in a marine atmosphere, J. Geophys. Res., 82, 988-990, 1977.

Johansson, T. B., R. E. Van Grieken, and J. W. Winchester, Marine influences on aerosol composition in the coastal zone, J. Rech. Atmos., 8, 761-776, 1974.

Likens, G. E., and F. H. Bormann, Acid rain, a serious regional environmental problem, Science, 184, 1176-1179, 1974.

Martens, C. S., J. J. Wesolowski, R. C. Harriss, and R. Kaifer, Chlorine loss from Puerto Rican and San Francisco Bay area marine aerosols, J. Geophys. Res., 78, 8778-8792, 1973.

Meinert, D. L., and J. W. Winchester, Chemical relationships in the North Atlantic marine aerosol, J. Geophys. Res., 82, 1778-1782, 1977.

Overein, L. N., Sulphur pollution patterns observed, leaching of calcium in forest soil determined, Ambio, 1, 145-147, 1972.

Rahn, K. A., Sources of trace elements in aerosols--approach to clear air, Ph. D. Thesis, Univ. of Mich., Ann Arbor, 1971.

Ren Lixin, J. W. Winchester, Lü Weixiu, and Wang Mingxing, Elemental composition of Beijing aerosols (in Chinese), Daqi Kexue (Scientia Atmospherica Sinica), Vol. VI, 1982.

Wang Mingxing, J. W. Winchester, Ren Lixin, and Lü Weixiu, Sampling and chemical analysis techniques for atmospheric aerosols (in Chinese), Huanjing Kexue Congkan, 2, 1-10, 1981a.

Wang Mingxing, J. W. Winchester, Lü Weixiu, and Ren Lixin, Aerosol composition in a nonurban area near the Great Wall (in Chinese). Daqi Kexue (Scientia Atmospherica Sinica), Vol. V(2), 1981b.

Wilkniss, P. E., and D. J. Bressan, Fractionation of the elements F, Cl, Na, and K at the sea-air interface, J. Geophys. Res., 77, 5307-5315, 1972.

Winchester, J. W., Lü Weixiu, Ren Lixin, Wang Mingxing, and W. Maenhaut, Fine and coarse aerosol composition from a rural area in north China, Atmos. Environ., 15, 933-937, 1981a.

Winchester, J. W., Wang Mingxing, Ren Lixin, Lü Weixiu, H.-C. Hansson, H. Lannefors, M. Darzi, and A. C. D. Leslie, Nonurban aerosol composition near Beijing, China, Nucl. Instr. Meth., 181, 391-398, 1981b.

Winchester, J. W., Wang Mingxing, Ren Lixin, Lü Weixiu, M. Darzi, and A. C. D. Leslie, Aerosol composition in relation to air mass movement in north China, in Atmospheric Aerosol: Source/Air Quality Relationship, edited by E.S. Macias and P. K. Hopke, Am. Chem. Soc. Symposium Series, Vol. 167, 287-301, 1981c.

THE POSSIBLE ROLE OF HETEROGENEOUS AEROSOL PROCESSES IN THE CHEMISTRY OF CH_4 AND CO IN THE TROPOSPHERE

Cindy J. Luther and Leonard K. Peters

Department of Chemical Engineering, University of Kentucky, Lexington, Kentucky 40506

Abstract. The tropospheric chemistry of CO and CH_4 was analyzed to evaluate the role that heterogeneous interactions between trace gaseous species and aerosol particles covered by an aqueous layer could have on this oxidation sequence. In-cloud and below-cloud scavenging were not considered. The heterogeneous processes were represented by first-order rate expressions described by a heterogeneous loss constant based on classical mass transfer theory or gas-kinetic theory. Estimated global CO and CH_4 emission rates and measured CH_2O, H_2O_2, and HNO_3 concentrations were used as a basis for comparing the results of two purely homogeneous gas-phase chemistry models with the same models which also included heterogeneous removal. On the basis of CH_2O, H_2O_2, and HNO_3 concentrations, the results indicated that the homogeneous models describe the CO-CH_4 tropospheric chemistry more accurately, thereby suggesting that heterogeneous processes may not be significant in this sequence of chemical reactions. Estimates of the CO and CH_4 emission rates did not provide a conclusive basis on which to compare the significance of the homogeneous and heterogeneous processes. Finally, the time period during which an aerosol particle is effective in removing a trace gaseous species from the troposphere was found to vary according to the tropospheric concentration and solubility of the particular gaseous species.

Nomenclature

[A]	gas-phase concentration of species A
[A*]	gas-phase concentration of A if it is in equilibrium with the particle phase
$[A_a(D)]$	particle-phase concentration of species A
$[A_a^*(D)]$	particle-phase concentration of A if it is in equilibrium with the gas phase
A_s	particle surface area
D	particle diameter
D_o	smallest particle diameter
D_∞	largest particle diameter
$\mathcal{D}_{A,air}$	diffusivity of A in air
$\mathcal{D}_{A,H_2O}$	diffusivity of A in water
f(D)	normalized particle size distribution function
g(θ)	normalized particle age distribution function
H	equilibrium distribution coefficient
k	homogeneous or heterogeneous loss rate constant
k_g	gas-phase mass transfer coefficient
k_g^t	gas-phase mass transfer coefficient for the transition regime
k_g^c	gas-phase mass transfer coefficient for the continuum regime
k_g^n	gas-phase mass transfer coefficient for the noncontinuum regime
k_ℓ	liquid-phase mass transfer coefficient
K_g	overall mass transfer coefficient
Kn	Knudsen number (λ/D)
M	active aerosol mass, or third body species
M_a	mass of A in the particle phase
n(D)	number of particles per unit volume of size D
n(D,θ)	number of particles per unit volume of size D and age θ
n	total number of particles per unit volume
T	temperature
Δt	time required to saturate the particles
u	molecular velocity
V	particle volume
α	accommodation coefficient
β	capacity of active aerosol particles to remove species from the gas phase
λ	mean free path of air
τ_D	particle residence time
θ	solar zenith angle, or age of particle
——	averaged over time
< >	averaged over particle diameter

Introduction

The chemistry of trace species in the atmosphere is a complex area of active atmospheric research. From the viewpoint of the global scale, NO_x, O_3, CH_4, and CO are several important minor species in the troposphere. These species are principal ones in a significant subsystem of atmospheric chemistry, the conversion of CH_4 to CO.

The oxidation of CH_4 is a significant source of CO in the atmosphere, while the oxidation of CO represents a principal sink for that species. The estimated global budgets of tropospheric CH_4 and CO are shown in Table 1. The bulk of the data has been taken from Peters and Chameides (1980), except that estimated CO sources from biomass burning (Crutzen et al., 1979) and from terpene and isoprene oxidation (Zimmerman et al., 1978) have also been considered. One can see that the source and sink budgets are approximately in balance and that chemistry processes are an important aspect of that balance.

Tropospheric CO-CH_4 chemistry models have concentrated primarily on a series of homogeneous gas-phase chemical and photolytic reactions. In this paper, we will examine the role that heterogeneous processes might have on the overall CO-CH_4 chemistry in the troposphere. This objective will be pursued by employing gas-kinetic theory along with classical mass transfer theory to develop relationships for a loss rate constant describing the heterogeneous processes. The analysis will provide some insight into the parameters that are required for evaluation of such processes. Following this, heterogeneous loss constants will be calculated for chemical species involved in two different existing CO-CH_4 chemistry models. These heterogeneous loss constants will be incorporated into the reaction schemes to determine their effects. The general ability of aerosol particles to actively remove trace gas species found in the troposphere will also be discussed.

Cadle et al. (1975) made calculations which indicated that heterogeneous chemical reactions between trace gases and aerosol particles cannot always be ignored. They argued that the particle concentration in the lower stratosphere is such that collisions of molecules with a particle surface are relatively fast and quite probable. At about the same time, Graedel et al. (1975) made calculations which included the interaction of aerosols and free radicals. They concluded that aerosol effects are a significant part of atmospheric chemical kinetics. However, they made no effort to compare their predicted results with known concentrations or with results based on other studies. Recently, Baldwin and Golden (1980) suggested that heterogeneous radical reactions would need a collisional reaction probability of about 1 to have any significance in tropospheric chemistry. They measured the collisional reaction probability with sulfuric acid for various species present in the troposphere and found relatively low values (e.g., $\alpha \sim 5 \times 10^{-4}$ for the OH radical). Their results were probably affected strongly by the chemical nature of the two species involved.

Review of CO-CH_4 Chemistry

The oxidation of CH_4 to CO is initiated by the reaction of CH_4 with the OH radical to form the CH_3 radical. Figure 1 illustrates the chemical pathways that are currently believed to be important in the oxidation sequence. OH is produced principally from reactions of $O(^1D)$ with CH_4, CH_3OOH, H_2O, and H_2, and from the photolysis of H_2O_2, HNO_3, and HNO_2. The CH_3 radical begins a series of reactions which eventually produces the HCO radical and CH_2O. HCO reacts with O_2 to form

TABLE 1. Global Budgets of Tropospheric CH_4 and CO

CH_4 Sources	Strength (10^{14} g/yr)	CO Sources	Strength (10^{14} g/yr)
Oceans	0.04 - 0.2	Oceans	0.6 - 3
Soils	5 - 8	Anthropogenic	5 - 7
Anthropogenic	0.1 - 0.5	Biomass burning	2 - 16
		CH_4 + OH	0.7 - 6
TOTAL	5 - 9	NMHC oxidation	4 - 13
		TOTAL	12 - 45
Sinks	**Strength (10^{14} g/yr)**	**Sinks**	**Strength (10^{14} g/yr)**
CH_4 + OH	2 - 4	Soils	3 - 5
Transport to stratosphere	0.1 - 0.7	CO + OH	12 - 19
		Transport to stratosphere	0.1 - 0.6
TOTAL	2 - 5	TOTAL	15 - 25

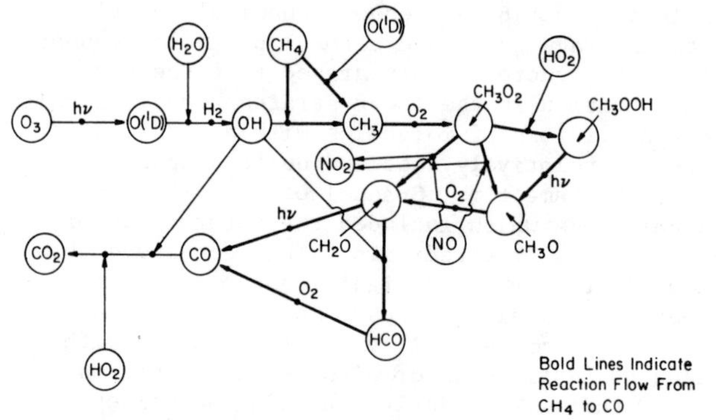

Fig. 1. The principal pathways for oxidation of CH_4 to CO and CO_2 following the chemical reaction scheme of Chameides (1978).

CO, which is also produced from the photolysis of CH_2O.

Several investigators have proposed mechanistic schemes to describe the CO-CH_4 chemistry, but the models developed by Kitada (Kitada and Peters, 1980) and Chameides (Chameides, 1978) were used in this study. These are shown in Table 2. In order to study the importance of heterogeneous processes on the CO-CH_4 chemistry, two simplifications were made in the solution of these kinetic models. The first was that CO and CH_4 are the only variables expressed by differential equations. The second assumption was that the concentrations of NO, NO_2, O_3, H_2O, and M are specified and are not affected by the system. All other species were described using steady-state assumptions.

A Scheme to Incorporate Heterogeneous Processes

Heterogeneous Removal

For the purposes of this study, the aerosol particles are assumed to be spherical and have the physical and chemical properties of water. This does not seem too unrealistic since particles in the troposphere can quite regularly be covered with a layer of water. Figure 2 is a representation of the aerosol particle for this model. Obviously, some species would be more likely than others to interact with aerosol particles. There are several limitations relative to this representation of the heterogeneous processes. First, in-cloud and below-cloud scavenging by rain are not considered. These phenomena could be analyzed in a similar manner. But since they are not always active, this intermittent nature must be incorporated. It is conceivable that these processes could dominate in regions where cloud formation and precipitation are occurring. Secondly, the model aerosol particle does not consider absorption by nonaqueous surfaces. While that phenomenon could be included, knowledge of the fraction of particles with surfaces of that nature would be necessary. Finally, adsorption-desorption phenomena are not included. Thus, processes such as reactant adsorption and product desorption or product incorporation into the particle are not accounted for.

It is assumed that all of the species present in the two CO-CH_4 chemistry models can potentially interact with the aerosol population. However, the extent to which this interaction takes place depends primarily upon the solubility of the species in the particle (water) and the concentration of the species surrounding the particle. The species concentration is important from two aspects. First, the heterogeneous removal rate varies directly as the species concentration. There is, however, a more subtle effect associated with the capacity of the particles to absorb or hold the species. For a species that is very soluble in the particle, the particle phase can rapidly become saturated with respect to that species. As a result, the particle activity toward that species will be greatly reduced, and the particle will only be active for part of its residence time in the atmosphere. This effect has apparently not been considered previously and will be discussed in more detail later.

The heterogeneous removal rate is described using an effective first-order loss rate constant. Two mass conservation balances are used to develop the loss rate constant expression. With n(D) particles of diameter D present, the rate of removal of A from the gas phase by these particles is

$$\frac{d[A]}{dt} = -K_g \pi D^2 n(D) \; ([A] - [A^*]) \qquad (1)$$

where K_g is the overall mass transfer coefficient based on the gas-phase concentration and $[A^*]$ represents the hypothetical gas-phase concentration of A if the gas and particle phases are in equilibrium. Assuming a linear relationship for the species distribution between the gas and particle phases,

$$[A^*] = H \; [A_a(D)] \qquad (2)$$

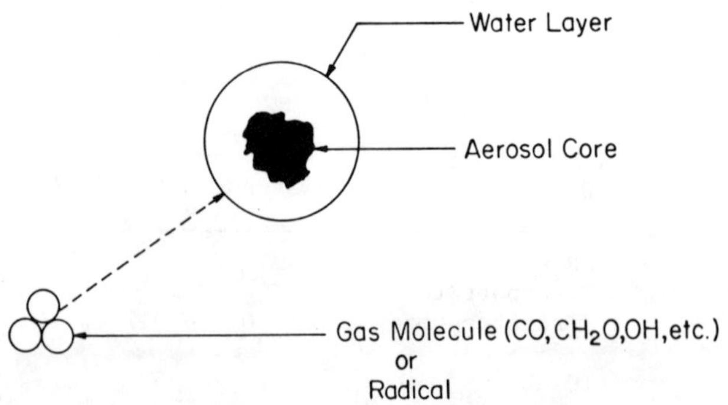

Fig. 2. Schematic representation of the aerosol model.

TABLE 2. Reaction Mechanisms for the CO-CH_4 Chemistry

Reaction	Rate Constant for Chameides Model[a]	Rate Constant for Kitada Model[a]
$CH_4 + OH \rightarrow CH_3 + H_2O$	$2.35 \times 10^{-12}exp(-1710/T)$	same
$CH_4 + O(^1D) \rightarrow CH_3 + OH$	1.3×10^{-10}	b
$CH_3 + M + O_2 \rightarrow CH_3O_2 + M$	3.0×10^{-32}	b
$CH_3O_2 + HO_2 \rightarrow CH_3OOH + O_2$	$3.0 \times 10^{-11}exp(-500/T)$	b
$2CH_3O_2 \rightarrow 2CH_3O + O_2$	2.6×10^{-13}	1.6×10^{-13}
$CH_3OOH + OH \rightarrow CH_3O_2 + H_2O$	$1.0 \times 10^{-11}exp(-750/T)$	b
$CH_3OOH + h\nu \rightarrow CH_3O + OH$	$1.812 \times 10^{-5}exp(-03297/\cos\theta)$	b
$NO + CH_3O_2 \rightarrow CH_2O + HNO_2$	0.0	b
$NO_2 + CH_3O_2 \rightarrow CH_2O + HNO_3$	0.0	b
$NO + CH_3O_2 \rightarrow CH_3O + NO_2$	$3.3 \times 10^{-12}exp(-500/T)$	same
$CH_3O + O_2 \rightarrow CH_2O + HO_2$	$1.6 \times 10^{-13}exp(-3300/T)$	same
$CH_2O + OH \rightarrow HCO + H_2O$	$3.0 \times 10^{-11}exp(-250/T)$	same
$CH_2O + h\nu \rightarrow HCO + H$	$3.921 \times 10^{-5}exp(-0.825/\cos\theta)$	same
$CH_2O + h\nu \rightarrow CO + H_2$	$1.225 \times 10^{-4}exp(-0.625/\cos\theta)$	same
$H + O_2 + M \rightarrow HO_2 + M$	$3.0 \times 10^{-32}(273/T)^{1.3}$	$2.08 \times 10^{-32}exp(290/T)$
$HCO + O_2 \rightarrow CO + HO_2$	1.0×10^{-13}	5.7×10^{-12}
$CO + OH \rightarrow CO_2 + H$	$2.1 \times 10^{-13}exp(-115/T) + 7.3 \times 10^{-33}[M]$	same
$CO + HO_2 \rightarrow CO_2 + OH$	1.0×10^{-20}	b
$H_2 + OH \rightarrow H + H_2O$	$6.8 \times 10^{-12}exp(-2020/T)$	b
$O(^1D) + H_2O \rightarrow 2OH$	2.3×10^{-10}	same
$O(^1D) + H_2 \rightarrow H + OH$	1.3×10^{-10}	b
$HO_2 + OH \rightarrow H_2O + O_2$	3.0×10^{-11}	same
$OH + OH \rightarrow H_2O + O$	$1.0 \times 10^{-11}exp(-550/T)$	b
$HO_2 + HO_2 \rightarrow H_2O_2 + O_2$	$3.0 \times 10^{-11}exp(-500/T)$	2.5×10^{-12}
$H_2O_2 + h\nu \rightarrow 2OH$	$1.812 \times 10^{-5}exp(-0.3297/\cos\theta)$	same
$H_2O_2 + OH \rightarrow HO_2 + H_2O$	$1.0 \times 10^{-11}exp(-750/T)$	same
$HO_2 + NO \rightarrow OH + NO_2$	8.0×10^{-12}	8.1×10^{-12}
$NO_2 + HO_2 + M \rightarrow HO_2NO_2 + M$	1.97×10^{-31}	b
$HO_2NO_2 \rightarrow HO_2 + NO_2$	$1.4 \times 10^{14}exp(-10350/T)$	$1.259 \times 10^{16}exp(-11700/T)$
$OH + NO_2 \rightarrow HNO_3$	1.25×10^{-11}	2.0×10^{-12}
$OH + NO \rightarrow HNO_2$	2.0×10^{-12}	2.0×10^{-12}
$HNO_3 + h\nu \rightarrow OH + NO_2$	$9.88 \times 10^{-7}\cos\theta$	same
$HNO_3 + OH \rightarrow NO_3 + H_2O$	8.0×10^{-14}	b
$HNO_2 + h\nu \rightarrow OH + NO$	$0.275 \times 10^{-2}\cos\theta$	same
$NO_2 + O_3 \rightarrow NO_3 + O_2$	$1.1 \times 10^{-13}exp(-2450/T)$	same
$NO + O_3 \rightarrow NO_2 + O_2$	$2.1 \times 10^{-12}exp(-1450/T)$	$9.0 \times 10^{-13}exp(-500/T)$
$NO_2 + h\nu \rightarrow NO + O$	3.8×10^{-3}	$1.55 \times 10^{-2}exp(-0.48/\cos\theta)$
$NO_3 + NO_2 \rightarrow N_2O_5$	3.8×10^{-12}	$1.48 \times 10^{-13}exp(861/T)$
$NO_3 + NO_2 \rightarrow NO + NO_2 + O_2$	$2.3 \times 10^{-13}exp(-1000/T)$	b
$NO_3 + NO \rightarrow 2NO_2$	8.7×10^{-12}	1.9×10^{-11}
$NO_3 + h\nu \rightarrow NO_2 + O$	$0.1065exp(-0.2097/\cos\theta)$	b
$NO_3 + h\nu \rightarrow NO + O_2$	$0.0334\ exp(-0.2193/\cos\theta)$	b
$NO + NO_2 + H_2O \rightarrow 2HNO_2$	6.0×10^{-37}	6.0×10^{-38}
$N_2O_5 + H_2O \rightarrow 2HNO_3$	1.0×10^{-20}	same
$N_2O_5 \rightarrow NO_3 + NO_2$	$5.7 \times 10^{14}exp(-10600/T)$	$1.24 \times 10^{14}exp(-10317/T)$
$N_2O_5 + h\nu \rightarrow NO_2 + NO_3$	2.0×10^{-5}	b
$O_3 + h\nu \rightarrow O_2 + O$	1.3×10^{-4}	b
$O + O_2 + M \rightarrow O_3 + M$	$1.1 \times 10^{-34}exp(510/T)$	same
$O_3 + h\nu \rightarrow O(^1D) + O_2$	$2.5 \times 10^{-4}exp(-0.66/\cos\theta)$	$1.89 \times 10^{-4}exp(-1.93/\cos\theta)$
$O(^1D) + M \rightarrow O + M$	3.2×10^{-11}	$2.0 \times 10^{-11}exp(107/T)$
$OH + O_3 \rightarrow HO_2 + O_2$	$1.5 \times 10^{-12}exp(-1000/T)$	b
$HO_2 + O_3 \rightarrow OH + 2O_2$	$1.0 \times 10^{-13}exp(-1250/T)$	$1.5 \times 10^{-12}exp(-1000/T)$
$2OH + M \rightarrow H_2O_2 + M$	c	$1.25 \times 10^{-32}exp(900/T)$
$CH_3 + O_2 \rightarrow CH_3O_2$	c	2.2×10^{-12}

[a]Rate constants in cm^3/molec·sec.
[b]This reaction is not included in the Kitada model (Kitada and Peters, 1980).
[c]This reaction is not included in the Chameides (1978) model.

where $[A_a(D)]$ is the concentration of A in aerosol particles of diameter D. A balance on the concentration of A in the particle phase is

$$\frac{dM_a}{dt} = \frac{d\{n(D)V[A_a(D)]\}}{dt} = n(D)\,\frac{\pi D^3}{6}\,\frac{d[A_a(D)]}{dt}$$

$$= K_g n(D)\pi D^2([A] - [A^*]) \qquad (3)$$

The linear distribution relationship can also be used to define $[A^*_a(D)]$, which represents the equilibrium, or maximum, concentration that can exist in the particle phase when in contact with a gas phase at a concentration of [A]; i.e., $[A] = H\,[A^*_a(D)]$. Rearranging equation (3) then yields the following:

$$\frac{d[A_a(D)]}{[A^*_a(D)] - [A_a(D)]} = \frac{6K_g H}{D}\,dt \qquad (4)$$

Before integrating equation (4), the limits of integration must be considered. Initially (at time zero), the concentration of species A in the particle is assumed to be zero. This assumes that the particle, when it is formed, has none of the species present. When the residence time of the particle in the atmosphere (τ_D) is reached, the concentration of A in the particle will be represented as $[A_a(D)]_f$. Therefore, equation (4) becomes

$$[A_a(D)]_f = [A^*_a(D)][1 - \exp(-\frac{6HK_g}{D}\tau_D)] \qquad (5)$$

Variation With Particle Residence Time

The particle population in the troposphere varies in size, age, and chemical composition. Let n be the total number of particles present and assume that the distribution can be represented as

$$n(D,\theta) = nf(D)g(\theta) \qquad (6)$$

where θ is the age of a particle. Equation (6) assumes that there are no composition variations, an obvious simplification. By taking an average of $[A_a(D)]$ over the particle age, the following expression is obtained:

$$\overline{[A_a(D)]} = \frac{1}{n(D)}\int_0^{\tau_D} [A_a(D,\theta)]\,nf(D)g(\theta)\,d\theta \qquad (7)$$

where τ_D represents the average residence time for particles of size D, and the bar superscript represents a particle age-averaged property. Consider as an approximation that

$$g(\theta) \simeq 1/\tau_D \quad \text{when } \theta \le \tau_D$$
and
$$g(\theta) \simeq 0 \quad \text{when } \theta > \tau_D \qquad (8)$$

This implies that for a given size, particles of all ages $0 \le \theta \le \tau_D$ are equally probable. Equation (8) also implies that particles of size D cannot have an age greater than τ_D The expression for $\overline{[A_a(D)]}$ thus becomes

$$\overline{[A_a(D)]} = \frac{1}{\tau_D}\int_0^{\tau_D} [A_a(D,\theta)]\,d\theta \qquad (9)$$

where n(D) = nf(D) has been used. Substituting for $[A_a(D,\theta)]$ from equation (5) for any time θ gives the average particle-phase concentration for particles of size D:

$$\overline{[A_a(D)]} = [A^*_a(D)]\,\{1 - \frac{D}{6HK_g\tau_D}[\exp(-\frac{6HK_g}{D}\tau_D) - 1]\} \qquad (10)$$

Particles of all sizes can of course remove the species from the gas phase. Thus, equation (1) is integrated over the entire size range; i.e.,

$$\frac{d[A]}{dt} = \int_{D_o}^{D_\infty} - K_g\pi D^2 n(D)\,([A] - H\overline{[A_a(D)]})\,dD \qquad (11)$$

where the age-averaged notation has been included. Substituting for $\overline{[A_a(D)]}$ gives

$$\frac{d[A]}{dt} = -\{\int_{D_o}^{D_\infty} K_g\pi D^2 n(D)\,\frac{D}{6HK_g\tau_D}[1 - \exp(-\frac{6HK_g\tau_D}{D})]\,dD\}\,[A] \qquad (12)$$

The quantity within the braces can thus be identified as the heterogeneous loss constant; i.e.,

$$k = \int_{D_o}^{D_\infty} \frac{\pi D^3 n(D)}{6H\tau_D}\,[1 - \exp(-\frac{6HK_g\tau_D}{D})]\,dD \qquad (13)$$

It is instructive to investigate the limiting cases of very insoluble gases ($H \to \infty$), very soluble gases ($H \to 0$), and very short and long particle residence times. These limits can be summarized as follows:

$$H \to 0 \text{ or } \tau_D \to 0, \quad k = \int_{D_o}^{D_\infty} \pi D^2 n(D) k_g\,dD$$

$$H \to \infty \text{ or } \tau_D \to \infty, \quad k = 0$$

These results show that the heterogeneous removal rate constant is controlled by the gas-phase exchange resistance for infinitely soluble gases or for particles with a very short residence time.

Neither limit is surprising. The case of $\tau_D \to 0$ corresponds to an instantaneous renewal of particle surface. On the other hand, the heterogeneous loss constant approaches zero for the insoluble species or for very long residence time particles.

Equation (13) shows that information on the equilibrium distribution of the species between the gas and particle phases (H) and the rate of transfer to the surface (K_g) must be known. These data are not always available. In addition, the residence times and size distribution of the particles must be known. The residence time distribution function for particles in the troposphere has been presented by Jaenicke (1978) as

$$\frac{1}{\tau_D} = \frac{1}{C}\left(\frac{D}{2R}\right)^2 + \frac{1}{C}\left(\frac{D}{2R}\right)^{-1} + \frac{1}{\tau_{wet}} \qquad (14)$$

where $C = 1.26 \times 10^8$ sec, $\tau_{wet} = 1.81 \times 10^6$ sec, and $R = 0.3$ μm. This distribution function is used in evaluating the heterogeneous rate constant and is shown in Figure 3 superimposed on the classical Junge size distribution (Junge, 1963) for a moderately polluted atmosphere.

Since the principal emphasis in this study is on background gas concentrations, an aerosol size distribution more typical of those conditions has been used. The clean continental background troposphere typically has a surface area around 27 $\mu m^2/cm^3$ (Willeke and Whitby, 1975) whereas that shown in Figure 3 has a surface area of 522 $\mu m^2/cm^3$. Therefore, the distribution shown in Figure 3 was reduced by two orders of magnitude to approximate a background troposphere and was used to evaluate the particle-gas system for background conditions.

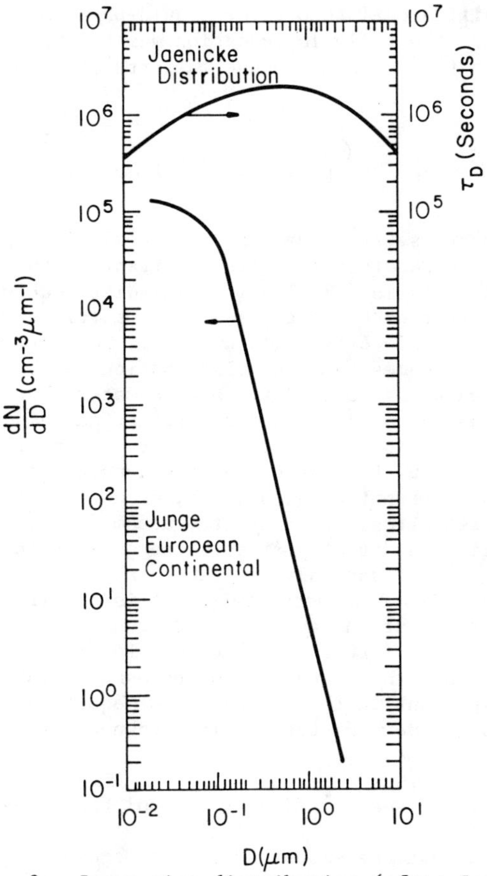

Fig. 3. Junge size distribution (after Seinfeld, 1975) with particle residence time distribution of Jaenicke (1978) superimposed.

Overall Mass Transfer Coefficient

The overall mass transfer coefficient used in equation (13) can be written for a linear equilibrium relationship as

$$K_g = \frac{1}{1/k_g + H/k_\ell} \qquad (15)$$

It is standard to divide the aerosol particles into size regimes according to the Knudsen number ($Kn \equiv \lambda/D$). The first is the noncontinuum regime ($Kn \gtrsim 10$), in which the rate of absorption of a species into the particle can be analyzed using gas-kinetic theory. The gas-phase mass transfer coefficient for this regime can be written as

$$k_g^n = \frac{\alpha u}{4} \qquad (16)$$

where α is an accommodation coefficient dependent on the fraction of collisions of molecules with the particles that are effective, and u is the gas-kinetic velocity. The upper limit on α is unity, which has been used to determine the maximum effect that the heterogeneous processes might have.

The second regime for relatively large particles is the continuum regime ($Kn \lesssim 0.1$), and the rate of absorption of a species into the particle can be calculated by classical diffusional mass transfer theory. Assuming the net velocity between the air and particles is small, the gas-phase mass transfer coefficient for this regime is simply

$$k_g^c = \frac{2\mathcal{D}_{A,air}}{D} \qquad (17)$$

The third size regime is the transition region ($0.1 \lesssim Kn \lesssim 10$) which matches the noncontinuum and continuum regimes in a smooth manner. Fuchs and Sutugin (1971) have used the results of Sahni (1966) to obtain

$$k_g^t = \left(\frac{1}{1+\ell Kn}\right) k_g^c \qquad (18)$$

where

$$\ell = \frac{4/3 + 0.71\ Kn^{-1}}{1 + Kn^{-1}} \qquad (19)$$

Using diffusional theory for liquid-phase mass transfer, the following equation was used to approximate the liquid-phase mass transfer coefficient.

$$k_\ell = \frac{2\mathcal{D}_{A,H_2O}}{D} \tag{20}$$

There are several physical and chemical properties and parameters that are important to the overall analysis. The equilibrium distribution of the gas species between the air and particle phases (described by H) is crucial. The current research assumes that the distribution can be described using solubility data in water, but additional data are needed for better calculations. It also seems apparent that the entire particle population is not active toward the gas species. Current data, however, do not permit better estimates. Finally, the accommodation coefficient α directly affects the gas-particle exchange rate, and only limited data are available to assess its role accurately. It is significant to observe that the quantity D/τ_D ranges from about 3×10^{-12} to 2×10^{-9} cm/s. If HK_g is greater than this range, then the exponential term is unimportant in the evaluation of equation (13), and H alone becomes the critical unknown parameter.

Effective Removal Time of Aerosol Particles

Assume that the amount of active aerosol in the troposphere is M and that it has the capacity for removing βM of species A from the gas phase. It would seem that, as a rough approximation, $\beta \sim 1$. If [A] is the gas-phase concentration and the concentration in the particle phase is negligible, then

$$\frac{d(\beta M)}{dt} \simeq \frac{\Delta(\beta M)}{\Delta t} \simeq \langle K_g A_s \rangle [A] \tag{21}$$

where A_s represents the particle surface area, and $\langle K_g A_s \rangle$ is the mass transfer coefficient times the particle surface area integrated over all particle sizes. Thus,

$$\Delta t \simeq \frac{\beta M}{\langle K_g A_s \rangle [A]} \tag{22}$$

where Δt is an estimate of the time required to saturate the aerosol particles. This characteristic time, Δt, is the period during which the aerosol particles are active in removing trace species from the gas phase and will be referred to as the effective removal time.

In Figure 4, the Jaenicke residence time distribution function is superimposed on a plot of effective removal time versus concentration. The concentrations shown are only representative for some of the species since their concentrations can

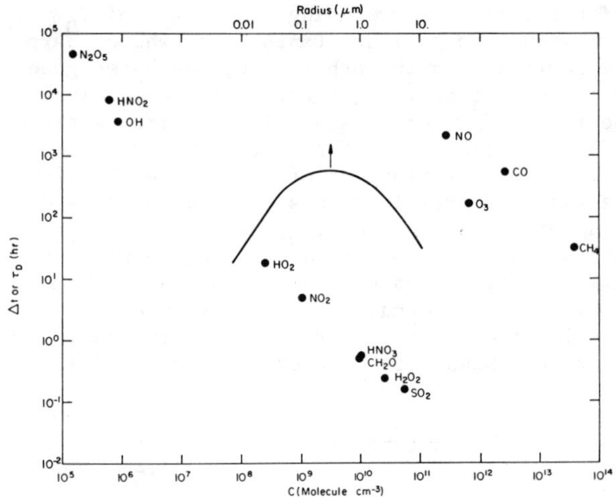

Fig. 4. Effective removal times of various gaseous species compared to the Jaenicke (1978) particle residence time distribution.

be quite variable. But this shows that for some species, such as CO or NO, the characteristic removal time is significantly longer than the particle residence time and thus is truly limited by the mass transfer rate. This is not surprising since these species are less soluble in the particle phase, and the overall mass transfer coefficient is a function of solubility. For other species such as H_2O_2 or SO_2 the removal is not limited by mass transfer since they saturate the particles in a very short period of time. In these cases, the particles only actively remove the species for a part of their residence time. This is understandable since these species are quite soluble in the particle phase, and the particle residence time is an important factor in estimating characteristic removal times of trace gaseous species by heterogeneous processes. For species such as O_3 the effect appears to be intermediate and some particles are saturated in a time shorter than their residence time while other particles are not saturated prior to their average lifetime. It must be emphasized that the results shown in Figure 4 depend on H, which in turn depends on the nature of the particles.

Effect of Heterogeneous Removal on the CO-CH_4 Chemistry Models

The effects of such heterogeneous removal on the two CO-CH_4 chemistry models were considered in light of important factors affecting the tropospheric abundance of these species. Table 3 lists the heterogeneous loss constants calculated according to equation (13). Many of these rate constants are quite large because of the values used for H. Some indicate that the heterogeneous rates could dominate the known homogeneous processes. The most striking example is for HNO_3, where the homogeneous loss processes are on the

TABLE 3. Heterogeneous Loss Constants

Species	Heterogeneous Loss Constant (s^{-1})
NO	1.41×10^{-8}
CO	6.25×10^{-9}
O_3	6.70×10^{-8}
CH_4	7.42×10^{-9}
CH_3O_2	8.18×10^{-4}
OH	1.11×10^{-3}
H_2O_2	6.20×10^{-4}
CH_2O	7.23×10^{-4}
CH_3OOH	6.07×10^{-4} [a]
CH_3O	7.03×10^{-4}
HNO_2	7.31×10^{-4}
HNO_3	6.53×10^{-4}
HCO	7.44×10^{-4}
N_2O_5	5.34×10^{-4}
NO_3	6.69×10^{-4}
$O(^1D)$, $O(^3P)$	1.19×10^{-3}
CH_3	8.32×10^{-4}
HO_2NO_2	5.98×10^{-4}
NO_2	7.56×10^{-4}
HO_2	8.52×10^{-5}
H	5.92×10^{-4}

[a] This loss constant is not required for the Kitada model.

order of 10^{-6} to 10^{-7} s^{-1} compared to 6.53×10^{-4} s^{-1} for heterogeneous processes. Other differences, although less substantial, are also noted for CH_2O and H_2O_2.

These first-order loss processes were added to the mechanisms shown in Table 2 and the results were compared with those from the homogeneous gas-phase mechanisms. Two factors used for comparison were the estimated global production rates of CO and CH_4 and the concentrations of key intermediates in the chemistry. The simulations assumed a pressure of 1 atm, a temperature of 298 K, and a solar zenith angle of 75°.

Comparison with CO and CH_4 Budgets

Table 4 is a summary of CO and CH_4 production rates predicted by each of the two homogeneous models and the two modified models including the heterogeneous processes. From this table it is evident that the original Kitada model predicted quite well the estimated CH_4 production rate. On the other hand, the modified Kitada model predicted a CH_4 production rate slightly higher than estimated. The CH_4 production rates predicted by both versions of the Chameides model were higher than the estimated production rate.

The original Kitada model predicted a CO production rate slightly higher than the estimated rate, while the modified Kitada model predicted a CO production rate at the lower end of the estimated range. Again, the Chameides models predict significantly higher CO production rates than estimated from other considerations.

Comparison with CH_2O, H_2O_2, and HNO_3 Concentrations

The second criterion used to evaluate the contributions of the heterogeneous processes was comparison of the model predictions with reported measurements of certain gaseous species in the troposphere. CH_2O, H_2O_2, and HNO_3 are key intermediate species in the atmospheric chemistry of CO-CH_4, and their concentrations were substantially reduced by incorporation of heterogeneous removal into both CO-CH_4 chemistry models. This well-defined difference between the original and modified models provides a basis on which conclusions can be drawn.

Table 5 is a summary of predicted and measured concentrations of CH_2O, H_2O_2, and HNO_3. In comparing the CH_2O concentrations predicted by the CO-CH_4 models with background measurements, it is evident that both of the original models predicted the background CH_2O concentration more accurately than the heterogeneous models. The same was true for background HNO_3 and H_2O_2 concentrations, with heterogeneous removal reducing these concentrations substantially below the observed values. This implies that the purely homogeneous chemical mechanisms are more representative of the overall CO-CH_4 tropospheric chemistry system. However, the homogeneous chemistry models may be predicting HNO_3 concentrations 10 to 10^2 times too large based on the data of Huebert and Lazrus (1978) and Huebert (1980). This could indicate that, although a heterogeneous loss constant for HNO_3 of 6.53×10^{-4} s^{-1} is too large, a value around 10^{-5} to 10^{-6} s^{-1} could be realistic.

TABLE 4. CO and CH_4 Production Rates

Study	CO Production Rate (molec/cm^3·sec)	CH_4 Production Rate (molec/cm^3·sec)
From Table 1	4.5×10^4 - 3.3×10^5	1.2×10^5 - 2.3×10^5
Kitada model	4.6×10^5	2.2×10^5
Modified Kitada model	3.75×10^4	2.8×10^5
Chameides model	1.95×10^6	9.0×10^5
Modified Chameides model	1.80×10^6	8.4×10^5

TABLE 5. Comparison of CH_2O, H_2O_2, and HNO_3 Concentrations

Species	Study	Location	Concentration (molec/cm^3)
CH_2O-predicted	Kitada original	Background	9.64×10^9
"	Kitada modified	Background	1.60×10^6
"	Chameides original	Background	1.58×10^{10}
"	Chameides modified	Background	4.91×10^7
CH_2O-observed	Zafiriou et al. (1980)	Enewetak Atoll (background)	1.06×10^{10}
"	W. Seiler (private communication, 1980)	Marine air masses (background)	1.34×10^{10}
H_2O_2-predicted	Kitada original	Background	2.46×10^{10}
"	Kitada modified	Background	1.42×10^5
"	Chameides original	Background	1.74×10^{11}
"	Chameides modified	Background	4.31×10^7
H_2O_2-observed	Kelly and Stedman (1979)	Rural U.S. air	$8.04 \times 10^9 - 8.04 \times 10^{10}$
HNO_3-predicted	Kitada original	Background	1.00×10^{10}
"	Kitada modified	Background	1.24×10^6
"	Chameides original	Background	8.27×10^{10}
"	Chameides modified	Background	6.05×10^8
HNO_3-observed	Huebert (1980)	Equatorial Pacific (background)	1.01×10^9
"	Huebert and Lazrus (1978)	Pacific marine air (background)	$<5.36 \times 10^8 - 8.04 \times 10^9$
"	Kelly and Stedman (1979)	Rural U.S. air	$<2.7 \times 10^{10} - 1.34 \times 10^{11}$

In order to determine which factors most influenced the concentrations of CH_2O, H_2O_2, and HNO_3, the modified models were solved with various heterogeneous loss constants set to zero. From these evaluations it was apparent that the dominant factor in reducing the CH_2O, H_2O_2, and HNO_3 concentrations in both modified models was the heterogeneous removal of the respective individual species and not the heterogeneous removal of the intermediate radicals which are involved in the formation of these species. This, of course, indicated that the assumptions involved in obtaining the heterogeneous loss constants for the intermediate free radical species were not very critical.

The effect of the total aerosol surface area on the concentrations predicted by the modified $CO-CH_4$ chemistry models was also studied. It was established that as the aerosol particle surface area in the troposphere was reduced, the concentrations predicted by the modified $CO-CH_4$ chemistry models approach those of the original homogeneous models. This is logical since there would be fewer aerosol particles to interact with the gaseous species. Reduction of the particle surface area is analogous to decreasing the accommodation coefficient.

Summary

The effect of particle-gas interactions on the $CO-CH_4$ oxidation sequence in the troposphere was studied by developing rate expressions for heterogeneous removal of gaseous species based on absorption of the species by aerosol particles. The analysis indicates that the equilibrium distribution of species between the gas and particle phases and the accommodation coefficient with the particle surface are important parameters to be measured. Those species with high solubilities have larger loss constants than those species with low solubilities. The heterogeneous removal of tropospheric gases depends upon the residence time of the aerosol particle also. Smaller particles with longer residence times in the troposphere, since they are exposed to the gaseous species longer, would have higher concentrations of these species when their residence times are reached than larger particles.

The global production rates of CO and CH_4 predicted by the chemistry models were used to evaluate the effects of heterogeneous removal by comparing the results to production rates estimated on other bases. Also, CH_2O, H_2O_2, and HNO_3 concentrations predicted by the homogeneous and modified models were compared to measured background concentrations of these gases reported in the literature. Comparisons on the basis of the CO and CH_4 production rates are not conclusive in establishing a preference for the honogeneous or heterogeneous models. On the basis of CH_2O, H_2O_2, and HNO_3 concentrations, it appears that the CO-CH_4 chemistry models which consist of only homogeneous gas-phase reactions represent more accurately the tropospheric chemistry of these species. Overall, these results indicate that one or a combination of the following factors is true: the accommodation coefficient for heterogeneous processes is substantially smaller than unity; only a small fraction of the total aerosol area is effective in removing intermediates in the CH_4-CO oxidation sequence; or other chemical processes producing CH_2O, H_2O_2, and/or HNO_3 are important.

Considerable work remains to be done in modeling CO-CH_4 chemistry in the troposphere. As measurement programs for monitoring levels of atmospheric pollutants develop, it is recommended that CO-CH_4 models be adapted to incorporate new information. However, our current knowledge indicates that heterogeneous removal processes do not significantly affect the homogeneous gas phase chemistry of CO and CH_4 in the troposphere.

Acknowledgement. This research was supported by the National Aeronautics and Space Administration under Research Grant NSG-1424.

References

Baldwin, A. C., and D. M. Golden, Heterogeneous atmospheric reactions: Atom and radical reactions with sulfuric acid, J. Geophys. Res., 85, 2888-2889, 1980.

Cadle, R. D., P. Crutzen, and D. Enhalt, Heterogeneous chemical reactions in the stratosphere, J. Geophys. Res., 80, 3381-3385, 1975.

Chameides, W. L., The effect of anthropogenic carbon monoxide on the methane budget of the troposphere, in Proceedings of the Fourth Joint Conference on Remote Sensing of Environmental Pollutants, American Chemical Society, 1978.

Crutzen, P. J., L. C. Heidt, J. D. Krasnec, W. H. Pollock, and W. Seiler, Biomass burning as a source of the atmospheric gases CO, H, N_2O, NO, CH_3Cl, and COS, Nature, 282, 253-256, 1979.

Fuch, N. A., and A. G. Sutugin, Highly dispersed aerosols, Topics in Current Aerosol Research, Pergamon Press, New York and London, 1971.

Graedel, T. E., L. A. Farrow, and T. A. Weber, The influence of aerosols on the chemistry of the troposphere, Int. J. Chem. Kin. Symposium No. 1, John Wiley and Sons, New York, 1975.

Huebert, B. J., and A. L. Lazrus, Global tropospheric measurements of nitric acid vapor and particulate nitrate, Geophys. Res. Lett., 5, 577-580, 1978.

Huebert, B. J., Nitric acid and aerosol nitrate measurements in the equatorial pacific region, Geophys. Res. Lett., 7, 325-328, 1980.

Jaenicke, R., Über die Dynamik atmospharischer Aitkenteilchen, Ber. Bunsenges. Phys. Chem., 82, 1198-1202, 1978.

Junge, C. E., Air Chemistry and Radioactivity, Academic Press, New York, 1963.

Kelly, T. J., and D. H. Stedman, Measurements of H_2O_2 and HNO_3 in rural air, Geophys. Res. Lett., 6, 375-378, 1979.

Kitada, T., and L. K. Peters, A model of CO-CH_4 global transport/chemistry: I. Chemistry model, J. Japan Soc. Air Poll., 15, 91-108, 1980.

Peters, L. K., and W. L. Chameides, The chemistry and transport of methane and carbon monoxide in the troposphere, Adv. Env. Sci. Eng., 111, 100-149, 1980.

Sahni, D. C., Effect of a black sphere on the flux distribution in an infinite moderator, J. Nucl. Eng. A/B, 20, 915, 1966.

Seinfeld, J. H., Air Pollution: Physical and Chemical Fundamentals, McGraw-Hill Book Co., New York, 1975.

Willeke, K., and K. T. Whitby, Atmospheric aerosols: Size distribution interpretation, J. Air Poll. Control Assoc., 25, 529-534, 1975.

Zafiriou, O. C., J. Alford, M. Herrera, E. T. Peltzer, and R. B. Gagosian, Formaldehyde in remote marine air and rain: Flux measurements and estimates, Geophys. Res. Lett., 7, 341-344, 1980.

Zimmerman, P. R., R. B. Chatfield, J. Fishman, P. J. Crutzen, and P. L. Hanst, Estimates on the production of CO and H_2 from the oxidation of hydrocarbon emissions from vegetation, Geophys. Res. Lett., 5, 679-682, 1978.

Acknowledgements

The editor wishes to thank the authors and referees of the papers presented in this monograph for their conscientious efforts as well as for their cooperation in making timely publication possible. Thanks are also due to Susan Caughlan and Sybil Watson for invaluable editorial assistance. Finally, I wish to acknowledge with gratitude the support and encouragement of Volker Mohnen, John Mugler, Ed Prior, and Bob Hess throughout the entire endeavor which has culminated in this document.

This monograph was prepared and published under the sponsorship of the National Science Foundation and the National Aeronautics and Space Administration, and their support is gratefully acknowledged.